CHEESE:
CHEMISTRY, PHYSICS AND MICROBIOLOGY

Volume 2
Major Cheese Groups

Second Edition

CHEESE:
CHEMISTRY, PHYSICS AND MICROBIOLOGY

Volume 2
Major Cheese Groups

Second Edition

Edited by

P.F. FOX

Department of Food Chemistry
University College
Cork, Ireland

CHAPMAN & HALL

London · Glasgow · New York · Tokyo · Melbourne · Madras

Published by Chapman & Hall, 2–6 Boundary Row, London SE1 8HN, UK

Chapman & Hall, 2–6 Boundary Row, London SE1 8HN, UK

Blackie Academic & Professional, Wester Cleddens Road,
Bishopbriggs, Glasgow G64 2NZ, UK

Chapman & Hall Inc., 29 West 35th Street, New York NY10001, USA

Chapman & Hall Japan, Thomson Publishing Japan,
Hirakawacho Nemoto Building, 6F, 1-7-11 Hirakawa-cho, Chiyoda-ku,
Tokyo 102, Japan

Chapman & Hall Australia, Thomas Nelson Australia, 102 Dodds Street,
South Melbourne, Victoria 3205, Australia

Chapman & Hall India, R. Seshadri, 32 Second Main Road, CIT East,
Madras 600 035, India

First edition 1987
Second edition 1993

© 1993 Chapman & Hall

Typeset in Times Roman by Variorum Publishing Limited, Rugby
Printed in Great Britain by Galliard (Printers) Ltd, Great Yarmouth

ISBN 0 412 53510 6   0 412 58230 9 (Set)

A catalogue record for this book is available from the British Library

Library of Congress Cataloging-in-Publication data
Cheese: chemistry, physics, and microbiology / edited by P.F. Fox. —
   2nd ed.
        p.   cm.
    Includes bibliographical references and index.
    Contents: v. 1. General aspects — v. 2. Major cheese groups.
    ISBN 0 412 58230 9 (set). — ISBN 0 412 53500 9 (v. 1). — ISBN
0 412 53510 6 (v. 2)
    1. Cheese.   2. Cheese — Varieties.   I. Fox, P.F.
SF271.C43   1993
637'.3 — dc20   92-42748
                                                          CIP

ISBN 0 412 53500 9 (v. 1)
ISBN 0 412 53510 6 (v. 2)
ISBN 0 412 58230 9 (set)

Preface to the Second Edition

The first edition of this book was very well received by the various groups (lecturers, students, researchers and industrialists) interested in the scientific and technological aspects of cheese. The initial printing was sold out faster than anticipated and created an opportunity to revise and extend the book.

The second edition retains all 21 subjects from the first edition, generally revised by the same authors and in some cases expanded considerably. In addition, 10 new chapters have been added: Cheese: Methods of chemical analysis; Biochemistry of cheese ripening; Water activity and the composition of cheese; Growth and survival of pathogenic and other undesirable microorganisms in cheese; Membrane processes in cheese technology, in Volume 1 and North-European varieties; Cheeses of the former USSR; Mozzarella and Pizza cheese; Acid-coagulated cheeses and Cheeses from sheep's and goats' milk in Volume 2. These new chapters were included mainly to fill perceived deficiencies in the first edition.

The book provides an in-depth coverage of the principal scientific and technological aspects of cheese. While it is intended primarily for lecturers, senior students and researchers, production management and quality control personnel should find it to be a very valuable reference book. Although cheese production has become increasingly scientific in recent years, the quality of the final product is still not totally predictable. It is not claimed that this book will provide all the answers for the cheese scientist/technologist but it does provide the most comprehensive compendium of scientific knowledge on cheese available.

Each of the 31 chapters is extensively referenced to facilitate further exploration of the extensive literature on cheese. It will be apparent that while cheese manufacture is now firmly based on sound scientific principles, many questions remain unanswered. It is hoped that this book will serve to stimulate further scientific study on the chemical, physical and biological aspects of cheese.

I wish to thank sincerely all the authors who contributed to the two volumes of this book and whose co-operation made my task as editor a pleasure.

P.F. Fox

Preface to the First Edition

Cheese manufacture is one of the classical examples of food preservation, dating from 6000–7000 BC. Preservation of the most important constituents of milk (i.e. fat and protein) as cheese exploits two of the classical principles of food preservation, i.e.: lactic acid fermentation, and reduction of water activity through removal of water and addition of NaCl. Establishment of a low redox potential and secretion of antibiotics by starter microorganisms contribute to the storage stability of cheese.

About 500 varieties of cheese are now produced throughout the world; present production is $\sim 10^7$ tonnes per annum and is increasing at a rate of $\sim 4\%$ per annum. Cheese manufacture essentially involves gelation of the casein via isoelectric (acid) or enzymatic (rennet) coagulation; a few cheeses are produced by a combination of heat and acid and still fewer by thermal evaporation. Developments in ultrafiltration facilitate the production of a new family of cheeses. Cheeses produced by acid or heat/acid coagulation are usually consumed fresh, and hence their production is relatively simple and they are not particularly interesting from the biochemical viewpoint although they may have interesting physicochemical features. Rennet cheeses are almost always ripened (matured) before consumption through the action of a complex battery of enzymes. Consequently, they are in a dynamic state and provide fascinating subjects for enzymologists and microbiologists, as well as physical chemists.

Researchers on cheese have created a very substantial literature, including several texts dealing mainly with the technological aspects of cheese production. Although certain chemical, physical and microbiological aspects of cheese have been reviewed extensively, this is probably the first attempt to review comprehensively the scientific aspects of cheese manufacture and ripening. The topics applicable to most cheese varieties, i.e. rennets, starters, primary and secondary phases of rennet coagulation, gel formation, gel syneresis, salting, proteolysis, rheology and nutrition, are reviewed in Volume 1. Volume 2 is devoted to the more specific aspects of the nine major cheese families: Cheddar, Dutch, Swiss, Iberian, Italian, Balkan, Middle Eastern, Mould-ripened and Smear-ripened. A chapter is devoted to non-European cheeses, many of which are ill-defined; it is hoped that the review will stimulate scientific interest in these minor, but locally important, varieties. The final chapter is devoted to processed cheeses.

It is hoped that the book will provide an up-to-date reference on the scientific aspects of this fascinating group of ancient, yet ultramodern, foods; each chapter is extensively referenced. It will be clear that a considerably body of scientific knowledge on the manufacture and ripening of cheese is currently available but it will be apparent also that many major gaps exist in our knowledge; it is hoped that this book will serve to stimulate scientists to fill these gaps.

I wish to thank sincerely the other 26 authors who contributed to the text and whose co-operation made my task as editor a pleasure.

P.F. Fox

Contents

List of Contributors

M.H. ABD EL-SALAM
Department of Food Technology and Dairying, National Research Centre, Dokki, Cairo, Egypt.

E. ALICHANIDIS
Laboratory of Dairy Technology, Department of Agriculture, Aristotelian University of Thessaloniki, 540 06 Thessaloniki, Greece.

V. ANTILA
Food Research Institute, Agricultural Research Centre, Jokioinen, Finland.

Y. ARDÖ
Swedish Dairies Association R & D, Lund, Sweden.

B. BATTISTOTTI
Istituto di Microbiologia, Università Cattolica, Via Emilia Parmense, 84-Piacenza, Italy.

J.O. BOSSET
Federal Dairy Research Institute, CH-3097 Liebefeld-Bern, Switzerland.

M. CARIĆ
Faculty of Technology, Institute of Meat, Milk Fat and Oil, Fruit and Vegetable Technology, University of Novi Sad, 2100 Novi Sad, V. Vlahovica 2, Yugoslavia.

C. CORRADINI
Istituto di Tecnologie Alimentari, Università di Udine, Via Marangoni, 97-Udine, Italy.

L.K. CREAMER
New Zealand Dairy Research Institute, Private Bag, Palmerston North, New Zealand.

P. EBERHARD
Federal Dairy Research Institute, CH-3097 Liebefeld-Bern, Switzerland.

M.A. ESTEBAN
Department of Food Science and Technology, University of Córdoba, E-14005 Córdoba, Spain.

N.Y. FARKYE
Dairy Products Technology Centre, California Polytechnic State University, San Luis, Obispo, CA 93407, USA.

P.F. FOX
Department of Food Chemistry, University College, Cork, Republic of Ireland.

J. GILLES
New Zealand Dairy Research Institute, Private Bag, Palmerston North, New Zealand.

T.J. GEURTS
Department of Food Science, Wageningen Agricultural University, Bomenweg 2, 6703 HD Wageningen, The Netherlands.

J.C. GRIPON
Laboratoire de Recherches Laitières, Institut National de la Recherche Agronomique, Jouy-en-Josas, France.

A.V. GUDKOV
Department of Microbiology, All-Union Research Institute for Butter and Cheese-Making Industry, Krasnoarmeiskii Boulevard, 15262 Uglich, Yaroslav District 19, Russia.

T.P. GUINEE
National Dairy Products Research Centre, Teagasc, Moorepark, Fermoy, Co. Cork, Republic of Ireland.

M. KALÁB
Food Research Centre, Agriculture Canada, Ottawa, Ontario, Canada.

G.C. KALANTZOPOULOS
Department of Food Science and Technology, Agricultural University of Athens, Athens, Greece.

P.S. KINDSTEDT
Department of Animal Sciences, University of Vermont, College of Agriculture, Carrigan Hall, Burlington, Vermont 05405-0044, USA.

R.C. LAWRENCE
New Zealand Dairy Research Institute, Private Bag, Palmerston North, New Zealand.

A. MARCOS
Department of Food Science and Technology, University of Córdoba, E-14005 Córdoba, Spain.

E.W. NIELSEN
Department of Dairy and Food Science, Royal Veterinary and Agricultural University, Copenhagen, Denmark.

A. NOOMEN
Department of Food Science, Wageningen Agricultural University, Bomenweg 2, 6703 HD Wageningen, The Netherlands.

J.A. PHELAN
Meat and Dairy Service, Animal Production and Health Division, Room C556, Via Terrme di Carracalla, Rome 00100, Italy.

P.D. PUDJA
Institute of Food Technology and Biochemistry, Faculty of Agriculture, University of Belgrade, Yugoslavia.

A. RAGE
Norwegian Dairies Research Centre, Voll, Norway.

J. RENAUD
Meat and Dairy Service, Animal Production and Health Division, Room C556, Via Terrme di Carracalla, Rome 00100, Italy.

A. REPS
Institute of Food Engineering and Biotechnology, University of Agriculture and Technology, 10-957 Olsztyn-Kortowo, BL. 43, Poland.

M. RÜEGG
Federal Dairy Research Institute, CH-3097 Liebefeld-Bern, Switzerland.

C. STEFFEN
Federal Dairy Research Institute, CH-3097 Liebefeld-Bern, Switzerland.

P. WALSTRA
Department of Food Science, Wageningen Agricultural University, Bomenweg 2, 6703 HD Wageningen, The Netherlands.

H. WERNER
Danish Dairy Board, Aarhus, Denmark.

G.K. ZERFIRIDIS
Laboratory of Dairy Technology, Department of Agriculture, Aristotelian University of Thessaloniki, 540 06 Thessaloniki, Greece.

1

Cheddar Cheese and Related Dry-Salted Cheese Varieties

R.C. Lawrence, J. Gilles & L.K. Creamer

New Zealand Dairy Research Institute, Palmerston North, New Zealand

1 INTRODUCTION

In the warm climates in which cheesemaking was first practised, cheeses would have tended to be of low pH as a result of the acid-producing activity of the lactic acid bacteria and coliforms in the raw milk. In colder climates, it would have been logical either to add warm water to the curds and whey to encourage acid production (the prototype of Gouda-type cheeses) or to drain off the whey and pile the curd into heaps to prevent the temperature falling. In the latter case, the piles became known as 'Cheddars', after the village in Somerset, England, where the technique is said to have been first used about the middle of the 19th century. The concept of cheddaring was quickly adopted also outside Britain. The first Cheddar cheese factory, as opposed to farmhouse cheesemaking, was in operation in the United States (N.Y. State) in 1861, followed by Canada (Ontario) in 1864 and by New Zealand and England in 1871.

1.1 Role of Cheddaring

Cheddar cheese was apparently made originally by a stirred curd process without matting, but poor sanitary conditions led to many gassy cheeses with unclean flavours.[1] Cheddaring was found to improve the quality of the cheese, presumably as a result of the greater extent of acid production. As the pH fell below about 5·4, the growth of undesirable, gas-forming organisms, such as coliforms, would have been increasingly inhibited. The piling and repiling of blocks of warm curd in the cheese vat for about two hours also squeezed out any pockets of gas that formed during manufacture. Cheesemakers came to believe that the characteristic texture of Cheddar cheese was a direct result of the cheddaring process. It is now clear, however, that cheddaring and the development of the Cheddar texture are concurrent rather than interdependent processes. As will be described later, recently developed methods for the manufacture of Cheddar cheese do not involve a cheddaring step but the cheese obtained is identical to traditionally-made Cheddar.

The development of the fibrous structure in the curd of traditionally-made Cheddar does not commence until the curd has reached a pH of 5·8 or less.[2] The changes that occur are a consequence of the development of acid in the curd and the loss of calcium phosphate from the protein matrix. It is important, therefore, to recognize that 'cheddaring' is not confined to Cheddar cheese. All cheeses are 'cheddared' in the sense that all go through this same process of chemical change. The only difference is of degree, i.e., the extent of flow varies due to differences in calcium levels, pH and moisture.[3,4] In addition, with brine-salted cheeses, flow is normally restricted at an early stage during manufacture by placing the curd in a hoop. If, however, Gouda curd is removed from a hoop, it flows in the same way as Cheddar curd. Similarly, the stretching that is induced in Mozzarella by kneading in hot water is best viewed as a very exaggerated form of 'cheddaring'. All young cheese, regardless of the presence of salt, can be stretched in the same way as Mozzarella provided that the calcium content and pH are within the required range.[5]

1.2 Development of Dry-Salting

In the early days of cheesemaking, the surface of the curd mass was presumably covered with dry salt in an attempt to preserve the cheese curd for longer periods. In localities where the salt was obtained by the evaporation of seawater, it would have been a rational step to consider using the concentrated brine rather than wait for all the liquid to evaporate. The technique of dry-salting, i.e. salting relatively small particles of curd before pressing, appears to have evolved in England, probably in the county of Cheshire, where rock salt is abundant. Cheshire has been manufactured for at least 1000 years and this is a more ancient cheese than Cheddar. Variants of Cheshire and Cheddar were developed in specific localities of Britain and have come to be known as British Territorial cheeses. Blue-veined cheeses such as Stilton, Wensleydale and Dorset are also dry-salted.

Dry-salting overcomes the major disadvantage of brine-salting, i.e. the 'blowing' of the cheese due to the growth of such bacteria as coliforms and clostridia, but introduces new difficulties since the starter organisms are also inhibited by the salt. This inhibition is not a problem when the pH of the curd granules is allowed to reach a relatively low value prior to the application of salt, as in Cheshire and Stilton manufacture. The manufacture of a dry-salted cheese in the medium pH range (5·0–5·4), such as Cheddar, is, however, more difficult than that of the Gouda-type cheeses in which the pH is controlled mainly by the addition of water. At the time of salt addition, a relatively large amount of lactose is still present in the Cheddar curd.[6] This is not, however, detrimental to the quality of the cheese provided that salt-in-moisture (S/M) level is greater than 4·5% and the cheese is allowed to cool after pressing.[7]

Differences obviously exist in the procedures used for the manufacture of dry- and brine-salted cheeses but these have relatively little effect on the finished cheeses; the production of dry-salted cheeses is in principle similar to that of brine-salted cheeses. Clearly, the rate of solubilization of the casein micelles and the activity of the residual rennet and plasmin in the curd will be affected more rapidly by dry-salting than by brining, but only during the first few weeks of ripening. There is no evidence to suggest that the mechanisms by which the protein is degraded are affected by the changes in salt concentration as the salt diffuses into the curd. Any differences between dry- and brine-salted cheeses of the same chemical composition will therefore decrease as the cheeses age.

Traditional Cheddar cheese is visually different from the common brine-salted cheeses such as the Gouda- and Swiss-type cheeses, which are more plastic in texture and possess 'eyes'. Both these characteristics, however, are a result of the relatively high pH and moisture of these cheeses and not of brine-salting *per se*. The texture of brine-salted cheese is closer than that of traditionally made Cheddar cheese since the curd is pressed under the whey to remove pockets of air before brining. As a close texture is a prerequisite for the formation of 'eyes', it has come to be believed generally that 'eyes' can be formed only in brine-salted cheese. The recently developed technique of vacuum pressing, however, also allows the removal of air from between the particles of dry-salted curd. This can result in a closeness of texture similar to that of Gouda-type cheeses. Therefore, it is now possible to manufacture dry-salted cheese with 'eyes' provided that the chemical composition is similar to that of traditional brine-salted cheeses and if the starter contains gas-producing strains.[5]

2 MANUFACTURE OF CHEDDAR CHEESE

Dramatic changes in the manufacture of Cheddar cheese have occurred during the past 25 years. The single most important factor has been the availability of reliable starter cultures because the successful development of continuous mechanized systems for Cheddar manufacture has depended upon the ability of the cheesemaker to control precisely both the expulsion of moisture and the increase in acidity required in a given time. This in turn has led to the recognition that the quality of cheese, now being made on such a large scale in modern cheese plants, can only be guaranteed if its chemical composition falls within predetermined ranges. Nevertheless, Cheddar cheese is still a relatively difficult variety to manufacture since the long ripening period necessary for the development of the required mature flavour can also be conducive to the formation of off-flavours. In addition, its texture can vary considerably. The intermediate position of Cheddar cheese in the total cheese spectrum[3] (Fig. 1) is particularly exemplified by its textural properties which lie between the crumbly nature of Cheshire and the plastic texture of Gouda.

The traditional manufacture of Cheddar cheese consists of: (a) coagulating milk, containing a starter culture, with rennet; (b) cutting the resulting coagulum into small cubes; (c) heating and stirring the cubes with the concomitant production of a required amount of acid; (d) whey removal; (e) fusing the cubes of curd into slabs by cheddaring; (f) cutting (milling) the cheddared curd; (g) salting; (h) pressing; (i) packaging and ripening. Although it is impossible to separate the

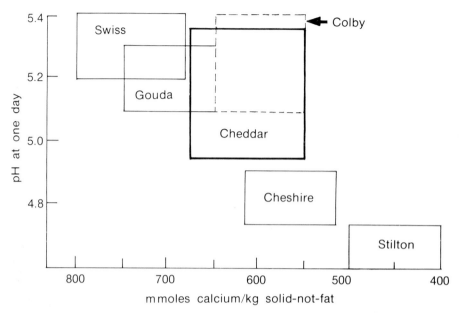

Fig. 1. Classification of traditionally manufactured cheese varieties by their characteristic ranges of ratio of calcium/solid-non-fat and pH.

combined effects of some of these operations on the final quality of the cheese, they will as far as possible be considered individually.

2.1 Effect of Milk Composition and Starter Culture

Cheesemaking basically consists of the removal of moisture from a rennet coagulum (Fig. 2). The three major factors involved are the proportion of fat in the curd, the cooking (scalding) temperature and the rate and extent of acid production.[4] In order to achieve uniform cheese quality in large commercial plants, the manufacturing procedures must be as consistent as possible. The first requirement is uniformity of the raw milk. This is achieved by bulking the milk in a silo to even out differences in milk composition from the various districts supplying milk to the cheese plant. Preferably, the milk should be bulked before use so that its fat content can be accurately standardized. For Cheddar cheese varieties, the milk is normally standardized to a casein/fat ratio between 0·67 and 0·72. The more fat present in the cheese milk, and therefore in the rennet coagulum, the more difficult it is to remove moisture under the same manufacturing conditions since the presence of fat interferes mechanically with the syneresis process.[8]

The manufacture of Cheddar cheese is more dependent upon uniform starter activity than that of washed curd cheeses, such as Gouda. The proper rate of acid development, particularly before the whey is drained from the curd, is essential if the required chemical composition of the cheese is to be obtained.[3,9]

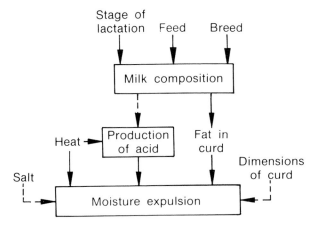

Fig. 2. Main factors in the expulsion of moisture from a rennet coagulum.

However, the curd is 'cooked' to expel moisture at a temperature which normally adversely affects the starter bacteria. The cheesemaker must therefore exert judgement to ensure that the desired acid development in the curd is reached at about the same time as the required moisture content. It is not surprising that the defined single strain starter systems developed in New Zealand[3,10-12] have been based upon the temperature sensitivity of the selected strains as well as their phage resistance and acid-producing ability. These defined starter systems have now also been widely adopted in the United States[13] and Ireland[14] and are replacing the undefined commercial mixed-strain cultures, of the type still almost universally used for the manufacture of Dutch-type cheeses.[15] If the cooking temperature is kept constant (for instance at 38°C) throughout the cheesemaking year and standardized milk is used, by far the most important factor in producing Cheddar cheese of uniform quality is the extent of acid production in the vats. To compensate for seasonal changes in milk composition it is normally only necessary to vary the percentage inoculum of starter to achieve the required acidity at draining.

2.2 Effect of Coagulant

The proportion of rennet added should be the minimum necessary to give a firm coagulum in 30 to 40 min. To achieve a similar firmness of coagulum throughout the season may involve the addition of calcium chloride and/or an increase in the temperature of the milk at which the rennet is added. The early stages of Cheddar cheese manufacture, specifically gel assembly and curd syneresis, have recently been reviewed in detail[16,17] (see Volume 1, Chapter 5). Electron microscopy studies[18-20] have shown that the casein micelles, which are separate initially, aggregate into a network, coalesce and finally form a granular mass. The fat globules, also separate at first, are gradually forced together as a result of shrinkage of the casein network. After the coagulum is cut, the surface fat globules are exposed and washed away as the curd is stirred. This leaves a thin layer depleted of fat at the curd granule surface. During matting, the layers of adjacent curd granules fuse, leading to the formation of fat-depleted junctions.[21] Starter bacteria are trapped in the casein network near the fat-casein interface, which has been shown to be the region of highest water content in the mature cheese.[18] In all cheese varieties, the outline of the original particles of curd formed when the rennet coagulum is cut can be readily distinguished by scanning electron microscopy.[22] In addition, in traditionally-made Cheddar cheese, the boundaries of the milled curd pieces can also be seen.[21] These curd granules and milled curd junctions in Cheddar cheese are permanent features which can still be distinguished in aged cheese.

The rennet coagulum consists of a continuous matrix of an aqueous protein mass interspersed with fat globules. The protein matrix, in turn, consists of small protein particles held together as a network by various binding forces. Several reports[23-25] conclude that the microstructure of the coagulum produced by the different types of milk coagulant is a major factor determining the structure and texture of Cheddar cheese. It has been suggested[24] that 'the structure of the protein network is laid down during the initial curd forming process and is not fundamentally altered during the later stages of cheesemaking and that the fibrous and more open framework of curd formed by bovine and porcine pepsins might be the reason for the softer curd associated with their use'.[23] This implies that different milk coagulants significantly affect the initial arrangement of the network of protein structural units. It is more likely, however, that the proportion of minerals lost from the rennet coagulum, as a result of the change in pH, largely determines the texture of a cheese. As one would expect, the type of rennet used affects the degree of proteolysis as the cheese ripens[20,26] (cf. Volume 1, Chapters 3 and 10).

2.3 Effect of Heating (Cooking) the Curd

During cooking, the curd is heated to facilitate syneresis and aid in the control of acid development. The moisture content of the curd is normally reduced from approximately 87% in the initial gel to below 39% in the finished Cheddar cheese. The expulsion of whey is aided by the continued action of rennet as well as the combined influence of heat and acid. The rate at which heat is applied was once considered to be important[27] but this is less relevant now that more reliable starter cultures are available. The temperature should be raised to 38–39°C over a period of about 35 min. The curd shrinks in size and becomes firmer during cooking.

2.4 Acid Production at the Vat Stage

The single most important factor in the control of Cheddar cheese quality is the extent of acid production in the vat (Fig. 3) since this largely determines its final pH[28,29] and the basic structure of the cheese.[4] As the pH of the curd decreases there is a concomitant loss of colloidal calcium phosphate from the casein

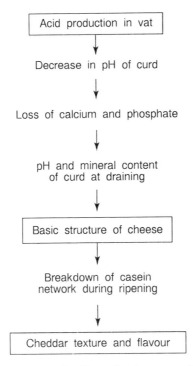

Fig. 3. Relationship between extent of acid production up to the draining stage and production of Cheddar cheese flavour and texture.

sub-micelles and, below about pH 5·5, a progressive dissociation of the sub-micelles into smaller aggregates.[30] As the amount of rennet added and the temperature profile are normally constant in the manufacture of Cheddar cheese, the pH change in the curd becomes the important factor in regulating the rate of whey expulsion.[3,27] In mechanized cheesemaking systems, the cheese is usually made to a fixed time schedule. In New Zealand, the time between 'setting' (addition of milk coagulant) and 'running' (draining of whey from curd; also called 'pump out') is normally 2 h 30 min ± 10 min (Fig. 4). The percentage of starter added determines the increase in titratable acidity between 'cutting' and 'running'. The extent of acidity increase at this stage is particularly important since it also controls the increase in titratable acidity from 'drying' (when most of the whey has been removed) onwards.[31] The actual increase in titratable acidity may need to be adjusted at intervals throughout the year to achieve the required pH in the cheese at one day. This depends on changes in the chemical composition of the milk, which in turn are determined by both the feed of the cow and the lactational cycle. The pH at draining also determines the proportions of residual chymosin (calf rennet) and plasmin in the cheese.[3,26,32] Chymosin plays a major role in the degradation of the caseins during ripening and in the consequent development of characteristic cheese flavour and texture.

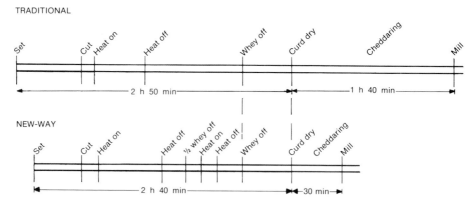

Fig. 4. Schedules for the manufacture of Cheddar cheese by the traditional and 'New-Way' processes.

While curd remains in the whey there is a continual transfer of lactose to the curd. The whey thus provides a reserve of lactose which prevents any great decrease in lactose concentration in the curd. After the whey has been removed this reserve is no longer available and the lactose content of the curd falls rapidly as the fermentation proceeds. Curd that has been left in contact with the whey for a longer period has a higher lactose content than curd of the same pH value from which the whey has been removed earlier.[31,33] Acid production can be under complete control only if defined starter systems, such as the 'multiple strain' and 'single pair' systems recently developed in New Zealand,[3] are used. Use of these cultures has allowed New Zealand cheesemakers to reduce the time from 'set' to 'salt' to about 4 h 30 min (Fig. 4). Even shorter times are potentially possible but these are limited by the rate at which moisture can be expelled from the curd in the traditional Cheddar process. Experience has shown that it is preferable to produce lactic acid relatively slowly during the early stages of curd formation and cooking, followed by a steady and increasing rate after draining the whey from the curd, since this retains more of the calcium phosphate in the curd. Whitehead & Harkness[34] pointed out that a major acceleration of the process makes considerable demands on the skill of the cheesemaker. They doubted whether a manufacturing time of less than 3 h 30 min would be practical for traditional Cheddar cheesemaking since the rate of moisture expulsion must be kept in balance with acid development.

A 'New-Way' method has, nevertheless, been developed[35-37] in Australia in which the time from 'set' to 'salt' has been reduced to about 3 h 10 min (Fig. 4). This method differs from traditional manufacture in four main respects: (1) a thermophilic starter is used in addition to the normal mesophilic strains; (2) half the whey is run off about 25 min after the first 'cook' (38°C) has been completed; (3) a second 'cook' stage, 10 min later, further raises the temperature of the curd to 43–46°C; (4) the remaining whey is then run off and the curd cheddared for 30 min only. While this short-time method has not been adopted commercially, the technique of draining off part of the whey at some stage after the completion of 'cooking' is relatively common.[38] The advantage of this practice is an increased rate of syneresis, probably due to a greater frequency of impact of the curd particles during stirring.

2.5 Effect of Cheddaring

The series of operations consisting of packing, turning, piling and repiling the slabs of matted curd is known as cheddaring. The curd granules fuse under gravity into solid blocks. Under the combined effect of heat and acid, matting of the curd particles proceeds rapidly. The original rubber-like texture gradually changes into a close-knit texture with the matted curd particles becoming fibrous. The importance attached to flow in the past varied markedly from country to country. In Britain, it was common for each Cheddar block to be made to spread into a thin, hide-like sheet covering an area of several square feet, whereas in New Zealand only moderate flow was induced, the final Cheddar block being little different in dimensions from when first cut. Investigations by Czulak and his colleagues initially led them to conclude[2,39-41] that extensive deformation and flow were essential in Cheddar cheesemaking. Further research in Australia,[42] New Zealand[43] and Canada,[21] however, has slowly led to the view that 'cheddaring' is not an essential step and serves no purpose other than to provide a holding period during which the necessary degree of acidity is developed and further whey can be released from the curd. This loss of whey is controlled by the acidity and temperature of the curd. The temperature is important both directly and indirectly since the rate of acid development is also influenced by temperature. In general, higher temperatures during cheddaring increase the expulsion of whey from the curd. In the traditional process, manual manipulations of the curd, i.e. the cutting of the matted curd into different size blocks, the height of piling and the frequency of turning the curd blocks, also aid in moisture control.

Mechanical forces—pressure and flow—have been shown[2,39] to be an important factor in the development of fibrous structure in the curd. This is clearly seen in the arrangement of fibres, which follow the direction of the flow. Fibrous structure cannot be brought about by pressure and deformation, however, unless the curd has reached a pH of 5·8 or less.[2] This suggested that pressure and flow serve to knit, join, stretch and orientate the network of casein fibres already partly formed in response to rising acidity. The readiness to flow, the type of fibres and the density of their network are also influenced by temperature and moisture. The warmer the curd and the higher its moisture content, the more readily it flows and the finer, longer and denser are the fibres. These investigators[2] also concluded that it is possible to influence curd structure by manipulating pH, pressure and temperature and that a direct relationship exists between the structure and the water-holding capacity of the curd. This was confirmed by Olson & Price[44] who showed that extension and rapid flow of curd during cheddaring produced a higher moisture content in the resulting cheese.

Fluorescence microscope studies have demonstrated the change of the casein from spherical granular particles to a fibrous network.[41] While some granular structure was evident in curd grains, the conversion to the fibrous form was complete in cheddared curd. The fibrous shreds of cheddared curd consist of flattened, elongated curd particles that overlap each other, forming a network-type structure with the protein as a continuous phase. The exact mechanism responsible for these observed changes in cheddared curd is not known with certainty, but the loss of minerals from the casein micelles in the curd is likely to be the major factor. The loss of calcium phosphate will destabilize the casein micelles, resulting in a change in the conformation of the caseins. The concomitant loss of moisture from the casein micelles may also possibly contribute to the conformational change.

Czulak[42] concluded that the characteristic close texture of Cheddar cheese could be obtained without cheddaring. He suggested, however, that in mechanizing the cheesemaking process it was probably most convenient, while holding the curd for acidity to develop, to allow the particles to mat together 'but to apply no labour or equipment for its fusing beyond that necessary for ready handling'. Almost all modern mechanized Cheddar cheesemaking systems are based upon these conclusions and involve little or no flow of the curd. This development was supported by the success achieved in the manufacture of cheese of normal Cheddar characteristics, particularly in the United States, by 'the stirred curd' process. This strongly indicated that flow and the cheddaring process itself are of little or no significance in the Cheddar cheesemaking process. Similar conclusions were also reached by research workers[43] in New Zealand.

2.6 Effect of Milling

The milling operation consists of mechanically cutting the cheddared curd in small pieces in order to: (a) reduce the curd block and so enable more uniform salt distribution into the curd; (b) encourage whey drainage from the curd; (c) assemble the curd in a convenient form for hooping. There is a practical upper limit to the cross-section of milled curd before salting for two reasons: (a) there is inadequate whey drainage after salting with large particles, (b) the larger the curd particles, the smaller is the surface/volume ratio. With larger sized particles, a higher salting rate is therefore required to achieve a given final level of salt-in-moisture (S/M) in the cheese. This increases the chance of seaminess[45] and gives higher salt losses in the whey.[46] The longer time required for salt penetration allows a greater development of acid in the centre of large curd particles than in smaller particles and this may result in a 'mottled' colour in the final cheese.

Gilbert[47] has pointed out that ideally the curd should be cut into spheres to obtain a uniform mass/surface area profile. The best that can be achieved,[47,48] however, is to use a curd mill that produces a shredded curd, flakes of curd or pieces of curd resembling a finger. The more uniform the ratio of surface area to curd mass after milling, the more uniform will be the rate of salt diffusion into the milled curd particles and the proportion of salt retained. It is worth noting that these conditions are more closely satisfied if the curd is not cheddared but is kept in the granular state prior to salting.

2.6.1 Mellowing Prior to Salting
In the traditional procedure for Cheddar cheese manufacture, the milled 'chips' were left until the newly cut surfaces glistened as a mixture of whey and fat exuded from them. The mellowing period provided time to produce sufficient surface moisture to dissolve the salt crystals when they were applied, and gave rise to higher salt retention. In a sense, therefore, the milled curd was being brine-salted.[47] The real purpose of the traditional mellowing period ('dwell time'), however, was to allow for further moisture release and acidity increase. In mechanized Cheddar cheese plants, a mellowing period of only 5 to 10 min has proved sufficient since the use of the improved culture systems now available has greatly simplified the control of moisture expulsion from the curd. In addition, the acidity at which curd is commonly salted these days has been considerably reduced. In some mechanized cheesemaking systems, salt is added to the curd immediately after milling and continuous agitation of the milled particles is used to encourage whey flow and brine development.

2.7 Effect of Salting

The major difficulty in achieving cheese of uniform quality in modern Cheddar cheese plants results from the relatively wide variation in salt-in-moisture (S/M) levels that occurs in the cheese.[5] Salt (and more specifically S/M) plays a number of roles in the quality of Cheddar cheese by controlling: (a) the final pH of the cheese,[28,49] (b) the growth of microorganisms, specifically starter bacteria and undesirable species such as coliforms, staphylococci and clostridia, and (c) the overall flavour and texture of the cheese. The S/M level controls the rate of proteolysis of the caseins by the rennet, plasmin and bacterial proteases. Proteolysis, and thus the incidence of bitterness and other off-flavours, decreases with an increase in salt concentration.[49,50] At S/M levels >5·0, bitter flavours are rarely encountered;[51] below this level there is more or less a linear relationship between S/M and the incidence of bitterness. General aspects of salt in cheese were considered in Volume 1, Chapter 7; some specific aspects in relation to Cheddar are considered below.

2.7.1 Salting of Milled Curd
The salt crystals dissolve on the moist surfaces of the milled curd particles and form a brine. This diffuses into the curd matrix through the aqueous phase, causing the curd to shrink in volume, and more whey is thereby released to dissolve more salt. The proportion of moisture in the curd and the amount of salt added both affect the rate of solution of the salt. The high salt content of the surfaces of the milled curd particles reduces the tendency of the particles to fuse together. The difference between dry-salting and brine-salting is, in effect, the

availability of water at the surface of the curd. With brine-salting, salt absorption begins immediately; release of whey occurs, as in dry-salting, but is not a prerequisite for salt absorption. The faster rate of salt absorption from brine has prompted the use of brine or salted whey in mechanized salting procedures[52] but none has yet met with commercial success.

While salt does promote syneresis, it should not be used in mechanized Cheddar cheesemaking as a means of making a significant adjustment to the moisture content of the curd. In modern cheese plants it is essential that the curd particles prior to salting are consistent from day to day with respect to moisture content, acidity levels, temperature, cross-sectional size and structure (degree of cheddaring), and that the application of salt is uniform. This gives the cheesemaker control over both the mean salt content and, equally important, the variation (standard deviation) within a day's manufacture. Cheese specifications normally require both moisture and S/M to be within specified ranges. This means in practice that variations in the moisture content of the curd prior to salting must not be greater than $\pm 1\%$.

It has been suggested[47,53] that the size of the salt crystals used is important for both salt uptake and moisture control. In practice, however, the major requirement in mechanized cheese plants is that the size range of the salt crystals should be narrow. If the range is variable, the delivery of salt from the equipment is erratic. The presence of high amounts of very fine crystals also results in excessive salt dust within the plant environment.

2.7.2 Mellowing after Salting

Sufficient time must be allowed after salting (the mellowing period) to ensure the required absorption of salt on the curd surfaces and continued free drainage of whey. With the better starter systems now available there is no longer any necessity for cheesemakers to attempt to control the moisture content of the cheese by varying the amount of salt added. It was earlier suggested[54] that the curd could be hooped as soon as it had been salted. This has, however, led to problems in cheese made by these shorter processes,[54] specifically to the entrapment of whey and consequently to excessive moisture and uneven colour in the cheese. As a result, a number of investigations have been carried out recently to determine the factors that influence the amount of salt absorbed and the speed of its absorption.[46-48,53]

The amount of salt absorbed by the curd and the rate of subsequent whey drainage are related to the availability of dissolved salt on the curd particle surfaces, and to the physical characteristics of the curd, e.g. fat-free curd allows faster diffusion.[55] Even when a holding time of more than 30 min is maintained and the rate of salt solution addition is uniform, large variations may still occur in the salt content of cheeses because other conditions that affect salt absorption are not controlled. For instance, the curd temperature, the depth of curd, the extent of stirring after salt addition, and the degree of structure development in the curd are also significant factors in the control of salt absorption and subsequent whey drainage.[46,53] It is not surprising, therefore, that there have been conflicting reports as to how long the mellowing period after salting should be. It is clear that at least 15 min holding is necessary to minimize loss of salt during pressing.[48] Other reports suggest that pressing of the salted curd should be delayed for at least 30 min[46] and preferably for 45–60 min.[48] Some loss of salt occurs even when the holding period is extended to 60 min. An increase in the time of holding, however, substantially reduces the proportion of whey expelled during pressing and greatly improves the degree of salt absorption.[53] Nevertheless, experience in mechanized cheese plants has shown that a mellowing period of only 15–20 min is usually adequate for satisfactory salt uptake and whey removal.[5]

The irregular effect of curd temperature on the extent of salt absorption was thought[48] to be caused by a protective layer of fat exuding from the surfaces of curd particles. Less fat was present on curd surfaces at 26°C than at 32°C. Above 38°C such fat was melted and dispersed in the brine solution that was present on the surface. In general, however, a decrease in the temperature of the curd at salting increases the S/M of the final cheese.[53] Curd salted at a high pH retains more salt[31] and is more plastic than curd salted at a low pH. Similarly, for a given salting rate, the S/M is high when the titratable acidity is low.[46] Salting the curd under the most favourable conditions for salt absorption reduces the proportion of salt required, and thereby reduces salt losses,[46] and also helps to overcome the defect of seaminess.[54,56]

2.7.3 Equilibrium of Salt within a Cheese

The rate of penetration of salt into cheese curd is very slow[55,57–59] and a mean diffusion rate of 0·126 cm²/day for salt in the water of Cheddar cheese has been reported.[55] This corresponds well with salt migration values for Gouda cheese of the same moisture content,[58] suggesting that the matrix structures of the two cheese types are similar. Despite the low salt diffusion rate it was nevertheless believed that the S/M concentration in Cheddar cheese was essentially uniform within a few days.[59] Reports[49,55,60–62] that wide variations in salt content occur between blocks from the same vat and even within a block now suggest that, if salt distribution is not uniform initially, equilibrium will not be attained during the normal period of ripening of that cheese. The appreciable variation in the salt and moisture contents of small plug samples taken from different cheeses from the same vat[49,55,60] demonstrates how difficult it is to manufacture cheese to a uniform S/M level. As the consistency of cheese flavour is directly related to the extent of variability in S/M, the need to produce a curd mass consisting of particles of uniform cross-section at the time of salting cannot be over-emphasized (Fig. 5).

2.7.4 Seaminess and Fusion

When curd particles are dry-salted, discrete boundaries are set up between the individual particles, in contrast to brine-salted cheeses where there is only one boundary, i.e. the cheese rind or exterior. The addition of dry salt causes shrinkage of the curd and a rapid rate of release of whey containing calcium and

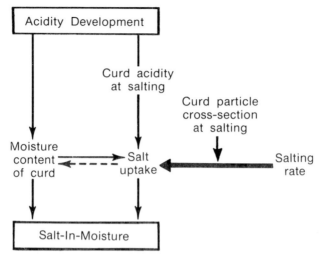

Fig. 5. Main factors that affect the salt uptake and S/M level in Cheddar cheese.

phosphate, particularly in the first few minutes of pressing. It has been suggested that the salted surface of the curd particle acts as a selective permeable membrane, thereby concentrating calcium and phosphate at the surface of the curd particle.[59] It is possible that this calcium gradient is also accentuated, under some circumstances, by the variations in pH between the surface and interior of the salted curd particle owing to inhibition of starter activity by the high salt concentration at each curd boundary. The establishment of a pH gradient leads, in turn, to a shallow calcium gradient,[63] the magnitude of which will depend upon the size of the curd particle and the proportion of salt added. In its most extreme form, the deposition of calcium phosphate crystals results in the phenomenon of seaminess in Cheddar cheese,[45,54,64] a condition in which the junctions of the milled curd particles are visible after pressing. Seaminess is more frequent and more marked with cheese of low moisture and high salt content and in some cases persists after the cheese has matured.[56] The binding between curd particles is usually weak, due to incomplete fusion. This often leads to crumbling when the cheese is sliced or cut into small blocks for packing.

Photomicrographs show that, in both seamy and non-seamy Cheddar cheese, crystals of calcium *ortho*phosphate dihydrate are dispersed throughout the cheese mass,[45] but in seamy cheese they are concentrated in the vicinity of the surfaces of the milled curd particles to which salt was applied. To a depth of about 20 μm below these surfaces, the protein appears to be denser than elsewhere, suggesting that severe dehydration of the surface occurs on contact with dry salt. The observation[27,54] that seaminess is reduced by washing the curd after milling and before salting can be explained by the removal of calcium phosphate from the surface layer. In addition, the provision of more water will lessen the dehydrating and contracting effect of salt on the surface layer.

Seaminess and poor bonding between the curd particles occur together and treatment with warm water corrects both defects. Poor fusion of the curd as a consequence of heavy salting results from irreversible changes in the protein at the surface, from poor contact between the hardened surfaces, from the physical separation brought about by the presence of salt crystals and, when these have disappeared, from the growth of the calcium *ortho*phosphate crystals.[45] Fusion of the particles is improved by an increase in the pH, temperature or moisture content of the curd.

2.8 Effect of Pressing

Traditionally, Cheddar cheese was pressed overnight using a batch method. The recently developed 'block-former' system[65,66] has two major advantages for modern cheesemaking plants: firstly it is a continuous process and secondly the residence time is reduced to about 30 min. The curd is fed continuously into an extended hoop (tower) under a partial vacuum and a very short period of mechanical pressure is applied at the base of the tower, usually for about 1 min.

In traditionally made Cheddar, the two common types of textural defect are mechanical- and slit-openness. Mechanical-openness (occurrence of irregularly shaped holes) is evident in very young cheese but decreases markedly during the first week or two after manufacture and thereafter changes little.[67-69] On the other hand, slit-openness is usually absent in freshly made cheese[70] but develops during maturation.[71] The extreme expression of this defect, known as fractured texture, is found only in mature cheese. A comprehensive survey of commercial Cheddar cheese on the UK market carried out during 1958–61 found that mechanically open cheese was usually almost free of fractures and conversely that badly fractured cheese usually had few mechanical openings.[72] The term 'fracture' is normally used to describe long slits, i.e. slits longer than about 3·5 cm. As a result of the growth of the cheese pre-packaging trade in recent years, the importance attached to fractures in cheese has greatly increased since fractures can result in the breaking up of cheese during pre-packaging.

From the observation[73] that cheese hooped under whey had a completely close texture and from their own studies of curd behaviour, Czulak & Hammond[39] concluded that air entrapped during compression of curd was responsible for mechanically open texture. They considered that during compression of the salted, granular curd, the spaces between the granules diminish until they form a complex of narrow channels filled with air. Under further pressure some of the air is forced out, the escape of the remainder being blocked by closure of the channels at various points and the high surface tension developed by traces of whey in the remaining narrowed outlets. The isolated pockets of trapped air form numerous small irregular holes in the cheese. Conventionally made Cheddar cheese has a significantly closer texture than Granular cheese. The effect of cheddar-ing on texture appears to be due to the presence of milled strips of curd (fingers) compared with the relatively small granules of uncheddared curd present in Granular cheese. The larger the fingers of curd, the fewer are the pockets of trapped air and the closer is the texture of the cheese. As mentioned previously, however, there is a practical limit to the size of milled curd since large curd fingers may result in inadequate whey drainage after salting.

During the last 30 years there has been a marked reduction in the incidence of both texture defects by: (a) the use of higher pressures during curd fusion;[74] (b) the change from the manufacture of rinded 36 kg cheeses to smaller 20 kg rindless cheeses; (c) the introduction of vacuum pressing; (d) the use of defined single strain cultures from which gas-producing strains have been omitted. The beneficial effects of these modifications are undoubtedly associated with a reduction in the gas content of the cheese. The production of carbon dioxide during ripening by non-starter bacteria has been associated with the development of slit-openess[71] but gas production is considered to be of secondary importance when compared with manufacturing conditions.[75] It is the relatively insoluble and biologically inactive nitrogen in the entrapped air which contributes to the ultimate openness of the cheese since the oxygen is rapidly metabolized during ripening.

2.8.1 Vacuum Pressing

It was a logical step to prevent the entrapment of air between the curd particles by pressing the curd under a vacuum, a procedure first[76] patented in Canada in 1959. A moderately high vacuum is required, approximately 33 kPa pressure. Vacuum-treated cheese is free, or almost so, of mechanical openings when two weeks old and remains free throughout maturation.[70] There was some disagreement among the various research groups as to the optimum conditions for vacuum pressing.[72] Initially, the cheddared and salted curd was pre-pressed under a vacuum for 30 min before dressing, followed by normal pressing.[67] Later work suggested that pressures greater than 180 kPa appeared to be required during and after vacuum pressing to achieve close texture.[72]

An important development in Australia was the hooping of granular, salted curd and pressing under vacuum.[42] It was found that the use of vacuum pressing ensured the characteristic close texture of Cheddar cheese, and thus eliminated the need for cheddaring. This observation was particularly significant for the complete mechanization of Cheddar cheese manufacture. Trials in New Zealand[70] quickly confirmed the Australian conclusions. Maximum reduction in openness was achieved with the combined use of vacuum pressing of granular curd and a homofermentative starter.[71] Presumably, air can be removed more readily by vacuum from granular curd than from the closer textured cheddared curd. The recently developed technique used in the 'block-former' system of filling hoops under a partial vacuum is particularly effective in achieving a close texture. Mechanical-openness is never found in cheese made by the 'block-former' system but slit-openness does develop if gas producing organisms are present.

A factor which formerly restricted the size of Cheddar cheese blocks was the tendency for large cheeses to show severe mechanical-openness. With the aid of vacuum pressing it has been found quite practicable to form curd into very large blocks[77] which, by extrusion into cutting equipment, can be sub-divided into 20 kg blocks.

3 CHEMICAL COMPOSITION AND CHEDDAR CHEESE QUALITY

Recent developments in cheese marketing have resulted in a demand for cheese of greater uniformity of composition than in the past. Such uniformity is best achieved by a grading system based on compositional analysis, since a relationship between the composition and quality of Cheddar cheese is now well established.[52,78–82] Lyall[81] briefly reported on a procedure for evaluating chemical analyses of cheese, points being assigned on the basis of composition. However, the only scheme in commercial use for assessing Cheddar cheese quality by compositional analysis appears to be that proposed by Gilles and Lawrence.[80] Suggested ranges of moisture in the non-fat substance (MNFS), salt-in-moisture (S/M), fat-in-the-dry matter (FDM) and pH for both first and second grade cheeses are given in Fig. 6. All New Zealand export Cheddar cheese is now subject to compositional grading to ensure that atypical cheese is segregated. In addition, a sensory flavour assessment is made to ensure that the cheese is free from flavour defects.[83] A recent review[84] has concluded that grading on the basis of composition may be a satisfactory method for deciding which cheese should be allowed to mature for the British market and which should be sold more quickly.

Any grading system based on compositional analysis will be relatively complex since a further factor, the rate and extent of acid production at the vat stage, must also be considered.[3,85] The point in the process at which the curd is drained from the whey is the key stage in the manufacture of Cheddar cheese since it controls to a large extent its mineral content, the proportion of residual chymosin in the cheese, the final pH and the moisture to casein ratio.[3] All of these influence the rate of proteolysis in the cheese. A relationship has also been found between the calcium content of the cheese, the concentration of residual chymosin and the protein breakdown during ripening[4] and between the rate of

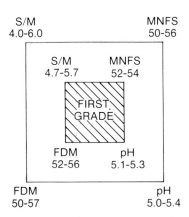

Fig. 6. Suggested ranges of salt-in-moisture (S/M), moisture in non-fat substance (MNFS), fat-in-dry matter (FDM) and pH for First grade (shaded) and Second grade Cheddar cheese. Analyses 14 days after manufacture.

acid development during the early stages of manufacture and proteolysis in the cheese.[86] The calcium level is therefore an index of the extent of acid production up to the draining stage and also an indication of the rate of proteolysis that is likely to occur during ripening. Significant differences in the calcium content of Cheddar cheese would suggest differences in the proportion of residual chymosin in the cheeses and thus differences in the rate of proteolysis and the development of flavour.

Variations in calcium content, however, have a much smaller effect on Cheddar cheese quality than MNFS, S/M and pH. It is important to recognize that these three parameters are interrelated[85] and must be controlled as a group to ensure first grade cheese. Nevertheless, the effect of each of these factors will, as far as possible, be examined separately.

3.1 Effect of MNFS

There is considerable circumstantial evidence that the main factor in the production of the characteristic flavour of hard and semi-hard cheese varieties is the breakdown of casein. This is supported by the finding that the ratios of moisture to casein and of salt to moisture are critical factors in cheese quality[80,85] since both parameters affect the rate of proteolysis in a cheese.[49] Traditionally, cheesemakers describe cheese in terms of its absolute moisture content but the ratio of moisture to casein is much more important since it is the relative hydration of the casein in the cheese that influences the course of the ripening process.[83] It is difficult, however, to measure the casein content of cheese accurately and most commercial plants analyse only for fat and moisture. A practical compromise, therefore, is to determine the ratio of moisture to non-fat substances (MNFS) rather than measure the moisture to casein ratio. The non-fat substance is not the same as the casein in the cheese but is equal to the moisture plus the solids-not-fat. Casein represents approximately 85% of the solids-not-fat. There is not, therefore, a strong relationship between the moisture to casein ratio and the MNFS. It is important to note, however, that changes in MNFS for any particular cheese variety correlate well with changes in the ratio of moisture to casein.

The MNFS value for cheese gives a much better indication of potential cheese quality than the moisture content of the cheese in the same way as the S/M ratio is a more reliable guide to potential cheese quality than is the salt content of the cheese *per se*.[83] In mechanized cheese plants, a significant relationship exists between the FDM and MNFS values in a cheese,[85] probably as a result of the relative inflexibility of the procedures available for the control of moisture. This is of commercial interest since changing the FDM is an effective way of controlling the MNFS in the cheese as the composition of the milk changes throughout the season. The actual MNFS percentage for which a cheesemaker should aim depends upon the storage temperatures used and when the cheese is required to reach optimum quality. Experience has shown that if Cheddar cheese is to be stored at 10°C, and the cheese is consumed after six to seven months, then the MNFS of the cheese should be about 53%. The higher the MNFS percentage, the faster is the rate of breakdown. Thus, if one anticipates that the cheese will be consumed after three to four months, the MNFS percentage can be increased to about 56%. However, the higher the MNFS the more rapidly Cheddar cheese will deteriorate in quality after reaching its optimum. The same is true for a Cheddar cheese with a relatively low S/M, i.e. less than 4%, or with a high acid content (i.e. a low pH, <4·95). Such cheeses tend to develop gas and sulphide-type off-flavours after they have reached maturity.

3.2 Effect of pH

Every cheese variety has a characteristic pH range,[3] within which the quality of the cheese is dependent upon both its composition and the way in which it is manufactured.[4] The pH value is important in that it provides an indication of the extent of acid production throughout the cheesemaking process. In normal manufacture, the curd acidity at salting is a key factor in determining the pH of dry-salted cheese (Fig. 7). The salting acidity is, however, to a large extent controlled in turn by the acidity developed at draining.[28] The potential for a further decrease in pH after salting depends upon the residual lactose in the curd and its buffering capacity. The residual lactose will be determined by the rate at which an inhibitory level of NaCl is absorbed by the cheese curd and salt tolerance of the starter strains used. The buffering capacity is largely determined by the concentrations of protein and phosphate present, and to a much lesser extent by ions such as calcium. The percentages of phosphate and calcium retained in the cheese are influenced mainly by the rate of acidification prior to the separation

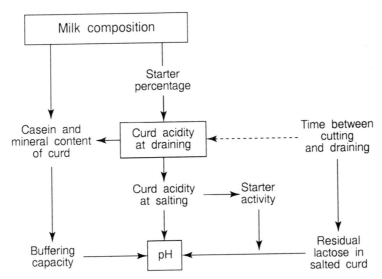

Fig. 7. Main factors that determine the pH of Cheddar cheese.

of the whey from the curd. The buffering capacity is also influenced by seasonal, regional and lactational factors.

Given reliable starter activity at the vat stage, the actual pH reached in dry-salted cheeses is determined by the S/M value since this controls the extent of starter activity after salting, the rate of lactose utilization in the salted curd and thus the final pH reached. A S/M concentration of 6% will inhibit the activity of all *Lactococcus cremoris* strains, the starter organisms of choice for Cheddar manufacture.[12] The proportion of residual lactose that remains unmetabolized in such cheese will be high even after 2 months.[6] In a cheese with a S/M of 4·5%, however, the starter will not be completely inhibited and the lactose will be rapidly metabolized. This explains why the pH values of one-day Cheddar cheese may range from 5·3 (which is about the pH of the curd at salting) to pH 4·9. In general, the higher the pH the greater is the amount of lactose left unmetabolized. Under normal circumstances, this residual lactose does not affect the quality of Cheddar cheese at maturity.[5]

The importance of measuring the pH at one day has been generally over-looked in the past, probably because it is relatively difficult to measure the pH of cheese accurately. This has led to a lack of appreciation of the significance of relatively small differences in pH. In addition, a pH value *per se* is sometimes difficult to interpret unless considered in conjunction with the calcium level in the cheese.[83]

3.3 Effect of S/M

The main factors that determine the S/M percentage in Cheddar cheese are summarized in Fig. 5. In young Cheddar cheese, the S/M ratio is the major factor controlling water activity. This in turn determines the rate of bacterial growth and enzyme activity in the cheese, specifically the proteolytic activity of chymosin,[50,87,88] plasmin[89] and starter proteinases.[90] If the S/M value is low (<4·5%), the starter numbers will reach a high level in the cheese and the chance of off-flavours due to starter is greatly increased.[91,92] For this reason, cheesemakers normally aim for a S/M value in Cheddar cheese between 4·5 and 5·5%.[28,83] Within this S/M range, the rate of metabolism of the lactose is controlled by a second factor, the temperature of the cheese during the first few days of ripening, since this controls the rate of growth of non-starter bacteria such as lactobacilli and pediococci.[7] While non-starter bacteria grow on substrates other than lactose in cheese, undoubtedly the presence of lactose encourages their rapid growth. This tends to result in a more heterolactic metabolism of lactose, usually with the production of acetate, ethanol and carbon dioxide and may lead to flavour and textural defects. Clearly, the initial numbers of non-starter bacteria in the salted curd should be controlled by hygiene during manufacture. Thereafter, their rate of growth, particularly after the first few days of ripening, should be kept to a minimum and this is largely controlled by the temperature of the cheese.[7]

The necessity to introduce compositional ranges into the grading system is well illustrated by the fact that it is not yet possible to produce a uniform line of cheese within a day's manufacture, particularly with respect to S/M values. This does not necessarily mean, however, that some of the cheese produced is of poorer quality than the rest. The rate of ripening will differ but all of the cheese is likely to be acceptable as long as its composition is within the required compositional range. For instance, variations in the moisture content and acidity of the curd before salting, in the accuracy of salt delivery by salting equipment and in the dimensions and structure of the milled curd will result in considerable variation in salt uptake.[28] Even within a single 20 kg block of cheese, S/M values vary by as much as 1%. Nevertheless, as long as the titratable acidity at salting is normal, S/M values between 4·5 and 6·0% tend to result in acceptable cheese.[83]

3.4 Effect of FDM

The FDM percentage in Cheddar cheese is of less importance than MNFS, S/M or pH, in that it normally only influences cheese quality indirectly through its effect on MNFS.[93] Nevertheless, the FDM percentage has more relevance to the cheesemaker than the fat content *per se* because moisture is volatile and legal limits for fat are usually specified in terms of FDM. Use of FDM has the further advantage that it can be controlled directly by altering the casein to fat ratio of the milk.

A prerequisite for the manufacture of cheese of a desired composition is the use of milk with a standardized casein/fat ratio. The actual ratio required to obtain a specified FDM must be found from experience for each cheese plant since the composition of the milk will vary significantly between different milk supply areas. Milk composition fluctuates as a result of many factors, some of which, such as seasonal changes, the stage of lactation and the cows' health, are not under the cheesemaker's control. An increase of about 0·02–0·04 in the casein/fat ratio will generally cause a decrease of 1% in the FDM.[94]

4 TEXTURE OF CHEDDAR CHEESE

The texture of Cheddar cheese is recognized as being important to the consumer, yet it has not been studied extensively. This is partly due to the complexity of the situation since several variables affect the texture[93,95,96] and the effect of any one variable depends on the magnitude of the others. It is difficult, if not impossible, to make cheeses that differ with respect to one variable only, leaving the other variables unaltered. A further problem is that individual teams of workers use different nomenclature for defining textural properties. For example, the properties that they claim to be measuring are variously described as body, structure, firmness, hardness, consistency, cohesiveness, crumbliness, toughness or shortness, yet the methods used may be almost identical. On the other hand, a single property, the 'firmness' of cheese, has been measured by various research groups in several different ways, i.e. using compression, penetration, tensile and breaking methods (see Volume 1, Chapter 8).

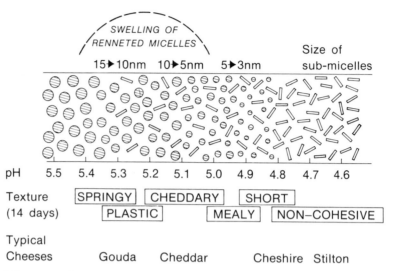

Fig. 8. Diagrammatic representation of the effect of pH on cheese microstructure and texture.

Cheddar cheese has a texture that is intermediate (Fig. 8) between those of the relatively high pH cheeses, which flow readily when a force is applied, and the low pH cheeses which tend to deform, by shattering, only at their yield point. Scanning electron microscopy has established that cheese consists of a continuous protein matrix but that this matrix is clearly different in the various cheese types.[97] The structural units in the protein matrix of Gouda are essentially in the same globular form (10–15 nm in diameter) as in the original milk. In contrast, the protein aggregates in Cheshire are much smaller (3–4 nm) and are apparently in the form of strands or chains, i.e. the original sub-micellar protein aggregates appear to have lost almost all their identity. Cheddar is intermediate between Gouda and Cheshire, i.e. much of the protein in Cheddar is in the form of smaller particles than in Gouda (Fig. 8). As the pH decreases towards that of the isoelectric point of paracasein, the protein assumes an increasingly more compact conformation and the cheese becomes shorter in texture and fractures at a smaller deformation.[96,98] The texture of Cheddar cheese has a wider range of consumer acceptability than other varieties as a consequence of its intermediate position in the cheese spectrum.

The high moisture and relatively high pH (5·2) of American Cheddar resulted traditionally in a more cohesive and waxy texture[1] than that of English and New Zealand Cheddar. In North America, relatively low levels of acid were developed in the curd up to the salting stage (less than 0·65% titratable acidity). In contrast, English cheesemakers strove for a high salting acidity (about 0·85%) with a consequently low final pH (about 4·9). New Zealand cheesemakers aimed for a final pH of 5·0 and a moisture content of about 35% in contrast to the 38 to 39% moisture level found in both American and English Cheddar. In recent times, however, most Cheddar-producing countries have tended towards the American style of 'sweet' Cheddar cheese with a final pH between 5·1 and 5·3 now being common. Care must be taken, however, not to exceed pH 5·3 since the cheese will then be curdy in texture and will take longer to reach the smoothness of texture normally associated with mature Cheddar.

4.1 Effect of pH, Calcium and Salt

Although the mineral content plays an important role in establishing the characteristic structure,[3,4] the texture of Cheddar cheese appears to be more dependent upon pH than on any other factor.[95] For the same calcium content, the texture at 35 days can vary from curdy (pH >5·3) to waxy (pH 5·3–5·1) to mealy (pH <5·1). Trials in New Zealand have shown that, for any given pH value, the calcium level of Cheddar can vary over a range of ±15 mmoles/kg with only a slight effect on the texture[5] although there is a general tendency for the cheese to become less firm as the calcium content decreases.

The dominant effect of pH on texture can, however, be modified by other compositional factors, particularly the moisture, salt and calcium contents. Between pH 5·5 and 5·1, much of the colloidal calcium phosphate and a considerable part of the casein is dissociated from the sub-micelles.[30] These changes in the size and characteristics of the sub-micelles significantly increase their ability to absorb water,[30,99–101] casein hydration reaching a maximum at about pH 5·35. More relevantly, Creamer[101] found that casein hydration in renneted milk greatly increased in the presence of NaCl between pH 5·0 and 5·4. Furthermore, at any given pH the extent of solubilization of the micelles by the NaCl decreased as the calcium concentration in the solution increased. This finding is in agreement with the effects of calcium in brine on the solubilization of the rind of Gouda-type cheese.[102] It also explains the observations that a higher Ca^{2+}/Na^+ ratio results in a firmer cheese,[98] and that Cheddar cheese made from milk to which calcium has been added has a reduced protein breakdown and is of poorer quality.[103,104] The high levels of calcium present in buffalo milk[105] may also account for the difficulty in manufacturing Cheddar cheese from buffalo milk. The extent of proteolysis is low,[106] presumably because the degree of solubilization of the casein micelles by the NaCl is reduced. As a result, the cheese needs to be stored for long periods before the characteristic Cheddar texture and flavour develop.

It is, therefore, not surprising that the texture of Cheddar cheese changes markedly as the pH varies between 5·4 and 4·9. A wide range of casein aggregates is present and differences in the sodium and calcium ion concentration, as well as the proportion of water to casein, markedly affect the extent of swelling of the sub-micelles (Fig. 8). Salt also has a more direct effect on the texture of Cheddar: excessive salting (i.e. a S/M >~6%) produces a firm-textured cheese which is drier and ripens at a slow rate,[27] whereas under-salting (i.e. a S/M <~4%) results in pasty cheese with abnormal ripening and flavour characteristics. Such factors as enzyme activity and the conformation of α_{s1}- and β-caseins in salt solutions,[87] solubility of protein breakdown products, hydration of the protein network,[58] and interactions of calcium with the paracaseinate complex in cheese[102] are all influenced by salt concentration.

4.2 Effect of Protein, Fat and Moisture

In dry-salted cheeses, water, fat and casein are present in roughly equal proportions by weight, together with small amounts of NaCl and lactic acid, Since protein is considerably denser than either water or fat, it occupies only about one-sixth of the total volume. Nevertheless, it is largely the protein matrix which gives rise to the rigid form of the cheese. Any modification of the nature or the amount of the protein present in the cheese will modify its texture. Thus, reduced-fat Cheddar (17% fat) is considerably firmer and more elastic than full-fat Cheddar (35% fat), even when the MNFS levels of the cheese are the same.[107] This difference was explained by the presence in the reduced-fat cheese of about 30% more protein matrix, which must be cut or deformed in texture assessments, but such a large reduction in fat must also affect the texture of the cheese.

Fat in cheese exists as physically distinct globules, dispersed in the aqueous protein matrix.[18] In general, increasing the fat content results in a slightly softer cheese, as does an increase in moisture content, since the protein framework is weakened as the volume fraction of protein molecules decreases. Relatively large variations in the fat content are, however, necessary before the texture of the cheese is significantly affected.[83] Commercial cheese with a high FDM usually has a high MNFS[85] and this causes a decrease in firmness. An inverse relationship between the fat content and cheese hardness has been reported.[93,108]

4.3 Effect of Ripening

Considerable changes in texture occur during ripening as a consequence of proteolysis. The rubbery texture of 'green' cheese changes relatively rapidly as the framework of α_{s1}-casein molecules is cleaved by the residual coagulant.[96] A group of Cheddar cheeses examined over a period of nearly a year increased in hardness and decreased in elasticity, with the age of the cheese, the greatest changes occurring during the first 30 days.[108] In part, this is caused by the loss of structural elements but another feature of proteolysis is probably important:[96] as each peptide bond is cleaved, two new ionic groups are generated, each of which will compete for the available water in the system. Thus, the water previously available for solvation of the protein chains becomes bound by the new ionic groups, making the cheese firmer and less easily deformed. This change, in combination with loss of an extensive protein network, gives the observed effect.

Clearly, the change in texture during ripening depends upon the extent of proteolysis which, for any individual cheese, is determined by the duration and temperature of maturation. The main factor that influences the rate of proteolysis appears to be S/M.[50,87,88] A direct relationship between S/M and residual protein was established whereas the correlation between moisture and residual protein was relatively weak. A cheese with a low S/M value has a higher rate of proteolysis and is correspondingly softer in texture than a cheese with a high S/M. The concentrations of residual rennet and plasmin in the cheese, together with the starter and non-starter proteinases present, must, however, also be important factors that determine the rate of proteolysis.[4]

5 FLAVOUR OF CHEDDAR CHEESE

Cheese ripening is essentially the controlled slow decomposition of a rennet co-agulum of the constituents of milk. Hydrolysis of the casein network, specifically α_{s1}-casein, by the coagulant appears to be responsible for the initial changes in the coagulum matrix.[96] The level of chymosin retained in the curd is pH-dependent.[4,26,32] In fresh milk, plasmin, the native alkaline milk proteinase, is associated with the casein micelles but it dissociates at low pH.[89,109] The activity of plasmin in cheese is reported to be dependent on cooking temperature[109] as well as on pH and the salt and moisture contents of the cheese.[89,109] The role of plasmin in Cheddar cheese flavour has yet to be elucidated but it has been reported that the rate and extent of characteristic flavour development in Cheddar cheese slurries appeared to be related directly only to the degradation of β-casein.[110] Plasmin may well therefore prove to be an enzyme of considerable importance in the development of cheese flavour.

As the original casein network is broken down, the desired balance of flavour and aroma compounds is formed. However, the precise nature of the reactions that produce flavour compounds and the way in which their relative rates are controlled are poorly understood. This has been due firstly to the lack of knowledge of the compounds that impart typical flavour to Cheddar cheese, and secondly to the complexity of the cheese microflora as the potential producers of flavour compounds. Any organism that grows in the cheese, whether starter or non-starter, and any active enzyme that may be present such as chymosin or plasmin, must have an influence on the subsequent cheese flavour (Fig. 9). Research in New Zealand has clearly established, however, that if the growth of starter and non-starter is controlled so as not to reach levels that give discernible off-flavours[4,7] and if as little chymosin as possible is used,[111,112] the flavour which develops in Cheddar cheese is likely to be acceptable to most consumers.

There appear to be only two 'facts' on which all research workers on Cheddar cheese flavour seem to agree. Firstly, that milk fat has to be present for the perception of flavour[113] and secondly that lactic starters must be used for the development of a typical balanced Cheddar flavour. However, the mechanisms by which fat and starter perform their functions are not at all well understood. Virtually every finding and conclusion reached by any research group have become subsequently a matter of controversy. Totally contradictory conclusions have been reached, for example, on the role of starter and non-starter organisms in Cheddar cheese.[114–117] This section is an attempt by the present authors to summarize what they consider to be relevant.

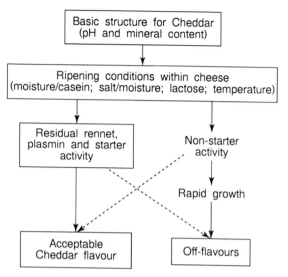

Fig. 9. Main factors that determine the development of flavour in Cheddar cheese.

5.1 Effect of Milkfat

It is well accepted that Cheddar cheese made from skim milk does not develop a characteristic flavour. Cheese with a FDM greater than 50% developed a typical flavour whereas cheese with a FDM less than 50% did not.[113] When, in this study, a series of batches of cheese was made from milk of increasing fat content (from 0 to 4·5%), the quality of the flavour improved as the fat content increased. If, however, the fat content was increased above a certain limit the flavour was not further improved. Substituting vegetable or mineral oil for milkfat still resulted in a degree of Cheddar flavour.[118] This suggests that the water-fat interface in cheese is important and that the flavour components are dissolved and retained in the fat. Clearly, although milk proteins and lactose are the most likely sources of the flavour precursors in Cheddar cheese, the fat plays an important but not yet defined role.

5.2 Effect of Proteolysis

As described earlier, it is likely that the breakdown of casein is involved in the production of Cheddar cheese flavour.[119] A further consequence of proteolysis could also be the release of flavour components that were previously bound to the protein.[120] Most research has concentrated on the aroma aspect of Cheddar flavour but it now appears to have been established that the non-volatile, water-soluble fraction makes the greater contribution to flavour intensity. It has been suggested[120,121] that the importance of low levels of such non-volatile compounds as peptides, amino acids and salts has been underrated in the past. This view is supported by the highly significant correlations found between the levels of proteolysis products and the extent of flavour development.[122] The level of phosphotungstic acid-soluble amino nitrogen was found to be a reliable indicator of flavour development. Above certain limits, however, the level of specific low molecular weight peptides results in bitterness. Cheddar cheeses made with temperature-insensitive starter strains were found to become bitter because large numbers of starter cells contributed excessive levels of proteinases. These released bitter-tasting peptides from high molecular weight peptides that had been produced mainly as a result of chymosin action.[91] The subject of bitterness, the single most common defect in Cheddar cheese, has been extensively reviewed (for instance Ref. 123).

Several reports strongly implicate the volatile sulphur compounds, specifically methanethiol, in Cheddar cheese flavour,[124,125] but an Australian report[126] concludes that none of these sulphur compounds is a reliable indicator of flavour development. It is conceivable, however, that although the volatiles do not make a measurable contribution to the intensity of Cheddar flavour, they may still be an essential factor in the quality of the flavour.[120] This is supported by the finding[127] that the quality of blocks of Cheddar cheese decreased, and off-flavours increased, with a decrease in block size. Headspace analysis showed that the concentrations of H_2S and CH_3SH, compounds that are extremely susceptible to oxidation, decreased as the quality of the cheese decreased.

5.3 Role of Starter

The absence of any Cheddar flavour in gluconolactone-acidified cheese and the development of typical, balanced Cheddar flavour in starter-only cheese[128] established that starter has a role in the development of cheese flavour. What exactly that role is has been much more difficult to determine. An indirect contribution of starters to the production of flavour compounds has been suggested by New Zealand research workers.[129] They considered that the main role of the starter is to provide a suitable environment which allows the elaboration of characteristic cheese flavour. Starter activity results in the required redox potential, pH and moisture content in the cheese that allows enzyme activity to proceed favourably. In addition, the temperature during manufacture and the S/M must be controlled so as to ensure that the net metabolic activity of the starter organisms is low[112,129] but nevertheless adequate to allow the required pH at one day to be reached. Should the starter reach too high a population or survive too long, flavour defects such as bitterness, which mask or detract from cheese flavour, are produced. A reduction in unpleasant flavour is associated with improved perception of the Cheddar flavour.[91,129]

It is likely, however, that the starter enzymes also have a direct role in flavour development.[130-132] Lactococci, for instance, can metabolize lactose and citric acid to diacetyl and acetic acid.[132] Cheddar cheese flavour therefore may well result from the presence of diacetyl,[132] -SH compounds,[125] acetic acid,[132-134] carbon dioxide,[134] lactic acid,[134] and NaCl, together with low levels of specific amino acids and peptides.[120–122] One research group has concluded that some key flavour compounds are produced non-enzymically.[135–137]

5.4 Role of Non-Starter

Cheddar cheese made under controlled bacteriological conditions and containing only starter lactococci develops balanced, typical flavour[128] but it is intriguing that cheeses made in open vats develop such flavour more rapidly.[138–140] This would suggest that a non-starter flora, present as a result of post-pasteurization contamination, is beneficial. There have, nevertheless, been reports that conclude that non-starter bacteria have little effect on normal Cheddar cheese flavour development.[136,141]

The role of non-starter bacteria in Cheddar cheese flavour has clearly yet to be elucidated.[142] However, as has been amply shown for starter organisms, it is probable that relatively low densities of non-starter bacteria are beneficial but high densities are not.[4] It appears that the rate of growth of the non-starter bacteria during the early days of ripening is important. The rate of cooling of the cheese, after pressing the curd, appears to be the single most significant factor in controlling the cheese flora[7] and appears to offer the easiest method of controlling cheese flavour.[143]

An alternative to actually determining the numbers of non-starter bacteria in cheese is to measure a compound such as acetate, since this provides an index of the degree of hetero-fermentative metabolism that may have taken place (T.F. Fryer, pers. comm., 1986). When such growth is extensive, cheese quality is often less than satisfactory because of fermented and sour off-flavours and one of the compounds responsible is undoubtedly acetic acid. More recently, it has been shown that the L-lactate originally present in the cheese is slowly converted to D-lactate.[144–148] Racemization, which normally takes about 90 days, represents a significant change during Cheddar cheese ripening since about half of the lactate (0·7% of the cheese mass) is transformed. Pediococci that have been isolated from Cheddar cheese[144,145] and some lactobacilli[144,146] have been shown to have racemizing activity. An interesting point is that the calcium salt of D-lactic acid is less soluble than that of L-lactic acid. Crystallization of calcium lactate on the surface of Cheddar cheese is a common and troublesome defect[149] and racemization will therefore encourage the formation of these undesirable white deposits in cheese.

5.5 Effect of Redox Potential

Since the redox potential determines the status of disulphide/sulphydryls in both the casein and the enzymes in cheese, the redox potential would be expected to affect the rate of proteolysis during cheese ripening. The development of characteristic cheese flavour is considered to be determined by the ability of protein-based sulphur groups to accept hydrogen resulting from oxidative ripening processes.[114,150] Whenever active-SH groups failed to appear early in ripening, as a consequence of atypical manufacturing conditions, the resulting cheese was found to lack flavour.

The unsuccessful attempts to duplicate the flavour of cheese made from unheated milk by the addition of 'flavour-producing' bacteria to heated milk is thought to result from the heat-induced interaction of sulphur-containing groups of milk proteins, which reduces the ability of the groups to accept hydrogen during fermentation.[114] A more recent finding,[151] however, suggests that non-starter bacteria vary in their ability to maintain the required low redox potential in cheese. This might account for the many conflicting reports concerning the effect of non-starter bacteria on cheese flavour.

6 GRADING OF CHEDDAR CHEESE

There is no absolute standard for measuring cheese quality. Young Cheddar cheese is judged on the basis of whether it has properties characteristic of its variety. Compositional analysis provides an objective method for detecting atypical cheese and is to be preferred to subjective grading methods. In the case of mature cheese, quality assessment is largely a matter of personal preference, with consumers differing considerably in their requirements with respect to cheese flavour.

The assignment of a grade to a consignment of cheese may be improperly influenced by the sample since differences may exist between blocks of cheese made from the same vat of milk and even within the same block of cheese. Flavour defects, such as fruitiness and sulphide off-flavours, have sometimes been located in particular areas within a cheese.[5] Such lack of uniform flavour usually results from variations in S/M.[28] Differences between cheeses have also been attributed to an uneven cooling of cheese blocks stacked closely on pallets while the cheese is still warm.[152] It is clear that a grade score is likely to be highly biased if the assessment of a whole vat depends on a single randomly-drawn sample.[55]

It has been long recognized that the texture of Cheddar cheese changes dramatically during the first few days of ripening. The simplest explanation for this observation is that the cheese microstructure consists of an extensive network of α_{s1}-casein and that cleavage by chymosin (or rennet substitute) of just a few peptide bonds of α_{s1}-casein greatly weakens this network.[96] This results in a relatively large change in the force necessary for deformation. It is differences in this force that a grader attempts to assess when he rubs down a plug of cheese between his thumb and forefinger. From this assessment of the texture, after the cheese has been allowed to ripen for at least seven days, the grader proceeds to predict what the quality of the cheese will be after it has matured.[4]

The sensory method of prediction traditionally used by graders has therefore some validity since the rate of change in cheese texture during the first few days of ripening is determined by the same factors, i.e. the pH at one day, the salt to moisture ratio and the moisture to casein ratio, which also influence the quality

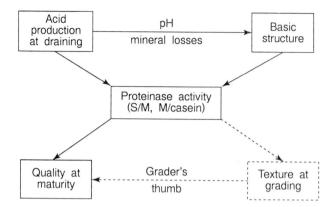

Fig. 10. An explanation for the general validity of traditional sensory testing of Cheddar cheese.

of the cheese at maturity (Fig. 10). Experience has long shown that a Cheddar cheese with an atypical texture seldom, if ever, develops a characteristic flavour but unfortunately the reverse is not true. A good-textured cheese does not always result in an acceptable flavour, since off-flavours can still be produced if unsuitable manufacturing and ripening procedures are used.[4] Essentially, the grader, using his thumb, is attempting to assess that acid production at draining and the cheese composition after salting were normal. However, this can be achieved more directly and objectively by compositional analysis, which would also partly overcome the confusion in nomenclature used in grading circles. 'Body', 'texture', 'firmness' and 'hardness' may all be synonymous if different experts are judging the same batch of cheese and there is a need to define and standardize the various terms.

7 VARIETIES OF CHEDDAR CHEESE

7.1 Low-fat Cheddar Cheese

Although still a relatively minor category, there is a growing demand for low-fat cheese in today's calorie-conscious society.[153] It is proving difficult, however, to produce a low-fat Cheddar cheese with the same flavour and texture characteristics as a full-fat cheese.[154,155] The flavour and acceptability at 3 months decreases with decreasing fat content.[156] Cheddar cheeses containing only 15 to 30% fat are noticeably more firm and less smooth when young than full-fat cheese. The differences in texture, although marked during the early stages of ripening, apparently narrow after the cheese has matured for one or two months.[155] Attempts to improve the quality of low-fat Cheddar cheeses by UF treatment of milk and addition of flavour-enhancing enzymes have not been successful,[157] but the addition of pediococci[145] and other bacterial cultures[158] may prove to be more useful.

7.2 Stirred Curd or Granular Cheeses

As discussed earlier, granular cheese preceded, historically, the manufacture of traditional Cheddar cheese. It is made as for normal Cheddar cheese except that the curd fusion or cheddaring step is omitted. More acid, therefore, has to be developed at the vat stage to compensate for the shorter total manufacturing time but starter systems are now available that allow very acceptable granular cheese to be made. Maintaining curd in the granular form, without the need for milling prior to salting, has obvious attractions. There is, however, a tendency for the curd to mat after drying unless it is agitated and continued stirring may lead to higher fat losses. Moisture expulsion is also faster than during cheddaring.

The salted curd particles take some time to fuse together, the rate of bonding depending largely upon the pH of the curd at salting. There are, however, advantages in mechanized cheesemaking systems in having the curd in a granular form. The salt readily mixes with the curd, the salted granules flow and can be easily hooped. Stirred curd cheesemaking is now widely used in the manufacture of 'barrel' (bulk pressed) cheese, although variations in moisture levels may occur as a consequence of different temperatures within the block.[159,160]

The pressing of granular curd gives rise to open-textured cheese as a result of air being entrapped within the cheese.[39] This has been overcome, however, by the development of methods of pressing the curd under a vacuum.[66] Granular cheese resembles Cheddar cheese in composition but it does mature somewhat differently because of the relatively low acidity at which the curd is salted. Curd hooped in the granular form gives a texture at 14 days which, although completely close, is just perceptibly different from normal Cheddar cheese.[42] This difference in texture, however, becomes increasingly less obvious as the cheese matures.

7.3 Washed Curd Varieties

There has been a substantial increase in recent years in the consumption of washed curd varieties of Cheddar cheese.[161] Varieties such as Colby and Monterey are milder in flavour and have a more plastic texture than Cheddar. They are relatively high-moisture cheeses (39–40%) and ripen rapidly.

7.3.1 Colby and Monterey

The recent improvements in the production of granular Cheddar for processing are also indirectly responsible for the production of Colby and Monterey since these varieties are in fact washed curd, granular cheeses. Traditionally, whey is drained off until the curd on the bottom of the vat is just breaking the surface and cold water is added to reduce the temperature of the curds/whey to about 27°C (Fig. 11). The moisture content of the cheese can be controlled by the temperature of the curd/whey/water mix. The moisture content decreases as the temperature is increased between 26 and 34°C. The pH of the cheese is determined

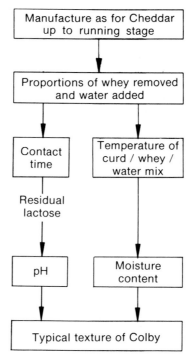

Fig. 11. Main factors that determine the characteristic texture of Colby cheese.

both by the proportions of whey removed and water added. The length of time the water is in contact with the curd is also important since this determines the level of residual lactose. Since salt penetration into the interior of the granular curd particles is rapid, no pH gradient occurs and seaminess is not a problem.

The calcium content of Colby tends to be slightly lower than that of Cheddar because of the higher starter percentage used and a further small loss of calcium at the washing stage. As discussed previously, however, it is the pH that determines to a large extent the texture of a cheese. The addition of water results in a small increase (about 0·1 to 0·2 units) in the pH of the finished cheese but this is sufficient to give the cheese a more plastic texture than Cheddar. Recent trials[29] have shown that the calcium level in Colby cheese can vary between 120 and 180 mmoles/kg cheese without influencing significantly the texture of the cheese as long as the pH is greater than about 5·2. The characteristic texture of Colby cheese is thus influenced almost entirely by its pH and moisture level (Fig. 11). Traditionally, Colby had a mechanically open texture but the use of short-time pressing systems,[65,66] in which the curd is transported to the press under a partial vacuum, results in a texture that is as close as that of Gouda-type cheese varieties. Monterey cheese has many similarities to Colby but is usually softer.[1]

8   British Territorial Cheeses

In 1983, about 161 000 tonnes of Cheddar and 74 000 tonnes of other dry-salted cheeses were manufactured in Britain.[162] Of the latter, Cheshire (25 900 t) was the most popular, followed by Double Gloucester (14 000 t), Leicester (13 700 t), Stilton (7100 t), Lancashire (4500 t), Wensleydale (3100 t) and Caerphilly (1500 t). About 4000 tonnes of Cheshire are still produced on farms but only very small amounts of other varieties.[163] Annatto is added to the cheese milk in varying quantities to give a special identity to Leicester (deep orange–red), Dunlop (orange–red), Cheshire (orange) and Double Gloucester (yellow–orange).

8.1 Cheshire Cheese

Cheshire, the oldest of the English named cheeses, differs markedly from Cheddar as a consequence of the high level of acidity developed prior to the formation of a rennet coagulum. This results in a low pH (less than 4·9 at one day) and a low mineral content (less than 170 mM calcium/kg cheese). The texture is therefore shorter and its flavour sharper than that of Cheddar. The higher acidity developed before the whey is run off also results in a higher chymosin retention and therefore in an increased level of proteolysis. The combination of higher moisture, low calcium content and increased proteolysis imparts good melting properties to the cheese. Small quantities of Cheshire are made in France, where it is known as Chester.

8.1.1 Variants of Cheshire Cheese

Although Lancashire is the major variant of Cheshire, it is virtually unknown outside its own county. It is an unusual variety in that the cheese is made from a mixture of fresh, one- and two-day old curd. The finished textural characteristics are controlled by changing the amounts of curd of differing ages. The origin of the procedure is presumably a consequence of the small quantities of milk that were left over for cheesemaking on the farms. Setting aside the curd in pans for 2 or 3 days would also have overcome the problem of variable acid production before commercial starters were available. Titratable acidities of the curd prior to blending vary between 0·2% for fresh curd and 1·5% for the two-day old curd. The target acidity for the blend is within the range 0·85 to 1%, depending on seasonal conditions. It is unlikely that modern plants have time for such niceties and most Lancashire these days is similar to a moist Cheshire. The melting properties of traditional Lancashire have a high reputation and the cheese used to be known as the 'Leigh Toaster'. It is softer than typical hard-pressed cheeses and can be spread with a knife after ripening for 2 or 3 months.

Caerphilly is a smaller version (about 3·5 kg) of Cheshire but with a higher moisture content.[164] It originated in the village of the same name in Wales but is now made almost entirely in the West of England. The major departure from Cheshire manufacture is the 24 h immersion in brine following overnight pressing of the lightly salted curd. Caerphilly is more granular than Cheshire but easily sliced. It is normally eaten unripened after about 7 days but may be ripened for up to 1 month, when it tends to develop a white mould which is considered desirable. Prior to sale, the cheese surface is sometimes rubbed with a mixture of rice flour and barley meal to give a uniform finish. Caerphilly and Lancashire are the only British semi-hard cheeses.

8.1.2 Variants of Cheddar Cheese

One hundred years ago, most farms throughout Britain made their own cheeses and numerous hard-pressed varieties were developed with properties between those of Cheddar and Cheshire. The rennet coagulum was heated at a lower temperature (34–36°C) than in the manufacture of Cheddar and the cheeses were typically more moist, softer and more acidic. They tended not to keep as well as Cheddar and were not, therefore, well known outside the regions where they were made. Four varieties, Leicester, Double Gloucester, Derby and Dunlop, have survived but have changed in type these days to suit modern tastes and conditions. They do not differ markedly from Cheddar except in their higher moisture and slightly lower calcium contents,[164] and the use of regional names has often been reduced to little more than a sales gimmick. The cheeses are not usually subject to rigorous grading and little information is available as to their compositions or modern methods of manufacture.

Traditional Leicester has a crumbly, moist texture and good melting characteristics, which makes it particularly suitable for use in cooking. Double Gloucester was originally made from the high-fat milk of Gloucester cows, a breed that is now almost extinct. The cheese probably derived its name from the use of full-cream milk rather than the partially skimmed milk used for Single Gloucester, which was also roughly half the thickness and weight of Double Gloucester.

Dunlop originated in Ayrshire in Scotland. The curd is cut into larger pieces than for Cheddar, which together with a lower cooking temperature, results in a cheese that can be considerably more moist and softer than Cheddar. Traditional Dunlop is, therefore, eaten when quite young, between 6 weeks and 4 months. The high moisture content makes it a good cheese for toasting, particularly when mature. Derby is similar to Dunlop but is smaller in size. A combination of sage and spinach leaves is used to give the green streaks or marbling effects in Sage Derby.

REFERENCES

1. Kosikowski, F.V., 1977. *Cheese and Fermented Milk Foods*, 2nd edn. Edwards Brothers Inc., Ann Arbor, Michigan.
2. Czulak, J., 1959. *Proc. 15th Int. Dairy Congress, London*, Vol. 2, p. 829.
3. Lawrence, R.C., Heap, H.A. & Gilles, J., 1984. *J. Dairy Sci.*, **67**, 1632.
4. Lawrence, R.C., Gilles, J. & Creamer, L.K., 1983. *N.Z. J. Dairy Sci. Technol.*, **18**, 175.
5. Gilles, J. & Lawrence, R.C., unpublished results.
6. Turner, K.W. & Thomas, T.D., 1980. *N.Z. J. Dairy Sci. Technol.*, **15**, 265.

7. Fryer, T.F., 1982. *Proc. 21st Int. Dairy Congress, Moscow*, Vol. 1, Book 1, p. 485.
8. Sammis, J.L., 1910. *Res. Bull. Univ. Wis. Agric. Exp. Stn.*, No. 7.
9. Whitehead, H.R. & Harkness, W.L., 1954. *Aust. J. Dairy Technol.*, **9**, 103.
10. Lawrence, R.C. & Pearce, L.E., 1972. *Dairy Ind.*, **37**, 73.
11. Lawrence, R.C., Heap, H.A., Limsowtin, G.K.Y. & Jarvis, A.W., 1978. *J. Dairy Sci.*, **61**, 1181.
12. Lawrence, R.C. & Heap, H.A., 1986. *Bull. Int. Dairy Fed.*, No. 199, 14.
13. Richardson, G.H., Ernstrom, C.A. & Hong, G.L., 1981. *Cult. Dairy Prod. J.*, **16**, 11.
14. Timmons, P., Hurley, M., Drinan, F., Daly, C. & Cogan, T.M., 1988. *J. Soc. Dairy Technol.*, **41**, 49.
15. Stadhouders, J. & Leenders, G.J.M., 1984. *Neth. Milk Dairy J.*, **38**, 157.
16. Fox, P.F., 1984. In *Developments in Food Proteins*, Vol. 3, ed. B.J.F. Hudson. Elsevier Applied Science Publishers, London, p. 69.
17. Green, M.L., 1984. In *Advances in the Microbiology and Biochemistry of Cheese and Fermented Milk*, eds F.L. Davies and B.A. Law. Elsevier Applied Science Publishers, London, p. 1.
18. Kimber, A.M., Brooker, B.E., Hobbs, D.G. & Prentice, J.H., 1974. *J. Dairy Res.*, **41**, 389.
19. Kalab, M., 1977. *Milchwissenschaft*, **32**, 449.
20. Stanley, D.W. & Emmons, D.B., 1978. *J. Inst. Can. Sci. Technol. Aliment.*, **10**, 78.
21. Lowrie, R.J., Kalab, M. & Nichols, D., 1982. *J. Dairy Sci.*, **65**, 1122.
22. Kalab, M., Lowrie, R.J. & Nichols, D., 1982. *J. Dairy Sci.*, **65**, 1117.
23. Eino, M.F., Biggs, D.A., Irvine, D.M. & Stanley, D.W., 1976. *J. Dairy Res.*, **43**, 113
24. Green, M.L., Turvey, A. & Hobbs, D.G., 1981. *J. Dairy Res.*, **48**, 343.
25. Green, M.L., Marshall, R.J. & Glover, F.A., 1983. *J. Dairy Res.*, **50**, 341.
26. Creamer, L.K., Lawrence, R.C. & Gilles, J., 1985. *N.Z. J. Dairy Sci. Technol.*, **20**, 185.
27. Van Slyke, L.L. & Price, W.V., 1952. *Cheese*, 4th edn. Orange Judd Publishing Co., New York.
28. Lawrence, R.C. & Gilles, J., 1982. *N.Z. J. Dairy Sci. Technol.*, **17**, 1.
29. Creamer, L.K., Gilles, J. & Lawrence, R.C., 1988. *N.Z. J. Dairy Sci. Technol.*, **23**, 23.
30. Roefs, S.P.F.M., Walstra, P., Dalgleish, D.G. & Horne, D.S., 1985. *Neth. Milk Dairy J.*, **39**, 119.
31. Dolby, R.M., 1941. *N.Z. J. Sci. Technol.*, **22**, 289A.
32. Holmes, D.G., Duersch, J.W. & Ernstrom, C.A., 1977. *J. Dairy Sci.*, **60**, 862.
33. Czulak, J., Conochie, J., Sutherland, B.J. & van Leeuwen, H.J.M., 1969. *J. Dairy Res.* **36**, 93.
34. Whitehead, H.R. & Harkness, W.L., 1959. *Proc. 15th Int. Dairy Congress, London*, Vol. 2, p. 832.
35. Czulak, J., Hammond, L.A. & Meharry, H.J., 1954. *Aust. Dairy Rev.*, **22**, 18.
36. Czulak, J. & Hammond, L.A., 1956. *Aust. Dairy Rev.*, **24**, 11.
37. Hammond, L.A., 1979. *Proc. 1st Biennial Marschall Int. Cheese Conf.*, p. 495.
38. Patel, M.C., Lund, D.B. & Olson, N.F., 1972. *J. Dairy Sci.*, **55**, 913.
39. Czulak, J. & Hammond, L.A., 1956. *Aust. J. Dairy Technol.*, **11**, 58.
40. Czulak, J., 1958. *Dairy Eng.*, **75**, 67.
41. King, N. & Czulak, J., 1958. *Nature*, **181**, 113.
42. Czulak, J., 1962. *Dairy Eng.*, **79**, 183.
43. Harkness, W.L., King, D.W. & McGillivray, W.A., 1968. *N.Z. J. Dairy Technol.*, **3**, 124.
44. Olson, N.F. & Price, W.V., 1970. *J. Dairy Sci.*, **53**, 1676.
45. Conochie, J. & Sutherland, B.J., 1965. *J. Dairy Res.*, **32**, 35.
46. Gilles, J., 1976. *N.Z. J. Dairy Sci. Technol.*, **11**, 219.
47. Gilbert, R.W., 1979. *Proc. 1st Biennial Marschall Int. Cheese Conf.*, p. 503.
48. Breene, W.M., Olson, N.F. & Price, W.V., 1965. *J. Dairy Sci.*, **48**, 621.
49. Thomas, T.D. & Pearce, K.N., 1981. *N.Z. J. Dairy Sci. Technol.*, **16**, 253.

50. Pearce, K.N., 1982. *Proc. 21st Int. Dairy Congress, Moscow*, Vol. 1, Book 1, p. 519.
51. Lawrence, R.C. & Gilles, J., 1969. *N.Z. J. Dairy Technol.*, **4** 189.
52. Robertson, P.S., 1966. *J. Dairy Res.*, **33**, 343.
53. Sutherland, B.J., 1974. *Aust. J. Dairy Technol.*, **29**, 86.
54. Czulak, J., 1963. *Aust. J. Dairy Technol.*, **18**, 192.
55. Sutherland, B.J., 1977. *Aust. J. Dairy Technol.*, **32**, 17.
56. Czulak, J., Conochie, J. & Hammond, L.A. 1964. *Aust. J. Dairy Technol.*, **19**, 157.
57. Morris, H.A., Guinee, T.P. & Fox, P.F., 1985. *J. Dairy Sci.*, **68**, 1851.
58. Geurts, T.J., Walstra, P. & Mulder, H., 1974. *Neth. Milk Dairy J.*, **28**, 102.
59. McDowall, F.H. & Dolby, R.M. 1936. *J. Dairy Res.*, **7**, 156.
60. McDowall, F.H. & Whelan, L.A., 1933. *J. Dairy Res.*, **4**, 147.
61. Fox, P.F., 1974. *Irish J. Agric. Res.*, **13**, 129.
62. Morris, T.A., 1961. *Aust. J. Dairy Technol.*, **16**, 31.
63. Le Graet, Y., Lepienne, A., Brule, G. & Ducruet, P., 1983. *Le Lait*, **63**, 317.
64. Al-Dahhan, A.H. & Crawford, R.J.M., 1982. *Proc. 21st Int. Dairy Congress, Moscow*, Vol. 1, Book 1, p. 389.
65. Wegner, F., 1979. *Proc. 1st Biennial Marschall Int. Cheese Conf.*, p. 213.
66. Brockwell, I.P., 1981. *Proc. 2nd Biennial Marschall Int. Cheese Conf.*, p. 208.
67. Czulak, J., Freeman, N.H. & Hammond, L.A., 1962. *Aust. J. Dairy Technol.*, **17**, 22.
68. Irvine, O.R. & Burnett, K.A. 1962. *Can. Dairy Ice Cream J.*, **41**(8), 24.
69. Price, W.V., Olson N.F. & Grimstad, A., 1963. *J. Dairy Sci.*, **46**, 604.
70. Robertson, P.S., 1965. *Aust. J. Dairy Technol.*, **20**, 155.
71. Hoglund, G.F., Fryer, T.F. & Gilles, J., 1972. *N.Z. J. Dairy Sci. Technol.*, **7**, 150.
72. Robertson, P.S., 1965. *Dairy Ind.*, **30**, 779.
73. Walter, H.E., Sadler, A.M., Malkames, J.P. & Mitchell, C.D., 1953. *U.S. Dept. Agric. Bur. Dairy Ind.*, BDI-Inf., 158.
74. Whitehead, H.R. & Jones, L.J., 1946. *N.Z. J. Sci. Technol.*, **27A**, 406.
75. Hoglund, G.F., Fryer, T.F. & Gilles, J., 1972. *N.Z. J. Dairy Sci. Technol.*, **7**, 159.
76. Smith, A. B., Roberts, M.J. & Wagner, D.W., 1959, *Canadian Patent* No. 578 251.
77. Robertson, P.S., 1967. *Dairy Ind.*, **32**, 32.
78. O'Connor, C.B., 1971. *Irish Agric. Creamery Rev.*, **24**(6), 5.
79. Fox, P.F., 1975. *Irish J. Agric. Res.*, **14**, 33.
80. Gilles, J. & Lawrence, R.C., 1973. *N.Z. J. Dairy Sci. Technol.*, **8**, 148.
81. Lyall, A., 1968. *Aust. J. Dairy Technol.*, **23**, 30.
82. Pearce, K.N. & Gilles, J., 1979. *N.Z. J. Dairy Sci. Technol.*, **14**, 63.
83. Lawrence, R.C. & Gilles, J., 1980. *N.Z. J. Dairy Sci. Technol.*, **15**, 1.
84. Burton, J., 1989. *Dairy Ind. Int.*, **54**(4), 17.
85. Lawrence, R.C. & Gilles, J., 1986. *Milk. The Vital Force*. In *Proc. 22nd Dairy Congress, The Hague*, p. 111.
86. O'Keeffe, R.B., Fox, P.F. & Daly, C., 1975. *J. Dairy Res.*, **42**, 111.
87. Fox, P.F. & Walley, B.F., 1971. *J. Dairy Res.*, **38**, 165.
88. Fox, P.F., 1987. *Dairy Ind. Int.*, **52**(9), 19, 21–22.
89. Richardson, B.C. & Pearce, K.N., 1981. *N.Z. J. Dairy Sci. Technol.*, **16**, 209.
90. Martley, F.G. & Lawrence, R.C., 1972. *N.Z. J. Dairy Sci. Technol.*, **7**, 38.
91. Lowrie, R.J. & Lawrence, R.C., 1972. *N.Z. J. Dairy Sci. Technol.*, **7**, 51.
92. Breheny, S., Kanasaki, M., Hillier, A.J. & Jago, G.R., 1975. *Aust. J. Dairy Technol.*, **30**, 145.
93. Whitehead, H.R., 1948. *J. Dairy Res.*, **15**, 387.
94. Dolby, R.M. & Harkness, W.L., 1955. *N.Z. J. Sci. Technol.*, **37A**, 68.
95. Lawrence, R.C., Creamer, L.K. & Gilles, J., 1987. *J. Dairy Sci.*, **70**, 1748.
96. Creamer, L.K. & Olson, N.F., 1982. *J. Food Sci.*, **47**, 631.
97. Hall, D.M. & Creamer, L.K., 1972. *N.Z. J. Dairy Sci. Technol.*, **7**, 95.
98. Walters, P. & van Vliet, T., 1982. *Bull. Int. Dairy Fed.*, No. 153, 22.

99. Snoeren, T.H.M., Klok, H.J., van Hooydonk, A.C.M. & Dammam, A.J., 1984. *Milchwissenschaft*, **39**, 461.
100. Tarodo de la Fuente, B. & Alais, C., 1975. *J. Dairy Sci.*, **58**, 293.
101. Creamer, L.K., 1985. *Milchwissenschaft*, **40**, 589.
102. Geurts, T.J., Walstra, P. & Mulder, H., 1972. *Neth. Milk Dairy J.*, **26**, 168.
103. Ernstrom, C.A., Price, W.V. & Swanson, A.M., 1958. *J. Dairy Sci.*, **41**, 61.
104. Babel, F.J., 1948. *Nat. Butter Cheese J.*, **39**, 42.
105. Rajput, Y.S., Bhavadasan, M.K. & Ganguli, N.C., 1983. *Milchwissenschaft*, **38**, 211.
106. Neogi, S.B. & Jude, T.V.R., 1978. *Proc. 20th Int. Dairy Congress, Paris*, Vol. E, p. 810.
107. Emmons, D.B., Kalab, M., Larmond, E. & Lowrie, R.J., 1980. *J. Texture Studies*, **11**, 15.
108. Baron, M., 1949. *Dairy Ind.*, **14**, 146.
109. Farkye, N.Y. & Fox, P.F., 1990. *J. Dairy Res.*, **57**, 413.
110. Harper, W.J., Carmona, A. & Kristoffersen, T., 1971. *J. Food Sci.*, **36**, 503.
111. Lawrence, R.C. & Gilles, J., 1971. *N.Z. J. Dairy Sci. Technol.*, **6**, 30.
112. Lawrence, R.C., Creamer, L.K., Gilles, J. & Martley, F.G., 1972. *N.Z. J. Dairy Sci. Technol.*, **7**, 32.
113. Ohren, J.A. & Tuckey, S.L., 1969. *J. Dairy Sci.*, **52**, 598.
114. Kristoffersen, T., 1973. *J. Agric. Food Chem.*, **21**, 573.
115. Law, B.A., 1981. *Dairy Sci. Abstr.*, **43**, 143.
116. Law, B.A., 1984. In *Advances in the Microbiology and Biochemistry of Cheese and Fermented Milk*, ed. F.L. Davies & B.A. Law. Elsevier Applied Science Publishers, London, p. 187.
117. Aston, J.W. & Dulley, J.R., 1982. *Aust. J. Dairy Technol.*, **37**, 59.
118. Foda, E.A., Hammond, E.G., Reinbold, G.W. & Hotchkiss, D.K., 1974. *J. Dairy Sci.*, **57**, 1137.
119. Fox, P.F., 1989. *J. Dairy Sci.*, **72**, 1379.
120. McGugan, W.A., Emmons, D.B. & Larmond, E., 1979. *J. Dairy Sci.*, **62**, 398.
121. Aston, J.W. & Creamer, L.K., 1986. *N.Z. J. Dairy Sci. Technol.*, **21**, 229.
122. Aston, J.W., Durward, I.G. & Dulley, J.R., 1983. *Aust. J. Dairy Technol.*, **38**, 55.
123. Crawford, R.J.M., 1977. *Ann. Bull. Int. Dairy Fed.*, No. 97.
124. Green, M.L. & Manning, D.J., 1982. *J. Dairy Res.*, **49**, 737.
125. Lindsay, R.C. & Rippe, J.K., 1986. *ACS Symposium Series*, **317**, 286.
126. Aston, J.W. & Douglas, K., 1983. *Aust. J. Dairy Technol.*, **38**, 66.
127. Manning, D.J., Ridout, E.A., Price, J.C. & Gregory, R.J., 1983. *J. Dairy Res.*, **50**, 527.
128. Reiter, B., Fryer, T.F., Sharpe, M.E. & Lawrence, R.C., 1966. *J. Appl. Bact.*, **29**, 231.
129. Lowrie, R.J., Lawrence, R.C. & Peberdy, M.F., 1974. *N.Z. J. Dairy Sci. Technol.*, **9**, 116.
130. Law, B.A. & Wigmore, A.S., 1983. *J. Dairy Res.*, **50**, 519.
131. Farkye, N.Y., Fox, P.F., Fitzgerald, G.F. & Daly, C., 1990. *J. Dairy Sci.*, **73**, 874.
132. Lawrence, R.C. & Thomas, T.D., 1979. In *Microbial Technology: Current State, Future Prospects*, eds A.T. Bull, D.C. Ellwood & C. Ratledge. Cambridge University Press, Cambridge, p. 187.
133. Forss, D.A. and Patton, S., 1966. *J. Dairy Sci.*, **49**, 89.
134. Morris, H.A., 1978. *J. Dairy Sci.*, **61**, 1198.
135. Law, B.A., Castanon, M.J. & Sharpe, M.E., 1976. *J. Dairy Res.*, **43**, 301.
136. Law, B.A. & Sharpe, M.E., 1977. *Dairy Ind. Int.*, **42**(12), 10.
137. Law, B.A. & Sharpe, M.E., 1978. *J. Dairy Res.*, **45**, 267.
138. Law, B.A., Castanon, M. & Sharpe, M.E., 1976. *J. Dairy Res.*, **43**, 117.
139. Law, B.A., Hosking, Z.D. & Chapman, H.R., 1979. *J. Soc. Dairy Technol.*, **32**, 87.
140. Reiter, B., Fryer, T.F., Pickering, A., Chapman, H.R., Lawrence, R.C. & Sharpe, M.E., 1967. *J. Dairy Res.*, **34**, 257.

141. Law, B.A. & Sharpe, M.E., 1978. *Proc. 20th Int. Dairy Congress, Paris*, Vol. E, p. 769.
142. Peterson, S.D. & Marshall, R.T., 1990. *J. Dairy Sci.*, **73**, 1395.
143. Miah, A.H., Reinbold, G.W., Hartley, J.C., Vedamuthu, E.R. & Hammond, E.G., 1974. *J. Milk Food Technol.*, **37**, 47.
144. Thomas, T.D. & Crow, V.L., 1983. *N.Z. J. Dairy Sci. Technol.*, **18** 131.
145. Bhowmik, T., Riesterer, R., van Boekel, M.A.J.S. & Marth, E.H., 1990. *Milchwissenschaft*, **45**, 230.
146. Johnson, M.E., Severn, D., Ito, O. & Olson, N.F., 1986. *J. Dairy Sci.* **69**(Suppl. 1), 75.
147. Johnson, M.E., Riesterer, B. & Olson, N.F., 1989. *J. Dairy Sci.*, **72**(Suppl. 1), 125.
148. Dybing, S.T., Wiegand, J.A., Brudvig, S.A., Huang, E.A. & Chandan, R.C., 1988. *J. Dairy Sci.*, **71**, 1701.
149. Pearce, K.N., Creamer, L.K. & Gilles, J., 1973. *N.Z. J. Dairy Sci. Technol.*, **8**, 3.
150. Kristoffersen, T., 1967. *J. Dairy Sci.*, **50**, 279.
151. Thomas, T.D., McKay, L.L. & Morris, H.A., 1985. *Appl. Environ. Microbiol.*, **49**, 908.
152. Conochie, J. & Sutherland, B.J., 1965. *Aust. J. Dairy Technol.*, **20**, 36.
153. O'Donnell, J., 1990. *Dairy Foods*, **91**(7), 53.
154. Olson, N.F., 1980. *Dairy Field*, **163**(2), 64.
155. Olson, N.F., 1984. *Dairy Record*, **85**(10), 115.
156. Banks, J.M., Brechany, E.Y. & Christie, W.W., 1989. *J. Soc. Dairy Technol.*, **42**, 6.
157. McGregor, J.U. & White, C.H., 1990. *J. Dairy Sci.*, **73**, 571.
158. McGregor, J.U. & White, C.H., 1988. *J. Dairy Sci.*, **71**(Suppl. 1), 116.
159. Olson, N.F., 1984. *Dairy Record*, **85**(11), 102.
160. Reinbold, R.S. & Ernstrom, C.A., 1988. *J. Dairy Sci.*, **71**, 1499.
161. Olson, N.F., 1981. *J. Dairy Sci.*, **64**, 1063.
162. Anon., 1985. *Dairy Ind. Int.*, **50**(2), 21.
163. Wade, O., 1982. *J. Soc. Dairy Technol.*, **35**, 138.
164. Florence, E., Milner, D.F. & Harris, W.M., 1984. *J. Soc. Dairy Technol.*, **37**, 13.

2

Dutch-Type Varieties

P. Walstra, A. Noomen and T.J. Geurts

Department of Food Science, Agricultural University, Wageningen,
The Netherlands

1 Description

We define Dutch-type varieties of cheese as those that:

(a) are made of fresh cows' milk, the milk being at most partly skimmed (generally leading to at least 40% fat in the dry matter of the cheese);
(b) are clotted by means of rennet (usually extracted from calves' stomachs);
(c) use starters consisting of mesophilic lactococci and usually leuconostocs, that generally produce CO_2;
(d) have a water content in the fat-free cheese below 63% (ratio of water to solids-not-fat $<1\cdot70$);
(e) are pressed to obtain a closed rind;
(f) are salted after pressing, usually in brine;
(g) have no essential surface flora;
(h) are at least somewhat matured (a few weeks) and thus have undergone significant proteolysis.

Consequently, the cheese usually has a semi-hard to hard consistency and a smooth texture, usually with small holes; the flavour intensity varies widely.

So defined, Dutch-type varieties constitute one of the most important (if not the most important) types of cheese produced in the world (in terms of tonnage), comparable in that respect to Cheddar and the group of white, fresh cheeses.

Variation within the type is considerable:

(a) loaf size may be between 0·2 and 20 kg;
(b) loaf shape may be a sphere (Edam), a flat cylinder with bulging sides (Gouda), a block, like a loaf of bread, etc.;
(c) fat content in the dry matter ranges from 40 to over 50%;
(d) water content in the fat-free cheese ranges from 53 to 63%;
(e) salt content in the cheese water ranges from 2 to 7%;
(f) pH may be anywhere from 5·0 to 5·6;
(g) maturation may take from 2 weeks to 2 years.

39

Generally, a larger loaf is likely to have a lower water content (initially) and is matured for a longer time. The character (taste, consistency) ranges almost from that of a typical St Paulin to Parmesan, even within a type of the same designation, e.g. Gouda. Further variation occurs because of the use of different starters, different degrees of acidification during curd making, whether the cheese milk is pasteurized or not, contamination with different microorganisms (lactobacilli, in particular, can grow in the cheese) and different conditions during maturation. Finally, herbs or spices are sometimes added, particularly cumin (i.e. the seeds of *Cuminum cyminum*, not of *Carum carvi*—caraway or 'Kümmel'—as applied in other varieties).

Traditionally, two main types of cheese were made in the Netherlands: Gouda and Edam. Gouda cheese (Dutch: Goudse kaas) was made in fairly large loaves of flat cylindrical shape (mostly 4–12 kg) from fresh unskimmed milk, and was matured for variable periods (6–60 weeks); it is still made on some farms in much the same way ('Goudse boerenkaas'). Edam (Edammer kaas), a sphere of, for example, 2 kg, was made from a mixture of skimmed evening milk and fresh morning milk, leading to about 40% fat in the dry matter; the cheese had a somewhat shorter texture than Gouda, and was usually matured for 6 months or more. Later, a greater range of cheeses, differing in shape, body and taste, evolved from these types. Most modern types have a somewhat higher pH and moisture content than the cheeses used to have; one reason for change was to obtain better sliceability. The same or similar types are made in several countries, either having evolved in the country itself or as an imitation of cheese as made in the Netherlands; the phylogeny is not always clear. Table I lists several of these; the table is not complete and may even contain some inaccuracies as it is difficult to obtain reliable data from all over the world. Other derived types have, for example, a higher (Dutch: roomkaas) or a lower fat content. The method of making the cheese has altered greatly. Some Italian varieties are similar to Dutch-type varieties, although higher scalding temperatures are applied; even varieties like Colby and Monterey are rather similar to Gouda, although they are made in a different way, e.g. with dry salting. After all, it is the composition of the cheese, more than the way in which it is made, that determines its properties, as was, for example, pointed out by Lawrence *et al.*[3]

Some cheese types, although clearly outside our definition of Dutch-type varieties, have, nevertheless, evolved from traditional Dutch cheeses. These include:

(a) 'Meshanger', originally an Edam cheese that failed to develop sufficient acidity and so obtained a high water content, a soft body and a flat shape.[4]

(b) Tilsiter, originally derived from Gouda, from which it differs mainly in having a red slime on the surface.[2]

(c) 'Maasdammer', a cheese very much like Jarlsberg, and thus containing propionic acid bacteria (considered to be a defect in true Gouda).

For the sake of completeness, we mention also some other varieties that were traditionally made in the area which now constitutes the Netherlands.

TABLE I
Dutch-Type Cheeses[a]

Country	Designation	Weight (kg)	Maturation (months)	Fat in DM[b] (%)	Wff[c] (%)	Remarks
The Netherlands	Amsterdammer kaas	2·5-5	0·7-1·5	49	62	Gouda shape
	Goudse kaas	2·5-30	1-20	49	59	Gouda shape
	Lunchkaas (baby Gouda)	0·2-1·1	0·7-2	49	62	
	Edammer kaas	1·7-2·5	1-15	41	59	
	Baby-Edammer	0·9-1·1	0·7-2	41	61	Sphere
	Commissie-kaas	3-4·5	4-12	41	59	Sphere
	Middelbare kaas	5-6·5	4-12	41	59	Sphere
Argentina	Pategrás Argentino	5		40	58	
Brazil	Bola (Prato Esférico)	2		45	56	Sphere
	Prato Estepe	4		46	60	
	Reino	1·5-1·8		45	53	
Czechoslovakia	Javor	12		50	57	
Denmark	Danbo	1-14	1·5	47	61	Loaf
	Elbo	5·5	1·5	47	61	Gouda shape
	Fynbo	7	≥1·5	47	61	Part of salt added to curd
	Maribo	14	2-3	47	59	
	Molbo	1·5-3	1·5	47	60	Sphere
	Samsø	14	1·5-3	47	59	
	Tybo	3	1·5	47	61	Loaf
Egypt	Memphis (Menfis)	4	2·5	48	52	
Finland	Kartano	4·5	1·5	42-47	59	
	Lappi (Pehtori, Vouti)	1·8-3	1-1·5	42-47	58	
	Turunmaa (Korsholm)	6	2	52	61	
France	Mimolette	2·5-4	>1·5	41	59	Sphere
Germany	Brotedamer	2·5-4·5	≥1	41	58	Loaf
	Geheimratskäse	0·5	1	45	61	
Hungary	Balaton sajt	9-12	1	46	55	

(continued)

Table I—*contd.*
Dutch-Type Cheeses[a]

Country	Designation	Weight (kg)	Maturation (months)	Fat in DM[b] (%)	Wff[c] (%)	Remarks
Ireland	Blarney	11–13		48	57	
Italy	Fontal	1·5–2	1·5	46	60	
Jugoslavia	Trapist sir		1–1·5	45	63	
Norway	Norvegia	4–12	3	46	58	
	Nøkkelost	4–15	4	46	56	Spiced
Poland	Mazurski	18		50	58	
	Salami	1·2		40–45	58	Sausage shape
	Warmiński	4·5		40–45	58	
Sweden	Drabantost	4–12	2–4	45–50	59	Block
	Herrgårdsost	12–15	3–4	43–46	55	
	Hushållsost	1–2	2	45–50	60	
	Prästost	12–14	4–5	52	58	
	Sveciaost	12–15	4	47	57	Part of salt added to curd
	Västerbottensost	18	8	50	52	
Former USSR	Jaroslavskij syr	3–10	2	45–50	59	
	Kostromskoj syr	5–12	2·5	45	55	
	Pošechonskij syr	5–6	1·5	45	59	
	Rossiskij syr	5–18	3·5	50	60	Part of salt added to curd
	Stepnoj syr	5–10	2·5	45	58	
	Uglickij syr	3	2	45	59	

[a] Partly after Refs 1 and 2. The numerous varieties of which the designation includes words like Dutch, Gouda or Edam have been deleted, although the specifications for such cheeses often differ somewhat from those of the original types.
[b] Fat content in the dry matter.
[c] Water content of fat-free cheese, usually at the minimum time of maturation.

(a) Leyden cheese (Dutch: Leidse kaas), still made on a few farms from raw milk, skimmed a few times and thus pre-acidified. Shape: flat cylindrical; weight: 8–12 kg; fat content in the dry matter: about 30%; dry and hard and containing cumin seeds; matured for 6–12 months. A type with 40% fat-in-dry matter, but otherwise much the same as Gouda (the shape is slightly different), is now being produced industrially.

(b) Friesian cheese (Dutch: Friese kaas), made from raw milk, skimmed several times which was, therefore, rather acid at the time of renneting. A kind of 'cheddaring' was applied and it was salted at the curd stage. It contained about 20% fat-in-dry matter, was very hard and dry, and was matured for a long time. It was made in three varieties: without spices ('kanterkaas'), with cumin ('kruidkaas') and with cumin and cloves ('nagelkaas'). 'Friese nagelkaas' is still produced, but its texture and composition are more like a traditional Leyden cheese, and it is made from fresh, pasteurized, partly skimmed milk.

(c) Limburger (Dutch: Limburgse kaas) made in Dutch Limburg, is the same as the 'Hervekaas' or 'Fromage d'Herve', made in the Belgian provinces Limburg and Liège.

(d) 'Witte meikaas', a soft but pressed, white cheese consumed when a few days old.

(e) Ewes' milk cheese, e.g. from the Texel (made like a small Gouda) or from Friesland (somewhat like Brynzda).

At the present time, several other types of cheese are being produced in the Netherlands.

2 MANUFACTURE

The authors assume that the reader is familiar with the general principles of cheese-making, e.g. as outlined in Volume 1 of this book.

2.1 Treatment of Milk

Treatment of milk aims at improving or maintaining the quality of the milk for cheesemaking, with respect to cheese quality and composition, yield and ease of manufacture.

Milk quality may be defined so as to include composition. The fat and casein content of the milk naturally affect cheese yield and fat content; lactose content affects cheese acidity (Section 3.2). Off-flavours, particularly if associated with the fat, may be carried over into the cheese. In a well-ripened cheese, such flavours may be masked and flavour due to lipolysis may even be desirable, but in the milder varieties this is not so. Physical dirt should be absent as it shows up in the cheese, but can be removed easily by filtering and/or centrifuging.

The bacteriological quality of cheese milk is of great importance. Pathogenic organisms may survive in cheese, which may be a problem in raw-milk cheese. Pathogenic enterobacteriaceae and staphylococci may even grow in cheese, although not if the cheese contains no sugar. Proper manufacturing procedures ensure that everywhere in the cheese, either the sugar is fully and rapidly converted into acid by the starter organisms,[5] or the salt content is already high enough to prevent the growth of pathogens. *Listeria monocytogenes* is killed by normal pasteurization and can, moreover, grow only on the cheese surface if it is sufficiently moist; problems with this organism have never been observed with Dutch cheeses. *Staph. aureus* can grow somewhat if salt is present, but then normally does not produce toxin. Several bacteria present can cause defects in the cheese: coliforms, *Lactobacillus* spp., *Str. thermophilus*, faecal streptococci, propionic acid bacteria, *Clostridium tyrobutyricum* (see further Section 3.5). Growth of psychrotrophic bacteria in the raw milk may lead to the production of sufficient thermostable lipolytic enzymes to cause undesired lipolysis in the cheese;[6] bacterial proteinases do not seem to cause undesirable effects.[6,7] Milk to

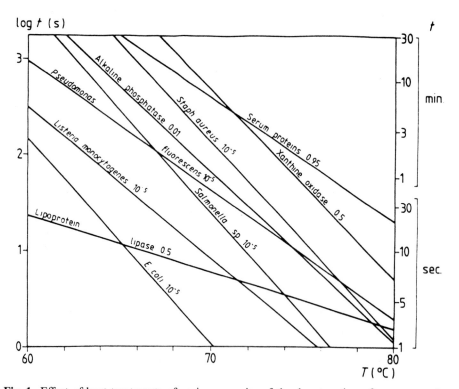

Fig. 1. Effect of heat treatments of various severity of the denaturation of serum proteins, the inactivation of some enzymes, and the killing of some microorganisms, important for the cheese quality or safety. The figures indicate the fraction left unchanged in the milk after the treatment. Approximate average results after various sources.

be stored for long periods before cheesemaking is usually given a mild heat-treatment ('thermalization'), sufficient to kill several types of bacteria, including most psychrotrophs, but not to greatly alter the milk otherwise, e.g. 10 s at 65°C.

The milk is commonly pasteurized, usually HTST, e.g. 20 s at 72°C. Figure 1 gives time–temperature relations for effects that are important for cheesemaking; it should be realized that these are only examples since there is considerable variation in the thermolability of microorganisms, etc. Pasteurization serves the following functions:

(a) Killing of pathogens; those that possibly occur are killed if heating is sufficient to inactivate alkaline phosphatase. Toxin produced by *Staphylococcus aureus* is not inactivated by such treatment.[8]

(b) Killing of spoilage organisms. Spores of *Clostridium tyrobutyricum* survive but enterobacteriaceae, propionic acid and most lactic acid bacteria are killed. Some species of *Lactobacillus* and *Streptococcus* are not fully killed, but they are seldom present in great numbers in the milk. However, *S. salivarius* ssp. *thermophilus* may grow in the regeneration section of the heat exchanger[9,10] and thus attain high numbers if it is continuously used for a long time (e.g. 10 h); this may lead to undesirable flavour and texture of the cheese.

(c) Inactivation of milk enzymes. Probably, this may be useful only with regard to lipoprotein lipase (EC 3.1.1.34), but even the usefulness of this is variable (see below). Several lipases and proteinases produced by psychrotrophic bacteria are not inactivated.[6]

(d) If the milk is pasteurized shortly before renneting it also serves to undo, more or less, the adverse effects of cold storage on the casein micelles and salt equilibrium (which in turn affects renneting properties, Volume 1, Chapter 3), to melt the fat in the globules and to bring the milk to renneting temperature.

Heat treatment can also have undesirable effects, particularly if its intensity is more than that needed to inactivate alkaline phosphatase:

(a) Denaturation of serum proteins leads to slow renneting, a weak curd and poor syneresis (Volume 1, Chapters 3–5). It may also easily cause a cheese of poorer quality, particularly the development of a bitter flavour; the explanation is not clear. Since heat denaturation also causes a profitable increase in cheese yield, fairly rigorous control is exerted in some countries, e.g. via the nitrogen content of the whey (which, for example, should be at least 95% of that of the whey made from raw milk).

(b) Useful milk enzymes may be inactivated, especially xanthine oxidase (EC 1.2.3.2). This enzyme is needed to slowly convert (added) NO_3^- into NO_2^-, which is essential for the desired action of nitrate against clostridia (Section 3.5).

Although pasteurization of cheese milk is widespread, and has certainly helped to considerably improve average cheese quality, it is often held to be responsible for a certain lack of flavour, especially in well-matured varieties. This may be due to inactivation of lipoprotein lipase or to killing of bacteria that may impart in a raw-milk cheese some flavour that is desirable to some people and that anyway may be more variable than the flavour of cheese made from pasteurized milk. A less severe pasteurization may improve flavour. If the bacteriological quality of the milk is very good, cheese can be made from raw milk or from mixtures of raw and pasteurized milk. An old-fashioned way of improving milk quality is to let the fresh milk cream at low temperatures (5–10°C); in this way, most bacteria accompany the cream because of agglutination.[11] By pasteurizing the cream, but not the skim-milk, most bacteria are killed without greatly affecting milk enzymes.

Another way to enhance bacteriological quality is by bactofugation. This may be applied when cheese is made from raw milk. Another purpose is the removal of spores of *Clostridium tyrobutyricum*. This treatment reduces the number of spores drastically, even to about 3%.[12,13] The sediment obtained contains the spores but also casein and its removal would cause a significant reduction in cheese yield (about 6%). Consequently, the sediment is commonly UHT-heated to kill the spores and added again to the milk; the concomitant denaturation of serum proteins involves a small enough quantity to be acceptable. Double bactofugation increases the efficacy of spore removal, but is, of course, costly.

Homogenization of cheesemilk is rarely practised. It may enhance lipolysis, depending on conditions, and this may be either desirable or not, but generally not. However, it causes the cheese to attain an undesirable, sticky texture. Damage to the fat globules, e.g. by foaming, may also enhance lipolysis. Splashing milk from a height into the cheese vat may even cause some 'churning', leading to significant creaming in the vat; this happens if cold milk is warmed to 30°C and then brought into the vat. The remedy is to warm the milk sufficiently to melt the fat and then cool it to 30°C.[14]

In most cases, the milk is standardized so as to yield the desired fat content in the dry matter of the cheese (Section 3.1). This generally implies some skimming of the milk. Usually, part of the milk is passed through a separator (which also removes dirt particles) and sufficient cream is removed.

Substances added to the milk may include:

(a) CaCl$_2$ to speed up and particularly to diminish variability in renneting and syneresis.

(b) Nitrate, to prevent early blowing by coliforms and growth of *C. tyrobutyricum*. Nowadays, nitrate is often added later, i.e. to the curd–whey mixture after about half of the whey has been removed. This is both to save on nitrate and to avoid producing large quantities of whey that contains nitrate.

(c) Colouring matter, either β-carotene or annatto (an extract of the fruits of *Bixa orellana*), for obvious reasons. Its use appears to be waning and is often omitted. Some types are highly coloured, e.g. Mimolette.

Spices, if any, are commonly added to the curd.

2.2 Main Process Steps

Figures 2 and 3 are examples of flow-sheets for the production of Edam and Gouda cheese. Some essential process steps in the transformation of milk into

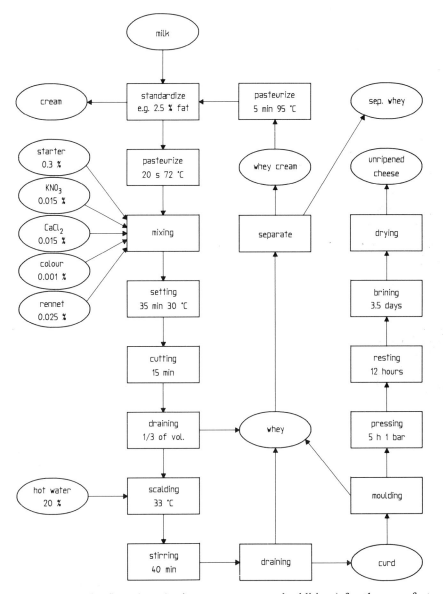

Fig. 2. Example of a flow sheet (main process steps and additions) for the manufacture of Edam cheese (until curing) in a fairly traditional way. Wooden moulds and cheese cloth were used during pressing.

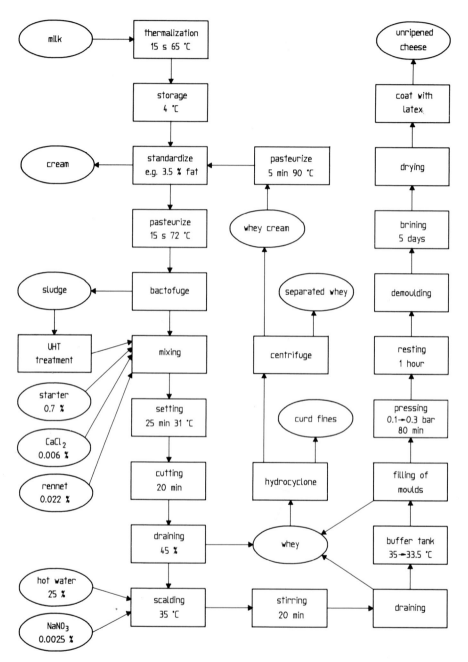

Fig. 3. Example of a flow-sheet for the manufacture of 12 kg Gouda cheese (until curing) by a modern method. Time from start of cutting to start of filling: 60 min. A curd filling machine with vertical columns and plastic moulds with a nylon gauze lining are used. NaCl content of brine: 18%, brine temperature: 14°C, pH <4·6. If bactofugation is omitted, about 0·015% NaNO$_3$ is added.

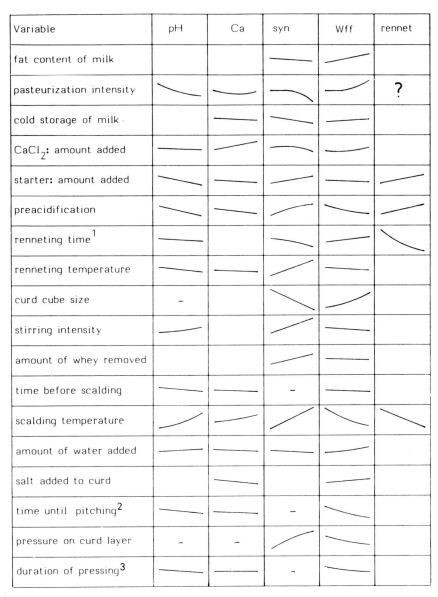

Fig. 4. The effects of milk properties and conditions during curd making on the pH of the curd at the end of curd making (moment of separation of curd and whey), the amount of Ca retained in the curd, the rate of syneresis (syn), the water content of the fat-free (Wff) cheese (before brining) and the quantity of rennet retained in the curd. The figure is meant only to illustrate trends for cheese of average water content. If no curve is given, the relation is unknown but probably is weak; a dash implies no relation. Notes: (1). This generally implies reciprocal of rennet quantity. (2). Total time between cutting and pitching. (3). On the curd layer (not in the moulds). After various sources, e.g. Refs 15–20, and authors' observations.

fresh cheese are discussed below. Those steps largely determine cheese composition and the efficiency of production. Important aims are to obtain maximum yield, to control cheese composition (hence quality), and to keep the process as short as possible, while following a fixed time schedule. The main points to be considered in relation to cheese composition are: final water content, pH of the cheese, quantity of calcium phosphate remaining in the cheese, and the quantity of rennet retained in the curd. To regulate the process, syneresis rate is paramount. The effects of several process parameters on these variables are summarized in Fig. 4; the relationships are, of course, only very approximate and they often depend on the level of other parameters.

Another essential variable is the bacterial population in the fresh cheese, as it greatly affects ripening (Section 3.4) and the possible development of defects (Section 3.5). This concerns starter bacteria (Section 3.3) as well as contaminating organisms. Strictly enforced hygienic measures must be taken to prevent undesirable growth of lactobacilli[21] and the propagation of bacteriophages.[22]

Pre-acidification (which is rarely practised today), quantity of starter added, rate of growth of starter bacteria and the time needed until the curd is separated from the whey, all affect the pH of the curd at the latter moment and consequently, the concentrations of calcium and inorganic phosphate left in the cheese. Lower concentrations of these give a somewhat lower cheese yield, a lower buffering capacity and consequently a slightly lower pH, and may somewhat affect cheese texture, the consistency becoming slightly softer and shorter. The concentration of rennet retained in the cheese increases markedly with a lower pH and a lower temperature at the moment of whey separation.

Renneting is usually done at about 30°C and cutting starts 20–30 min afterwards. About 20 ml rennet with a specific activity of 10 800 Soxhlet units and about 7 g $CaCl_2$ are usually added to 100 litres of cheese milk. The aim is to produce a curd that can be cut easily and stirred without undue losses of 'fines' in the whey and that shows rapid syneresis. If the milk is drawn from a very large quantity and moreover if the calving pattern of the cows is fairly evenly spread throughout the year, milk composition is generally sufficiently constant to always give good results with fixed quantities of rennet and $CaCl_2$.

The curd is cut, usually in cubes of some 8–15 mm size. Stirring, at first gently (to minimize loss of fines, etc.) and later more vigorously, is done either with the knives used for cutting or with special stirrers. After a while, part of the whey is removed so that stirring becomes more effective, i.e. the forces acting on the curd grains are higher and thus promote syneresis. The temperature is increased (scalding), also to speed up syneresis, but not to temperatures high enough to injure the starter organisms, which usually implies keeping below 38°C. Scalding can be done by indirect heating, by adding heated whey, or by adding hot water. The latter practice is the most common, since water usually has to be added anyway to regulate pH. The higher the moisture content of the cheese, the higher its ratio of lactose to buffering substances, and hence the lower its pH becomes, since, ideally, all lactose is converted into lactic acid. Independent control of the water content and the pH of cheese can thus be achieved by more or less diluting the cheese moisture with added water or 'washing' the curd. These aspects are further discussed in Section 3.2.

For some varieties, part of the salt is added to the curd–whey mixture at the end of the curd making to hopefully inhibit growth of undesired microorganisms.

After the curd has lost enough moisture (the water content being, for example, 65% in the case of Gouda cheese and the pH 6·5), stirring is stopped and the curd grains are allowed to sediment. Partial fusion of curd grains now occurs and a continuous mass of curd is formed that can be cut into blocks and taken out of the whey. Considerable loss of whey from the curd occurs during these stages[23] and this is promoted by applying some pressure (e.g. 400 Pa) by placing perforated metal plates on top of the curd or by the curd layer itself being deep enough; pressure also promotes curd fusion. If a very low water content is desired, the drained curd may be stirred or worked; this causes considerable additional syneresis, but also loss of fines and fat and a cheese with a rather open texture (many irregular, small holes).

The whey obtained at various stages may be collected separately, because it differs in pH, added water and added nitrate and, rarely, salt. The whey is usually separated and the cream obtained is pasteurized, e.g. 5 min at 90°C, to fully destroy any bacteriophages that may be present; it is used to adjust the fat content of the next lot of cheese milk. Curd fines are sometimes separated from the whey by means of hydrocyclones or filters, but are not recycled to the curd because of the danger of contamination (bacteria and phages).

The blocks of curd obtained are put into moulds and pressed. Originally, wooden moulds were used, and the curd was wrapped in cloth; a pressure of 50–100 kPa was applied for several hours and the cheese developed a very distinct, firm rind. Nowadays, cheese loaves are formed in perforated metal or plastic moulds, sometimes lined with a gauze or some kind of cloth to promote drainage and rind formation. The pressure is usually much lower and is applied for a shorter time. Consequently, only a weak rind is formed, although the rind should be fully 'closed' (i.e. free of visible openings). Closing of the rind is due to complete fusion of the outermost layer of curd grains. Pressing as it is applied nowadays is generally insufficient to cause complete fusion throughout the mass of the freshly pressed loaf. This implies that some moisture can still move fairly freely through the cheese mass, possibly leading to uneven moisture distribution. Within 24 h, however, curd fusion is complete.

The cheese loses considerable moisture during pressing, but moisture loss is slight once a closed rind has formed. This implies that starting the pressing earlier and applying a higher pressure lead to less moisture loss, and hence to a cheese with a higher water content. By varying the moment of applying pressure, water content can thus be regulated to some extent, especially to ensure the same water content in cheese loaves of the same batch. During stirring in the whey, syneresis proceeds faster than in a block of curd. Consequently, the blocks formed last have the lowest water content immediately after shaping and pressure should be applied to those loaves directly to 'keep in the moisture', while the blocks

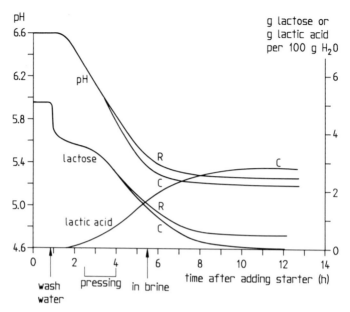

Fig. 5. Acid production during Gouda cheese manufacture (conditions comparable to those in Figure 3). Quantity of lactose and lactic acid, and the pH of curd and cheese as a function of the time after starter inoculation of the cheese milk. C = centre of the loaf; R = rind portion. After Northolt & Van den Berg, NIZO.

formed earlier should be left for a much longer time before pressing[24] (see also Section 3.2).

In traditional cheesemaking, the freshly pressed loaves were turned in the moulds and left there till the next day for 'shaping' (Dutch: omlopen), i.e. to attain a symmetrical shape. The main change occurring was, however, complete conversion of lactose to lactic acid. This is important because during brining the fermentation by the starter bacteria is slowed down and even effectively stopped in the rind, owing to the combined effects of low temperature and high salt content. Nowadays, a greater quantity of a faster growing starter is usually added, which implies that much more lactic acid has already been produced a few hours after adding starter. These aspects are illustrated in Fig. 5. Often, the cheese is put into brine within 1 h after pressing when some lactose is not yet converted to acid. (Incidentally, this means that even in a 3-week-old cheese, the outermost layer may contain about 0·2% lactose in the cheese moisture and also that the brine contains lactose.) The use of pasteurized milk, strict hygiene and adequate rind treatment (Section 2.4) are needed to prevent undesirable growth of micro-organisms.

Brining is primarily done to provide the cheese with the necessary salt. More-over, it serves to cool the loaves rapidly to below 15°C (to stop further syneresis, and prevent or slow down the growth of undesirable bacteria), and to give them a certain rigidity (due to the high salt content in the rind) during the necessary

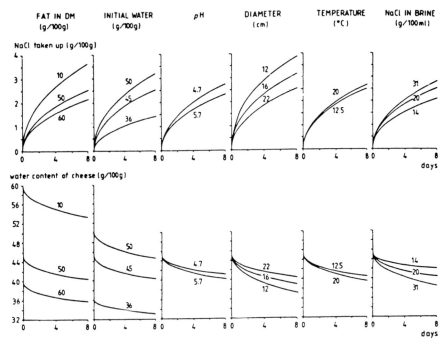

Fig. 6. Salt uptake and water content of cheese (spherical loaves) as a function of the duration of brining. Unless mentioned otherwise, a cheese of 50% fat in dry matter and 45% water, diameter 22 cm, pH 5·0 and a brine of pH 5·0, containing 20 g NaCl/100 ml, temperature 12·5°C, were used (from Ref. 25).

handling shortly after brining. Brining further causes a considerable loss of water (2–3 times the quantity of salt taken up) and loss of a little soluble matter (<0·2% of the cheese mass). Factors affecting these processes are discussed in Chapter 7, Volume 1. Figure 6 summarizes the effects of several variables on salt uptake and water loss. The average salt content is generally 3–5 g/100 g water in the cheese; of course, it takes considerable time (up to some weeks in a large loaf) for the salt to become more or less evenly distributed throughout the cheese.

The brine generally contains about 18% NaCl, but weaker brines are sometimes used to allow brining the loaves for exactly one week. This poses the problem of growth of salt-tolerant lactobacilli in the brine that may even to some extent penetrate the cheese and cause flavour defects.[26,27] Regular cleaning of the equipment is desirable. The brine should contain enough Ca (0·3%) and acid (pH 4·5) to prevent dissolution of cheese protein in the brine, which would cause a slimy rind.[28]

2.3 Mechanization

The past 35 years especially have witnessed drastic changes in the cheese industry. Refrigeration of the milk at the farm (~4°C) and every second or third day collection of this milk has become the accepted system in many countries. Rigorous control of the hygienic quality of the milk leads to far smaller variation in composition, thus facilitating the introduction of systems for process control. Thermalization of raw milk (e.g. 10 s at 65°C), which prevents the growth of psychrotrophic spoilage organisms, enables further storage of milk at the factory for some days. Moreover, cheese factories have been modernized and merged into plants with high capacity. Plants with an annual production of 10 000 tonnes of cheese, manufactured in a 5-day working week, are not exceptional. These plants are highly mechanized, automated and computerized, producing cheese of the desired quality at relatively low labour costs but with very costly equipment. In the majority of cases, individual plants specialize in the manufacture of a single cheese variety.

The main developments in the mechanization and automation of cheese production can be described briefly as follows.

Curd making and moulding. The size of cheese vats has been increased very considerably. This initially became possible with the introduction of specific curd-sedimentation vats (strainers), with a moving, perforated belt. At one end of this vat, the sedimented and drained curd layer is mechanically cut into pieces of desired dimensions, which are put into moulds, pressed, etc. This method allows for the production of cheese with less variation in composition, moisture content in particular, and loaf weight. The capacity of cheese vats has been increased from 5000 to 12 000 litres. This system is still applied, but in many cases has been superseded by the following technique. Curd is made batch-wise in vats of up to 20 000 litres and the time schedule for successive batches (vats) is programmed in such a way that finished curd–whey mixtures can be pumped separately into a buffer tank, whence the mixture is fed into continuously working machines that separate the whey from the curd, shape the curd blocks and fill them into moulds. (Most machines operate with a downward curd stream, but machines with an upward stream can be used also.) Technologically, these machines have the advantage that the weight of the loaves can be controlled accurately, the relative standard deviation being 0·5–1·5%; this is especially important for small loaves, e.g. Edam cheese. During the mould filling process, syneresis of the curd proceeds and would cause the water content of cheese to decrease if no precautions were taken. Therefore, syneresis during the moulding operation is slowed down by stirring the curd–whey mixture in the buffer tank only gently and by gradually decreasing the temperature by 0·5 to 2°C. In this way, the moisture content of the cheese can be controlled fairly accurately.

Pressing. Instead of pressing cheese loaves in stacks, it has become increasingly common to press them in a single layer. Presses are filled and emptied continuously. Continuous pressing, however, is not applied because differences still exist in water content between loaves of a batch at the beginning and the end of moulding, due to differences in the extent of syneresis of the curd. Consequently, to obtain a uniform water content in all loaves, special attention is given to the order in which loaves are set tight under the press. With respect to demoulding, systems have been designed which prevent damage to the relatively weak cheese rind.

Brining. Two brining systems are in use. One employs racks composed of several horizontal compartments; the racks can be moved up and down in a deep brine basin (2–3 m). Filling and, after brining, emptying of successive compartments is realized by circulation of the brine, which also enhances the rate of salt uptake by the cheese. In the other system, the loaves are salted while floating in a shallow layer of circulating brine (~0·4 m); brine sprinkling devices or rollers that periodically immerse the loaves take care of salting of the top of the loaves.

Storage. Treatments in curing rooms also have been highly mechanized: transport, plastifying and turning of cheese, cleaning of shelves, etc. Much progress has been made to control the temperature, relative humidity and velocity of air, in order to approximate the ideal situation in which each loaf is stored under identical conditions.

2.4 Rind Treatment and Curing

For Dutch-type cheese varieties according to our definition (Section 1), development of microorganisms on the cheese surface is undesirable, because they may negatively affect cheese quality. In particular, the growth of moulds must be prevented, since some species may produce mycotoxins, e.g. sterigmatocystine by *Aspergillus versicolor*. In former days, cheese was pressed in such a way as to obtain a thick and very tough rind; mould development was reduced by regular rubbing of the cheese rind with a dry cloth and unboiled linseed oil. After hardening, the coat formed also reduced water evaporation from the cheese rind, thus permitting it to remain relatively supple and springy. Mineral oils, e.g. paraffin, were applied as well. The practice was abandoned after the introduction of latexes, mostly called plastic emulsions, which offer superior protection and permit production of cheese with a much weaker rind. Without this, mechanization and marked speeding up of cheesemaking would not have been possible. These polymer latexes form, on drying, a coherent plastic film that offers protection against mechanical damage and slows down, but does not prevent, evaporation of water. The film mechanically hinders mould growth, but it may also contain fungicides, e.g. natamycin (pimaricin), an antibiotic produced by *Streptomyces natalensis*, or calcium and sodium sorbate. In the Netherlands, only natamycin is allowed while in some other countries only sorbates are permitted. When compared to sorbates, natamycin offers the advantages that its migration into the cheese is generally limited to the outer 1 mm of the rind and that it does not adversely affect the appearance, taste and flavour of the cheese.[29] Moreover, natamycin is much more effective than sorbates; for comparable protection from mould growth, the amount of sorbate needed is about 200 times that of natamycin. With respect to public health, an acceptable daily intake of

0·3 mg natamycin per kg body weight per day has been established.[30] Dutch cheese regulations limit the quantity to 2 mg per dm^2 of cheese surface when the cheese is sold.

In practice, successive treatments (2–3 times) with latex are applied to all sides of the cheese shortly after manufacture. Care is taken that the cheese surface is sufficiently dry before each treatment. Treatment is repeated during long curing.

Generally, the cheese is cured at 12–16°C and 85–90% RH. The conditions must allow the latex to dry quickly, otherwise undesired organisms like coryneform bacteria and yeasts may develop and cause off-flavours. However, if the latex dries too quickly, cracks may form in the plastic layer, allowing microbial growth. Particularly at the beginning of ripening, the cheese inevitably expels a little moisture, causing a high humidity between the loaf and the shelf, which favours bacterial growth. To prevent this and to allow the cheese to retain a good shape, loaves are turned frequently during this period. Upon prolonged ripening, this frequency is reduced. Regular cleaning and drying of the shelves and control of the microbial condition of the air in curing rooms form part of a general programme on hygiene.

Just before they are put on the market, cheeses may be treated with paraffin, generally after they have been treated with latex; in the Netherlands, this especially concerns Edam-type cheese. Before waxing, the loaves must have a very clean and dry surface, since a high humidity between the cheese and the wax layer favours bacterial growth, causing off-flavours and gas formation. Consequently, wax is applied predominantly to mature cheese.

Some cheese is made in rectangular or square loaves for curing while wrapped in saran foil. This particularly suits the processed cheese industry and those customers who prefer this type of cheese when it is to be sold in prepacked portions or slices. Compared to normal cheese, important differences are the lack of a firm rind, the almost complete absence of moisture loss and the consequently more homogeneous composition. The cheese should have a lower (1–2%) water content immediately after manufacture, because this content must meet the standards for normal cheese after ripening. Prolonged curing at the usual temperature, say 14°C, however, tends to produce cheese of poor flavour. Therefore, the cheese is kept at low temperatures (<8°C); the resulting flavour may be rather flat. A starter with low CO_2-producing capacity is used to prevent loosening of the wrapping. It is not necessary to turn the cheese, which may be ripened at a lower relative humidity (e.g. 70%).

3 IMPORTANT TOPICS

3.1 Standardization and Yield

Standardization of milk for cheesemaking means adjustment of its fat content to ensure that the cheese being made contains the legally required percentage of fat in the dry matter (FDM). The yield is the mass of cheese obtained from a certain quantity of this milk.

It is of importance to calculate precisely the desired fat content of the cheese milk. The yield of cheese should also be predictable. Comparison with the ultimate analytical results may enhance understanding of the cheesemaking process.

Usually, all cheese made from one vat of milk is weighed. If this is always performed in much the same way it may be a valuable help, since it gives a first indication of cheese composition.

3.1.1 Standardization

Under practical cheesemaking circumstances, establishing the correct fat content of the milk causes specific problems. Firstly, the cheese mass is always inhomogeneous, causing difficulties in establishing its real fat in dry matter content. For this reason, borer samples may give considerable bias and the whole loaf, a sector from it, or a quarter from a square-shaped cheese, is ground. Secondly, one has to take into account that, generally, the fat content of different loaves from one batch is not identical, the standard deviation often amounting to about 0·5% FDM. Moreover, FDM decreases during ripening, since proteolysis involves 'conversion' of some water into dry matter.

For these reasons, a safety margin is taken into account, i.e. the initial fat content is adjusted to a somewhat higher level than is required. As a rule, a surplus value of ~1·5% FDM is taken. The plus sign in notations like 40+ or 60+ refers to this margin.

Difficulties in standardization are also caused by the multiplicity of variables affecting the ratio of fat to dry matter in the finished product:

1. The composition of the milk changes with season and shows short-term fluctuations. Moreover, changes may occur during prolonged cold storage. Previously, the fat content of the whole milk was taken as the basis of standardization, based on an assumed constant fat/protein ratio. The higher the fat content of the whole milk, the higher the fat content of the cheese milk has to be. The ratio between the fat and protein contents of the whole milk is, however, not constant. Hence, this method is not very precise. Much greater certainty is obtained if the protein content of the milk is estimated, or still better, its paracasein content. Almost fixed proportions of the fat and the Ca-paracasein Ca-phosphate complex are carried over from the milk into the cheese.

2. The method of making the cheese. Important aspects are:
   (a) The pasteurization of the milk: denatured serum proteins are incorporated in the curd, increasing its SnF content.
   (b) The cutting of the curd, which affects fat losses into the whey and the amount of curd fines. (The latter fraction has a lower fat content than the curd itself.)
   (c) The quantity of wash water used, which affects the SnF content of the moisture in the cheese.
   (d) The amount of acid produced in the curd, and thereby the loss of calcium phosphate into the whey.
   (e) The quantity of salt absorbed by the cheese.

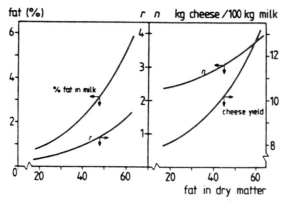

Fig. 7. The factors r (Eqn 1) and n (Eqn 3). Examples of the fat content of the cheese milk (3·4% protein) and of the yield of cheese (12 days old, 58% water in the fat-free part, 4% salt in dry matter).

3. The maturation of the cheese; the quantity of fat hardly changes but the SnF does, since water is converted into dry matter during the hydrolytic processes.

To standardize the cheese milk, the ratio between its fat content (v) and (crude) protein content (p) may simply be used as a basis. Suppose that F is the fraction of the fat that is transferred from the milk to the cheese, and that K kg fat-free dry cheese, including added salt, originated from 1 kg of milk protein, then Fv/Kp represents the ratio between fat and SnF in the cheese. As far as the making of Dutch-type cheese is concerned, both F and K approximate 0·9.[18] Hence, the ratio of fat to protein in the milk may be adjusted to the ratio that is desired between fat and SnF in the cheese. Schulz and Kay[31] accordingly arranged their 'Käse-Tabellen'. If p is known, the appropriate value of v may be found in the table for any cheese being made.

In the Dutch cheesemaking industry, more detailed formulae are in use, e.g. (Ref. 32):

$$v = rp + q \tag{1}$$

Under normal conditions of Gouda and Edam cheesemaking, r depends primarily on the desired fat in dry matter content of the cheese. For 40% FDM, $r \approx 0.67$, for 48% FDM, $r \approx 0.91$ (see also Fig. 7). The last factor, q, refers to the fat lost in the first and second whey. The loss increases more than proportionally with v. For cheese with 20% FDM, $q \approx 0.05$, for 40% FDM, $q \approx 0.14$, for 48% FDM, $q \approx 20$, for 60% FDM, $q \approx 0.40$. In Fig. 7, v is shown as a function of FDM; note the strong increase of v with increasing FDM. Some cheesemakers use more elaborate calculations to arrive at v.

3.1.2 Yield
The yield of cheese can be defined as kg of product (y) obtained from 100 kg of cheese milk.

$$y = \text{kg (fat + protein + other solids + water)} \qquad (2)$$

Most factors affecting the ratio of fat to dry matter (Section 3.1.1) also influence cheese yield.

An important variable is undoubtedly the water content of the cheese and, hence, the loss of whey (syneresis) during curd making and pressing. Many factors affect the water content (Section 3.2). Its standard deviation between loaves of one batch often amounts to 0·5–1%.[24,33] It has been observed that y increases by ~0·2 kg if the water content increases by 1 percentage unit.[32]

If we consider the water content as given, additional factors affecting y are:

Season. Under Dutch conditions, y is relatively high in autumn and low in spring, the discrepancy amounting to over 10%.

Mastitis. Severe mastitis leads to the production of milk with a reduced casein content and a reduced casein/total N ratio.[34,35] Actually, since large quantities of bulk milk are used, real problems are seldom met.

Genetic variants of milk proteins. These may affect cheese yield,[36,37] presumably because milk composition is somehow correlated with specific genetic variants, especially of β-lactoglobulin.

Cold storage of the milk. The extent of a possible effect is not quite clear. Unequivocal results are difficult to obtain from experiments, since purely physico-chemical changes must be separated from bacterial effects. It is, however, clear that proteolysis generally causes a loss of yield.

Pasteurization of the milk. A higher heating temperature will increase y (Section 2.1).

Rennet type. Differences in proteolysis, other than the hydrolysis of κ-casein, would reduce y but they are usually negligible, unless some microbial rennets are used.[37a]

Starter. A change in the amount added introduces several other changes. Firstly, the incorporation into the curd of denatured serum proteins may increase with the starter quantity Banks *et al.*[38,39] reported such an effect with Cheddar cheese. The higher yield is caused by the increased retention of serum proteins from the starter, since it is prepared from severely heated milk. Secondly, in Dutch-type cheese manufacture the use of more starter inevitably requires more curd wash water (Section 3.2) which increasingly dilutes the whey and hence reduces the yield. The net result of both factors may be almost nil. Increasing the rate of acidification (more or more active starter) decreases the pH of milk and curd, inducing more dissolution of Ca and inorganic phosphate. Probably, the subsequent loss in y is small, say 30 g for 10 kg cheese produced if the pH at

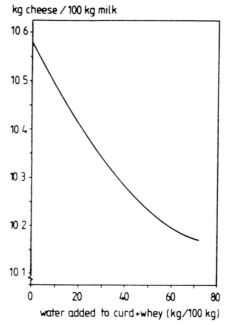

Fig. 8. Yield of Gouda cheese (12 days' old, 41% water) as a function of the quantity of 'curd wash' water used. Water content and pH of the cheese are assumed to remain constant (recalculated from Ref. 42).

separation of curd from whey is as low as 6·25 instead of 6·5. Moreover, more proteose-peptone will leave the micelles at a lower pH, and will be lost in the whey. A lower yield (a higher protein content of the whey) might be expected since O'Leary *et al.*[34] and Richardson[40] found more proteolysis when more starter was used. The findings of van den Berg and de Vries[16] do not, however, point to such an effect.

CaCl₂. Addition to the milk causes some accumulation of colloidal calcium phosphate in the micelles.[41] Usually, about 1 mmol/litre $CaCl_2$ is added to enhance the clotting, presumably yielding ~30 g 'cheese' from 100 kg milk.

Washing. The dilution of whey with water at scalding affects y. Increasing the quantity of added water, e.g. from 30 to 40% (expressed per mass of curd and whey after part of the whey has been let off) reduces y by 0·5 to 1%. The effect is illustrated in Fig. 8. In the calculation of y, the efficiency of decreasing the concentration in the curd particles by the washing was taken to be 90% for lactose and other low-molecular weight substances, and 50% for the serum proteins.

Ultrafiltration. An appreciable increase in cheese yield is obtained because of accumulation of serum proteins in the curd.[43,43a] In view of the results of Van Boekel and Walstra[43b] on steric exclusion of serum proteins with respect to (para-)casein micelles, it is likely that Dutch-type (and many other types of) cheese produced in the usual manner (no ultrafiltration) contains no significant quantities of serum proteins, but quantitative experimental results seem to be lacking. Also the quantitative effect of partial concentration by ultrafiltration has yet to be established.

Salting. Absorption of NaCl obviously causes a gain in weight of the cheese. Against this profit there is usually a greater loss of moisture,[25] hence a net loss of weight. The quantity of salt absorbed varies, e.g. from 1 to 3% and the net weight loss may vary from, say, 0·02 to 0·06 kg/kg cheese produced. The loss of solids-not-salt ranges from 1 to 3 g/kg; this may include losses caused by mechanical damage during salting.

Strictly speaking, *y* refers to the ultimate product, excluding curd remnants, fines, and any rind trimmings which have to be discarded; *y* includes the latex coating. Clearly, *y* can be in no way predicted very precisely, the cheesemaking process being too complex. Even the random variation in water content (see above) makes exact prediction difficult. To predict the yield of cheese from a given vat of standardized milk one obviously will proceed on its protein content. In the tables of Schulz and Kay,[31] the *y* value can be looked up if cheese of certain fat and water content and with 5% salt-in-moisture has to be made, starting with milk of known protein content. The yield of 12-day old cheese may be calculated[32] from:

$$y = np - 0 \cdot 084 \qquad (3)$$

where *n* depends mainly on FDM. In the case of full-cream cheese, $n \approx 3 \cdot 1$, and *y* amounts, for example, to 10·7 kg (see also Fig. 7).

An example of calculations with respect to the standardization of cheese milk and to cheese yield is given in Ref. 24, which also deals with applications under practical manufacturing conditions.

3.2 Control of pH and Water Content

Very few quantitative data have been published on this subject. Process control has made considerable progress during the last decades but some variations in the cheese composition cannot yet be explained satisfactorily. The degree of acidification in curd and freshly made cheese still shows unexplained fluctuations, although most variables have been identified and can be controlled.[43c]

It is not easy to adjust the water content and pH of cheese independent of each other. In Section 3.2.3, interrelationships under varying conditions are considered.

3.2.1 Control of the Water Content
The basic information with respect to this subject is outlined in Chapter 5, Volume 1. Moreover, the effects of several factors involved are illustrated in

Fig. 4. In fact we have to deal here with the amount of water in the fat-free cheese (Wff), rather than with the absolute water content of the cheese which, within one cheese type, decreases fairly proportionally with increasing fat content.[44] Wff is characteristic for the type of cheese considered (see also Table I).

As a matter of fact, numerous factors affect the water content. Under normal manufacturing conditions, however, the number of process parameters available to really adjust Wff turns out to be restricted. Important are:

(a) Cutting of the curd. The smaller the grains, the higher the syneresis rate, causing a lower water content. Cutting the curd very finely, however, seems to increase Wff, and it causes a greater loss of fines and fat into the whey. A very inhomogeneous cheese mass may result if the initial size of the grains differs widely.

(b) Stirring of the curd–whey mixture. This concerns the intensity of stirring, which increases with the stirring rate and on any removal of part of the whey; the duration of stirring; and (scalding) temperature. Figure 16 in Chapter 5, Volume 1 shows, among other things, that extended stirring causes a lower ultimate water content. If the temperature of the curd mass is kept constant after separation of the whey, then the time of separation should affect the final water content only slightly. This illustrates that in practical cheesemaking the lowering of the temperature after the whey drainage rapidly restrains syneresis.

(c) The above process steps also play a part in the acidification rate, since the temperature affects the virulence of the starter bacteria, and since stirring can be stopped earlier at a higher syneresis rate, hence at a higher pH. Figure 16 in Chapter 5, Volume 1 illustrates the great importance of acidification.

Other variables occurring at renneting, pressing, shaping and salting have effects also but they are not suitable process parameters through which to adjust the water content.

3.2.2 Control of the pH

Here, we deal with the pH at one or a few days after making the cheese; after this, the pH gradually but slowly increases as a result of maturation.

The pH of the cheese results mainly from the amount of lactic acid on the one hand, and that of the buffering compounds on the other. The acid is produced by the starter bacteria, metabolizing the available lactose. The main buffering substance in curd and cheese is the Ca-paracasein Ca-phosphate complex, of which calcium phosphate contributes roughly one third of the buffering capacity. Lactic acid itself is a buffer in cheese at low pH (pK ≈ 3.9). The other salts in the curd moisture presumably play a minor part; after about 2 weeks, citrate is virtually absent due to conversion by starter organisms.

It is worthwhile to distinguish between different situations with respect to pH control. Some lactose may remain in the finished cheese, because the acidification process is hindered or stopped. Adding salt at an early stage, as is done when making Meshanger cheese,[4] may cause this. If Dutch-type cheese is brined shortly after pressing, some lactose is usually left in the outer rind portion (Section 2.2, Fig. 5). The lactic acid fermentation may also be restrained by cooling, which is practised in the manufacture of Quarg, Cottage cheese and Bel Paese. Lactic acid produced by starter bacteria may be metabolized by organisms of the surface flora on soft cheese, thereby increasing the pH. In the majority of cheese types, including the Dutch, all lactose is converted, mainly into lactic acid. If we now presume that the water content of the fat-free cheese (Wff) is adjusted to its desired value, important compositional characteristics of the milk in relation to the final pH are:

(a) The lactose content of the milk serum (rather than the lactose to casein ratio in the milk).
(b) The quantity (and composition?) of calcium phosphate in the casein micelles; a changed buffering capacity of the curd is due predominantly to a different calcium phosphate concentration.

As soon as the cheese milk has been collected and bulked, these variables (a and b) are fixed. To make the desired type of cheese from this milk, important process parameters involved are:

(1) Factors affecting the water content of the cheese. The higher the water content, the more lactose, or its equivalent, lactic acid, is present in the cheese, and the lower the pH will be. In other words, from the moment the cheese loaf is formed, the ratio between incorporated lactose and buffering substances controls the pH. It has been observed that, *ceteris paribus*, increasing the water content of Gouda cheese by one percentage unit, decreases the pH by 0·1 unit.

(2) The decrease in pH during curd making, and the ensuing loss of Ca and phosphate into the whey, may play a part too. These phenomena depend on the buffering capacity of the curd, on the amount of added $CaCl_2$ (which decreases the pH slightly), and on the degree of acid production which is, in turn, affected by the amount and type of starter added, the temperature, any pre-acidification, infection with bacteriophage and the presence of inhibiting components: antibiotics and disinfectants, agglutinins (active in milk but not in curd and cheese), and the peroxidase-H_2O_2-thiocyanate system. In many cases, however, the pH will have changed by only ~0·2 units at moulding and this drop causes dissolution of very little calcium phosphate and hence it would have only a minor effect on the final pH. On the other hand, a decrease in pH increases the syneresis rate, which affects the water content and hence the pH (via point (1) above). If the water content is kept constant by other means, a small effect still remains, since now a slightly smaller quantity of the buffering calcium phosphate is incorporated into the curd, causing a lower final pH and a lower cheese yield (Section 3.1).

(3) The best process parameter through which to adjust the pH, independently of the water content, is washing. After the addition of the water to the whey–curd mixture, lactose diffuses from the grains into the whey to equalize the lactose concentrations inside and outside the particles, although equilibrium is rarely reached. When the size distribution of the particles is normal and the contact time with the wash-water is 25 min, the efficiency of reducing the lactose concentration in the curd is ~90%.[45] More water causes a lower yield (Section 3.1) as well as a less valuable whey.

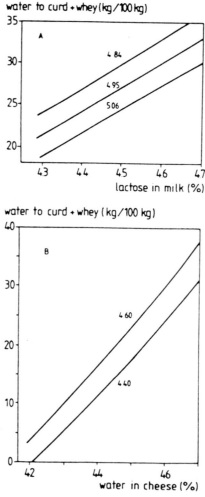

Fig. 9. Amount of curd wash water to be used in relation to the lactose content of the milk (A), and the water content of the cheese before brining (B). Figures near the curves indicate lactose (%) in the fat-free dry matter of the finished cheese (A) and in the milk (B). In A, water content of the cheese is 46%; in B, lactose content in the fat-free dry cheese is 4·85% (from Ref. 15).

TABLE II
Feasibility of Making Dutch Type Cheese of Desired pH and Water Content; Very Qualitative

		Low	Normal	High
	High	++	+	?
pH	Normal	+	++	+
↑	Low	?	+	++

Low   Normal   High
→ water content

++ = easy
+ = possible
? = hardly feasible

Figure 9 illustrates the quantities of water to be added under conditions related to normal Gouda cheese manufacture. The 'lactose in the fat-free dry matter' represents the ratio between lactose and buffering substances (see point (1)). Note that increasing the water content before salting from 44 to 46%, increases the amount of water to be added by 15 percentage units. A 0·2% higher lactose content of the milk necessitates 5 percentage units more added water.

3.2.3 Interrelations
Table II tentatively illustrates the feasibility of adjusting the pH and the water content at various levels. If a cheese with a high water content is desired, e.g. 47%, and a normal pH of 5·1, then in addition to gentle cutting and stirring, much water at a relatively low temperature must be added. To obtain a normal content and a high pH, whey drainage and water addition may be repeated.

If the water content is to be normal and the pH rather low (~5·0), addition of water should be omitted. Scalding can preferably be achieved by means of hot water in the vat jacket, or by adding heated whey (which formerly was common practice in Edam cheese manufacture). If an extremely low pH is desired, giving a typical short consistency, the milk may be pre-acidified.

To obtain a low water content and a normal pH, the curd should be cut rather finely and, after removal of part of the whey, it should be stirred vigorously at a rather high scalding temperature. Heating should, again, be indirect and slow to prevent the formation of curd particles with a 'skin'. Moreover, a high amount of starter should be added, and possibly some pre-acidification applied. A further lowering of the water content can be obtained by 'working' the drained curd (Section 2.2).

3.3 Starters: Composition and Handling

Use is made of mesophilic starters, usually composed of strains of *Lactococcus lactis* spp. *cremoris* and/or *Lactococcus lactis* ssp. *lactis* as acid-producing microorganisms and citric-acid fermenting organisms, either *Leuconostoc lactis* and/or *Leuc. cremoris* (L-starters) or both the leuconostocs and *Lactococcus lactis* ssp.

lactis biovar *diacetylactis* (DL-starters). Formerly, L and DL starters were denoted as B and BD starters, respectively. Whether L- or DL-starters are selected depends largely on the desired degree of eye formation in the cheese: DL-starters ferment citric acid more rapidly and produce more CO_2 than L-starters. Detailed information on these organisms, for example as to their taxonomy, physiology, biochemical characteristics, phages and phage resistance, and on the composition of starters and their propagation can be found in various publications (e.g. Refs 46–48; see also Chapter 6, Volume 1).

In many European countries spontaneously-developed mixed-strain starters have been used traditionally. These so-called Practice (P)-starters[49] are phage-carrying and partially phage resistant, contrary to 'traditional' single-strain starters. Mechanisms involved in the protection of P-starters against phages have been identified.[22] Daily propagation of these starters in cheese factories is possible without any controlled protection against air-borne bacteriophages. They do not show complete failure of acid production when they become contaminated with disturbing phages, but the strain composition of the starter is greatly affected and the rate of acidification may vary considerably.

Modern large-scale cheese factories require the use of starters with constant activity. Acid production in cheese must proceed fairly quickly and at a constant rate, the latter being essential for the control of syneresis and the water content of the cheese. Therefore, traditional methods for starter production are increasingly being replaced by the use of concentrated starters, enabling more uniform bacterial composition of starters and controlling their rate of acidification when they are propagated under complete protection from phage. In the Netherlands, P-starters in use at cheese factories have been selected according to their taste and flavour formation properties, rate of acidification, capability to induce eye formation, and phage-resistance. They are kept as inoculated milk in a frozen condition and are rarely transferred to preserve their P-properties, phage-resistance in particular. These starters serve for the production of concentrates for bulk starter preparation (thus eliminating the use of mother cultures at the factory), and the concentrates are distributed to the cheese factories in a frozen state.[22] The manufacture of bulk starter concentrates is not yet common.

The most common procedure now is as follows. Bulk starter milk is pasteurized, e.g. for 10 min at 90°C (batch-pasteurization) or 1 min at 95°C (HTST-pasteurization). The intensity of the heat treatment should be at least equivalent to that for 3 min at 90°C to destroy any phages present. Specially designed bulk starter equipment offers an effective barrier against air-borne contamination with phages: generally, tanks provided with HEPA (High Efficiency Particulate Air) filters and a special device enabling decontamination of the outer side of boxes of starter concentrate with hypochlorite solution before the starter is introduced into the tank, are used.[50,51] Additional precautions should be taken to avoid accumulation of disturbing phages in the factory, which especially could affect the rate of acidification of the curd in the vat. These measures include: the manufacture of bulk starters in separate rooms; use of closed equipment, cheese vats in particular; and frequent cleaning and disinfection of all installations. Cheese whey is a specially dangerous source of phage contamination. Starters are propagated for 18–24 h at ~20°C. In almost all modern factories, the starter is automatically metered and added to the cheese vat. Starters may be kept for a limited time (e.g. 24 h) below 5°C without loss of activity.

The activity of the bulk starter should be the same on successive days of manufacture. Activity is usually tested in a standardized activity test[52] performed with a standard, pasteurized, reconstituted, high-quality skim-milk powder, and also with the pasteurized cheese milk, which ought to be skimmed. The activity of the starter in either of these milks should be constant. Any change in activity may indicate either: contamination of the starter with disturbing bacteriophages, a decreased activity of the starter (e.g. if it had been kept too long at a low temperature), the presence of antibiotics and/or disinfecting agents in the cheese milk, or gross variations in the composition of the milk. To a certain extent, variation in activity may be corrected by adjusting the quantity of starter added to the cheese milk, or by adjusting other conditions during curd making, e.g. the scalding temperature. It must be remarked that results of the activity test and the acidification rate of cheese do not always precisely agree because of different conditions in milk and fresh cheese, notably phage concentration. According to practical standards for the Dutch cheese industry, the pH of cheese should be 5·7–5·9 after 4 h from the start of manufacture, and 5·3–5·5 after 5·5 h.[53]

3.4 Maturation

Maturation is the result of numerous changes occurring in the cheese. Some changes start during curd making, but most become manifest during storage. The structure and composition of cheese alter greatly and so do organoleptic properties. Biochemical, microbiological, chemical and physical aspects are involved. Development of cheese properties is due particularly to the conversion of lactose, protein, fat and, in Dutch-type varieties, of citric acid.

3.4.1 Fermentation of Lactose and Citric Acid

Formation of lactic acid by the starter bacteria is paramount for the preservation of cheese. By their action they:

(a) Ferment lactose quickly and almost completely; consequently, the cheese soon lacks available carbohydrate.

(b) Produce lactic (and a little acetic) acid and reduce the pH of the cheese to 5·1–5·2. At the end of fermentation (after about 10 hours) the lactic acid concentration in the cheese moisture amounts to about 3%. Part (mostly 4–7%) of the lactic acid is present in its undissociated (i.e. bacteriostatic) form, the more so as the pH is lower.

(c) Reduce the redox potential of the cheese to about −140 to −150mV at pH ≈ 5·2,[53,54] as measured with a normal hydrogen electrode.

All these changes aid in inhibiting the growth of undesired microorganisms; salt uptake by the cheese, the presence of a protective cheese rind and adequate

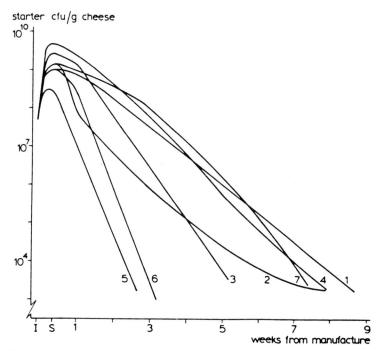

Fig. 10. Populations of different starters (1–7) during the manufacture and ripening of aseptically made Gouda cheese. Plate counts of samples I (milk after inoculation with starter) and S (cheese immediately before brining) are expressed per gram of finished cheese: cfu = colony forming units (from Ref. 55).

treatment of this rind contribute also (Section 2.4). However, microbial defects cannot always be prevented (Section 3.5).

Starter bacteria may differ greatly as to growth rate, the maximum number to which they grow in cheese and the rate at which they lose viability during cheese ripening (Fig. 10). Cheese milk is commonly inoculated at a level of 10^7 starter bacteria/ml of milk. Mechanical inclusion in the curd leads already to $\sim 10^8$ cfu/g of curd, where they grow to, at most, $\sim 10^9$; this implies that starter bacteria generate (divide) only a few times in the fresh cheese. After growth, fermentation is far from complete (pH of cheese $\sim 5\cdot 7$), and during further conversion of lactose, growth and fermentation are uncoupled.

Fermentation of citric acid is of particular importance to eye formation in Dutch-type cheese. The DL- and L-starters used in the manufacture of Dutch-type varieties ferment citric acid, but DL-starters do so more rapidly and produce more CO_2; they are, therefore, selected if more extensive eye formation is desired. The rate of decrease of the citric acid content in the young cheese may be used as an indication of the capability to induce eye formation;[53] the rate of citric acid fermentation is, however, not the only factor involved in eye formation (see Section 3.4.5).

3.4.2 Proteolysis

Protein breakdown in Dutch-type varieties is due mainly to the action of (calf) rennet enzymes, enzymes of the starter bacteria and, to a much lesser extent, milk proteinases. Basic information about these proteolytic systems is given in Chapters 10 and 14, Volume 1. The separate and combined actions of these systems in Gouda cheese have been studied intensively by Kleter[56-59] and Visser[55,60-63] making use of aseptic milking and cheesemaking techniques.

Effectively, the action of calf rennet is determined predominantly by the amount remaining in the curd. Dutch-type cheese contains approximately 0·3 ml rennet (strength 10 000 Soxhlet units) in 1 kg of cheese. The quantity depends primarily on:[20,63a]

(a) the quantity of rennet used in cheesemaking;
(b) the pH during cheese manufacture; the lower the pH, the more calf rennet becomes associated with the paracasein. The amount of starter used, its rate of acidification and the initial pH of the cheese milk and its composition are, therefore, of importance;
(c) the scalding temperature of the curd; the higher this temperature, the less active rennet is included;

See also Fig. 4. Whether an increased moisture content of the cheese and a higher heat treatment of the milk (i.e. some denaturation of serum proteins) enhance the quantity of rennet included in the cheese, is not certain.

The action of rennet enzymes, predominantly chymosin, is characterized by the rapid degradation of α_{s1}-casein at the onset of maturation, about 80% being hydrolysed within 1 month. β-Casein is degraded far more slowly, about 50% remaining even after 6 months.[62] Rapid breakdown of α_{s1}-casein is particularly favoured by the pH of the cheese being near to the optimum (about 5) for rennet action, and a moderate NaCl content (~4%) in the cheese moisture. β-Casein degradation is slowed down considerably, even at this low NaCl content.[64] Calf rennet appears to be responsible for the formation of the greater part of soluble N and the liberation of high and low molecular weight (MW) peptides, but only very low amounts of amino acids (Table III).

When acting separately in cheese, starter bacteria are able to decompose paracasein gradually, proteolytic activity becomes manifest only after several weeks of ripening. Therefore, this capacity, in particular the ability to degrade α_{s1}-casein, seems of minor importance because in the meantime, the casein will have been largely degraded by rennet.[62] Contrary to rennet, starter bacteria predominantly produce low-MW peptides and amino acids (see Table III). Starters may vary greatly in this respect.[61] Rennet and/or starter bacteria may cause bitterness in cheese; for the underlying mechanisms reference is made to Chapter 10, Volume 1.

When acting alone in cheese, milk proteinases may hydrolyse α_{s1}-, α_{s2}- and β-caseins to some extent during prolonged ripening. It can be seen from Table III that small amounts of low-MW peptides and amino acids are liberated.[61] The pH and NaCl content of Dutch-type varieties are not very favourable for the activity of most enzymes; in particular, plasmin activity is reduced greatly.[65]

TABLE III
Quantity of Soluble N Compounds as Produced in Aseptically-made Gouda Cheese by the Combined and Separate Actions of Rennet, Starter Bacteria and Milk Proteinase (after Ref. 61). Data to Illustrate Trends; the Blank Values Varied Slightly

Ripening time (months)	Proteolytic system	Soluble nitrogen, as percentage of total nitrogen				
		Total	*As peptides of MW*			*As amino acids*
			>14 000	14 000– 1400	<1400	
1	Rennet	6·7	2·7	2·7	1·2	0·1
	Starter	2·5	0·2	0·6	0·4	1·3
	Milk proteinase	2·0	0·2	0·4	1·3	0·1
	All systems (cheese)	12·2	1·8	2·3	6·1	2·0
3	Rennet	12·7	3·6	5·2	3·7	0·2
	Starter	4·7	0·3	0·7	1·4	2·3
	Milk proteinase	3·3	0·4	0·7	1·9	0·3
	All systems (cheese)	19·5	2·3	3·3	9·1	4·8
6	Rennet	17·3	4·4	4·1	8·4	0·3
	Starter	7·6	0·9	0·3	2·4	4·0
	Milk proteinase	4·7	0·5	1·0	2·7	0·5
	All systems (cheese)	26·0	5·5	2·3	10·8	7·4

In normal cheeses, where all enzyme systems act together, no clear mutual stimulation or inhibition of the systems in the formation of soluble N components is observed. From Table III it can, however, be deduced that the action of rennet clearly stimulates the starter bacteria to accumulate amino acids and low-MW peptides, which is most likely due to the progressive degradation by starter peptidases of the higher MW products of rennet action.

Contents of soluble N compounds reflect the 'width' of ripening. The 'depth' of ripening is defined as the ratio between the amount of degradation products of low MW, e.g. amino acids or peptides with MW < 1400, and the total amount of soluble breakdown products. In that sense, the width of ripening of Dutch-type cheese varieties is predominantly determined by rennet action, and the depth by the action of starter bacteria.

Serum proteins seem to be hardly degraded in cheese.[66]

Up to now, attempts to accelerate proteolysis in, e.g., Gouda cheese, aimed at shortening its ripening period, failed to produce cheese with characteristics corresponding to those of the normally made cheese. The application of specific starter bacteria may, however, result in cheese which ripens more quickly and develops an agreeable flavour, albeit clearly different from that of Gouda cheese. An example is the Pardano cheese, which has recently been introduced commercially in the Netherlands.[66a] In addition to mesophilic starter bacteria, a mixture of thermophilic lactobacilli and streptococci is used in its production, which otherwise follows the normal process for Gouda cheese making.

3.4.3 Lipolysis

In Dutch-type varieties some lipolysis usually occurs and is even desired, but it should be limited. Factors affecting lipolysis have been intensively studied by Stadhouders and co-workers.[67-73] Cheese made from raw milk shows distinct action of milk lipase; if made from aseptically-drawn milk containing a negligible number of lipolytic bacteria, fat acidity increases gradually.[67,68] HTST-treatment of milk, e.g. 15 s at 72°C, largely but not completely inactivates milk lipase[6] (see also Fig. 1). Cheese made from aseptically drawn, low-temperature pasteurized milk still shows an increase in fat acidity during maturation, although this increase is slight or even scarcely noticeable.[58,59] One may question as to what extent variations in results were caused by differences in susceptibility of the milk to lipolysis, e.g. due to mechanical damage of the fat globules.

Under well-controlled conditions during making and curing of cheese made from pasteurized milk, lipolysis in cheese will result predominantly from the action of starter bacteria, residual milk lipase and, possibly, heat-stable lipases of psychrotrophic organisms. Such conditions in particular imply a good bacteriological quality of the milk prior to thermalization or pasteurization (especially psychrotrophs should be virtually absent), and the absence of microbial growth on the cheese surface during maturation. Enzymes of the starter bacteria hardly decompose triglycerides, but are able to produce free fatty acids from mono- and diglycerides formed by milk lipase and/or other microbial lipases.[73] The activity of milk lipase is reduced by the NaCl in the cheese.[67] Milk lipase action is also affected by the pH of the cheese. Although this action has been found to decrease markedly with decreasing pH when assayed on substrates greatly different from cheese (e.g. Ref. 67), the acidity of cheese fat has been reported to increase faster in cheese of low pH.[67,74] The explanation is unclear, though it should be noted that at a lower pH a higher proportion of any short-chain fatty acids present will be in the fat phase. Lipase activity in cheese increases markedly with temperature.[67,75] In cheese made from milk containing high numbers of psychrotrophic bacteria (or their heat-stable lipases), lipolysis may be increased to undesirable levels. Also, growth of organisms on the cheese surface, e.g. moulds, coryneform bacteria, yeasts,[67,70] may contribute to increased acidity of the fat. Growth of such organisms, however, is usually minimized but cannot be fully prevented; consequently, the rind portion of the cheese generally acquires a somewhat higher fat acidity. Homogenization of the cheese milk greatly enhances lipolysis in the cheese, but is seldom practised.

The free fatty acid content of cheese milk normally amounts to, say, 0·5 mmol per 100 g of isolated fat. In a 6-weeks old Gouda cheese made from HTST pasteurized milk, this value averages about 1 mmol, which increases to about 3 mmol in a well-matured cheese. Cheese made from raw milk usually shows higher FFA contents.

3.4.4 Flavour

Flavour is defined as the complex sensation comprising aroma, taste and texture (see Ref. 76). Little is known about the nature of the aroma compounds in

Dutch-type cheese, but it is clear that the breakdown products of lactose and citric acid (lactic acid, diacetyl, CO_2, etc.), of paracasein (peptides and amino acids), and of lipids (free fatty acids) are essential for the flavour. A correct balance must exist between the various flavour substances.[77] Lactic acid causes the refreshing acid taste, which is particularly noticeable in young cheese; an excess of lactic acid renders the cheese sour, a shortage insipid. Indirectly, lactic acid exerts influence on the texture of cheese. Large changes in flavour develop during maturation. Numerous secondary products formed during the fermentation of lactose and the subsequent partial transformation of lactic acid affect aroma and taste (e.g. aldehydes, ketones, alcohols, esters, organic acids, CO_2). Proteolysis is also crucial in flavour formation. Paracasein is tasteless, but many degradation products are not; for example, peptides may be bitter and many amino acids have specific tastes, sweet, bitter or brothlike, in particular.[77] Short peptides and amino acids contribute at least to the basic flavour of cheese.[77,78] Higher temperatures, a higher pH, a higher water content and a lower NaCl content in cheese appear to favour proteolysis, in particular the formation of amino acids.[79] Mature cheese contains numerous volatile compounds, usually in small amounts. These are predominantly degradation products of amino acids, e.g. NH_3, various amines, H_2S, phenylacetic acid.

Protein degradation greatly affects cheese consistency and thereby probably flavour perception. Carbon dioxide, although tasteless, appears to influence the flavour; loss of CO_2 may contribute to the rapid loss of typical flavour from grated cheese. NaCl accentuates the flavour; saltiness of cheese is governed by the NaCl content of the cheese rather than that in the cheese moisture. In mature cheese, free fatty acids may render the cheese flavour very piquant. However, in cheese lacking sufficient basic flavour, free fatty acids soon produce a rancid flavour.

3.4.5 Texture

Cheese consistency is discussed in Chapter 8, Volume 1 of this book (see also Ref. 80) and has recently been studied in detail.[79a] The main factors affecting the consistency in Dutch-type varieties probably are moisture content, extent of proteolysis, pH, NaCl and fat contents, any inhomogeneity of these variables throughout the cheese mass and, of course, temperature.

During maturation, several changes occur that may be important to texture:

(a) Structure and composition become more uniform, particularly in the early stages due to further fusion of curd grains, and reduction of salt, moisture and pH gradients.

(b) The cheese loses water by evaporation and ongoing syneresis (especially near the rind) and due to proteolysis.

(c) Maturation primarily implies breakdown of the paracaseinate network; it also causes a slight increase in pH (formation of alkaline groups by proteolysis, degradation of lactic acid).

(d) Gas is formed.

The common result is that during maturation the apparent elastic modulus of the cheese increases, the deformation at which fracture occurs decreases and the fracture stress at first decreases and subsequently increases again.[81,79a] The modulus seems, for the same pH and NaCl content, to depend on the water content in the fat-free matter only.[79a] The only rheological parameter that appears to correlate well with the degree of maturation of the cheese is the deformation at fracture, as was, for instance, found in a study of several, widely different cheeses ranging in age from 4 to 20 weeks.[82] The relative deformation (Hencky strain) at fracture is, say, 1·6 for unsyneresed curd, higher than 1 shortly after salting and about 0·5 after 3 months of maturation.[79a]

In several Dutch-type varieties, the number, size and shape of holes is considered an important texture characteristic. Conditions allowing hole formation have recently been studied in some detail by Akkerman et al.[82a] Holes can be formed if gas pressure exceeds saturation and if sufficient nuclei are present. The gas is commonly N_2 already present in the milk (any O_2 is consumed by the starter organisms) and CO_2 produced by starter organisms. CO_2 and possibly H_2 produced by undesirable bacteria (Section 3.5) can also contribute to formation of the holes. The supersaturation needed for hole formation (by approximately 0·3 bar) can be achieved when the rate of CO_2 production is relatively fast (which depends on temperature, type and number of bacteria and citrate content), its rate of diffusion ($D \approx 3 \cdot 10^{-10}\,\text{m}^2 \cdot \text{s}^{-1}$) out of the cheese is slow (mainly depending on loaf size and shape) and if the partial pressure of N_2 is high. Quantitative relations have been given.[82a] Nuclei are usually small air bubbles, either incorporated as such in the curd when the curd mass is 'worked' after draining off all whey, or present in the milk. The latter presumably exist as tiny air bubbles adhering to dirt particles and very small granules or partially coalesced fat globules; they can remain only if the milk is (almost) saturated with air. Nucleation predominantly determines the number of holes and their shape depends on cheese consistency, while both characteristics also depend on the rate of gas production. If the latter is not too fast and the cheese consistency allows for viscous flow of the cheese material, eyes (i.e. spherical holes) develop. If the consistency is short, or, more precisely, if the breaking stress of the material at slow deformation is low, slits may develop because the cheese mass fractures in the vicinity of the holes. Such may be the case for a cheese of low pH, low calcium phosphate content and considerable proteolysis at the time of gas production, but quantitative relations cannot be given yet. The problem is that most variables causing the fracture stress to be low (implying easy slit formation), also cause the extensional viscosity of the cheese to be low (implying a low overpressure in the hole and thus less possibility of fracture). See further Ref. 82b.

3·5 Microbial Defects

Whether or not deteriorative microorganisms will develop in cheese is determined by the microbial composition of the cheese milk, any contamination during cheesemaking, the composition of the cheese and curing conditions. Most undesirable organisms are killed by HTST-treatment of milk, e.g. 15 s at 72°C. Consequently, the nature and the extent of the defects may differ greatly between cheese made from raw milk and pasteurized milk. However, in spite of rigorous hygienic standards, contamination of milk or curd with undesired organisms cannot fully be prevented; moreover, conditions of modern cheesemaking, e.g. the almost continuous use of moulding machines, may allow considerable growth. Some defects may even originate specifically from the introduction of modern technology. Nevertheless, to guarantee the microbial quality of cheese, practical GMP standards for the Dutch cheese industry have been drawn up.[53] It must be noted that any organisms in the milk are almost completely entrapped in the curd, thus raising their number per gram of fresh cheese by a factor of 10, as compared to their number per gram of curd (gel) at the end of renneting.

3.5.1 Coliform Bacteria

These bacteria require lactose for growth in cheese. They grow only during the early stages of cheesemaking and can multiply very rapidly during the first few hours when other conditions, such as pH and temperature, are also favourable. According to the species or strains involved, varying amounts of metabolites are formed; most important are lactic acid, acetic acid, formic acid, succinic acid, ethanol, 2,3-butyleneglycol, hydrogen and carbon dioxide. In cases of excessive growth, cheeses develop off-flavours (yeasty, gassy, unclean) and show 'early blowing' due to the formation of CO_2 and, in particular, of H_2, which has very low solubility in cheese. Strains may differ in their potency to produce free H_2 in cheese; those that ferment citric acid readily show a reduced risk for early blowing, because hydrogenation of intermediate metabolites occurs.[83] HTST-treatment of milk, e.g. 15 s at 72°C, kills the coliform organisms and consequently, early blowing is unlikely to occur in cheese made from pasteurized milk, but in industrial practice a slight recontamination of this milk with coliform bacteria of non-faecal origin, e.g. *Enterobacter aerogenes*, is inevitable and growth during cheesemaking must be well controlled to limit their number in cheese. Apart from hygienic conditions, rapid acidification of the curd, which decreases the pH and the amount of lactose, is of great importance.

Formation of H_2 by coliform bacteria is inhibited by nitrate which suppresses the formation of the hydrogen lyase system normally involved in the production of hydrogen from lactose via formate under sufficiently anaerobic conditions, and induces the formation of a formate dehydrogenase/nitrate reductase system.[84,85] The nitrite formed induces the production of a formate dehydrogenase/nitrite reductase system.[86] Nitrate and nitrite act as terminal hydrogen acceptors and their nitrogen is used for bacterial growth or for ammonia production.[83,87] Hence, no H_2 is produced from formate. Growth of coliform bacteria is not prevented (e.g. Ref. 88), CO_2 is normally produced and development of off-flavour is not inhibited.

Nitrate is, however, used primarily to prevent the 'late blowing' defect caused by butyric acid bacteria.

3.5.2 Butyric Acid Bacteria

Butyric acid fermentation is characterized by the breakdown of lactic acid principally to butyric acid, CO_2 and H_2:

$$2CH_3CHOHCOOH \rightarrow CH_3CH_2CH_2COOH + 2CO_2 + 2H_2$$

Consequently, growth of anaerobic, spore-forming, lactate-fermenting butyric acid bacteria, especially of *Clostridium tyrobutyricum*, may cause 'late blowing' of cheese due to excessive production of CO_2 and H_2, and a very bad off-flavour. Silage, used as a feed in winter, represents the main source of contamination of the milk, especially when it is insufficiently preserved. Such silage contains large numbers of *Cl. tyrobutyricum* spores, which survive passage through the digestive tract of the cow and concentrate in dung. The degree of contamination of the milk with spores, therefore, strongly depends on hygienic conditions during milking,[89,90] but even with modern methods of milking, a slight contamination with dung present on the udder cannot be prevented. This problem is the more serious since the spores fully survive the HTST-treatment normally applied to the cheese milk.

Dutch-type varieties are especially vulnerable to butyric acid fermentation, the more so for larger cheese loaves. Because of the serious nature of the defect, much research has been undertaken to find ways to reduce the number of spores in milk and to prevent their germination and growth in cheese. Factors studied include: bactofugation of the milk; addition to the milk of nitrate, hydrogen peroxide and other oxidizing substances, or lysozyme; the use of a nisin-producing starter; the salt content and pH of the cheese; cheese ripening temperature; amount of (undissociated) lactic acid (for relevant literature information, see Refs 12, 91, 91a).

Nitrate may be used effectively to prevent butyric acid fermentation and has been used for this purpose for about 160 years.[91b] The mechanism of inhibition requires the presence of xanthine oxidase (EC 1.2.3.2) which reduces nitrate to nitrite.[92] Nitrite is considered to delay the germination of spores for a certain period after brining (but the actual mechanism may well be more complicated[91a]). Later on, the inhibitory action is taken over by NaCl when it has become evenly distributed throughout the cheese and if it is present at sufficient concentration.[93] If nitrite is the only factor involved in the initial inhibition, it must be very effective, since it is present at only very small concentrations (see also Fig. 11).

At a given curing temperature, usually about 14°C, the combined effect of several factors determines whether growth of *Cl. tyrobutyricum* is possible or not. Important factors promoting growth are a large number of spores in the cheese milk, a low content of undissociated lactic acid (hence usually a high pH), a low nitrate content in the cheese and a low level of NaCl in the cheese moisture. The rate at which salt becomes homogeneously distributed throughout the cheese mass, its final concentration and the initial nitrate content of the cheese are, therefore, crucial.[95] For example, a cheese with a high pH requires a higher than normal final salt concentration to inhibit growth. Any discrepancy

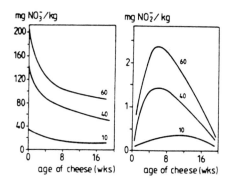

Fig. 11. Contents of nitrate and nitrite per kg of Gouda cheese, made from HTST-pasteurized milk (5 s at 76°C) during storage at 13°C. Parameter is the amount of NaNO₃ added to the cheese milk (g/100 litres) (after Ref. 94).

in conditions may allow growth of *Cl. tyrobutyricum* to start. Since the pH of the cheese is increased by the butyric acid fermentation, growth conditions for the organism then become more and more favourable and consequently the rate of fermentation is accelerated.

Increased numbers of coliform bacteria, e.g. >10⁵/g in a one-day old cheese, may cause nitrate to disappear too quickly from the cheese, increasing the risk of butyric acid fermentation. Growth of some particular strains of mesophilic lactobacilli also may have this effect.[96]

Heat treatments of milk that inactivate xanthine oxidase render cheese made from such milk very vulnerable to late blowing.

If legally accepted, lysozyme may be used as an alternative to nitrate. Its use is somewhat more expensive than bactofugation for a similar protective effect. In the amounts recommended (e.g. 500 I.U. per ml of cheese milk), it usually is less effective than nitrate, according to experience with Gouda cheese. Some spores are quite resistant to lysozyme, whereas others are readily inhibited or are even more sensitive to lysozyme than to nitrate.[96a]

Low numbers of spores in the cheese milk, which can also be achieved by bactofugation, permit the amount of nitrate to be reduced considerably. A certain amount remains necessary, because without nitrate the critical concentration of spores in milk capable of causing the butyric acid fermentation is extremely low, 5–10 spores/litre, which generally cannot be achieved.[13]

3.5.3 *Lactobacilli*

Growth of mesophilic normal or salt-tolerant lactobacilli may introduce flavour and texture defects, especially in mature cheese. Even when initially present in small numbers, e.g. 10 per ml of cheese milk, some strains of common lactobacilli (*L. plantarum, L. casei, L. brevis*) may grow slowly in cheese to more than 2×10^7/g in 4–6 weeks,[21] causing gassy and putrid flavours and an excessively open texture. Probably, amino acids are used as a carbon source.[97] The organisms are killed by adequate pasteurization of milk, e.g. 15 s at 72°C. In industrial practice, continuously working curd-drainage machines are the main source of contamination.

Especially when the salting of cheese is carried out in brine of reduced strength, there is a risk of defects caused by salt-tolerant lactobacilli, some strains being able to survive even in the presence of >15% NaCl. Furthermore, they differ from normal lactobacilli by their continuing growth in cheese and their active amino acid metabolism, causing phenolic, putrid, mealy and H_2S-like flavours in 4–6 months-old cheese. Some strains also produce excessive quantities of CO_2, causing the formation of holes, either eyes or cracks according to the consistency of cheese.[98] More than 10^3 of these gas-forming lactobacilli per ml of brine is considered to be dangerous. The lactobacilli enter the cheese through penetration of the rind during brining, this being facilitated if the cheese is insufficiently pressed and the rind not well closed.[99] Of course, contamination of the cheese milk with these bacteria must be prevented. If a weak brine (e.g. 14% NaCl) is kept sufficiently acid (pH ≤ 4·6) and cold (~13°C), growth of the organisms usually does not occur, and they die gradually. Increased numbers in brine originate from their growth in deposits, which are often present on the walls of basins just above the brine level, on racks and other equipment, and so contaminate the brine. Growth conditions for the lactobacilli are more favourable in these deposits as a result of the action of salt-tolerant yeasts increasing pH, a lower NaCl concentration (due to absorption of water), and a somewhat higher temperature than that of the brine, which usually is ~13°C. Measures to keep the number of lactobacilli low in brine include good hygiene in the brining room with removal of deposits, adjustment of the NaCl content of the brine to at least 16% and of its pH to ≤4·5.[100]

3.5.4 Thermoresistant Streptococci

These bacteria are normally present in raw milk. In particular, strains of *S. salivarius* ssp. *thermophilus* may be responsible for cheese of inferior quality. Contrary to the mesophilic streptococci, they can grow at 45°C and survive thermalization (e.g. 10 s at 65°C) and HTST-pasteurization (e.g. 15 s at 72°C) of milk. During such heat treatments, after some time a few organisms may become attached to the walls of the cooling section of the heat exchanger and may then start to multiply very rapidly (minimum generation time ≈15 min); this may depend on their initial number in the milk. Continuous use of heat exchangers for too long a period without cleaning may cause heavy contamination of the cheese milk (about 10^6 per ml). As a result of their concentration in curd and growth during the early stage of cheesemaking, their number may increase to more than 10^8/g of cheese. They render the flavour of cheese 'unclean' and 'yeasty'. Moreover, CO_2-production by these bacteria may yield cheese with an excessively open texture, especially if a starter with high CO_2-producing capacity is used for cheesemaking.[101,102]

3.5.5 Propionic Acid Bacteria

Very considerable growth of these organisms in cheese results in the development of a sweet taste and a very open texture, due to excessive gas formation. Propionic acid bacteria can convert lactates into propionic acid, acetic acid and CO_2, according to:

$$3CH_3CHOHCOOH \rightarrow 2CH_3CH_2COOH + CH_3COOH + CO_2 + H_2O$$

Consequently, the pH of the cheese does not change significantly. Because the bacteria develop very slowly in cheese at the commonly applied ripening temperatures, any serious defects occur only after prolonged ripening. Several conditions determine their growth in cheese. The pH is decisive, significant growth starting only from 5·1 and increasing at higher values. Increasing concentrations of NaCl retard their growth, but the effect of the usual variation in NaCl content in Dutch-type cheeses (4–5% NaCl in the water) is small. Higher salt concentrations, e.g. near the rind of the cheese, may be inhibitory. Higher storage temperatures favour the growth of propionic acid bacteria. Nitrate hinders their development. When conditions allow growth of these bacteria in cheese, the development of butyric acid bacteria (if present) may also be expected, provided that growth of the latter is not prevented otherwise. Propionic acid bacteria are killed by low-temperature treatment of milk, e.g. 15 s at 72°C. Therefore, they are predominantly of interest in the manufacture of cheese made from raw milk, farm-made cheese in particular.

3.5.6 Coryneform Bacteria, Yeasts and Moulds

Abundant growth of yeasts and coryneform bacteria on the cheese surface may lead to a somewhat slimy rind and a part-coloured or pink appearance. Growth of these organisms is favoured by: insufficient acidification of the cheese leading to a significant lactose content in the rind; salting of cheese in brine with a low NaCl content and a high pH; inadequate drying of the cheese rind after brining (this is the main factor in practice); and the use of insufficiently cleaned shelves. Growth of moulds causes discoloration and may under extreme conditions produce a health hazard because of mycotoxin formation. To prevent their development, special attention must be paid to treatment of the cheese rind and the hygienic conditions in curing rooms (see Section 2.4).

3.5.7 Minor Defects

Growth of certain lactobacilli (e.g. *L. büchneri*) and/or faecal streptococci (e.g. *Str. durans*) occasionally cause undesired levels of amines, histamine and tyramine in particular.[103] This problem is most frequently encountered in cheese made from raw milk.

Various microorganisms may sometimes cause flavour defects, predominantly in cheese made from raw milk. These include *Str. lactis* var. *maltigenes* (burned flavour), *Str. faecalis* var. *malodoratus* (H_2S flavour), and yeasts (yeasty, fruity flavour). Increased levels of psychrotrophic organisms or of their thermostable lipases in the cheese milk may cause the cheese to become rancid.

REFERENCES

1. Burkhalter, G. (Rapporteur), 1981. IDF-Catalogue of cheeses. *FIL-IDF Bulletin*, Doc. 141, p. 3.
2. Mair-Waldburg, H. (ed.), 1974. *Handbuch der Käse*, Volkswirtschaftlicher Verlag, Kempten.
3. Lawrence, R.C., Gilles, J. & Creamer, L.K., 1983. *New Zealand J. Dairy Sci. Technol.*, **18**, 175.
4. Noomen, A. & Mulder, H., 1976. *Neth. Milk Dairy J.*, **30**, 230.
5. van Schouwenburg-van Foeken, A.W.J., Stadhouders, J. & Witsenburg, W.W., 1979. *Neth. Milk Dairy J.*, **33**, 49.
6. Driessen, F.M., 1983. Lipases and proteinases in milk. Occurrence, heat inactivation, and their importance for the keeping quality of milk products. Doctoral thesis, Agricultural University, Wageningen.
7. Law, B.A., Andrews, A.T., Cliffe, A.J., Sharpe, M.E. & Chapman, H.R., 1979. *J. Dairy Res.*, **46**, 497.
8. Bergdoll, M.S., 1979. In *Food Borne Infections and Intoxications*, 2nd Edn, ed. H. Riemann & F. Bryan. Academic Press, New York, pp. 444–95.
9. Hup, G., Bangma, A., Stadhouders, J. & Bouman, S., 1979. *Zuivelzicht*, **71**, 1014.
10. Driessen, F.M. & Bouman, S., 1979. *Zuivelzicht*, **71**, 1062.
11. Stadhouders, J. & Hup, G., 1970. *Neth. Milk Dairy J.*, **24**, 79.
12. van den Berg, G., Hup, G., Stadhouders, J. & de Vries, E., 1980. *Rapport Nederlands Instituut voor Zuivelonderzoek*, **R112**.
13. Stadhouders, J., 1983. *Neth. Milk Dairy J.*, **37**, 233.
14. Mulder, H. & Walstra, P., 1974. *The Milk Fat Globule. Emulsion Science as Applied to Milk Products and Comparable Foods*, Pudoc, Wageningen.
15. van den Berg, G. & de Vries, E., 1976. *Zuivelzicht*, **68**, 878, 924 (NIZO-nieuws, 1976, No. 6, 7).
16. van den Berg, G., & de Vries, E., 1975. *Neth. Milk Dairy J.*, **29**, 181.
17. Birkkjaer, H.E., Sørensen, E.J., Jørgensen, J. & Sigersted, E., 1961. *Beretn. Statens Forsøgsmejeri*, No. 128.
18. Lolkema, H. & Blaauw, J., 1974. *Kaasbereiding*, Landelijke Stichting Beroepsopleiding Levensmiddelenindustrie, Apeldoorn.
19. Monib, A.M.M.F., 1962. *Meded. Landbouwhogeschool Wageningen*, **62**(10).
20. Stadhouders, J. & Hup, G., 1975. *Neth. Milk Dairy J.*, **29**, 335.
21. Stadhouders, J., Kleter, G., Lammers, W.L. & Tuinte, J.H.M., 1983. NIZO-nieuws M.9; *Voedingsmiddelentechnologie*, **26**, 20; *Zuivelzicht*, **75**, 1118.
22. Stadhouders, J. & Leenders, G.J.M., 1984. *Neth. Milk Dairy J.*, **38**, 157.
23. Walstra, P., van Dijk, H.J.M. & Geurts, T.J., 1985. *Neth. Milk Dairy J.*, **39**, 209.
24. Wilbrink, A., 1979. *Neth. Milk Dairy J.*, **33**, 202.
25. Geurts, T.J., Walstra, P. & Mulder, H., 1980. *Neth. Milk Dairy J.*, **34**, 229.
26. Wilbrink, A., Spoelstra, T. & Strampel, J., 1981. *Zuivelzicht*, **73**, 16.
27. Stadhouders, J., Hup, G. & Hassing, F., 1974. *Voedingsmiddelentechnologie*, **27**, 17.
28. Geurts, T.J., Walstra, P. & Mulder, H., 1972. *Neth. Milk Dairy J.*, **26**, 168.
29. de Ruig, W.G. & van den Berg, G., 1985. *Neth. Milk Dairy J.*, **39**, 165.
30. Het Additievenboekje. Een overzicht van toevoegingen aan drink- en eetwaren. Staatsuitgeverij, 's-Gravenhage, 1981.
31. Schulz, M.E. & Kay, H., 1957. *Käse-Tabellen*, Milchwirtschaftlichter Verlag Th. Mann KG, Hildesheim.
32. Posthumus, G., Klijn, C.J. & Booy, C.J., 1967. *Off. Orgaan Kon. Ned. Zuivelbond FNZ*, **59**, 712, 740, 749, 769.
33. Straatsma, J., de Vries, E., Heijnekamp, A. & Kloosterman, L., 1984. *Zuivelzicht*, **76**, 956 (NIZO-nieuws, 1984–8).

34. O'Leary, J., Hicks, C.L., Aylward, E.B. & Langlois, B.E., 1983. *Res. Food Sci. Nutr.*, **1**, 150.
35. Steffen, C. & Rentsch, F., 1981. *Schweiz. Milchztg*, **107**, 129, 137.
36. van den Berg, G., de Koning, P.J., Escher, J.T.M. & Bovenhuis, H., 1990. *Voedingsmiddelentechnologie*, **23**, 13.
37. van den Berg, G., Escher, J.T.M., de Koning, P.J. & Bovenhuis, H., 1992. *Neth. Milk Dairy J.*, **46**, 145.
37a. Emmons, D.B., Beckett, D.C. & Binns, M., 1978. *Proc. 20th Intern. Dairy Congr., Paris*, Vol. E, p. 491.
38. Banks, J.M. & Muir, D.D., 1985. *Milchwissenschaft*, **40**, 209.
39. Banks. J.M., Tamime, A.Y. & Muir, D.D., 1985. *Dairy Ind. Internat.*, **50**, 11.
40. Richardson, G.H., 1984. *Cultured Dairy Products J.*, **19**, 6.
41. Walstra, P. & Jenness, R., 1984. *Dairy Chemistry and Physics.* John Wiley, New York.
42. Posthumus, G., Booy, C.J. & Klijn, C.J., 1963. *Off. Orgaan Kon. Ned. Zuivelbond FNZ*, **55**, 986.
43. Maubois, J.L., 1984. In *Le Fromage*, A. Eck (ed.), Lavoisier, Paris, p. 157.
43a. Lawrence, R.C., 1989. *Bulletin of the International Dairy Federation*, No. 240.
43b. van Boekel, M.A.J.S. & Walstra, P., 1989. *Neth. Milk Dairy J.*, **43**, 437.
43c. Straatsma, J. & Heijnekamp, A., 1988. *NIZO-mededeling* M21; and 1987. *Zuivelzicht*, **79**(21), 24.
44. Pearce, K.N., 1978. *N.Z. J. Dairy Sci. Technol.*, **13**, 59.
45. van den Berg, G. & de Vries, E., 1974. *Milchwissenschaft*, **29**, 214.
46. Davies, F.L. & Law, B.A. (eds), 1984. *Advances in the Microbiology and Biochemistry of Cheese and Fermented Milk.* Elsevier Applied Science Ltd, London.
47. Daly, C., 1983. *Antonie van Leeuwenhoek*, **49**, 297.
48. Stadhouders, J., 1974. *Milchwissenschaft*, **29**, 329.
49. Galesloot, T.E., Hassing, F. & Stadhouders, J., 1966. *Proc. 17th Int. Dairy Congr.*, Munich, Vol. D, p. 491.
50. Lankveld, J.M.G., 1984. *Voedingsmidellentechnologie*, **17**, 33.
51. Stadhouders, J., Bangma, A. & Driessen, F.M., 1976. *Zuivelzicht*, **68**, 180.
52. Starters in the manufacture of cheese. 1980. *FIL-IDF Bulletin*, Doc. 129.
53. Northolt, M.D. & Stadhouders, J., 1985. *Zuivelzicht*, **77**, 324, 488.
54. Langeveld, L.P.M. & Galesloot, T.E., 1971. *Neth. Milk Dairy J.*, **25**, 15.
55. Visser, F.M.W., 1977. *Neth. Milk Dairy J.*, **31**, 120.
56. Kleter, G. & de Vries, Tj., 1974. *Neth. Milk Dairy J.*, **28**, 212.
57. Kleter, G., 1975. *Neth. Milk Dairy J.*, **29**, 295.
58. Kleter, G., 1976. *Neth. Milk Dairy J.*, **30**, 254.
59. Kleter, G., 1977. *Neth. Milk Dairy J.*, **31**, 177.
60. Visser, F.M.W., 1977. *Neth. Milk Dairy J.*, **31**, 188.
61. Visser, F.M.W., 1977. *Neth. Milk Dairy J.*, **31**, 210.
62. Visser, F.M.W., 1977. *Neth. Milk Dairy J.*, **31**, 247.
63. Visser, F.M.W., 1977. *Neth. Milk Dairy J.*, **31**, 265.
63a. Geurts, T.J. *et al.*, unpublished.
64. Noomen, A., 1978. *Neth. Milk Dairy J.*, **32**, 49.
65. Noomen, A., 1978. *Neth. Milk Dairy J.*, **32**, 26.
66. Youssef, Y.B. 1992. PhD thesis, Wageningen Agricultural University, Wageningen.
66a. Annual Report, NIZO, Ede, The Netherlands, 1989.
67. Stadhouders, J., 1956. De hydrolyse van vet bij de kaasrijping in verband met de smaak van kaas, Doctoral thesis, Agricultural University, Wageningen.
68. Stadhouders, J. & Mulder, H., 1957. *Neth. Milk Dairy J.*, **11**, 164.
69. Stadhouders, J. & Mulder, H., 1958. *Neth. Milk Dairy J.*, **12**, 238.
70. Stadhouders, J. & Mulder, H., 1959. *Neth. Milk Dairy J.*, **13**, 291.

71. Stadhouders, J. & Mulder, H., 1960. *Neth. Milk Dairy J.*, **14**, 141.
72. Driessen, F.M. & Stadhouders, J., Lipolysis in hard cheese made from pasteurized milk, *FIL-IDF Bulletin*, Doc. **86**, p. 101.
73. Stadhouders, J. & Veringa, H.A., 1973. *Neth. Milk Dairy J.*, **27**, 77.
74. Raadsveld, C.W. & Mulder, H., 1949. *Neth. Milk Dairy J.*, **3**, 222.
75. Raadsveld, C.W. & Mulder, H., 1949. *Neth. Milk Dairy J.*, **3**, 117.
76. Badings, H.T., 1984. Flavors and off-flavors. In *Dairy Chemistry and Physics*, ed. P. Walstra & R. Jenness. John Wiley, New York, p. 336.
77. Mulder, H., 1952. *Neth. Milk Dairy J.*, **6**, 157.
78. Ali, L.A.M., 1960. The amino acid content of Edam cheese and its relation to flavour, Doctoral thesis, Agricultural University, Wageningen.
79. Raadsveld, C.W., 1952. *Neth. Milk Dairy J.*, **6**, 342.
79a. Luyten, H., 1988. The rheological and fracture properties of Gouda cheese. Doctoral thesis, Agricultural University, Wageningen.
80. Walstra, P. & van Vliet, T., 1982. Rheology of cheese, *IDF-Bulletin*, Doc. **153**, p. 22.
81. Mulder, H., 1945. *Versl. Landbouwk, Onderz.*, **51C**, 467.
82. Oortwijn, H., 1984. Unpublished report 84.80 of the Rijkskwaliteitsinstituut voor Land- en tuinbouwprodukten (RIKILT), Wageningen.
82a. Akkerman, J.C., Walstra, P. & van Dijk, H.J.M., 1989. *Neth. Milk Dairy J.*, **43**, 453.
82b. Walstra, P., 1991. In *Rheological and fracture properties of cheese*, eds T. van Vliet & P. Walstra. *IDF Bulletin* **268**, 65–6.
83. Galesloot, T.E., 1946. Over de vroeg beginnende gasvorming in kaas. Doctoral thesis, Agricultural University, Wageningen; Publication L.E.B. Fonds, No. 29, Wageningen.
84. Ruiz Herrera, J. & De Moss, J.A., 1969. *J. Bact.*, **99**, 720.
85. Ruiz Herrera, J. & Alvarez, A., 1972. *Antonie van Leeuwenhoek*, **38**, 479.
86. Abou-Jaoudé, A., Chippaux, M., Pascal, M.C. & Casse, F., 1977. *Biochem. Biophys. Res. Comm.*, **78**, 579.
87. Cole, J.A. & Brown, C.M., 1980. *FEMS Microbiol. Lett.*, **7**, 65.
88. Galesloot, T.E. & Hassing, F., 1983. *Neth. Milk Dairy J.*, **37**, 1.
89. de Vries, Tj. & Stadhouders, J., 1977. *Zuivelzicht*, **69**, 196.
90. de Vries, Tj. & Brouwer, J., 1980. *Boerderij*, **64**, 50.
91. Kleter, G., Lammers, W.L. & Vos, E.A., 1984. *Neth. Milk Dairy J.*, **38**, 31.
91a. Stadhouders, J., Hup, G. & Nieuwenhof, F.F.J., 1983. *Mededelingen Ned. Inst. voor Zuivelonderzoek*, **M19**.
91b. Peereboom, J., 1830. Beantwoording der prijsvraag, houdende De oorzaak van het zoogenaamde knijpen of breken der Noord-Hollandsche of Edammer Kaas, met opgave der middelen om dit kwaad, ten dienste van Landbouw en Koophandel, te verhoeden. In *Beschrijving en gebruik van middelen ten algemeenen nutte*. Haarlem, Loosjes, pp. 2–5.
92. Galesloot, T.E., 1961. *Neth. Milk Dairy J.*, **15**, 31.
93. Galesloot, T.E., 1961. *Neth. Milk Dairy J.*, **15**, 395.
94. Goodhead, K., Gough, T.A., Webb, K.S., Stadhouders, J. & Elgersma, R.H.C., 1976. *Neth. Milk Dairy J.*, **30**, 207.
95. Galeshoot, T.E., 1964. *Neth. Milk Dairy J.*, **18**, 127.
96. Nieuwenhof, F.F.J., 1977. *Neth. Milk Dairy J.*, **31**, 153.
96a. Lindblad, O.T., 1990. *Brief Communications and Abstracts of Posters of the XXIII International Dairy Congress*, Montreal, Canada, 483.
97. Kristoffersen, T. & Nelson, F.E., 1955. *Appl. Microbiol.*, **3**, 268.
98. Stadhouders, J., Hup, G. & Hassing, F., 1974. *Voedingsmiddelentechnologie*, **27**, 17.
99. Hup, G., Stadhouders, J., de Vries, E. & van den Berg, G., 1982, *Zuivelzicht*, **74**, 270.

100. Stadhouders, J., Leenders, G.J.M., Maessen-Damsma, G., de Vries, E. & Eilert, J. G., 1985. *Zuivelzicht*, **77**, 892.
101. Hup, G., Bangma, A., Stadhouders, J. & Bouman, S., 1979. *Zuivelzicht*, **71**, 1014.
102. Driessen, F.M. & Bouman, S., 1979. *Zuivelzicht*, **71**, 1062.
103. Joosten, H.M.L.J., 1988. Conditions allowing the formation of biogenic amines in cheese. Doctoral thesis, Agricultural University, Wageningen.

3

Swiss-Type Varieties

C. STEFFEN, P. EBERHARD, J.O. BOSSET AND M. RÜEGG

Federal Dairy Research Institute, Liebefeld-Bern, Switzerland

1 GENERAL

There is no internationally recognized definition of Swiss-type cheeses that differentiates them from other varieties. Swiss-type cheeses have round regular eyes which vary in size from medium to large (Fig. 1). The body and texture correspond to those of hard or semi-hard cheeses. Swiss-type cheeses were originally manufactured in the Emmen valley in Switzerland, their precursors were mountain cheeses[1]. Emmentaler is probably the best-known Swiss-type cheese and is frequently referred to simply as 'Swiss cheese'. Gruyère and Appenzeller are other Swiss cheeses belonging to the Swiss-type group.

The characteristics of Swiss-manufactured Emmentaler are:

(a) cylindrical shape;
(b) firm dry rind;
(c) weight: 60–130 kg;
(d) 1000–2000 round eyes, diameter 1–4 cm, caused by propionic acid fermentation;
(e) flavour: mild, slightly sweet, becoming more aromatic with increasing age;
(f) cheese body: ivory to light-yellow, slightly elastic.

Today, Swiss-type cheeses are manufactured in many countries by methods differing from traditional Swiss procedures. Thus, the treatment of milk, the extent of mechanization, the weight, shape, ripening time and shelf life of foreign Swiss-type cheeses are often different from the original.

Various authors[2,3,4] classify different cheeses as being Swiss type: these include the following: Samsoe, Jarlbergost, Herregardsost, Maasdamer, Comté and Iowa-style Swiss cheese.

Table 1 shows the main constituents of ripe Emmentaler, Gruyère and Appenzeller cheeses.

2 FERMENTATION

The quality of Swiss-type cheese depends principally on the course of lactic acid fermentation, propionic acid fermentation (essential for Emmentaler), proteolysis and smear ripening (valued for Gruyère and Appenzeller).

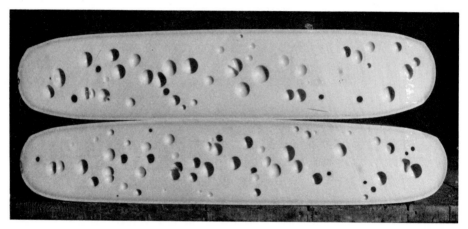

Fig. 1. Emmentaler cheese.[77]

TABLE I
Constituents of Ripe Emmentaler, Gruyère and Appenzeller (N = 60) [34,68,74,75]

Constituent	Unit	Emmentaler $\bar{x}$	s	Gruyère $\bar{x}$	s	Appenzeller $\bar{x}$	s
Protein	g/kg	291	5·8	268	6·4	249	7·0
Fat	g/kg	314	9·8	322	12·0	318	8·0
Water	g/kg	351	7·9	358	8·6	392	11·0
NaCl	g/kg	4·4	1·5	15·9	2·4	16·2	1·9
Calcium	g/kg	10·1	0·5	8·9	0·3	7·4	0·4
Manganese	mg/kg	0·3	0·01	0·3	0·08	0·3	0·06
Copper[a]	mg/kg	14·5	3·6	13·1	3·8	14·0	3·7
Iron	mg/kg	3·4	1·1	3·9	1·3	2·7	0·7
Volatile fatty acids	mmol/kg	128	19·5	33·0	12·1	74·0	24·9
—Acetic acid	mmol/kg	44·0	9·3	18·0	8·3	42·0	11·0
—Propionic acid	mmol/kg	84·0	15·8	10·0	9·0	24·0	13·5
Lactic acid	mmol/kg	34·0	22·0	106·0	14·0	45·0	16·7
Free amino acids	g/kg	30·0	9·6	39·0	8·8	39·0	8·6
Age	d	134	13	195	8	150	9

[a] Values for cheese manufactured in copper kettles.

2.1 Lactic Acid Fermentation

Every cheese type firstly undergoes homofermentative lactic acid fermentation, with over 90% of the lactose being converted to lactic acid. Controlled lactic acid fermentation is obtained by adding thermophilic and sometimes mesophilic lactic acid bacterial cultures to the cheese milk. Lactic acid bacteria are morphologically dissimilar[5,6] and consist of long and short rods (lactobacilli) as well as cocci (streptococci). They are all Gram-positive. Lactic acid bacteria have complex nutritional requirements for growth and milk is an ideal medium. They tolerate a high degree of acidity. The starter cultures used in Swiss cheese manufacture contain both thermophilic species such as *Streptococcus thermophilus, Lactobacillus lactis, Lactobacillus helveticus, Lactobacillus bulgaricus* and mesophilic species, e.g. *Lactococcus cremoris, Lactococcus lactis* (Fig. 2). Table II shows the properties of species of lactic acid bacteria frequently used in cheesemaking.

Lactose breakdown normally follows the fructose-1,6-diphosphate pathway described by Emden-Meyerof-Parnas.[6] Its catabolism starts with processing in the vat, and 4–6 h after dip filling of the moulds, the sugar is entirely hydrolysed. Whereas free glucose is never detected at high concentrations in the cheese, free galactose is found during the early stages. The complete lactic acid fermentation lasts about 24 h (Fig. 3).

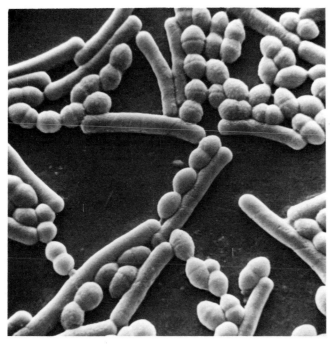

Fig. 2. Scanning electron micrograph of a culture of lactic acid bacteria (streptococci and lactobacilli).[77]

TABLE II
Characteristics of Lactic Acid Bacteria Used for Cheesemaking[6,76,77]

Characteristics	Units	S. lactis	S. thermophilus	Lb. helveticus	Lb. lactis
Generation time (optimal conditions)	min	15–20	15-20	35–45	35–45
Lactic acid production in milk (total acid possible)	g/l	5–8	8–10	20–30	10–25
Optimum growth temperature	°C	20–30	38–42	38–45	38–45
Optimum pH		6·0	6·0–6·5	5·0–5·5	5·0–5·5
Proteolysis		medium	weak	strong	strong
Lactic acid pattern		L(+)	L(+)	L(+)/D(−)	D(−)

The lactic acid produced in the cheese loaf influences the quality of cheese in several ways. It acts as a preservative since the lower pH values (milk 6·6–6·8, old cheese 5·1–5·3) inhibit the growth of strongly proteolytic and other bacteria. Moreover, it removes calcium from the paracasein-calcium phosphate complex; this influences synersis, body characteristics and proteolysis. Finally, lactic acid is an appropriate substrate for the subsequent propionic acid fermentation or for the surface flora of smear-ripened cheese (Gruyère, Comté, Appenzeller).

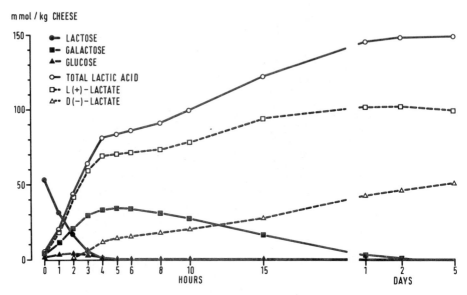

Fig. 3. Concentrations of lactose, glucose, galactose and D- and L-lactate in Emmentaler cheese.[78]

2.2 Propionic acid fermentation

Most Swiss-type cheeses undergo a more or less pronounced propionic acid fermentation, which is brought about by Gram-positive short-rod propionic acid bacteria. They grow under anaerobic conditions, at low oxygen concentrations only, and occur naturally in the rumen and intestine of ruminants.[5] Lactate is transformed into propionate, acetate and carbon dioxide in the following proportions:

$$3CH_3 CHOHCOO^- \rightarrow 2CH_3CH_2COO^- + CH_3COO^- + CO_2 + H_2O$$

Hygienic milk production calls for the inoculation of cheese milk in the vats with a culture of propionic acid bacteria, mostly *Propionibacterium freudenreichii* subsp. *shermanii*) before processing (Fig. 4). To initiate propionic acid fermentation, the ripening temperature for the cheese must be raised to approximately 18–25°C for a certain period of time. As a result, these lactate fermenters, originally at low concentrations in cheese milk, develop to levels of 10^8–10^9 organisms/g cheese. Emmentaler cheese undergoes propionic acid fermentation 20–30 days after the start of manufacture. The resulting metabolites, propionic acid and acetic acid, essentially contribute to the development of the characteristic flavours of Swiss-type cheese, whereas the carbon dioxide produced is responsible for eye formation. As soon as the development of sufficient eyes is accomplished, the propionic acid is usually retarded by storing the cheese at a lower temperature (Fig. 5).

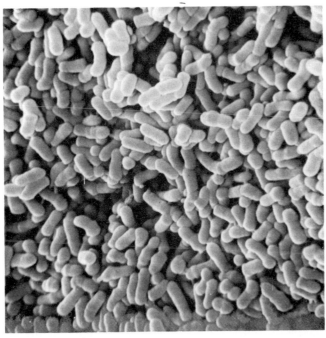

Fig. 4. Scanning electron micrograph of a culture of *Propionibacterium freudenreichii.*[79]

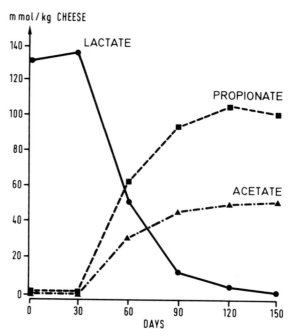

Fig. 5. Lactate breakdown and production of proprionate and acetate during ripening of Emmentaler cheese.[77]

2.3 Proteolysis

In Swiss-type cheese with a dry rind, proteolysis is, apart from propionic acid fermentation, the most important factor for ripening and flavour development. In Cheddar and Gouda type cheeses, rennet plays an important role in proteolysis. In Swiss-type cheeses, however, rennet is inactivated during the heating of the curd and does not play a significant role in proteolysis. In these cheeses, indigenous milk proteinase and the proteolytic enzymes of lactic acid bacteria are mainly responsible for protein breakdown.[7,8] Generally, thermophilic lactobacilli exert a stronger proteolytic effect than mesophilic lactococci, whereas thermophilic streptococci have very little influence on protein breakdown. The proteolytic activity of propionic acid bacteria is insignificant. When raw milk is processed, certain microorganisms of the wild (indigenous) flora of milk (e.g. enterococci, micrococci) may possibly be involved in proteolysis. The proteolytic enzymes of psychrotrophs from milk after prolonged cold storage sometimes influence ripening and flavour development. Proper selection of strains of lactic acid bacteria for starter cultures and the application of appropriate measures during manufacture in order to obtain the desired number of lactobacilli in the young cheese are the best means of controlling proteolysis. The activities of proteolytic enzymes in cheese further depend on the water content, lactic acid concentration, pH, storage temperature, storage time, NaCl concentration, water activity (a_w) and copper content.

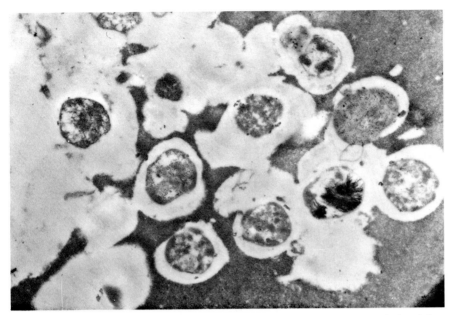

Fig. 6. Electron micrograph of a 12-month old Emmentaler cheese. Note casein breakdown in the neighbourhood of microorganisms: the white area around the bacteria represents the casein matrix degraded by proteolytic enzymes.[80]

Casein in unripened cheese is odourless, tasteless and water-insoluble and has a tough and rubber consistency. The proteolytic enzymes of microorganisms in cheese split this casein network into short-chain water-soluble compounds, e.g. peptones, polypeptides, peptides and amino acids (Fig. 6). These metabolites also contribute to the development of the characteristic flavour, body and texture of Swiss-type cheese. Common indices of proteolysis are the concentration of the water-soluble nitrogen (WSN), non-protein nitrogen (NPN), and different proteolytic enzymes (e.g. leucine aminopeptidase). In recent years, the measurement of peptides, free amino acids and amines has also been used in order to better determine protein breakdown (Table III). The amino acids may be further decomposed enzymatically by decarboxylation, deamination and transamination, but non-enzymatic reactions are also involved. The products arising from amino acid breakdown are: amines, ammonia, α-keto-acids, aldehydes, aromatic acids, mercaptans and fatty acids.

Some Swiss-type cheeses, such as Gruyère, undergo a smear ripening process. The surface of the cheese is first deacidified by yeasts. Early in the ripening process various salt-tolerant microorganisms develop (e.g. *Micrococcus*, *Enterococcus*, *Brevibacterium linens*). These microorganisms degrade the lactic acid and influence the flora and the colour of the surface of the cheese.[9]

In cheese with surface ripening, proteolysis is further affected, although the proteolytic enzymes of the surface flora, such as the leucine aminopeptidase,

TABLE III
Proteolysis in Emmentaler Cheese (Values per kg Dry Matter)[74,77,81]

Age in days	Total N g/kg	Water-soluble N g/kg	Non-protein N g/kg	Free amino acids	
				Total g/kg	Glutamic acid mmol/kg
1	69·0	3·7	1·3	1·61	1·8
30	68·2	8·8	3·1	4·91	4·2
60	68·9	14·9	6·0	11·40	11·1
90	69·2	15·5	7·7	17·70	17·5
120	68·9	16·2	8·5	22·80	21·8
150	68·5	17·0	9·6	27·50	25·2

arginine aminopeptidase, phenylalanine aminopeptidase and proline amino-peptidase do not penetrate, or penetrate only slightly, into the cheese mass.[10] An indirect, but strong proteolytic effect arises from the accelerated increase of the pH in the outer zones of the cheeses due to deacidification of the surface by the smear, which gives a wetter rind. Moreover, sapid substances from the smear flora diffuse into the cheese mass.

Insufficient proteolysis may cause different cheese defects. If the level of protein breakdown is too low, the taste is flat and the body consistency too 'long'. Sometimes, uneven openings also result. Excessive proteolysis gives an overripe and sharp taste and a shorter body consistency. This defect becomes particularly evident when a large amount of casein is decomposed into low-molecular compounds (high non-protein nitrogen level). Additional carbon dioxide production by decarboxylation clearly reduces the keeping quality of the cheese and leads to

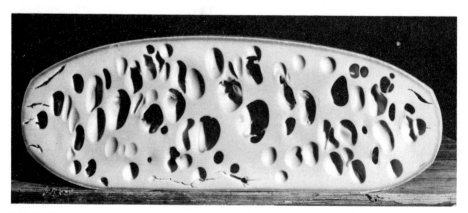

Fig. 7. Emmentaler cheese with secondary fermentation: note cracks due to excessive gas pressure from CO_2 produced by amino and decarboxylases.[77]

oversized eye formation. The cheese body often cannot withstand the pressure of the gas and cracks appear. This defect is called late or secondary formation.[11,12]

Frequently, the course of proteolysis in a cheese loaf varies from one zone to the other, a phenomenon that is due to temperature changes in the cheese loaf during lactic acid fermentation. Since the outer zone cools faster, it often develops a bacterial flora which is proteolytically more active than the microorganisms in the centre of the loaf. This usually leads to cheese defects such as short and firm body, or sharp taste, or the development of white colour under the rind.

A particularly serious defect results from the eventual presence of *Clostridium sporogenes*. This sporeformer brings about non-specific proteolysis, which may be local and very intensive. The cheese mass then shows putrid spots and is inedible.

2.4 Other Fermentation Processes

Besides the desired fermentation process, there are other possible metabolic processes caused by microbial activities, which may even lead to cheese defects.

Lipolysis is unwanted because of the atypical rancid taste it produces. However, it is sometimes supposed that a slight amount of fat hydrolysis in milk contributes to the development of the characteristic aroma of raw milk Swiss-type cheese.

Butyric acid fermentation is totally undesirable, since lactate decomposition by *Clostridium butyricum* and *Cl. tyrobutyricum* into butyric acid, acetic acid, carbon dioxide and hydrogen causes the cheese loaf to blow. Even in small amounts, butyric acid is unfavourable to flavour development. Therefore, in Switzerland, silage feeding of cows is prohibited for cheese production. This prohibition, together with hygienic milk production, helps prevent contamination with these butyric acid bacteria. In other countries, spores are either eliminated by bactofugation of milk or germination is restricted by additives, e.g. nitrates.

Heterofermentative lactic acid fermentation is of minor importance. In Emmentaler cheese it provokes abundant eye formation and in other cheese varieties such as Gruyère and Appenzeller it may contribute to the development of the eyes.

Mixed acid fermentation may occur due to the growth of *Enterobacteriaceae*, leading to an excess of eye formation in Swiss-type cheese due to the low water solubility of hydrogen produced by these bacteria.

2.5 Eye Formation

The characteristic eye formation of Emmentaler cheese is due mainly to the presence of carbon dioxide produced by propionic acid bacteria during lactate breakdown.[13] We can follow carbon dioxide production by measuring continually the carbon dioxide diffusing out of the cheese loaf.[14,15]

As shown in Fig. 8, carbon dioxide diffusion begins before propionic acid fermentation since small quantities of carbon dioxide are already produced during lactic acid fermentation. The steep rise in the production of carbon dioxide, however, coincides with the onset of the propionic acid fermentation. At the peak, carbon dioxide production and carbon dioxide diffusion rates are identical. The

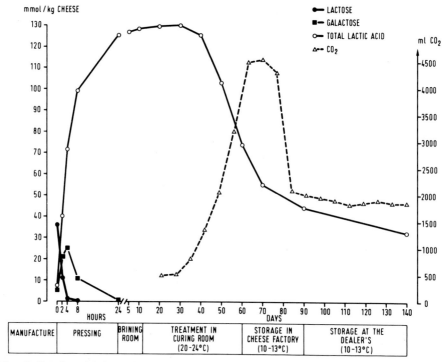

Fig. 8. Lactic acid fermentation, lactate breakdown and CO_2 loss during ripening of Emmentaler cheese.[12,14,77]

diffusion rate drops as soon as the cheese loaves are transferred from the ripening room (22°C) to the cold room (12°C). A new equilibrium then develops at a lower level between carbon dioxide production and carbon dioxide diffusion.

The different stages of carbon dioxide production and eye formation may be schematized as follows:

CO_2 Production and Eye Formation

CO_2 production and CO_2 diffusion
(starts at the beginning of the lactic acid fermentation)
↓
Accumulation of CO_2 in the cheese body
(proprionic acid fermentation)
↓
Oversaturation at the centres of future eye formation
(proprionic acid fermentation)
↓
Onset of eye formation at these centres
(after approximately 20–30 days)
↓
Increase in the number of eyes and their enlargement
(proprionic acid fermentation + decarboxylation of amino acids)

The development of eye formation depends mainly upon:

(a) time, quantity and intensity of CO_2 production;
(b) number and size of the areas of future eye formation;
(c) CO_2 pressure and diffusion rates;
(d) body texture and temperature.

Eye formation is a lengthy process. At the beginning, i.e. about 30 days after manufacture, only a few eyes appear; thereafter, the number of new holes increases progressively. The maximum rate is attained after about 50 days, which is also the time of rapid eye enlargement. The appearance of new eyes declines with decreasing carbon dioxide production and the simultaneous hardening of the cheese body. Nevertheless, eye formation sometimes continues in the cold room.

The quantity and distribution of the eyes also depend on other factors such as those mentioned above. The number of eyes is increased by the inhomogeneity of the curd, physical openness and hydrogen-forming microorganisms. Centrifugation and thermalization of the milk or application of vacuum after filling of the curd and during pressing of the cheese are performed in order to obtain a larger number of eyes (overset).[13,16] In a cheese loaf of approximately 80 kg, total carbon dioxide production is about 120 litres before the cheese is sufficiently aged for consumption. About 60 litres remain dissolved in the cheese body, approximately 20 litres are found in the eyes and approximately 40 litres diffuse out of the loaf.

The actual carbon dioxide production in Emmentaler cheese is higher than the amount calculated according to the equation of Fitz[17] on the basis of lactate breakdown and production of volatile fatty acid (acetate, propionate). Therefore, there must be further sources of carbon dioxide besides propionic acid fermentation. It has been demonstrated that lactic acid fermentation and protein breakdown are also sources of carbon dioxide.

In contrast to Emmentaler, Gruyère and Appenzeller cheeses do not strictly need propionic acid fermentation for eye formation, though it may be of some importance. In these cheese types, eye formation starts with lactic acid fermentation, probably because of the activity of heterofermentative lactic acid bacteria. Some carbon dioxide arises from oxidative decarboxylation of 6-phosphogluconate during the formation of lactate and acetate or ethanol from glucose[5]. Enlargement of the eyes, which occurs at a later stage, is probably due to a slight propionic acid fermentation and particularly to the carbon dioxide liberated by decarboxylation of amino acids.

Carbon dioxide pressure passes through two major phases (Fig. 9). The first covers the period of proper eye formation in the ripening room. During this period, the carbon dioxide pressure remains relatively low, between 1500 and 2500 Pa, because of the low resistance of the soft cheese mass to gas compression at 22–24°C. During storage, i.e. the second stage, the carbon dioxide pressure increases to 4000–8000 Pa. The differences in pressure between the various loaves are higher in the second stage than in the first. The pressure increase in the second stage is explained by the higher resistance to gas compression of

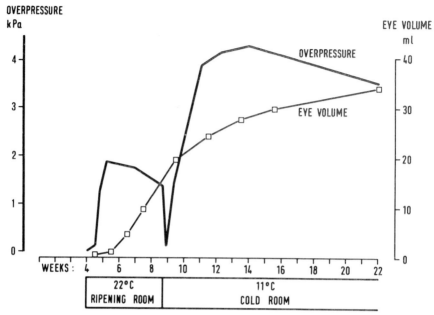

Fig. 9. Eye formation, volume and CO_2 overpressure in Emmentaler cheese.[14,82]

the cheese mass, which is due to a decrease in temperature from 22 to 12°C, and by continued gas production. During the first stage there is a marked pressure increase within the eyes.

3   FLAVOUR

Cheese flavour depends firstly on the properties of the cheese milk. Certain plants and feed-stuffs such as bulbous plants, leeks, vegetable wastes, herb mixtures and different mineral salt mixtures fed to dairy cows can influence the taste of milk and produce off-flavours. The chemical composition of milk also varies throughout lactation and is thus another factor of influence. Certain milk enzymes can induce flavours, e.g. lipase can cause rancidity. The microbial composition of milk plays an important role.[18,19,20] Whilst the wild flora of milk is generally composed of unwanted microorganisms,[21,22,23] which can influence the flavour directly by their fermentative activities or indirectly by other enzymatic reactions, the desired lactic acid bacteria must be added to the cheese milk in the form of starter cultures. Rennet is also a flavour influencing factor.

Cheese flavour depends secondly on the different operations involved in cheese-making and cheese ripening. In Switzerland, Swiss-type cheese is made from raw milk. Certain sapid compounds in milk are in fact lost and others are produced when it is subjected to thermalization or pasteurization before processing. The high temperatures applied during the early stages of manufacture and pressing

TABLE IV

Mean Volatile Fatty Acid and Lactic Acid Contents (mmol/kg) of Ripe Swiss-type Cheeses[12,34,74,75]

Components	Emmentaler Age: approx. 5 months			Gruyère Age: approx. 6·5 months			Appenzeller Age: approx. 4·5–5·5 months		
	$\bar{x}$	s	N	$\bar{x}$	s	N	$\bar{x}$	s	N
Tot. vol. fatty acids	124·0	17	60	33·1	12·1	62	73·7	24·9	59
Acetic acid	41·0	8·8	60	18·3	8·3	62	41·8	11·0	59
Propionic acid	83·7	15·8	60	9·9	9·0	62	24·1	13·5	59
i-butyric acid	—	—	60	—	—	62	0·4	1·8	59
n-butyric acid	—	—	60	3·1	2·8	62	4·4	4·3	59
i-valeric acid	—	—	60	0·9	0·7	62	2·2	2·2	59
i-caproic acid	—	—	60	—	—	62	0·9	1·4	59
n-caproic acid	—	—	60	—	—	62	—	—	59
Lactic acid	29·0	15·1	25	106	14·0	62	44·9	16·7	59

$\bar{x}$ = arithmetic mean (formally calculated although some distributions are asymmetric).
s = standard deviation.
N = number of samples.
— = traces or data missing.

of Swiss cheeses are essential for flavour development. Other important factors are the fermentation and ripening processes (pH, water activity), the conditions prevailing at the surface of the loaves, and even their size and shape. As regards brining and smearing, the salt concentration in these solutions affects the water activity in cheese and hence bacterial growth, thus indirectly contributing to cheese taste. The principal pathways of biosynthesis of cheese flavours are dealt with in many publications.[24-38] For analytical reasons, the flavour components are generally divided into two major groups; the volatile and the non-volatile compounds.

The *volatile compounds* include volatile fatty acids (Table IV), primary and secondary alcohols, methyl ketones, aldehydes, esters, lactones, alkanes, aromatic hydrocarbons and different sulphur- and nitrogen-containing compounds. Typical fingerprints of the three mature cheese-types considered are shown in Fig. 10. The GC-MS chromatograms were obtained using the so-called 'purge and trap' dynamic head space technique.[39]

The *non-volatile* group is composed of peptides, peptones, free amino acids, amines, organic and inorganic salts and fat (Tables V–VII).[40] Figure 11(a) shows the large number of peptides separated by RP-HPLC from mature Emmentaler, Gruyère and Appenzeller cheeses.[41] Polyacrylamide gel electrophoretograms of the same cheeses are shown in Fig. 11(b). Some compounds have been determined quantitatively, others have been identified qualitatively only, but a comprehensive study of flavours in Swiss-type cheese is lacking.

In a recent review, Ney[26] proposed a key which suggests the possible contribution of each of these components to cheese flavour. This key is used here to present the available data for Swiss-type cheese.

As regards other volatile flavour components not mentioned above, Swiss Gruyère as well as Emmentaler have been studied thoroughly. A large number of different volatile compounds could be identified by GC-MS analysis of Swiss Gruyère[42-45] and Emmentaler[39] cheese. The figures in brackets in the following list indicate the number of components found in Gruyère and Emmentaler, respectively: hydrocarbons [7, 31], primary alcohols [16, 16], secondary alcohols [6, 9], ketoalcohols [5, 1], aldehydes [6, 21], ketones [19, 26], acids [7, 7], esters [7, 22], lactones [1, 4], pyrazines [8, 6] and miscellaneous components [9, 8]. When

TABLE V

Mean Free Amino Acid Contents in Ripe Swiss-type Cheeses (mmol/kg)[34,81,84]

Free amino acids	Emmentaler	Gruyère	Appenzeller
Phosphoserine	4·23	5·85	5·00
Aspartic acid	1·07	6·32	5·82
Threonine	6·56	8·54	6·86
Serine	5·73	9·32	2·41
Asparagine	0	17·40	14·16
Glutamic acid	33·65	43·61	46·38
Glutamine	1·44	5·37	5·53
Proline	18·33	30·81	28·66
Glycine	6·14	9·54	9·98
Alanine	9·21	9·71	10·39
Citrulline	2·89	5·02	2·04
Aminobutyric acid	0	0	0·71
Valine	17·08	20·33	26·16
Methionine	4·68	5·82	6·18
Isoleucine	6·99	14·50	15·22
Leucine	24·24	28·71	31·84
Tyrosine	3·01	5·48	4·88
Phenylalanine	11·51	17·19	15·43
β-alanine	0	0·53	0
γ-aminobutyric acid	0·43	7·52	2·09
Ethanolamine	0·91	0·25	0
Ornithine	6·98	4·96	8·56
Lysine	20·38	27·63	31·29
Histidine	3·90	8·23	6·97
Arginine	0	0·91	0

TABLE VI

Concentrations of Amines in Ripe Swiss-type Cheeses (mg/kg, Median Values)[85]

Amines	Emmentaler	Gruyère	Appenzeller
Histamine	22	66	173
Tyramine	42	37	57
Phenylethylamine	0	0	0
Putrescine	1	5	2
Cadaverine	2	25	18

TABLE VII
Mean Contents of Water and Major Non-volatile Components in Ripe Swiss-
type Cheeses (g/kg)[34,74,75]

Component	Emmentaler	Gruyère	Appenzeller
Water	350·0	358·0	392·0
Fat	315·0	322·0	318·0
Total N	45·0	43·5	39·0
Water-soluble N	12·0	14·0	16·5
Non-protein N	7·0	8·5	9·0
NaCl	6·5	14·5	16·0

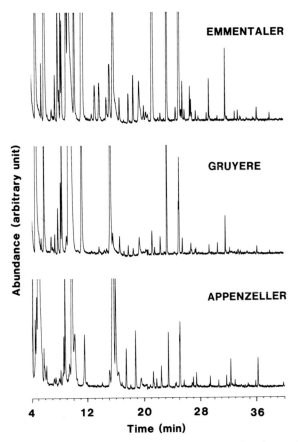

Fig. 10. Volatile compounds of cheese by capillary GC-MS using a 'Purge & Trap' extraction (trap of Supelco # 2-0295) followed by a cryo-focusing (Tekmar LSC-2000) coupled with a GC model 5890 of H.-P. *Chromatographic conditions*: DB Wax column of J.&W.: 60 m, 0·25 mm, 0·25 μm. Carrier gas: He. Inlet pressure: 150 kPa. Temp. programme: 13 min at 45°C, 5°C/min to 220°C, 10 min at 220°C. Detection with a MSD model 5970 of H.-P. in scan mode from 19 to 250 amu at 1·85 scans/s, ionisation by EI at 70 eV and 0·8 mA. For more details.[39]

comparing the numbers, it must be considered that the analytical techniques used were not exactly the same for both cheeses. For certain components, the changes in their concentrations were followed over 12 months of ripening.[44,45] Some of these volatile and non-volatile flavour components have also been found in Emmentaler and Gruyère-type cheese manufactured abroad.[46–55]

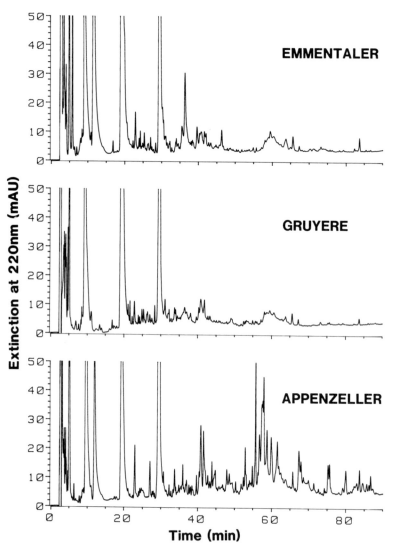

Fig. 11(a). Peptide chromatograms of cheese extracts by RP-HPLC with UV detection at 220 nm. *Chromatographic conditions*: Nucleosil 120-C_{18} column: 250 × 4 mm, 5 mm. Solvent A: 0·05% TFA; solvent B: 90% acetonitrile + 10% 0·5% TFA with a linear gradient from 10 to 90 min (0–50% B) and from 90 to 100 min (50–100% B). Flow rate: 1·00 ml/min. Temp.: 45°C. Inj. vol.: 20 μl. For more details see Ref. 41.

A   B   C

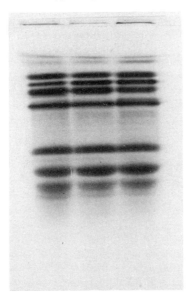

Fig. 11(b). Polyacrylamide-gel electrophoretograms of different Swiss cheese extracts. A—Emmentaler, B—Gruyère, C—Appenzell. 15% uniform PAA gel, 3% stacking gel, 5 M urea. Reservoir buffer : 30 mM Tris, 190 mM glycine, pH 8·3. The gel was run for 7 h at 25 mA, at 15°C. Gradual iso-propyl-alcohol-acetic acid-Coomassie blue staining. A protein load of 25 μg was analysed. (Courtesy Dr. P. Bican & A. Spahni, FAM.)

4 BODY CHARACTERISTICS

Cheese body and texture are very important qualities for both dealers and consumers. Variations from what is considered normal in body and texture within the same cheese variety are not tolerated since there is a close relationship between the body and texture and other qualities such as eye formation, taste and shelf-life. By texture we understand the structure and the consistency of the cheese body. The structure depends to a great extent on the microstructure inside the curd particles, whereas the body consistency is characterized by the reaction of the cheese mass to compression. Table VIII shows data on the size and shape of curd and granules in some traditional Swiss-type varieties. The measurements were made on vertical and horizontal sections of mature cheeses. An example of the prepared sections is shown in Fig. 12. This figure also shows the deformation of the granules around eyes.[56]

The structure of the cheese body can be firm or soft and the consistency short (coarse, brittle) or long (tough, elastic). The quality of the cheese body is evaluated in sensorial assessments by skilled panellists or scientifically analysed by objective determination of chemical and physical parameters. Universal testing machines, like INSTRON, are used to determine some important body characteristics of

TABLE VIII
Average Size and Elongation of Curd Granules in Horizontal and Vertical Sections of Some Swiss-type Cheeses (Median Values, x̄, and interquartile ranges r_x, (79) = 56)[a]

Parameter			Emmentaler	Gruyère	Appenzeller
a)	**Horizontal**				
	Area, mm^2	A	0·15	0·97	1·31
		r	0·56–1·27	0·48–2·04	0·65–2·67
	Major axis, mm	a	1·68	1·56	1·95
		r_a	1·23–2·38	1·05–2·35	1·33–2·85
	Minor axis, mm	b	0·93	0·84	0·93
		r_b	0·62–1·32	0·55–1·23	0·61–1·33
	Form factor, b/a	f	0·557	0·559	0·497
		r_f	0·44–0·69	0·43–0·67	0·37–0·62
b)	**Vertical**				
	Area, mm^2	A	1·00	0·89	1·02
		r	0·54–1·61	0·54–1·46	0·53–1·91
	Major axis, mm	a	1·69	1·62	1·78
		r_a	1·21–2·40	1·19–2·26	1·22–2·68
	Minor axis, mm	b	0·77	0·72	0·76
		r_b	0·57–1·02	0·54–0·93	0·54–1·04
	Form factor, b/a	f	0·448	0·436	0·436
		r_f	0·34–0·60	0·34–0·58	0·32–0·60

[a] 800–1000 measurements for each variety (5 to 8 different loaves, 6 (hard cheeses) or 12 zones (semi-hard cheeses) per loaf).

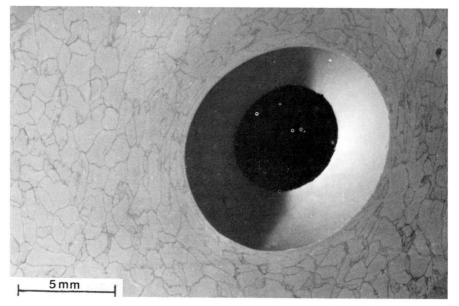

5 mm

Fig. 12. Deformation of curd granules around eyes in Emmentaler cheese.[56]

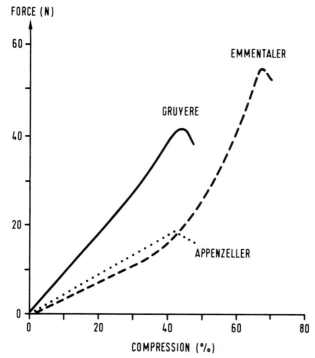

Fig. 13. Typical force–compression diagram for some cheese varieties.[58]

Swiss hard and semi-hard cheese: cylindrical cheese samples are subjected to compression.[57] Figure 13 shows typical force-compression diagrams resulting from such measurements.[58] These graphs permit the determination of:

(a) the force at breaking point which causes the body to crack;
(b) the compression at breaking point, expressed as a percentage of the original height of the sample; this is a description of the length of the body;
(c) the force required to compress the sample by 33%; this is characteristic of the firmness of the body.

The force at breaking point therefore depends on the length and firmness of the cheese body. The force required to cause cracking is higher for Emmentaler than for Gruyère (Fig. 13). The body of Emmentaler is soft and long, whereas a Gruyère has firmer and shorter body consistency. Differences in texture, within the same cheese variety, are partly due to differences in chemical composition. The correlations between body characteristics and chemical composition have been calculated for 195 5-month-old Emmentaler cheeses (Table IX).

Firmness depends on the total nitrogen content, which is the main support of the cheese body structure, as well as on the water content, which lends softness to the body. There is a negative correlation between the length (compression) of the cheese body and its lactate content. In hard and semi-hard cheese, the body

TABLE IX
Correlations Between Analytical Results for 195 150-Day Old Emmentaler Cheeses[83]

Parameter	1	2	3	4	5	6	7	8
1 Compression	1·00	—	—	—	—	—	—	—
2 Force	0·63	1·00	—	—	—	—	—	—
3 Firmness	−0·26	0·52	1·00	—	—	—	—	—
4 Water	−0·11	−0·35	−0·35	1·00	—	—	—	—
5 Fat	0·13	−0·04	−0·19	−0·69	1·00	—	—	—
6 Total N	0·19	0·63	0·62	−0·30	−0·29	1·00	—	—
7 Lactate	−0·36	−0·38	−0·07	0·36	−0·30	−0·19	1·00	—
8 pH	0·19	0·33	0·14	−0·02	−0·06	0·14	−0·26	1·00

characteristics change during ripening, mainly as a result of proteolysis (Fig. 14).

In Emmentaler, the body becomes longer during the first 30 days of ripening. Continued curd fusion changes the cheese body so that a higher compression is required to crack the samples. Comparison with Gruyère and Appenzeller (Fig. 14) shows that this increase in compression is less pronounced in these

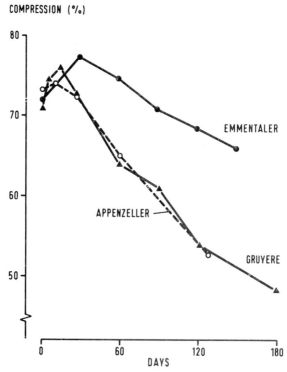

Fig. 14. Changes in the compression at the breaking point (cracking of the cheese body) throughout ripening.[83]

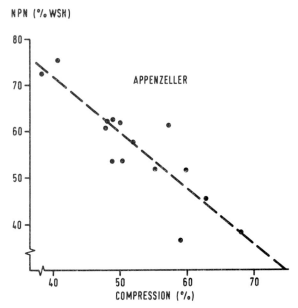

Fig. 15. Linear correlation between compression at the breaking point and protein breakdown (non protein nitrogen as % of water-soluble nitrogen); fifteen 130-day-old Appenzeller cheese loaves were studied.[83]

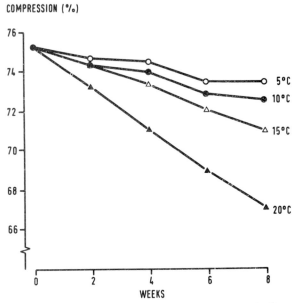

Fig. 16. Influence of different storage temperatures after the main fermentation on compression at the breaking point of Emmentaler cheese.[83]

varieties than in Emmentaler. After the beginning of eye formation, at the age of approximately 35 days, the cheese body again becomes shorter. In Gruyère and Appenzeller, the body becomes much shorter during the course of ripening.

These differences are due to divergent intensities of proteolysis. In smear-ripened cheeses, such as Gruyère and Appenzeller, the protein breakdown 'in depth', i.e. from peptones and polypeptides into smaller peptides and free amino acids, is greater than in dry-ripened Emmentaler.

The good correlation between protein breakdown 'in depth' and compression at the breaking point is evident from Fig. 15. The intensity of proteolysis is represented by the increasing quantity of water soluble NPN. In this example of an Appenzeller cheese, the slope value of linear regression was −0·87. Similar relationships have been found in other cheese varieties.

It is possible to accelerate proteolysis by increasing the storage temperature. Figure 16 shows the influence of temperature on the development of cheese body. Blocks of the same Emmentaler loaf were stored at 5, 10, 15 or 20°C after eye formation had terminated, i.e. after approximately 70 days of ripening. It is evident that the storage temperature clearly influences the compression at breaking point. The greatest difference was found between 15 and 20°C; the greatest difference in protein breakdown was also found at these temperatures. Thus, it appears that the typical body and texture of a variety can be obtained more rapidly by increasing the storage temperature. This, however, generally gives rise to pronounced off-tastes.[59]

Because of their low water content, Swiss-type cheeses melt at relatively high temperatures. The average softening points, determined with an automatic dropping point apparatus, are 68, 71 and 74°C for Appenzeller, Gruyère and Emmentaler, respectively.

The results presented are mainly based on samples taken from the central part of the cheese loaves. It must be considered, however, that most parameters show a gradient from the rind to the centre.[60,61]

5 INFLUENCES OF CHEESEMAKING OPERATIONS

The characteristics of eyes, flavour, body and texture and shelf-life of Swiss-type cheese originate mainly from the starter cultures, the quality of the milk and the different cheesemaking operations. Figure 17 shows the relationship between the factors that determine cheese quality.

The milk used for cheese manufacture should contain as few bacteria as possible so that the added starter cultures have an optimum effect. If raw milk is processed, the bacteriological requirements are particularly stringent. The microbial and hygienic state of farm milk, of course, also depends on the duration and temperature of storage and secondary contamination before or after processing must be avoided.

The temperature-time relationship during manufacture, pressing and ripening determines the development of lactic acid bacteria of the starters added. Most

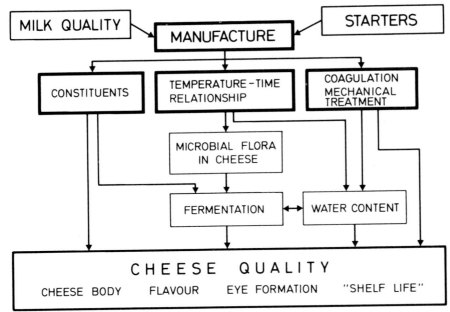

Fig. 17. Influence of factors throughout manufacture on the quality of Swiss-type cheese.[86]

Swiss-type cheese varieties need high cooking or scalding temperatures. Emmentaler is heated to 52–54°C after cutting. During pressing, the temperature remains at around 50°C for many hours (Fig. 18). At this temperature, the curd dries and most of the undesirable microorganisms are eliminated.

The fat, protein, water and salt contents are essential for body texture and taste. All Swiss-type varieties are rennet cheeses. This means that the milk coagulation is induced by animal rennet or rennet substitutes. The water content of the curd can be adjusted by controlled coagulation and mechanical treatment of the whey-curd mixture.[62] As regards eye formation, proper dip filling of the moulds is imperative since air inclusions can lead to undesirable openness (Fig. 19).

Swiss-type cheeses are very often manufactured in copper vats. The copper content of the cheese inhibits the formation of lactic and propionic acids, and influences the ripening, formation of aroma as well as the distribution of the eyes.[63–67]

For many decades, Swiss-type cheeses were manufactured in vats (Kessi) with a capacity of 800–1400 litres. Many manufacturing processes had to be mechanized or rationalized when the volume of production increased. Today, up to 15 000 litres of milk are transformed at once in 'Käserfertiger'. Cheese loaves are then formed in automatic presses and ripened in mechanically equipped cellars. In various countries, Swiss-type cheeses are manufactured in the form of blocks and ripened in plastic bags.[4,68,69] The development of the texture and flavour is further controlled by the storage conditions (temperature, storage time, relative air humidity, curing).

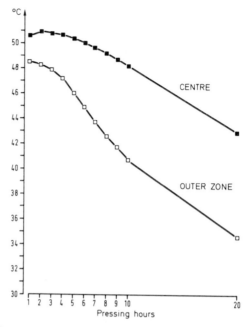

Fig. 18. Temperature profile during pressing of Emmentaler cheese.[87]

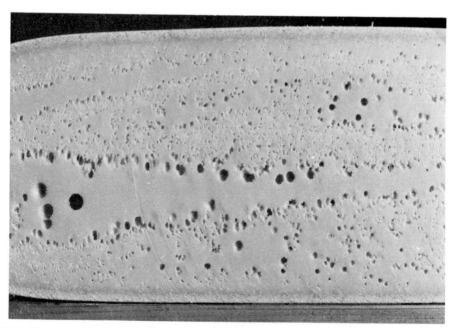

Fig. 19. Emmentaler cheese with air inclusion after dipping of the curd; note openness of texture.[77]

TABLE X
Typical a_w Values for Appenzeller, Emmentaler and Gruyère Cheeses[70,71,88]

Type	Age months	Average a_w	Standard deviation	a_w range in rind
Appenzeller	4–5	0·962	0·006	0·97–0·98
Emmentaler	4–5	0·972	0·007	0·90–0·95
Gruyère	6–7	0·948	0·012	0·92–0·98

6 CONTROL OF RIPENING REACTIONS

As can be seen from the preceding data, the ripening processes in Swiss-type cheese are influenced by different factors, among which the water activity (a_w) is important (Table X). It directly influences the growth and metabolism of the microorganisms as well as the enzymatic activities in cheese.[70,71] The rather low a_w in Swiss-type cheese is not only a characteristic of its ripening process, but is also responsible for its good keeping quality. The a_w value varies with the concentration of low-molecular water-soluble cheese components, i.e. the salt and soluble products of proteolysis as well as with the pH. It decreases with the progress of proteolysis during ripening (Fig. 20). Most cheeses show zonal differences.

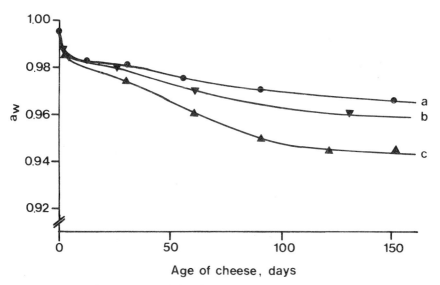

Fig. 20. Influence of the degree of ripening on a_w of Emmentaler (a), Gruyère (b) and Appenzeller (c) cheese (a_w at time 0, 0·995, corresponds to the value for milk estimated from its freezing point).[70]

Both the salt concentration and its uniform distribution throughout the curd are important for cheese quality. Salt penetration into cheese during and after brine salting is due mainly to diffusion. The diffusion coefficient of salt in Sbrinz-type cheese was found to increase from $1 \cdot 1 \cdot 10^{-10}$ at 7°C to $1 \cdot 9 \cdot 10^{-10}$ $m^2 s^{-1}$ at 20°C.[72,56,73]

REFERENCES

1. Peter, A. & Zollikofer, E., 1959. *Lehrbruch der Emmentalerkäserei*. Verlag K.J. Wyss Erben, AG, Bern.
2. Kosikowski, F.V., 1977. *Cheese and Fermented Milk Foods*. Edwards Brothers Inc., Michigan.
3. Reinbold, G.W., 1972. *Swiss Cheese Varieties*. Pfizer, New York.
4. Mair-Waldburg, H., 1974. *Handbuch der Käse*. Volkswirtschaftl. Verlag, Kempten.
5. Schoen, G., 1978. Mikrobiologie, Verlag Werder, Freiburg.
6. Carr, J.G., Cutting, C.V. & Whiting, G.C., 1973. *Lactic Acid Bacteria in Beverages & Food*. Academic Press, London.
7. Richardson, B.C. & Pearce, K.N., 1982. *N.Z. J. Dairy Sci. Technol.*, **16**, 209.
8. Casey, M., Gruskovnjak, J. & Furst, M., 1987. *Schweiz. milchw. Forschung*, **16**(1), 21.
9. Keller, S. & Puhan, Z., 1985. *Schweiz. milchw. Forschung*, **14**(2), 3.
10. Meyer, J., Casey, M. & Gruskovnjak, J., 1985. *Schweiz. milchw. Forschung*, **14**(2), 11.
11. Steffen, C., 1979. *Schweiz. milchw. Forschung*, **8**(3), 44.
12. Steffen, C. & Nick, B., 1981. *Schweiz. milchw. Forschung*, **10**(2), 32.
13. Steffen, C., 1973. *Schweiz. Milchztg*, **99**(63), 478, 488.
14. Flückiger, E., 1980. *Schweiz. Milchztg*, **106**(71), 473(72), 479.
15. Flückiger, E., Montagne, D.H. & Steffen, C., 1978. *Schweiz. milchw. Forschung*, **7**(4), 73.
16. Gehriger, G., Kurmann, J.L. & Kaufmann, H., 1974. *Schweiz. Milchztg*, **100**(84), 573(85), 581.
17. Clark, W.M., 1912. *US Dept. of Agr. Bulletin*, p. 151.
18. Gallmann, P. & Puhan, Z., 1981. *Schweiz. milchw. Forschung*, **11**(1 and 2), 3.
19. Gallmann, P. & Puhan, Z., 1982. *Schweiz. milchw. Forschung*, **11**(4), 64.
20. Forster, Inge, Grand, M. & Glättli, H., 1987. *Schweiz. milchw. Forschung*, **16**(4), 79.
21. Kielwein, G., 1975. *Dte Molk.-Ztg*, **96**, 1112.
22. Bolliger, O. & Zahnd, L., 1959. *Proc. XV Int. Milchwirtschaftskongress*, London, Vol. 2, Section 3, p. 751.
23. Dellaglio, F. & Bottaazzi, V., 1972. *Dairy Sci. Abstr.*, **34**, 703.
24. Adda, J., 1984. In *Le Fromage*, ed. A. Eck. Diffusion Lavoisier, Paris, p. 330.
25. Adda, J., Grippon, J.-C. & Vassal, L., 1982. *Food Chemistry*, **9**, 115.
26. Ney, K.H., 1981. In *The Quality of Foods and Beverages: Chemistry and Technology*, Vol. 1, eds G. Charalambous and G. Inglett. Academic Press, New York, p. 389.
27. Law, B.A., 1981. *Dairy Sci. Abstr.*, **43**, 143.
28. Moskowitz, J.J., 1980. In *The Analysis and Control of Less Desirable Flavors in Foods and Beverages*, ed. G. Charalambous. Academic Press, New York, p. 53.
29. Behnke, U., 1980. *Die Nahrung*, **24**, 71.
30. Badings, H.T. & Neeter, R., 1980. *Neth. Milk Dairy J.*, **34**, 9.
31. Dumont, J.-P. & Adda, J., 1979. In *Progress in Flavour Research*, eds D.G. Land & H.E. Nursten. Elsevier Applied Science, London, p. 245.
32. Forss, D.A., 1979. *J. Dairy Sci.*, **46**, 691.
33. Adda, J., Roger, S. & Dumont, J.-P., 1978. In *Flavour of Foods and Beverages: Chemistry and Technology*, eds G. Charalambous & G.E. Inglett. Academic Press, New York, p. 65.

34. Panouse, J.J., Masson, J.D. & Truong Thanh Tong, 1972. *Industr. Alim. Agr.*, **83**, 133.
35. Adda, J., 1986. In *Developments in Food Flavours*, eds G.G. Birch & M.G. Lindley. Elsevier Applied Science, London, p. 151.
36. Marshall, V.M.E., 1984. In *Advances in the Microbiology and Biochemistry of Cheese and Fermented Milk*, eds L.F. Davies & B.A. Law. Elsevier Applied Science, London, p. 153.
37. Law, B.A., 1982. In *Fermented Foods*, ed. A.H. Rose. Academic Press, London, p. 148.
38. Law, B.A., 1984. In *Advances in the Microbiology and Biochemistry of Cheese and Fermented Milk*, eds L.F. Davies & B.A. Law. Elsevier Applied Science, London, p. 187.
39. Bosset, J.O., Klein, B. & Gauch, R., *Food Sci. & Technol.* (in preparation).
40. Steffen, C., Glättli, H., Steiger, G., Flückiger, E., Bühlmann, C., Lavanchy, P. & Nick, B., 1980. *Schweiz. milchw. Forschung*, **9**(2), 19.
41. Bütikofer, U., Baumann, E. & Bosset, J.O., *J. Chromatogr.* (in preparation).
42. Liardon, R., Bosset, J.O. & Blanc, B., 1982. *Lebensmittelwissenschaft u.-technol.*, **15**, 143.
43. Bosset, J.O. & Liardon, R., 1984. *Lebensmittelwissenschaft u.-technol.*, **17**, 359.
44. Bosset, J.O. & Liardon, R., 1985. *Lebensmittelwissenschaft u.-technol.*, **18**, 178.
45. Bosset, J.O., *Lebensmittelwissenschaft u.-technol.* (in preparation).
46. Mocquot, G., 1979. *J. Dairy Res.*, **46**, 133.
47. Langsrud, T. & Reinhold, G.W., 1973. *J. Milch Food Technol.*, **36**, 593.
48. Langler, J.E. & Day, E.A., 1966. *J. Dairy Sci.*, **49**, 91.
49. Langler, J.E., Libbey, L.M. & Day, E.A., 1967. *J. Agr. Food Chem.*, **15**, 386.
50. Mitchell, G.E., 1981. *Aust. J. Dairy Technol.*, **36**, 21.
51. Biede, S.L., 1977. A Study of the Chemical and Flavour Profiles of Swiss cheese, Ph.D. Dissertation, Iowa State University.
52. Biede, S.L. & Hammond, E.G., 1979. *J. Dairy Sci.*, **62**, 227, 238.
53. Kiermeier, F., Mayr, A. & Hanusch, E.G., 1966. *J. Dairy Sci.*, **49**, 91.
54. Vamos-Vigyazo, L. & Kiss-Kutz, N., 1974. *Acta Alimentaria*, **3**, 309.
55. Schormueller, J., 1968. *Adv. Food Res.*, **16**, 231.
56. Rüegg, M. & Moor, U., 1987. *Food Microstructure*, **6**, 35.
57. Eberhard, P. & Flückiger, E., 1978. *Schweiz. Milchztg*, **104**(4), 24.
58. Eberhard, P. & Flückiger, E., 1981. *Schweiz. Milchztg*, **107**(5), 23.
59. Flückiger, E. & Eberhard, P., 1982. *Proc. XXI Intern. Milchwirtschaftskongress*, Moscow, Vol. I(1), p. 48.
60. Eberhard, P., Flückiger, E. & Puhan, Z., 1986. *Schweiz. milchw. Forschung*, **15**(4), 97.
61. Eberhard, P. & Flückiger, E., 1978. *Schweiz. Milchztg*, **104**(33), 253.
62. Schwartz, M.E., 1973. *Cheese-Making Technology*, Noyes Data Corporation, London.
63. Burkhalter, G., 1956. *Schweiz. Milchztg*, **82**, 41.
64. Kiermeier, F. & Weiss, G., 1970. *Z. f. Lebensm.-unters. u. Forschung*, **142**, 397.
65. Oehen, V., Haenni, H. & Bolliger, O., 1962. *Schweiz. Milchztg*, **88**, 423.
66. Ritter, W, 1967. *Schweiz. Milchztg*, **93**, 115.
67. Maurer, L., 1972. *Oesterr. Milchwirtsch.*, **27**(4), 249.
68. Steffen, C., Glättli, H., Steiger, G., Flückiger, E., Bühlmann, C., Lavanchy, P. & Nick, B., 1980. *Schweiz. milchw. Forschung*, **9**(2), 19.
69. Scott, R., 1981. *Cheesemaking Practice*. Elsevier Applied Science, London, p. 274.
70. Rüegg, M. & Blanc, B., 1981. In *Water Activity: Influences on Food Quality*, eds L.B. Rockland & G.F. Stewart. Academic Press, New York, p. 791.
71. Rüegg, M., 1985. In *Properties of Water in Food*, eds D. Simatos & J.L. Multon. Martinus Nijhoff, Dordrecht, p. 603.
72. Gros, J.B. & Rüegg, M., 1987. In *Physical Properties of Foods—2*, eds R. Jowitt, F. Escher, M. Kent, B. McKenna & M. Roques. Elsevier Applied Science, p. 71.

73. Rüegg, M. & Moor, U., 1988. *Schweiz. milchw. Forschung*, **17**(4), 69.
74. Steiger, G. & Flückiger, E., 1979. *Schweiz. milchw. Forschung*, **8**(3), 39.
75. Steffen, C., Glättli, H., Steiger, G., Flückiger, E., Bühlmann, C., Lavanchy, P., Nick, B., Schnider, J. & Rentsch, F., 1981. *Schweiz. milchw. Forschung*, **10**(3), 51.
76. Steffen, C., Nick, B. & Blanc, B., 1973. *Schweiz. milchw. Forschung*, **2**(4), 37.
77. Unpublished data from Dairy Research Institute (FAM), CH-3097 Liebefeld.
78. Steffen, C., 1975. *Schweiz. milchw. Forschung*, **4**(3), 16.
79. Rüegg, M., Moor, U. & Blanc, B., 1980. *Milchwissenschaft*, **35**, 383.
80. Rüegg, M. & Blanc, B., 1973. *Schweiz. milchw. Forschung*, **1**(1), 1.
81. Lavanchy, P., Buhlmann, C. & Blanc, B., 1979. *Schweiz. milchw. Forschung*, **8**(1), 9.
82. Kurmann, J.L. & Wüthrich, A., 1975. *Schweiz. Milchztg*, **101**(1), 1.
83. Eberhard, P., 1976. *Schweiz. milchw. Forschung*, **14**(4), 3.
84. Lavanchy, P. & Bühlmann, C., 1983. *Schweiz. milchw. Forschung*, **12**(1), 3.
85. Sieber, R., Collomb, M., Lavanchy, P. & Steiger, G., 1988. *Schweiz. milchw. Forschung*, **17**(1), 9.
86. Steffen, C., 1985. In *Käsereitechnologischer Sonderlehrgang '85,* Landesverband Bayerischer Molkereifachleute und Milchwirtschaftler E.V., Füssen, Jahresbroschüre, 151.
87. Steffen, C. & Schnider, J., 1978. *Scheiz. Milchztg*, **104**, 383.
88. Blanc, B., Rüegg, M., Baer, A., Casey, M. & Lukes, A., 1982. *Schweiz. milchw. Forschung*, **11**, 22.

4

Mould-Ripened Cheeses

J.C. GRIPON

Station de Recherches Laitières, Institut National de la Recherche Agronomique, Jouy-en-Josas, France

1 INTRODUCTION

Mould-ripened cheeses represent a small proportion of world cheese production. However, these cheeses are becoming increasingly popular with consumers and there is an increasing demand for them. Blue-veined cheeses have long been produced in various countries; Roquefort, Gorgonzola, Stilton and Danish Blue are typical examples. The production of surface mould-ripened soft cheeses, such as Camembert, was limited to France for a long time, but in recent years, many countries have developed the production of such cheeses.

The presence of mould within the cheese (*Penicillium roqueforti*) or on the surface (*P. camemberti*) gives these cheeses a different appearance and the high biochemical activities of these moulds produce a very typical aroma and taste. These moulds also lead to more complex ripening than in other varieties of cheese with simple flora.

The present chapter, after reviewing briefly the composition of the flora of these cheeses, considers the various biochemical changes occurring during their ripening, their aroma and textural properties and, finally, the control of their ripening. The more technological aspects have been treated in other reviews[1-4] and so are not discussed here. Coghill[5] briefly reviewed the microbiology and biochemistry of the ripening of blue-veined cheeses. Seth & Robinson[6] summarized the literature on the aroma compounds of surface mould-ripened soft cheeses and Desfleurs[7] reviewed the origin and history of traditional Camembert.

2 FLORA OF MOULD-RIPENED CHEESES

The composition and evolution of the flora of mould-ripened cheeses are complex, particularly when raw milk is used. Traditional Camembert is a good example: starters used are primarily lactococci (*Lactococcus lactis*, spp. *lactis* and *cremoris*) and after 24 h, a firm, drained demineralized curd is obtained with a pH of

4·5–4·6. The flora is then essentially composed of lactococci.[8] After curd-making, yeasts grow on the surface,[8–10] forming a dense layer about 200 μm thick;[11] *Kluyveromycces lactis, Saccharomyces cerevisiae* and *Debaryomyces hansensi* are the most common yeast species.[9,10] The mould, *Geotrichum candidum*, appears at the same time as the yeasts but its growth is limited by salting. After 6 or 7 days of ripening, the growth of *Penicillium camemberti* is observed and a white felt covers the entire surface of the cheese. After 15–20 days, when the *Penicillium* has consumed the lactic acid and de-acidified the cheese, an aerophilic acid-sensitive bacterial flora becomes established on the surface. This flora is formed of micrococci and coryneform bacteria (a high percentage are *Brevibacterium linens*).[8,12,13] Coliform bacteria can also be observed,[8,12–15] *Hafnia alvei* being the dominant species.[13] Inside the cheese, lactococci are clearly dominant, the yeast population remaining lower than on the surface (about 10^6 cells/g instead of 10^8 cells/g),[9] as well as the coliform population.[12] In the production of these cheeses from pasteurized milk (which is the case for the large majority of Camembert cheeses), the flora is less diverse, containing mostly organisms added as starters, i.e. lactococci and *P. camemberti*. The populations of other microorganisms are reduced and the cheese obtained is different from that made from raw milk: its taste and aroma are more neutral and less accentuated.

Blue-veined cheeses made from raw milk also have a complex flora. In 1966–1968, Devoyod[16–18] showed that besides *P. roqueforti* and lactic bacteria, the flora of Roquefort cheese (which is made with raw ewes' milk) also contains yeasts,

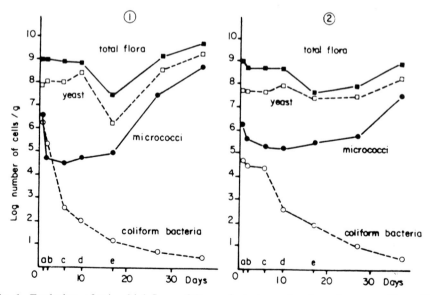

Fig. 1. Evolution of microbial flora of Roquefort cheese (surface: 1; centre: 2). a: 21 h after rennet; b: in warm room (18°C); c: in cold room (10°C); d: before salting; e: after salting (from Devoyod and Bret[18]).

lactobacilli, micrococci, staphylococci and coliform bacteria (Fig. 1). Lactococci are always clearly dominant inside the cheese but yeasts and leuconostocs are also present from the beginning of ripening.[19,20] Lactobacilli (mainly *Lactobacillus casei* and *L. plantarum*) reach a maximum just before salting.[17] The staphylococci decrease markedly during the first 48 h[21] and the coliform bacteria decrease regularly during the first month.[16] After salting, the surface flora contains primarily yeasts and micrococci.

Piercing the cheese admits enough air to allow the growth of *P. roqueforti* whose sporulation is visible inside the curd of 'bleu d'Auvergne' 2–3 weeks after manufacture.[22]

It is clear that in both surface mould and blue-veined cheeses, *Penicillium* spp. are the major components of the flora and the cheeses are marked by their characteristics. However, in traditional cheese made from unheated milk, the secondary flora plays an essential complementary role in the development of its organoleptic properties.

The morphological and physiological properties of *P. roqueforti* and *P. camemberti* have been reviewed by Moreau[23,24] and Choisy *et al.*[25] Up to the last few years, two species of the latter were distinguished: *P. caseicolum* and *P. camemberti*. As *P. caseicolum* is considered to be a white mutant of *P. camemberti*,[26] only the name *P. camemberti* is now used.[26,27] Moreau distinguished four different forms of *P. camemberti*:

(a) a form with fluffy mycelium, first white becoming grey-green (*P. camemberti sensus stricto*);
(b) a form with 'short hair', rapid growth; white, dense, close-napped mycelium;
(c) a form with 'long hair', rapid growth; white, loose, tall mycelium;
(d) 'neufchatel' form: vigorous, rapid growth, giving a thick white-yellow mycelium.

The last three forms correspond to the old name of *P. caseicolum*. The 'neufchatel' form has higher lipolytic and proteolytic activities than the others.[28–30] Only the white forms are actually used as starters in cheesemaking. Conditions for the production and utilisation of spores of *P. camemberti* have been described.[30–33]

Mould-ripened cheeses are generally thought to be limited to surface mould-ripened cheeses and blue-veined cheeses. In fact, there are a small number of other, less known, varieties of mould-ripened cheese that are produced in more limited quantities. In France, these are semi-hard cheeses called 'Saint-Nectaire' and 'Tome de Savoie'. The surface of these cheeses is covered by a complex fungal flora containing *Penicillium, Mucor, Cladosporium, Geotrichum, Epicoccum* and *Sporotrichum*.[34,35] Mould has also been reported on the surface of Italian cheeses such as Taleggio (*Penicillium, Mucor*) and Robiola (*Geotrichum*).[36,37]

Gammelost is a semi-hard cheese made in Norway from skimmed milk. After acidification, the milk is heated to 65°C and the casein precipitated and collected. After moulding, the curd is held for about 2 h in boiling whey, and the next day a suspension of *Mucor* is sprayed on the surface. The brown colour of the cheese is due to the development of this mould.[38] The inside of the cheese is brown–

yellow and this colour intensifies as the cheese ripens. Gammelost has a piquant, aromatic taste. *P. roqueforti* has also been reported to develop inside Gammelost. In this case, the cheese is pierced with metal needles covered with *P. roqueforti* spores to permit the mould to develop.[39]

The physico-chemical data on these latter varieties of mould-ripened cheese are, unfortunately, very limited so we shall treat here only blue-veined cheeses and surface mould-ripened soft cheeses.

3   PROTEOLYSIS AND AMINO ACID CATABOLISM

3.1 Intensity of Proteolysis

Proteolysis is very intense in blue-veined cheese. More than 50% of the total nitrogen (TN) is soluble at pH 4·6 in ripe Roquefort[16] and about 65% in Danish Blue.[40] This soluble fraction contains a large number of small peptides and the non-protein nitrogen (NPN: nitrogen soluble in 12% trichloroacetic acid) is about 30% of TN.[41] Free amino acids are also abundant, representing 10% of the total nitrogen in Danish Blue.[42] Equivalent or higher amounts (280–500 mg/10 g of cheese) were reported by Kosikowski & Dahlberg.[43] Godinho & Fox[41] noted that proteolysis is more limited in the outer parts than in the centre of the cheese and suggested that NaCl limits *Penicillium* growth and its proteolytic action.

In the outer part of a ripe raw-milk Camembert, about 35% of the nitrogen is soluble,[44] within the cheese, there is less breakdown and only 25% of the nitrogen is soluble. These values are lower than for blue-veined cheese but nevertheless show extensive proteolysis. The soluble nitrogen fraction contains many small peptides (the NPN is about 20% of the total N).[44,45] In traditional Camembert cheeses, ammonia represents 7–9% of total N[45,46] and results from extensive deamination of amino acids. The profile of free amino acids obtained by Do Ngoc *et al.*[46] is different from that of whole casein hydrolysate: alanine, leucine and phenylalanine occur in higher proportions and aspartic acid, tyrosine and lysine in lower proportions; arginine and serine occur in particularly low amounts.[46] This shows preferential release of amino acids during proteolysis and also the catabolism of free amino acids.

Electrophoretograms of blue-veined cheese show evident breakdown of α_{s1} and β-caseins;[22,40,47] at the end of ripening, these caseins have almost disappeared and the major products on the electrophoretograms show low mobility.[22,47] Trieu-Cuot & Gripon[22,48] noted that the electrophoretic profiles of surface mould-ripened cheese are very similar to those of blue-veined cheese but reveal less proteolysis.

Compared to proteolysis in other cheeses, that in mould-ripened cheeses is higher (particularly in blue-veined cheeses), and β-casein is degraded more extensively than in other varieties.[49] As in all cheeses, this breakdown results from coagulating enzymes, indigenous milk proteinases and microbial proteinases among which, in our case, enzymes synthesized by *Penicillium* ssp. have a dominant role.

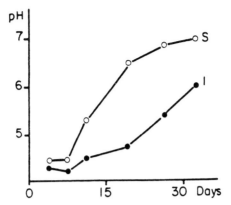

Fig. 2. Changes in pH during Camembert cheese ripening (S: surface; I: interior) (from Lenoir[52]).

3.2 Effect of Rennet

The action of rennet is quickly detectable in cheese by α_{s1}-casein hydrolysis and the production of α_{s1}-I peptide. Mould-ripened cheeses are no exception: α_{s1}-I peptide is seen in Camembert by electrophoresis after 6 h of draining and the concentration of this peptide increases during ripening.[48] Rapid production of α_{s1}-I peptide has also been observed in blue-veined cheeses.[41] Besides α_{s1}-I, Hewedi & Fox[40] detected other peptides typical of rennet action (α_{s1}-V and α_{s1}-VII) in blue-veined cheese. However, peptides resulting from α_{s1}-casein hydrolysis by *P. roqueforti* proteinases have similar or identical[49,50] electrophoretic mobilities and could also explain the intensification of the α_{s1}-V and α_{s1}-VII bands during ripening. In simulated cheese, Noomen[51] showed that rennet breakdown of casein was optimal at pH 5·0. Inside blue-veined cheese, the pH rises quickly to this value and after one month of ripening, reaches 5·5 in Danish Blue cheese[41] or 6·0 in 'bleu d'Auvergne'.[22] The pH in the outer part of Camembert increases more quickly, reaching more than 5·0 after two weeks and can attain 7·0 after 4–5 weeks (Fig. 2).[44] Under these conditions, one may suppose that rennet action decreases at the end of ripening when cheese pH has increased. As in other cheeses, β-casein is not attacked by rennet in mould-ripened cheeses or at least this action is not detectable by electrophoresis; β-I (the peptide obtained quickly *in vitro* by rennet hydrolysis of β-casein) has not been observed during Camembert ripening.[48]

3.3 Effect of Milk Proteinases

Milk proteinases are less active than rennet or microbial enzymes during cheese ripening.[53,54] However, in Camembert, Trieu-Cuot & Gripon[48] found a clear increase in γ-caseins at the end of ripening, showing increased activity of plasmin, the main milk proteinase. This augmented activity, first suggested by

Noomen,[55] is not surprising because the pH of the outer region of Camembert at the end of ripening (about 7·0) is not far from the optimum pH of plasmin (about 8·0). At the end of ripening, especially in the outer part of Camembert, this enzyme is probably much more active than in semi-hard cheeses where the pH remains at about 5·0–5·2.

3.4 Effect of *Penicillium*

Studies on controlled-flora rennet curds, in which *P. roqueforti* (or *P. camemberti*) develops alone with no other microorganism, have shown and permitted the definition of intensity of the proteolysis caused by these two moulds.[56] After 40 days of ripening, nitrogen soluble at pH 4·6, NPN, and nitrogen soluble in phosphotungstic acid in these curds represented about 50, 30 and 10%, respectively, of the total nitrogen. These values were very much higher than those for the control curds (where only rennet and plasmin were active) and show that there is extensive production of high- and low-molecular weight peptides as well as of free amino acids. These moulds thus have both an endopeptidase and an exopeptidase action. While rennet, plasmin or other flora have an effect, it is clear that *Penicillium* spp. play a major proteolytic role in the mould-ripened cheeses.

The extracellular proteolytic systems of *P. roqueforti* and *P. camemberti* are somewhat similar. Both synthesize a metalloproteinase[57–60] and an aspartate proteinase[61–64] as well as an acid carboxypeptidase[52,65,66] and an alkaline aminopeptidase.[67,68] Moreover, *P. roqueforti* synthesizes one or more alkaline carboxypeptidases[69] and some strains also produce an alkaline proteinase.[69]

Strains of *P. camemberti* have very similar enzyme potentials. In almost 110 strains studied, Lenoir & Choisy[28] observed that production of proteolytic enzyme (measured at pH 6·0) varied only by a factor of 2. In contrast, the levels of proteinases produced by *P. roqueforti* varied greatly from one strain to another.[70,71]

3.4.1 Properties of Proteinases
The aspartate proteinases (also called acid proteinases) of *P. roqueforti* and *P. camemberti* have similar properties.[58,62,64] Their pH optima are in the acid range (4 on haemoglobin as substrate and 5·5 on casein), and they are stable between pH 3·5 and 5·5 (*P. camemberti*) or pH 3·5–6·0 (*P. roqueforti*).

The *P. camemberti* enzyme hydrolyses α_{s1}-casein better than β- and κ-caseins (relative activity ratio, 1:0·7:0·6).[72] The acid proteinases of both species have the same action on β-casein and three bonds of the Lys-X type (Lys_{97}-Val_{98}, Lys_{99}-Glu_{100} and Lys_{29}-Ile_{30}) are cleaved more rapidly than others.[49,73] Of the three corresponding N-terminal fragments, β_{ap1}-peptide (Val_{98}-Val_{209}) has a lower electrophoretic mobility than β-casein and is easily detected in the electrophoretograms of mould-ripened cheeses.

Gripon & Hermier[57] and Lenoir & Auberger[59] also found that the metalloproteinases of *P. roqueforti* and *P. camemberti* had similar properties. Their pH optima are at 5·5–6·0 and they are stable between pH 4·5 and 8·5, which suit the conditions found in mould-ripened cheeses. Their specificity is wide and does not follow a clear rule.[60] *In vitro*, the *P. camemberti* enzyme cleaves α_{s1}-casein better than β- and κ-caseins (ratios of relative activity, $1:0\cdot4:0\cdot6$).[72] An electrophoretic study[50] showed that the metalloproteinases of both species have similar modes of action on α_{s1}- and β-caseins. Among the peptides of β-casein hydrolysate, β_{mp1}-peptide is clearly seen in the electrophoretograms of mould-ripened cheeses and can be used as a marker for the action of this enzyme *in situ*.[22,48,50] The Lys_{28}-Lys_{29}, Pro_{90}-Glu_{91} and Gly_{100}-Ala_{101} bonds of β-casein are cleaved by the *P. camemberti* enzyme.[50]

Our knowledge of the intracellular endopeptidases of *P. roqueforti* and *P. caseicolum* is very limited. Crude extracts of *P. roqueforti*, probably containing extra- and intracellular enzymes, have an optimum pH of $5\cdot0$–$6\cdot0$;[73-75] Lenoir & Choisy[28] observed an optimum pH of $6\cdot0$ for *P. camemberti* extract on casein. Two fractions with pH optima of $6\cdot0$ and $6\cdot5$ were separated by Takafuji & Yoshioka;[76] they hydrolysed β-casein, giving products with low electrophoretic mobilities.[77]

3.4.2 Action of Proteinases in Cheese

When the aspartate proteinase of *P. roqueforti* or the metalloproteinase of *P. camemberti* was added to aseptic control curds with no flora,[72,78] the electrophoretograms obtained after ripening were very similar to those of normal cheese, showing that these proteinases play a major role in mould-ripened cheeses. In these control curds, the enzymes very markedly increased the level of pH $4\cdot6$-soluble nitrogen as well as NPN but they released no free amino acids.[78]

The evolution of proteolytic activity in curds has been studied during Camembert ripening.[79-81] At the centre of the cheese, this activity is very low and hardly changes during ripening. However, in the outer region, it increases suddenly after 6–7 days of ripening, i.e. when *Penicillium* begins to grow (Fig. 3). Aspartate proteinase and metalloproteinase are both synthesized and their concentrations are maximal after about 15 days and then decrease slowly.[80,52] These two enzymes are thus fairly stable in cheese.

Trieu-Cuot & Gripon[48] studied the role of these enzymes *in situ* by following electrophoretic changes in the β_{ap1}- and β_{mp1}-peptides which are the respective markers for aspartate proteinase and metalloproteinase. β_{ap1} appeared shortly after *P. camemberti* developed and it intensified regularly during ripening, indicating the continuous action of aspartate proteinase during maturation, although the pH conditions of the curd were not favourable to its action. β_{mp1} decreased after 10–14 days of ripening, suggesting either that the action of metalloproteinase decreased or that β_{mp1} was actively degraded by another proteinase. The fact that the proteolytic activity in the curd is very low at the centre of Camembert, although it is high at the surface during ripening, suggests that migration of *Penicillium* proteinases in the curd is limited.[79,81] However, β_{ap1} has been detected inside Camembert at a depth of more than 7 mm at the end of ripening;[48] this may be due to the migration of either the aspartate proteinase or β_{ap1}. Lenoir[79] noted that the difference in the degree of proteolysis between the centre and the surface of Camembert was proportionally lower than the difference in proteolytic activity and suggested that the peptides migrated towards the centre of the cheese.

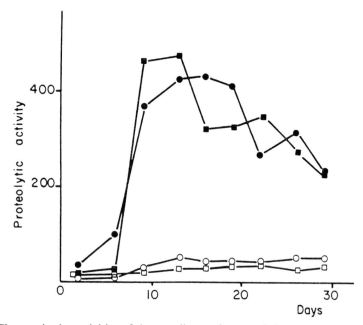

Fig. 3. Changes in the activities of the metalloproteinase and the aspartate proteinases of *P. camemberti* during Camembert cheese ripening (from Lenoir[52]). Aspartate proteinase: ●——● surface; ○——○ interior; metalloproteinase: ■——■ surface; □——□ interior.

Scanning electron microscopy studies of Camembert cheese[11] show mycelium alterations at the end of ripening, and the action of intracellular proteinases cannot be discarded. However, electrophoretograms of cheese do not show the appearance of new hydrolytic products, and the soluble nitrogen fraction increases little after 20 days of ripening, indicating that intracellular proteinases play a more limited role than extracellular proteinases.

3.4.3. Properties of peptidases
Some trials on controlled-flora curd have shown that *P. roqueforti* and *P. camemberti* can produce large amounts of free amino acids,[56] of the same order of magnitude as those observed in normal cheeses.[40] As mentioned earlier, *P. roqueforti* synthesizes several extracelluar exopeptidases. Acid carboxypeptidase[65] is a serine enzyme with an optimum pH of 3·5. It has broad specificity and releases acidic, basic or hydrophobic amino acids. Alkaline aminopeptidase is a metalloenzyme with an optimum pH of 8·0. It releases apolar amino acids but glycine in the penultimate or N-terminal positions causes low activity.[67] Paquet & Gripon[82] have observed intracellular exopeptidase activities (acid and alkaline carboxy-peptidases, alkaline aminopeptidase) and Ichishima et al.[83] have characterized an intracellular acid carboxypeptidase with properties similar to the corresponding extracellular enzyme.[65]

P. camemberti also produces several peptidases. The properties of the acid carboxypeptidase studied by Ahiko *et al.*[66] also correspond to those of serine carboxypeptidases. This enzyme reduces the bitterness of a casein hydrolysate[84] by releasing hydrophobic amino acids. Due to their broad specificity, the acid carboxypeptidase of *Penicillium* spp. may clearly contribute to the hydrolysis of the bitter peptides in mould-ripened cheese. Intracellular carboxypeptidase activity with an optimum pH of 6·5,[52] as well as alkaline aminopeptidase activities (intra- and extracellular), are also synthesized by *P. camemberti*.[68]

3.5 Effect of Other Flora

As in all cheeses, lactic bacteria participate in proteolysis, mainly by producing small peptides and free amino acids. However, the pH optima of the peptidases of these bacteria are usually closer to neutrality than to the pH of cheeses with a mainly lactic flora.[85] The higher pH of mould-ripened cheeses could thus favour their action.

Geotrichum candidium synthesizes extracellular and intracellular proteinases (pH optima: about 6·0).[86] Enzyme production varies greatly from one strain to another.[87,88] Proteolysis by *G. candidum* is difficult to determine but is clearly lower than that of *P. camemberti*; in fact, the proteolytic activity in the outer region of traditional Camembert does not increase during *Geotrichum* growth but only during that of *Penicillium*.[79] Also, *Geotrichum* seeded alone on the surface of the curds caused lower proteolysis than *P. camemberti* (L. Vassal, pers. comm.).

Yeast also have proteolytic activity which appears to be uniquely intracellular. Schmidt[89] observed caseinolytic activity with an optimum pH of about 6·0 in 165 strains isolated from Camembert cheese. The intracellular exopeptidases of *Kluyveromyces lactis* (a carboxypeptidase and three aminopeptidases) were partially purified and characterized.[90] *B. linens* secretes extracellular proteolytic enzymes: a proteinase has been demonstrated[91] and an aminopeptidase isolated and characterized;[92-94] several intracellular peptidases have been detected.[95] Although this bacterium plays a role in the development of the typical flavour of traditional Camembert cheese, its role in proteolysis at the end of ripening is difficult to determine.

3.6 Amino Acid Breakdown

The various breakdown reactions which amino acids undergo during ripening have been reviewed by Hemme *et al.*[96] Citrulline and ornithine, found in blue-veined cheese and in Camembert,[42-45] result from arginine breakdown, as shown for Gorgonzola.[97] γ-aminobutyric acids resulting from glutamic acid decarboxylation, is found in mould-ripened cheeses.[42,45] Decarboxylation also leads to the production of non-volatile amines. Tyramine is usually observed in higher amounts than tryptamine or histamine.[98] Concentrations of amines in mould-ripened cheeses vary greatly with the sample[99,100] and seem to decrease at the end of ripening.[98] Phenylalanine and tyrosine catabolism by coryneform bacteria isolated from Camembert have been studied by Lee *et al.*[101–102] *B. linens* degrades the amino acids by transamination and cleavage of the benzene ring by 3,4-dihydrophenylacetate-2,3-dioxygenase. This enzyme, as well as aminotransferase, is inducible.[96] Amino acid breakdown in mould-ripened cheeses also leads to the production of volatile compounds such as ammonia, aldehydes, acids, alcohols, amines or other products such as methanethiol resulting from the breakdown of the amino acid side chains (see Section 7).

4 LIPOLYSIS

4.1 Degree of Lipolysis in Cheese

As with proteolysis, lipolysis in mould-ripened cheese is much more extensive than in other cheeses (Table I). While the extent of lipolysis should not exceed 2% of the triglycerides in cheeses such as Gouda, Gruyère or Cheddar, it is usually between 5 and 20% in mould-ripened cheeses. Data in the literature vary greatly, probably depending on the degree of ripening. Extensive lipolysis can be attained in mould-ripened cheeses without any rancid taste occurring, probably due to neutralization of fatty acids on elevation of the pH. Anderson & Day[103] found even higher levels of free fatty acids in blue-veined cheese: about 65–100 meq/100 g of fat, i.e. 18–25% of the total fatty acids. Other workers report lower lipolysis in Danish Blue (about 45 meq/100 g of fat).[104] The extent of lipolysis in Roquefort is 8–10% of the total fatty acids. Godinho & Fox[105] noted a lower free fatty acid level in the outer part of blue cheese due to higher NaCl concentrations that limited the production of lipases and possibly their activity. Morris *et al.*[106] noted a regular increase in free fatty acid levels during ripening, while others[105] observed that these levels decreased at the end of ripening.

In Norman Camembert made from raw milk, lipolysis reaches 6–10% of total fatty acids,[107] but other, probably less typical, samples had lower values of 3–5%.[104] Lipolysis is always highest towards the surface.[108] Most of the free fatty acids arise from lipolysis: short-chain fatty acids, resulting from the breakdown of lactose or some amino acids, represent only 5% of the total free fatty acids.[109]

TABLE I
Fat Acidity of Different Cheese Varieties

Cheese variety	meq acid/100 g of fat
Gouda	6·14 ± 0·50
Camembert	22·27 ± 13·73
Danish Blue	45·34 ± 14·93
Roquefort	27·55 ± 12·16

Mean values of six samples.
Adapted from Van Belle *et al.*[104]

TABLE II
Relative Proportions in Camembert of Free Fatty Acids and of Fatty Acids in Glycerides (Expressed as % of Total)

Fatty acids	Glycerides	Free fatty acids
$4:0 + 6:0 + 8:0$	2·8	2·2
$10:0 + i\ 10:0$ to $12:0$	3·8	3·0
$12:0 + i\ 12:0$ to $14:0$	4·5	4·7
$14:0 + i\ 14:0$ to $16:0$	16·1	14·3
$16:0 + i\ 16:0$ to $18:0$	28·6	23·3
$18:0$	10·2	7·3
$18:1 + 18:2 + 18:3$	34·0	45·2

i = sum of intermediate peaks between the indicated fatty acids.
From Kuzdzal and Kuzdzal-Savoie.[107]

In Camembert, the relative proportions of free fatty acids are different from those of milk triglycerides since the former have a higher concentration of oleic acid[107] (Table II).

4.2 Properties and Effect of Mould Lipases

The essential lipolytic agents in mould-ripened cheeses are *Penicillium* spp. The natural lipase of milk is not very active, even in raw milk cheeses: however, its effect has been shown in blue-veined cheeses made from homogenized milk.[106] Except for *Geotrichum candidum*, microorganisms other than *P. roqueforti* or *P. camemberti* have very low lipolytic activity in mould-ripened cheeses. The particularly high proportion of free oleic acid in Camembert has been attributed to *G. candidum* lipase[107] which preferentially releases this fatty acid.[110]

Lamberet & Lenoir[111,112] noted that *P. camemberti* produces only one extracellular lipase which has optimal activity on tributyrin at pH 9·0 and 35°C. The production level of this enzyme varies from 1 to 10 (relative scale), depending on the strain.[113] At pH 6·0, this lipase retains 50% of its maximal activity and remains very active between 0 and 20°C. It is more active when calcium ions are present.[111] The production of lipase has been studied during the ripening of raw-milk Camembert;[114] activity appears after 10 days of ripening, during or shortly after mycelium growth, is maximal at 16 days and then decreases slightly until the 30th day when it increases again on the lysis of mycelia. In 10 cheeses of different origins, the lipase activity in the outer region of the cheeses varied from 1·2 to 4·45 units/g of cheese.

P. roqueforti has been reported to produce two lipases, one with an optimum at acid pH values, the other with an optimum in the alkaline pH range.[71,115–117] Several authors[71,115,117,118] report that the alkaline lipase is optimally active at pH 7·5–8·0 but optimum pH values at 9·0–9·5 have also been reported.[116,119] This enzyme still retains 20% of its activity at pH 6·0.[119] The activity of acid lipase is maximal at pH 6·0–6·5.[120,121] The specificities of the two enzymes are different, the acid lipase being more active on tricaproin and the alkaline lipase on tributyrin.[119] The activity of both enzymes has been measured in cheese.[122] More acid lipase was synthesized in six out of seven samples. However, in spite of the favourable pH of the cheese, it may not always play the more important role since alkaline lipase has higher activity on milk fat.[119,120] Samples in which alkaline lipase activity is high may have a slightly piquant taste or soapy aroma.[122] This would reflect the relative activity levels of the two lipases[111,120] and probably their different specificities.

5 LACTOSE BREAKDOWN

5.1 Lactose and Lactic Acid Content

The lactic starters used to make Camembert are homofermentative mesophilic lactococci, and lactose breakdown leads mainly to lactic acid production by the hexose diphosphate pathway. Rennet is added to the milk after ripening when the pH is about 6·4. Intense acidification occurs mainly during draining and the pH of the curd when taken from the moulds is about 4·6. The amount of residual lactose in the curd decreases very quickly between 5 and 10 days after processing, i.e. when *P. camemberti* grows; thereafter the lactose level continues to decrease more slowly.[123] After 20–30 days of ripening, the lactose has completely disappeared, this clearance being more rapid on the surface than at the centre of the cheese. Galactose and glucose concentrations in Camembert are very low.[123]

Besides homofermentative lactococci, the lactic flora of Roquefort cheese includes leuconostocs, heterofermentative bacteria which convert lactose into lactic acid, acetic acid, ethanol and CO_2.[16,20] The CO_2 causes small openings in the curd that favour the implantation of *P. roqueforti*. In this same cheese, yeasts also metabolize lactose[16,19] and contribute to the formation of openings in the curd.

In mould-ripened cheese, lactic acid produced by starters is utilized by moulds and yeasts. In Camembert, Berner[124] observed initial L-lactate levels of 2·9% (of dry matter) at the surface and 3·6% at the centre of the cheese. A rapid decrease in these levels between days 5 and 10 of ripening coincides with *Penicillium* growth and the amount of L-lactate in ripened cheese is <0·02%. Puhan & Wanner[125] noted levels of 0·06% on the surface and 0·02% at the centre of ripe Camembert. D-lactate is present at low concentrations[124,126] and has almost disappeared by the end of ripening.[126] Lactic acid breakdown leads to neutralization of the curd; the surface of a traditional Camembert reaches about pH 7·0 at the end of ripening[44] and the pH at the centre is about 6·0 (Fig. 2). The rise in pH in blue-veined cheese is less spectacular than on the surface of Camembert; however, the pH reaches about 6·2 in 'bleu d'Auvergne'[22] or 6·5 in Danish Blue.[35]

5.2 Consequences of pH Increase

This neutralization in cheese plays a considerable role in the ripening process. Due to it, acid-sensitive bacteria, including micrococci and coryneform bacteria, establish on the surface of mould-ripened soft cheese and contribute to their traditional taste qualities. Neutralization also favours the activity of various ripening enzymes whose pH optima are often closer to neutrality than to the acid zone. Furthermore, as we shall see in Section 8, it clearly influences the rheological properties of cheese. Besides these three main results, neutralization of cheese causes the minerals in surface mould cheeses to migrate. Metche & Fanni[127] and Le Graet et al.[128] showed considerable calcium and phosphate migration towards the exterior of Camembert during mould implantation on the surface. The rind of surface-mould cheese attains high concentrations of calcium and inorganic phosphorous (17 and 9 g/kg, respectively),[128] while the levels of these decrease at the centre. Le Graet et al.[128] showed that the high pH of the surface causes the formation of insoluble calcium phosphate, immobilizing this salt at the rind. Electron microscopy studies of the rind revealed the presence of crystals which tentatively were identified as calcium phosphate.[129] This is of nutritional interest as far as the mineral supply in surface mould-ripened cheese is concerned, depending on whether the rind is eaten or not, since at the end of ripening, the rind contains about 80% of the calcium and 55% of the phosphorus of the cheese.[128]

6 OTHER METABOLISMS

Changes in the concentrations of niacin and vitamin B6 have been studied during the ripening of blue-veined cheeses. The niacin content increases during the first six weeks and then decreases slowly; vitamin B6 concentration increases between the 3rd and 6th month of ripening.[130] In Camembert, the presence of ergosterol, which is a precursor of vitamin D, has been reported. Synthesis is carried out by P. camemberti and this substance does not migrate to the interior of the curd.[131]

A strain of P. camemberti, previously used as a fungal starter, was able to produce styrene in surface mould-ripened soft cheese, leading to the appearance of a celluloid taste.[132] Production of 1,3-pentadiene (which has a kerosene-like odour) by decarboxylation of sorbic acid was observed with a sorbate-resistant strain of P. roqueforti.[133]

Like other moulds, P. roqueforti and P. camemberti can synthesize toxic substances called mycotoxins. Their proportions and presence in cheese have been reviewed.[134–136] P. roqueforti synthesized roquefortine,[137] isofumigaclavine A and B,[137,138] PR-toxin,[139] mycophenolic acid,[140,141] patuline,[142] penicillic acid[142] and siderophores.[143] P. camemberti produces cyclopiazonic acid.[144] Patuline and penicillic acid are not found in cheese[142] nor is PR-toxin which is unstable under curd conditions.[145,146] The fungal metabolites that can be detected in cheese are roquefortine,[147] isofumigaclavine A,[147] mycophenolic acid,[148] siderophores[143] and cyclopiazonic acid[144] Considering the low levels (ppm) and biological activity of these substances, it has been concluded that there is no risk to human health, even if large amounts of cheese are eaten.[149–152] Cyclopiazonic acid production in cheese varies greatly according to the strain of P. camemberti used and is undetectable in cheese made with weakly-producing strains.[153]

7 AROMA COMPOUNDS

7.1 Blue-Veined Cheeses

Although the nature of the molecules that determine the aroma of some cheeses is well known, it is often difficult to define the role of one molecular species or a class of molecules in the aroma perception of the consumer. Blue-veined cheeses are an exception because it has been shown that methyl ketones have a key role in the typical flavour of these cheeses. 2-Heptanone and 2-nonanone are the most abundant; 2-pentanone, 2-propanone, 2-undecanone, 2-tridecanone are also present but in lower amounts[154-158] (Table III).

There are wide variations between samples as to total or individual concentrations of methyl ketones, and it is difficult to produce norms.[155-158] Dartey & Kinsella[159] observed that the concentration of methyl ketones increases regularly up to day 70 and then decreases. Comparable kinetics have been reported by Sato et al.[160]

The mechanism of methyl ketone formation in blue-veined cheeses has been thoroughly reviewed by Kinsella & Hwang.[161] Methyl ketones are produced by P. roqueforti from fatty acids via the β-oxidation pathway. β-Ketoacid-CoA, obtained by β-hydroxyacyl-CoA dehydrogenation, is deacylated to β-ketoacid by a β-ketoacyl-CoA deacyclase or thiohydrolase. A decarboxylase converts β-ketoacyl-CoA to a methyl ketone and CO_2.

Mycelium, as well as spores, can produce methyl ketones.[161-163] Spores oxidize fatty acids containing 2–12 carbon atoms; however, octanoic acid is the most rapidly converted substrate. Mycelium oxidizes fatty acids within a wide pH range but the reaction is optimal between pH 5 and 7,[164,165] i.e. at a pH similar to that of ripe cheese. A decarboxylase that converts β-ketododecanoic acid to 2-undecanone has an optimum pH of 6·5–7·0.[161]

There is a positive correlation between the free fatty acid level and the amount of methyl ketones produced,[71,160,161] and cheeses with limited lipolysis do not have a strong aroma.[161] This might be a result of a great excess of substrate and

TABLE III
Methyl Ketone Content of Blue-Veined Cheese

	mg ketone/kg cheese						
	A	B	C	D	E	F	G
2-Propanone	3·4	2·8	3·9	1·7	2·7	2·7	0·0
2-Pentanone	18·4	7·2	20·9	6·5	17·5	19·2	3·6
2-Heptanone	40·8	19·0	71·8	17·9	39·1	69·9	17·6
2-Nonanone	28·0	22·3	88·3	19·8	42·5	78·9	18·9
2-Undecanone	6·4	6·0	29·9	4·9	12·3	6·7	2·4
Total	97·0	57·3	214·8	50·8	114·1	177·4	42·5

Samples A–E: American blue cheese, samples F and G: Roquefort
(from Anderson & Day[156]).

adding lipases to the curd favours the appearance of methyl ketones.[166] Addition of fatty acids to a slurry system inhibits lipolysis but increases the concentrations of methyl ketones.[167] Mycelium physiological stage[168] and pH[162] play a role too. The intensity of *P. roqueforti* development is important; high salt concentrations limit mould growth, retard lipolysis and reduce methyl ketone production.[105]

The volatile fraction of blue-veined cheeses includes many other compounds besides methyl ketones (see review of Adda & Dumont[169]). Secondary alcohols (2-pentanol, 2-heptanol and 2-nonanol) are produced by *Penicillium* due to reduction of methyl ketones. Their concentrations are lower than those of methyl ketones but are still appreciable.[156,158,170] Other alcohols,[156] as well as many esters and aldehydes,[156,157] have also been identified. Jolly & Kosikowski[171] reported the presence of γ-lactones. Ney & Wirotama[172] showed the presence of volatile amines and Spettoli[99] quantified some of these in two samples of Gorgonzola. Fatty acids have been identified by Anderson.[157]

All of these data have contributed to the development of formulae for the composition of synthetic blue-veined cheese aroma.[151,173] The production of this aroma by fermentation (*P. roqueforti* culture on a fat-rich culture medium) has also been developed for use in salad dressing or processed cheeses and has been reported in many studies.[174–180]

7.2 Surface Mould-Ripened Soft Cheeses

The work of Dumont *et al.*[181,182] and Moinas *et al.*[182–184] has defined the aroma of Camembert and revealed its complexity. Table IV summarizes the composition of the volatile products found.

TABLE IV
Volatile Compounds Isolated from Camembert Cheese

1-Alkanols	C2, 3, 4, 6, 2-methylpropanol, 3-methylbutanol, oct-1-en-3-ol, 2-phenylethanol
2-Alkanols	C4, 5, 6, 7, 9, 11
Methyl ketones	C4, 5, 6, 7, 8, 9, 10, 11, 12, 13, 15
Aldehydes	C6, 7, 9, 2 and 3-methylbutanal
Esters	C2, 4, 6, 8, 10-ethyl, 2-phenylethylacetate
Phenols	phenol, p-cresol
Lactones	C_9, C_{10}, C_{12}
Sulphur compounds	H_2S, methyl sulphide, methyldisulphide, methanethiol, 2,4-dithiapentane, 3,4-dithiahexane, 2,4,5-trithiahexane 3-methylthio 2,4-dithiapentane 3-methylthiopropanol
Anisoles	anisole, 4-methylanisole, 2,4-dimethylanisole
Amines	phenylethylamine, $C_{2,3,4}$, diethylamine, isobutylamine, 3-methylbutylamine
Miscellaneous	dimethoxybenzene, isobutylacetamide

From Adda.[185]

Methyl ketones and their corresponding secondary alcohols are the most abundant neutral compounds in the volatile fraction[180,186] and contribute to the flavour of surface-mould cheeses. The work of Lamberet et al.[187] shows that the ability of *P. camemberti* to produce methyl ketones varies greatly among strains. Production of methyl ketones by *P. camemberti* mycelium in a model system containing milk lipids has been described.[188]

1-Octen-3-ol plays a particular role because it is responsible for the mushroom note in the characteristic flavour of Camembert.[189,190] When the concentration of 1-octen-3-ol is too high, the aroma is faulty. Production of 1-octen-3-ol and related compounds has been observed when *P. camemberti* is grown on potato dextrose or Czapeck agar.[191]

Sulphur compounds also play an important part; they cause the garlic note that is clearly present in ripe traditional Camembert. Compounds such as methylsulphide, methyldisulphide and 3-methylthiopropanol, which are present in other cheeses also, have a basic cheesy note, while 2,4-dithiapentane, 2,4,5-trithiahexane and 3-methylthio-2,4-dithiapentane are observed in typical Camembert.[185] Coryneform bacteria are usually thought to be key contributors to the formation of sulphur products in surface-mould cheeses. Strains of *B. linens* isolated from dairy products have been shown to produce methanethiol,[192–195] and these bacteria have a demethiolase which releases methanethiol from methionine. Methanethiol and sulphur compounds are also produced by *P. camemberti*[196] and *Geotrichum candidum*.[197] Several sulphur compounds develop from methanethiol by nonenzymatic reactions.[198]

Substantial amounts of 2-phenylethanol and lower amounts of 2-phenylethylacetate and 2-phenylethylpropionate are produced in traditional Camembert.[177] 2-Phenylethanol has a pleasant rose-like odour; its maximal concentration is reached after the first week of ripening and then decreases.[199] Lee & Richard[200] noted that yeasts can produce 2-phenylethanol from phenylalanine, while *G. candidum* and *B. linens* cannot. It is probable that phenylethanol is produced by yeasts that develop at the beginning of ripening. Phenylethanol would be obtained by the breakdown of phenylalanine by the Ehrlich–Neubauer pathway with phenylpyruvate as the intermediate product.[200] The mean detection thresholds of phenylethanol in a curd-type substrate are 7·6 ppm (aroma) and 9 ppm (flavour). The levels of phenylethanol found in cheese are slightly lower but correspond to the most sensitive taster thresholds.[199] This compound may therefore cause the floral note that some tasters can distinguish in traditional Camembert.

N-isobutylacetamide has been identified in Camembert.[201] This compound has a bitter taste which may contribute to bitterness.[189] It may originate from the Val-Gly dipeptide after decarboxylation and deamination. However, the Val-Gly sequence is not found in caseins and the presence of N-isobutylacetamide would imply that a transamination reaction occurs.[198] Another mechanism involving amine acetylation has been proposed.[198]

It must be recalled that ammonia is an important element in the aroma of traditional Camembert and results from the deamination activity of the microbial flora on amino acids. The various elements in the surface flora probably play a role; the action of *G. candidum*[202] and *B. linens*[96] has been emphasized.

Free fatty acids are found in large amounts in the non-volatile fraction (see Section 4). They contribute to the basic flavour of the cheese and are the precursors methyl ketones and secondary alcohols.

The effect of milk fat oxidation on Camembert flavour has been studied by adding small amounts of copper to the milk used to make the cheese,[203] but this did not lead to substantial fat oxidation in the curd nor to the appearance of an oxidized taste in ripe cheese.

8   CHANGES IN TEXTURE

The outer part of Camembert undergoes considerable modification of texture and the curd which is firm and brittle at the beginning of ripening, later becomes soft. Softening is visible in a cross-section of cheese and gradually extends towards the centre. The water content of Camembert is about 55% and, if it is too high, the outer part tends to flow when the ripe cheese is cut. These changes are usually attributed to the high level of proteolysis created by *P. camemberti*.[204,205] However, as seen earlier, the diffusion of fungal protease is limited,[80,81] and can affect only the outer few millimetres. Another important change caused by *P. camemberti* and the surface flora is the establishment of a pH gradient from the surface to the centre due to the consumption of lactic acid and the production of ammonia. This pH gradient can be simulated by incubating young Camembert (3 days of ripening without *Penicillium* seeding) in an ammoniacal atmosphere. The ammonia dissolves in the curd and, after equilibration, the pH gradient established is expressed by cheese softening; this process is more evident near the surface where the pH is higher[206] (Fig. 4). Increasing the pH, therefore, plays an important role by causing the cheese to soften. This may be explained by the fact that the pH increase augments the net charge on casein and modifies protein–protein interactions. It also changes protein–water interactions and thus the water absorption capacity of the caseins.[207] Indeed, during ripening, the outer part of Camembert has a higher water content than the centre, in spite of surface evaporation, which is inevitable.[128] According to Noomen,[81] the physico-chemical conditions (water content and pH) in Camembert alone cannot explain softening which could be related to rennet action. Indeed, experimental cheeses, containing no rennet and incubated in an ammoniacal atmosphere do not soften but become hard and springy, while cheese with rennet does soften.[81] The softening of Camembert could thus be explained by two processes: (1) α_{s1}-casein breakdown by rennet and (2) a rise in the pH caused by the surface flora. The importance of rennet action on α_{s1}-casein and of the physico-chemical conditions have also been reported for Meshanger cheese, a soft cheese with no surface flora.[208,209]

Few data are available on the texture of blue-veined cheeses. The degree of mineralization and the dry matter vary (50–58%), depending on the way the

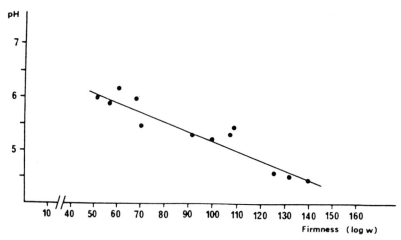

Fig. 4. Camembert firmness in relation to pH of the curd. Cheeses about 3 days old and without *Penicillium* were incubated for 1 h 30 min in an ammoniacal atmosphere. 24 h later a pH gradient was established and cheese firmness was measured at various distances from the surface by penetrometry. A metal blade was allowed to penetrate the curd, at a constant speed and perpendicularly to a section of cheese (i.e. parallel to cheese surface). The pH of the measured area varied depending on its distance from the surface. Firmness was determined by the logarithm of the work (*w*) necessary to insert the blade to a depth of 5 mm (Monnet *et al.*, unpublished data).

cheese is made, and this would explain the various textures of the different varieties. Because of their higher dry matter content, the rise in pH does not lead to softening as for surface mould cheeses. The modification of blue-veined cheese texture, therefore, mainly results from the extensive proteolysis.

9   CONTROL OF RIPENING

The intensity of the biochemical activity of *P. roqueforti* varies considerably among strains[70,210] and the choice of strains has a major effect on the quality of blue-veined cheese.[71,160,211,212] The characteristics of the desired strains also vary from one variety of cheese to another; 'bleu d'Auvergne' requires highly proteolytic strains which have low lipolytic activity, while 'fourme d'Ambert' needs strains with low proteolytic and low lipolytic activities (G. Pradel, pers. comm.). The control of strain growth is another crucial factor. If mould growth is too extensive or too limited, the aroma is faulty or weak. *P. roqueforti* develops within the curds because it tolerates a low O_2 level and a high CO_2 level.[24,213] Aeration and space for mould growth depend on whether the curd is more or less pierced. When heterofermentative leuconostocs are added to the milk during Roquefort production, CO_2 is produced, causing the formation of spaces which enhance mycelial development.[20]

Salt content, which is particularly high (6–8% of the liquid phase) in blue-veined cheeses, also affects *P. roqueforti* growth. Depending on the strain, such concentrations may retard growth.[214] Prolonging salting causes the rate of proteolysis and lipolysis to decrease.[41,105] Besides acting on *Penicillium*, salting also influences the implantation of micrococci and yeasts of the surface of Roquefort.[21]

The choice of the *P. camemberti* strain is also important in the production of surface mould soft cheeses. However, the proteolytic activity of the different strains does not vary[28] as much as their lipolytic and fatty acid oxidative activities.[29,187] The choice of a *P. camemberti* strain is also guided by the growth rate, colour, density and height of the mycelium which play a part in the appearance and attractiveness of surface mould cheeses.[23]

Salting has a selective effect on the moulds of surface mould cheeses. Too much salting limits or prevents *G. candidum* growth, while *P. camemberti* growth is much less affected. In whey culture, *P. camemberti* growth is slowed down when 10–15% of salt is present.[25] Inversely, too little salt, combined with insufficient draining, causes too much growth of *G. candidum* and hinders implantation of *Penicillium*, giving defective cheeses; this defect is called 'toad skin'. Under-salting may also favour the surface implantation of *Mucor*, altering the appearance of the cheese; this defect is called 'cat hair'. Reducing the water activity by higher salting[25,215] and using a *Penicillium* strain that implants quickly,[216] helps to correct this defect. Salting also influences the activity of *Penicillium* enzymes and at 4% it reduces the degree of proteolysis in Camembert (25 versus 40% in an unsalted control).[217] The effect of humidity and temperature conditions in the ripening room on the growth of *P. camemberti* and the quality of Camembert-type cheese have been described.[218]

The production of surface mould soft cheeses using milk highly contaminated with psychrotrophic bacteria leads to organoleptic defects. The lipolytic activity of these bacteria is expressed by increased lipolysis and a rancid taste; bitterness has also been reported.[219]

It is difficult to control the coliform flora of Camembert. Even a low level of coliform contamination of the milk can result in a high multiplication rate later.[14] At the beginning of cheesemaking, there is already some development before the pH has decreased sufficiently. Acidification destroys a large part of this flora. *E. coli* decrease during ripening[220] but in the case of *Hafnia alvei* the bacteria multiply again when the pH rises, resulting, in some cases, in high numbers in the cheese.[14,221] This can be avoided in pasteurized milk cheese through good hygienic practices.

Listeria monocytogenes is able to survive the Camembert cheesemaking process and to grow during ripening of the cheese.[222] Control of *L. monocytogenes* (no detection in 25 g of cheese) can be obtained by adequate pasteurization and avoiding contamination during cheesemaking through good hygienic practices (good equipment design and appropriate cleaning and disinfection). Bacteriocin-producing lactic acid bacteria can also be used for cheesemaking. *L. monocytogenes* content in curd can be very much reduced by using strains of *L. lactis* that produce nisin.[223]

As mentioned above, uncontrolled development of *G. candidum* produces defects in the appearance and taste of cheese, even though this mould probably contributes notably to the taste qualities of Camembert. Some strains of *G. candidum* clearly improve the taste and aroma of pasteurized Camembert cheese. Their controlled growth results in a more typical flavour, close to that of traditional Camembert.[197]

As in all cheeses, acidification plays an essential role by controlling syneresis and the degree of mineralization of surface mould cheese. When acidification is too high, the Camembert curd is too dry and brittle and enzyme activities are limited; insufficient acidification results in a cheese that is too moist at the end of ripening. The last 20 years have witnessed an increasing interest in 'stabilized' cheeses. Washing the curd permits a higher pH to be obtained at the end of draining. This gives a less demineralized cheese that seems more ripe than traditional cheese of the same age. These cheeses made from pasteurized milk have a milder taste and keep better. This could be due to more limited *P. camemberti* activity, perhaps because of the lower levels of available lactose and lactate. Due to their higher pH, these products are more sensitive to coliform bacteria.

An investigation in France by Pelissier *et al.*[224] showed that mould-ripened cheeses are more sensitive to bitterness than other varieties and the intensity of this defect may cause considerable damage. Cheboratev *et al.*[225] have selected strains of *P. roqueforti* for their ability to clear the bitterness of a protein hydrolysate, and they recommend using such strains to avoid the defect in blue-veined cheeses. *P. camemberti* has a crucial role in the appearance of bitterness in Camembert. A too abundant growth of the mycelium can lead to the defect; if *Penicillium* growth is limited by the presence of *G. candidum* or by incubating the cheese in an ammoniacal atmosphere, proteolysis is reduced and the defect does not occur.[226] Therefore, this defect could occur when there is too much proteolysis by *Penicillium* proteases. The rennet dose used and its augmentation does not seem to cause bitterness, perhaps because the pH of Camembert does not favour the action of rennet proteinases at the end of ripening.[226] Lactic bacteria and their proteinases have also been reported to affect the occurrence of bitterness: the defect appears when high populations of lactic bacteria are present in the curd; on the other hand, if these populations are reduced (for example, by phages), bitterness does not appear.[227] This seems to be related to the degree of curd acidification since the probability of bitterness is increased if the pH is low at the end of draining (L. Vassal, pers. comm.). Bitterness might not result directly from high amounts of lactic bacteria but could be related to *Penicillium* whose growth and protease production might be higher in very acid curds.

10 CONCLUSION

The particular characteristics of *P. roqueforti* and *P. camemberti* are expressed in mould-ripened cheese, giving the cheeses their characteristic appearance and contributing to the development of the rheological and gustative qualities. However, the secondary flora contributes to the attainment of traditional quality products. Great progress has been made during the last 20 years in our knowledge of the mechanisms of ripening in blue-veined and surface mould cheeses. However, the processes are very complex and no close relationship can yet be seen between the composition and the quality of mould-ripened cheese.

While studies on traditional mould-ripened cheeses should not be abandoned, it is to be recalled that more cheese is now being produced in large, automated factories. The good quality of these products must be maintained, taking into account consumer taste, which often favours rather mild products. Improving the storage life of surface mould-ripened soft cheese should also make it easier to distribute and to develop its production.

ACKNOWLEDGEMENTS

The author wishes to thank Drs M.J. Desmazeaud, G. Lamberet and L. Vassal for making useful comments and Mrs Alice Daifuku for translation of this manuscript.

REFERENCES

1. Morris, H.A., 1981. *Blue-veined Cheeses*. Pfizer Cheese Monographs, Vol. VII, Pfizer Inc., New York.
2. Kosikowski, F., 1977. In *Cheese and Fermented Milk Food*. Edwards Brothers, Michigan, p. 137.
3. Bakalor, S., 1962. *Dairy Sci. Abstr.*, Parts I and II, **24**, 529 and 583.
4. Pernodet, G., 1984. In *Le Fromage*, ed. A. Eck. Lavoisier, Paris, pp. 229 and 234.
5. Coghill, D., 1979. *Aust. J. Dairy Technol.*, **34**, 72.
6. Seth, R.J. & Robinson, R.K., 1988. In *Developments in Food Microbiology*—**4**, ed. R.K. Robinson. Elsevier Applied Science Publishers, London, p. 23.
7. Desfleurs, M., 1968. *Lait*, **48**, 493.
8. Lenoir, J., 1963. *Lait*, **43**, 262.
9. Schmidt, J.L. & Lenoir, J., 1978. *Lait*, **58**, 355.
10. Schmidt, J.L. & Lenoir, J., 1980. *Lait*, **60**, 272.
11. Rousseau, M., 1984. *Milchwissenschaft*, **39**, 129.
12. Richard, J. & Zadi, H., 1983. *Lait*, **63**, 25.
13. Richard, J., 1984. *Lait*, **64**, 496.
14. Mourgues, R., Vassal, L., Auclair, J., Mocquot, G. & Vandeweghe, J., 1977. *Lait*, **57**, 131.
15. Tolle, A., Otte, I., Suhren, G. & Heeschen, W., 1980. *Milchwissenschaft*, **35**, 21.
16. Devoyod, J.J., Bret G., & Auclair, J., 1968. *Lait*, **48**, 613.
17. Devoyod, J.J., 1970. *Lait*, **50**, 227.
18. Devoyod, J.J. & Bret, G., 1966. *Proc. 17th Intern. Dairy Congr.*, Munich, Vol. D, p. 585.
19. Devoyod, J.J. & Sponem, D., 1970. *Lait*, **50**, 524.
20. Devoyod, J.J. & Muller, M., 1969. *Lait*, **43**, 369.
21. Devoyod, J.J., 1969. *Lait*, **49**, 20.
22. Trieu-Cuot, P. & Gripon, J.C., 1983. *Lait*, **63**, 116.

23. Moreau, C., 1979. *Lait*, **59**, 219.
24. Moreau, C., 1980. *Lait*, **60**, 254.
25. Choisy, C., Gueguen, M., Lenoir, J., Schmidt, J.L. & Tourneur, C., 1984. In *Le Fromage*, ed. A. Eck. Lavoisier, Paris, p. 259.
26. Sanson, R.A., Eckardt, C. & Orth, R., 1977. *Antonie van Leuwenhoek*, **43**, 341.
27. Pitt, J.I., 1979. *The Genus Penicillium and its Teleomorphic States*. Academic Press, New York.
28. Lenoir, J. & Choisy, C., 1970. *Lait*, **51**, 138.
29. Lamberet, G., 1970. *Proc. 18th Intern. Dairy Congr.*, Sydney, Vol. 1F, 143.
30. Tourneur, C., 1982. *Proc. 21st Intern. Dairy Congr.*, Moscow, Vol. 1, Book 1, p. 380.
31. Engel, G., 1985. *Molkerei Zeitung Welt der Milch*, **39**, 700.
32. Engel, G., Prokopek, D. & Teuber, M., 1985. *Milchwissenschaft*, **40**, 661.
33. May, R., Grosse, H.H., Kuenkel, W., Kirchhuebel, W., Braun, I. & Scheunert, W., 1988. German-Democratic-Republic-Patent, 260837.
34. Delespaul, G., Gueguen, M. & Lenoir, J., 1973. *Revue Laitière Française*, **313**, 715.
35. Dale, G., 1972. *Revue Laitière Française*, **296**, 199.
36. Ottogali., G. & Galli, A., 1972. *Scienza e Technica Lattiero-casearia*, **23**, 363.
37. Carini, S., Kaderavek, G., De Gregori, A. & Invernizzi, F., 1971. *Il Latte*, **45**, 615.
38. Oterholm, A., 1984. *IDF Bulletin*, No. 171.
39. Chapman, H.R. & Sharpe, M.E., 1981. In *Dairy Technology*, ed. R.K. Robinson. Elsevier Applied Science Publishers, London, p. 157.
40. Hewedi, M.M. & Fox, P.F., 1984. *Milchwissenschaft*, **39**, 198.
41. Godinho, M. & Fox, P.F., 1982. *Milchwissenschaft*, **37**, 72.
42. Ismail, A. & Hanson, A.A., 1972. *Milchwissenschaft*, **27**, 556.
43. Kosikowski, F.V. & Dahlberg, A.C., 1954. *J. Dairy Sci.*, **37**, 167.
44. Lenoir, J., 1962. *C.R. Acad. Agric.*, **48**, 160.
45. Do Ngoc, M., Lenoir, J. & Choisy, C., 1971. *Revue Laitière Française*, **288**, 447.
46. Lenoir, J., 1963. *Ann. Technol. Agric.*, **12**, 51.
47. Marcos, A., Esteban, M.A., Leon, F. & Fernandes-Salguero, J., 1979. *J. Dairy Sci.*, **62**, 892.
48. Trieu-Cuot, P. & Gripon, J.C., 1982. *J. Dairy Res.*, **49**, 50.
49. Trieu-Cuot, P., Archieri-Hazé, M.J. & Gripon, J.C., 1982. *J. Dairy Res.*, **49**, 487.
50. Trieu-Cuot, P., Archieri-Hazé, M.J. & Gripon, J.C., 1982. *Lait*, **62**, 234.
51. Noomen, A., 1978. *Neth. Milk Dairy J.*, **32**, 49.
52. Lenoir, J., 1984. *IDF Bulletin*, No. 171, p. 3.
53. Visser, F.M.W., 1977. *Neth. Milk Dairy J.*, **31**, 210.
54. Visser, F.M.W. & Groot-Mostert, A.E.A., 1977. *Neth. Milk Dairy J.*, **31**, 247.
55. Noomen, A., 1978. *Neth. Milk Dairy J.*, **32** 26.
56. Desmazeaud, M.J., Gripon, J.C., Le Bars, D. & Bergere, J.L., 1976. *Lait*, **56**, 379.
57. Gripon, J.C. & Hermier, J., 1974. *Biochimie*, **56**, 1323.
58. Lenoir, J. & Auberger, B., 1977. *Lait*, **57**, 164.
59. Lenoir, J. & Auberger, B., 1977. *Lait*, **57**, 471.
60. Gripon, J. C., Auberger, B. & Lenoir, J., 1980. *Int. J. Biochem.*, **12**, 451.
61. Zevaco, C., Hermier, J. & Gripon, J.C., 1973. *Biochimie*, **55**, 1353.
62. Lenoir, J., Auberger, B. & Gripon, J.C., 1979. *Lait*, **59**, 244.
63. Modler, H.W., Brunner, J.R. & Stien, C.M., 1974. *J. Dairy Sci.*, **57**, 523.
64. Modler, H.W., Brunner, J.R. & Stien, C.M., 1974. *J. Dairy Sci.*, **57**, 528.
65. Gripon, J.C., 1977. *Ann. Biol. Anim. Biochem. Biophys.*, **17**, 283.
66. Ahiko, K., Iwasawa, S., Ueda, M. & Miyata, N., 1981. Reports of Research Laboratory, Snow Brand Milk Products Co. Ltd, Vol. 77, p. 127.
67. Gripon, J.C., 1977. *Biochimie*, **59**, 679.
68. Auberger, B., Bontals, M. & Lenoir, J., 1982. *Proc. 21st Intern. Dairy Congr.*, Moscow, Vol. 1(2), p. 276.
69. Gripon, J.C. & Debest, B., 1976. *Lait*, **56**, 423.

70. Fournet, G.P., 1971. Thèse, University of Montpellier, France.
71. Niki, T., Yoshioka, Y. & Ahiko, K., 1966. *Proc. 17th Intern. Dairy Congr.*, Munich, Vol. D, p. 531.
72. Lenoir, J. & Auberger, B., 1982. *Proc. 21st Intern. Dairy Congr.*, Moscow, Vol. 1(2), p. 335.
73. Le Bars, D. & Gripon, J.C., 1981. *J. Dairy Res.*, **48**, 479.
74. Nishikawa, I., 1957. Reports of Research Laboratory, Snow Brand Milk Products Co. Ltd, Vol. 36, p. 1.
75. Imamura, T., 1960. *J. Agr. Chem. Soc.*, **34**, 375.
76. Takafuji, S. & Yoshioka, Y., 1982. Reports of Research Laboratory, Snow Brand Milk Products Co. Ltd, Vol. 78, p. 63.
77. Takafuji, S. & Yoshioka, Y., 1982. Reports of Research Laboratory, Snow Brand Milk Products Co. Ltd, Vol. 78.
78. Gripon, J.C., Desmazeaud, M.J., Le Bars, D. & Bergere, J.L., 1977. *J. Dairy Sci.*, **60**, 1532.
79. Lenoir, J., 1970. *Revue Laitière Française*, **275**, 231.
80. Lenoir, J. & Auberger, B., 1982. *Proc. 21st Intern. Dairy Congr.*, Moscow, Vol. 1(2), p. 336.
81. Noomen, A., 1983. *Neth. Milk Dairy J.*, **37**, 229.
82. Paquet, J. & Gripon, J.C., 1980. *Milchwissenschaft*, **35**, 72.
83. Ichishima, E., Takeuchi, M., Yamamoto, K., Sano, Y. & Kikuchi, T., 1978. *Current Microb.*, **1**, 95.
84. Ahiko, K., Iwasawa, S., Ueda, M. & Miyata, N., 1981. Reports of Research Laboratory, Snow Brand Milk Products Co. Ltd, Vol. 77, p. 135.
85. Law, B.A. & Kolstad, J., 1983. *Antonie van Leeuwenhoek*, **49**, 225.
86. Gueguen, M. & Lenoir, J., 1976. *Lait*, **56**, 439.
87. Gueguen, M. & Lenoir, J., 1975. *Lait*, **55**, 621.
88. Gueguen, M. & Lenoir, J., 1975. *Lait*, **55**, 145.
89. Schmidt, J.L., 1982. *Proc. 21st Int. Dairy Congr.*, Moscow, Vol. 1(2), p. 365.
90. Desmazeaud, M.J. & Devoyod, J.J., 1974. *Ann. Biol. Anim. Biochim. Biophys.*, **14**, 327.
91. Friedman, M.E., Nelson, W.O. & Wood, W.A., 1953. *J. Dairy Sci.*, **36**, 1124.
92. Foissy, H., 1978. *Milchwissenschaft*, **33**, 221.
93. Foissy, H., 1978. *FEMS Microb. Lett.*, **3**, 207.
94. Foissy, H., 1978. *Z. Lebensmitt. Unters. Forsch.*, **166**, 164.
95. Torgersen, H. & Sorhaug, T., 1978. *FEMS Microb. Lett.*, **4**, 151.
96. Hemme, D., Bouillanne, C., Metro, F. & Desmazeaud, M.J., 1982. *Sciences des Aliments*, **2**, 113.
97. Cecchi, L. & Resmini, P., 1972. *Scienza et Technica Lattiero-casearia*, **23**, 389.
98. Colonna, P. & Adda, J., 1976. *Lait*, **56**, 143.
99. Spettoli, P., 1971. *Ind. Agr.*, **9**, 42.
100. Sen, N.P., 1969. *J. Food Sci.*, **34**, 22.
101. Lee, C.W., Lucas, S. & Desmazeaud, M.J., 1985. *FEMS Microb. Lett.*, **26**, 201.
102. Lee, C.W. & Desmazeaud, M.J., 1985. *Arch. Microbiol.*, **140**, 331.
103. Anderson, D.F. & Day, E.A., 1965. *J. Dairy Sci.*, **48**, 248.
104. van Belle, M., Vervack, W. & Foulon, M., 1978. *Lait*, **58**, 246.
105. Godinho, M. & Fox, P.F., 1981. *Milchwissenschaft*, **36**, 476.
106. Morris, H.A., Jeseski, J.J., Combs, W.B. & Kuramoto, S., 1963. *J. Dairy Sci.*, **46**, 1.
107. Kuzdzal, W. & Kuzdzal-Savoie, S., 1966. *Proc. 17th Intern. Dairy Congr.*, Munich, Vol. D2, p. 335.
108. Kuzdzal-Savoie, S., 1968. *Qualities Plantarum et Materiae Vegetabiles*, **16**, 312.
109. Kuzdzal-Savoie, S. & Kuzdzal, W., 1966. *Techn. Lait.*, **14**, 17.
110. Marks, T.A., Quinn, J.G., Sampugna, J. & Jensen, R.G., 1968. *Lipids*, **3**, 143.
111. Lamberet, G. & Lenoir, J., 1976. *Lait*, **56**, 662.

112. Lamberet, G. & Lenoir, J., 1976. *Lait*, **56**, 119.
113. Lamberet, G. & Lenoir, J., 1972. *Lait*, **52**, 175.
114. Lamberet, G. & Lopez, M., 1982. *Proc. 21st Intern. Dairy Congr.*, Moscow, Vol. 1(1), p. 499.
115. Imamura, T. & Kataoka, K., 1963. *Jap. J. Zootechn.*, **34**, 349.
116. Kman, I.M., Chandan, R.C. & Shahani, K.M., 1966. *J. Dairy Sci.*, **49**, 700.
117. Morris, H.A. & Jezeski, J.J., 1953, *J. Dairy Sci.*, **36**, 1285.
118. Eitenmiller, R.R., Vakil, J.R. & Shahani, K.M., 1970. *J. Food Sci.*, **35**, 130.
119. Menassa, A. & Lamberet, G., 1982. *Lait*, **62**, 32.
120. Lamberet, G. & Menassa, A. 1983. *J. Dairy Res.*, **50**, 459.
121. Lobyreva, L.B. & Marchenkova, A.I., 1981. *Mikrobiologiya*, **50**, 459.
122. Lamberet, G. & Menassa, A., 1983. *Lait*, **63**, 33.
123. Berner, G., 1970. *Milchwissenschaft*, **25**, 275.
124. Berner, G., 1971. *Milchwissenschaft*, **26**, 685
125. Puhan, Z. & Wanner, E., 1979. *Deutsche Molkerei-Zeitung*, **24**, 874.
126. Markwalder, H.U., 1982. *Lebensm. Wiss. u.Technol.*, **15**, 68.
127. Metche, M. & Fanni, J., 1978. *Lait*, **58**, 336.
128. Le Graet, Y., Lepienne, A., Brule, G. & Ducruet, P., 1983. *Lait.*, **63**, 317.
129. Brooker, B.E., 1987. *Food Microstructure*, **6**, 25.
130. Nilson, K.M., 1967. *Diss. Abst. Int. B*, **28**, 398.
131. Huyghebaert, A. & De Moor, H., 1979. *Lait*, **59**, 464.
132. Adda, J., Dekimpe, J., Vassal, L. & Spinnler, H.E., 1989. *Lait*, **69**, 115.
133. Liewen, M.B. & Marth, E.H., 1985. *Z. Lebensm. Unters. & Forsch.*, **180**, 45.
134. Scott, P.M., 1981. *J. Food Protection*, **44**, 702.
135. Orth, R. 1981. In *Mycotoxine in Lebensmitteln*, ed. J. Reiss. Gustav Fischer Verlag, Stuttgart, p. 273.
136. Sieber, R., 1978. *Z. Ernährungswissenschaft*, **17**, 112.
137. Scott, P.M., Merien, M.A. & Polonsky, J., 1976. *Experientia*, **32**, 140.
138. Ohmono, S., Sato, T., Utagawa, T. & Abe, M., 1975. *Agr. Biol. Chem.*, **39**, 1333.
139. Scott, P.M., Kennedy, B.P.C., Harwig, J. & Blanchfield, B.J., 1977. *Appl. Environ. Microbiol.*, **33**, 249.
140. Siriwardana, M.G. & Lafont, P., 1979. *J. Dairy Sci.*, **62**, 1145.
141. Engel, G., Von Milczewski, K.E., Prokopek, D. & Teuber, M., 1982. *Appl. Environ. Microbiol.*, **43**, 1034.
142. Engel, G. & Prokopek, D., 1980. *Milchwissenschaft*, **35**, 218.
143. Ong, S.A. & Neilands, J.B., 1979. *J. Agr. Food Chem.*, **27**, 990.
144. Le Bars, J., 1979. *Appl. Environ. Microbiol.*, **38**, 1052.
145. Scott, P.M. & Kanhere, S.R., 1979. *J. Assoc. Off. Anal. Chem.*, **62**, 141.
146. Piva, M.T., Guiraud, J., Crouzet, J. & Galzy, P., 1976. *Lait*, **56**, 397.
147. Scott, P.M. & Kennedy, B.P.C., 1976. *J. Agr. Food Chem.*, **24**, 865.
148. Engel, G., Von Milczewski, K.E., Prokopek, D. & Teuber, M., 1982. *Appl. Environ. Microbiol.*, **43**, 1034.
149. Scholch, U., Luthy, J. & Schlatter, C., 1984. *Milchwissenschaft*, **39**, 76.
150. Scholch, U, Luthy, J. & Schlatter, C., 1984. *Z. Lebensm. Unters. Forsch.*, **178**, 351.
151. Franck, H.K., Orth, R., Reichle, G. & Wunder, W., 1975. *Milchwissenschaft*, **30**, 594.
152. Franck, H.J., Orth, R., Ivankovic, S., Kuhlmann, M. & Schmahl, D., 1977. *Experientia*, **33**, 515.
153. Le Bars, J., Gripon, J.C., Vassal, L. & Le Bars, P., 1988. *Microbiologie-Aliments-Nutrition*, **6**, 337.
154. Schwartz, D.P. & Boyd, E.N., 1963. *J. Dairy Sci.*, **46**, 1422.
155. Schwartz, D.P. & Parks, O.W., 1963. *J. Dairy Sci.*, **46**, 989.
156. Anderson, D.F. & Day, E.A., 1966. *J. Agr. Food Chem.*, **14**, 241.
157. Anderson, D.F., 1966. *Diss. Abstr.*, **26**(6), 6636.

158. Svensen, A. & Ottestad, E., 1969. *Meieriposten*, **58**, 50, 77.
159. Dartey, C.K. & Kinsella, J.E., 1971. *J. Agric. Food Chem.*, **19**, 771.
160. Sato, M., Honda, T., Yamada, Y., Takada, A. & Kawanami, T., 1966. *Proc. 17th Intern. Dairy Congr.*, Munich, Vol. D, p. 539.
161. Kinsella, J.E. & Hwang, D.M., 1976. *CRC Crit. Rev. Food Sci. Nutr.*, **8**, 191.
162. Lawrence, R.C., 1966. *J. Gen. Microbiol.*, **44**, 383.
163. Larroche, C., Tallu, B. & Gros, J.B., 1988. *J. Indust. Microbiol.*, **3**, 1.
164. Dwivedi, B.K. & Kinsella, J.E., 1974. *J. Food Sci.*, **39**, 83.
165. Lawrence, R.C. & Hawke, J.C., 1968. *J. Gen. Microbiol.*, **51**, 289.
166. Jolly, R.C. & Kosikowski, F.V., 1975. *J Dairy Sci.*, **58**, 846.
167. King, R.D. & Clegg, G.H., 1979. *J. Sci. Food Agric.*, **30**, 197.
168. Fan, T.Y., Hwang, D.H. & Kinsella, J.E., 1976. *J. Agric. Food Chem.*, **24**, 443.
169. Adda, J. & Dumont, J.P., 1974. *Lait*, **54**, 1.
170. Martelli, A., 1989. *Rivista della Societa Italiana di Scienza dell'Alimentazione*, **18**, 251.
171. Jolly, R.C. & Kosikowski, F.V., 1974. *J. Dairy Sci.*, **57**, 597.
172. Ney, K.H. & Wirotama, I.P., 1972. *Z. Lebensm. Unters. Forsch.*, **149**, 275.
173. Ney, K.H., Wirotama I.P. & Freytag, W.G., British Patent, 1 381737.
174. Knight, S., 1963. US Patent, 3 100153.
175. Watt, J.C. & Nelson, J.H., 1963. US Patent, 3 072488.
176. Nelson, J.H., 1970. *J. Agric. Food Chem.*, **18**, 567.
177. Jolly, R.C. & Kosikowski, F.V., 1975. *J. Food Sci.*, **40**, 285.
178. Luksas, A.J., 1973. US Patent, 3 720520.
179. Revah, S. & Lebeault, J.-M., 1989. *Lait*, **69**, 281.
180. Pratt, N.G., 1989. US Patent, 4 882964.
181. Dumont, J.P., Roger, S. & Adda, J., 1976. *Lait*, **56**, 595.
182. Moinas, M., Groux, M. & Horman, I., 1973. *Lait*, **53**, 601.
183. Moinas, M., Groux, M. & Horman, I., 1975. *Lait*, **55**, 414.
184. Groux, M. & Moinas, M., 1974. *Lait*, **54**, 44.
185. Adda, J., 1984. In *Le Fromage*, ed. A. Eck. Lavoisier, Paris, p. 330.
186. Schwartz, D.P. & Parks O.W., 1963. *J. Dairy Sci.*, **46**, 1136.
187. Lamberet, G., Auberger, B., Canteri, G. & Lenoir, J., 1982. *Revue Laitière Française*, **406**, 13.
188. Kinsella, J.E. & Okomura, J., 1985. *J. Dairy Sci.*, **68**, 11.
189. Adda, J., Roger, S. & Dumont, J.P., 1973. In *Flavor of Food and Beverages*, ed. G. Charalambous & G.E. Inglett. Academic Press, New York, p. 65.
190. Karahadian, C., Josephson, D.B. & Lindsay, R.C., 1985. *J. Dairy Sci.*, **68**, 1865.
191. Karahadian, C., Josephson, D.B. & Lindsay, R.C., 1985. *J. Agric. Food Chem.*, **33**, 339.
192. Sharpe, M.E., Law, B.A., Phillips, B.A. & Pitcher, D.G., 1985. *J. Gen. Microbiol.*, **101**, 345.
193. Law, B.A. & Sharpe, M.E., 1978. *J. Dairy Res.*, **45**, 267.
194. Ferchichi, M., Hemme, D., Nardi, M. & Pamboukjian, N., 1985. *J. Gen. Microbiol.*, **131**, 715.
195. Cuer, A., Dauphin, G., Kergomard, A., Dumont, J.P. & Adda, J., 1979. *Appl. Microbiol.*, **38**, 332.
196. Tsugo, T. & Matsuoka, H., 1962. *Proc. 16th Intern. Dairy Congr.*, Copenhagen, Vol. IV, p. 385.
197. Mourgues, R., Bergere, J.L. & Vassal, L., 1983. *La Technique Laitière*, **978**, 11.
198. Adda, J., Gripon, J.C. & Vassal, L., 1982. *Food Chem.*, **9**, 115.
199. Roger, S., Degas, C. & Gripon, J.C., 1988. *Food Chem.*, **12**, 1.
200. Lee, C.W. & Richard, J., 1984. *J. Dairy Res.*, **51**, 461.
201. Dumont, J.P. & Adda, J., 1978. In *Progress in Flavour Research*, ed. D.G. Land & H.E. Nursten. Elsevier Applied Science Publishers, London, p. 245.
202. Greenberg, R.S. & Ledford, R.A., 1979. *J. Dairy Sci.*, **62**, 368.

203. Korycka-Dahl, M., Vassal, L., Ridadeau-Dumas, B. & Mocquot, G., 1983. *Sciences des Aliments*, **3**, 79.
204. Knoop, A.M. & Peters, K.H., 1971. *Milchwissenschaft*, **26**, 193.
205. Knoop, A.M. & Peters, K.H., 1972. *Milchwissenschaft*, **27**, 153.
206. Vassal, L., Monnet, V., Le Bars, J.C., Roux, C. & Gripon, J.C., 1986. *Lait*, **66**, 341.
207. Ruegg, M. & Bland, B., 1976. *J. Dairy Sci.*, **59**, 1019.
208. Noomen, A., 1977. *Neth. Milk Dairy J.*, **31**, 75.
209. De Jong, L., 1977. *Neth. Milk Dairy J.*, **31**, 314.
210. Farahat, S.M., Rabie, A.M. & Farag, A.A., 1990. *Food Chemistry*, **36**, 169.
211. Graham, D.M., 1968. *J. Dairy Sci.*, **41**, 719.
212. Zambrini, A.V., 1979. *Industria del Latte*, **15**, 91.
213. Golding, N.S., 1945. *J. Dairy Sci.*, **28**, 737.
214. Godinho, M. & Fox, P.F., 1981. *Milchwissenschaft*, **36**, 205.
215. Hardy, J., 1979. *Revue Laitière Française*, **377**, 19.
216. Von Weissenfluh, A., 1988. *Deutsche-Milchwirtschaft*, **39**, 590.
217. Kikuchi, T. & Takafuji, S., 1971. *Jap. J. Zootech. Sci.*, **42**, 276.
218. Von Weissenfluh, A. & Puhan, Z., 1987. *Schweiz. Milchwirtschaft. Forsch.*, **16**, 37.
219. Dumont, J.P., Delespaul, G., Miguot, B. & Adda, J., 1977. *Lait*, **57**, 619.
220. Parks, H.S., Marth, E.H. & Olson, N.F., 1974. *J. Milk Food Technol.*, **36**, 543.
221. Rutzinski, J.H., Marth, E.H. & Olson, N.F., 1979. *J. Food Protection*, **42**, 790.
222. Ryser, E.T. & Marth, E.H., 1987. *J. Food Protection*, **50**, 372.
223. Maisnier-Patin, S., Deschamps, N., Tatini, S.R. & Richard, J., 1992. *Lait*, **72**, 249.
224. Pelissier, J.P., Mercier, J.C. & Ribadeau-Dumas, B., 1974. *Revue Laitière Française*, **325**, 817.
225. Cheboratev, L.N., Bratsilo, T.E. & Rogoja, T.A., 1974. *Sovershenstovovanie teckhnologischkikh protsessov molochnoi promyshlennosti*, Tom 12, Chast II, 28 (Dairy Sci. Abst., 1976, **38**, 4406).
226. Vassal, L. & Gripon, J.C., 1984. *Lait*, **64**, 397.
227. Martley, F.G., 1975. *Lait*, **55**, 310.

5

Bacterial Surface-Ripened Cheeses

A. REPS

University of Agriculture and Technology, Olsztyn, Poland

1 INTRODUCTION

One of the most significant periods in cheese production is the curing (ripening) process; only a properly conducted ripening process, specific for a given type of cheese, ensures the production of a high quality product.

During ripening, microorganisms develop on the surface of cheeses. Quite often, this phenomenon is undesirable and periodic cleaning or wrapping of the cheese with protective coating prevents the growth of microorganisms on the surface. This is particularly the case with hard and semi-hard cheeses, ripened internally through the participation of coagulant, indigenous milk enzymes and of microbial enzymes present throughout the body of the cheese.

However, a group of cheeses exists for which the development of desirable microorganisms on the surface and, in effect, the formation of a viscous, red–orange (various shades) smear is necessary because it determines the organoleptic properties of these cheeses. They are soft cheeses, which ripen from the surface to the interior, mainly through the participation of enzymes secreted by microorganisms present in the smear.

There is also a group of semi-hard cheeses which ripen through the combined action of enzymes present inside the cheese and of enzymes secreted by microorganisms present in the smear.

Soft smear cheeses are characterized by a rich aromatic, piquant flavour; semi-hard smear cheeses are milder, with a pleasant, sweetish flavour.

The curing of cheeses without a smear considerably simplifies the ripening process and therefore their global production is increasing markedly. It is significant that the ripening process of non-smear cheeses may be automated, which favours the operation of large cheesemaking plants. Ripening of cheeses with the use of a smear is very laborious and therefore their level of production is considerably lower than that of internally-ripened cheese varieties. In the case of certain types of cheese, smear development is not used intentionally in the ripening process but it should be mentioned that during the ripening of certain types of hard cheeses, the presence of a smear on the surface positively affects their flavour properties.[1-3]

Depending on the production technology and the changes during ripening, the influence of the smear on the organoleptic properties of cheeses may be differentiated as:

(a) significant, e.g. Tilsiter, Gruyère, Beaufort, Appenzeller;
(b) of major importance, e.g. Trappist, Munster, Brick, Blue;
(c) essential, e.g. Romadour, Lederkranz, Saint Paulin.

The most popular soft smear cheeses in Europe are Limburger, Romadour, Livarot, Munster, Pont L'Eveque, and full-ripened Camembert and Brie. In the United States, the semi-hard, smear cheese, Brick, is very common. In the Soviet Union, many types of smear cheeses are produced; their organoleptic properties are similar to those of Backstein, a cheese of German origin.

Due to their short ripening period, relatively easy digestibility and high flavour properties, soft and semi-hard smear cheeses deserve attention.

2 FACTORS AFFECTING THE RIPENING PROCESS IN SMEAR CHEESES

A number of factors have an important influence on the more rapid (compared with other types of cheese) ripening process in smear cheeses and on the more intense flavour properties, namely:

(a) water content;
(b) size;
(c) manner of curing;
(d) development of desirable microorganisms on the surface of the cheese, i.e. microbiological composition of smear.

2.1 Water Content

The structure, consistency and the course of ripening of smear cheeses, which is the result of the development of a bacterial microflora on the surface, are affected, to a large extent, by the water content of the cheese.

During the manufacture of soft smear cheeses, cutting of the rennet coagulum may be delayed and, consequently, the resulting firm curd retains whey well, i.e. has poor syneresis properties. Furthermore, the curd is cut into large particles which also reduces exudation of whey. As a consequence, there is a lot of whey and, therefore, lactose, in the fresh cheese. The high lactose content favours the development of an acidifying microflora and the accumulation of high concentrations of lactic acid.

The pH of Brick cheese after manufacture is 5–5·2;[4] acidity increases further during the early stages of ripening, reaching a maximum of about pH 5 on the third day after production.[5] The acidity of Limburger cheese is higher and its pH reaches values much lower than 5.[6,7]

The correct water content, which ensures obtaining a product of high quality, is dependent on fat content. For example, in Romadour cheeses with fat contents of 20, 30, 40, 45 and 50% in dry matter (FDM), the water content should be 60·8, 57·8, 53·8, 51·6 and 49·2%, respectively, and in Limburger cheeses with 20 and 40% of FDM, it should be 58·8 and 51·7%, respectively. Cheeses with incorrect water contents have an irregular consistency and an atypical, often unclean, odour.

The high acidity and high salt concentration in the surface layer of these cheeses are inhibitory factors to the development of many microorganisms. The ripening process is inhibited initially and may commence only after neutralization of part of the lactic acid to a level at which the growth of bacteria and the activity of enzymes become possible.

2.2 Size of Cheese

Microorganisms that develop on the surface of smear cheeses synthesize much more active proteolytic and lipolytic enzymes than bacteria present inside the cheese. These enzymes penetrate the body of the cheese and participate in the ripening process.

Simultaneously, as a result of vital processes of smear microorganisms, many alkaline products are formed which also diffuse into the cheese, reducing the acidity of the cheese mass and creating conditions favourable for the development of other microorganisms on the surface of the cheese and also activating enzymatic processes within the cheese.

The influence of the smear on flavour development in the cheese is, therefore, affected by the size of the cheese—the bigger the cheese, the smaller is the effect of the smear enzymes. Due to this fact, smear cheeses are of small dimensions in order to ensure a high surface: volume ratio, and as a result there is a short distance for penetration of enzymes and metabolic products of smear microorganisms.

2.3 Curing of Cheeses During Ripening

The method of curing smear cheeses differs from that for other types of cheese. The curing procedure involves the creation of conditions favourable for the optimal development of desirable microorganisms on the surface of the cheese. The resultant, viscous, red-orange smear has a decisive effect upon the course of ripening and on the organoleptic properties of the cheese. The colour of smear is a result of the development of microorganisms, which form coloured colonies during growth.

Smear cheeses are ripened in rooms with a relative humidity above 95%, although certain types of smear cheese are maintained in a ripening room with a lower humidity for a short period to permit strengthening of the rind. The temperature of ripening depends on the type of cheese; for example, for Limburger cheese it is 15–20°C and for Trappist 12–13°C. During curing, the cheeses are inverted frequently and the surfaces massaged with a 2–3% NaCl solution.

After salting and in ripening rooms, on the walls and equipment, there are microorganisms which may develop on the surface of cheeses under favourable conditions. After a few days, a light yellow-coloured bloom having the consistency of pudding appears on the surface of the cheeses and during further ripening it changes colour to red–orange. Smear appears non-uniformly; coloured spots should be uniformly distributed over the whole surface of the cheese and smear should be rubbed into all irregularities on the cheese surface. In order to accelerate the development of smear, a pure culture of the desirable bacteria or 'good' smear collected from ripened cheeses, may be added to the salt solution used to massage the cheese surface.

The time of appearance of the smear and intensity of its development should be noted and followed. Weak or excessive development of smear has a negative effect on cheese quality. If the smear has a whitish colour it means that the cheese has been over-salted and when the smear dries too quickly it indicates that the cheese is under-salted. During incorrect curing, the layer of smear may become too thick which often causes a change of its colour, and simultaneously, beneath the layer of smear, under the anaerobic conditions, putrefaction of the cheese rind may occur. Absence of smear on the surface of cheeses may cause development of undesirable microorganisms, especially of moulds.

The duration of ripening of smear cheeses depends on the desired intensity of the cheese flavour. Certain cheeses, e.g. Limburger or Brick, are freed of smear after 2–3 weeks and after coating with protective layers, they are transferred to a ripening room at a lower temperature—about 10°C. For example, the ripening period for Limburger, Munster and Brick is 6–8 weeks and for Romadour is 3–4 weeks.

Paolo & Baćiková[8] investigated the influence of two different solutions of microelements on the ripening of Romadour cheese. They indicated that using the solution of microelements at concentration of 1:5000 stimulated the ripening of the cheese, but at higher concentrations of microelements an inhibition of ripening was observed. It was found that the investigated compositions of the microelements can be used to accelerate the ripening of Romadour cheese.

2.4 Microbiological Composition of the Smear

On the basis of studies conducted to date, it is very difficult to answer unequivocally the question: what is the microbiological composition of the smear?

First of all, the microorganisms found on the surface of cheese may be those which appear in a particular dairy plant. However, the most important factors affecting the composition of the smear microflora are:

(a) composition of the microflora in the brine and in rooms in which the cheeses are ripened;
(b) water activity, acidity and salt content in the surface layer of the cheese;
(c) humidity and temperature in the ripening room;
(d) rate of growth of the particular groups of microorganisms;
(e) symbiosis and anabiosis between the microorganisms;
(f) the regularity with which curing is conducted in a particular dairy plant.

During the initial period of ripening, microorganisms which tolerate the high acidity of the cheese as well as the high salt content in the surface layer at that time, may develop on the surface of the cheese.

In 1899, Laxa[9] isolated yeasts, lactic acid bacteria, yellow pigment-producing rods and the mould, *Oospora lactis* (now called *Geotrichum candidum*) from Backstein-type cheese. He expressed the opinion that the growth of *Geotrichum candidum* on the surface of the cheese lowers its acidity, creating favourable conditions for the development of other microorganisms and that the characteristic odour and flavour of these cheeses is a result of symbiosis between all microorganisms in the smear.

Kelly[10] examined microbiological changes in smear during the ripening of Limburger cheese in 14 plants. He stated that during the initial period of ripening, yeasts develop on the surface of the cheese, reaching maximum numbers in 4–5 days. From the 10th to 18th day of ripening, the number of yeasts decreased. From the 6th day of ripening, intensive growth of *Bacterium linens* (now *Brevibacterium linens*) was observed. The other microorganisms, including *Geotrichum candidum*, were sporadically present. He observed that the colour of the smear in various plants ranged from red-brownish to orange. The rate of cheese ripening was dependent on the water content, method of salting and the temperature of ripening.

In further studies, Kelly & Marquardt[6] confirmed that the first microorganisms which develop on the surface of Limburger cheeses are yeasts and, sporadically, small numbers of *Geotrichum candidum*. When the pH of the cheese surface is increased to 5·85 due to the growth of yeasts, the growth of *Br. linens*, which also tolerates high NaCl concentrations, is possible (Table I).

TABLE I
Changes in Microorganisms in Surface Smear (from Kelly & Marquardt[6])

Description of cheese	Age	Microorganisms in surface smear
	1 day	A few rods, evenly distributed
White slime	2 days	A few yeasts
	3 days	Fewer rods, yeasts well distributed
	4 days	Yeasts, some rods increasing
	6 days	Yeasts in masses, fewer rods
Gassy cheese	7 days	Yeasts more numerous, fewer rods
Good red colour	8 days	Yeasts same, rods in masses
Gassy cheese	10 days	Yeasts fewer, rods fewer. Geotrichum present
	11 days	Yeasts few, rods in masses, a few Geotrichum
	12 days	Yeasts fewer and smaller, rods in masses
Ready for storage	13 days	Yeasts in masses but small, rods in masses
	4 weeks	Yeasts small, rods in masses
	5 weeks	No yeasts, rods in masses
	6 weeks	Yeasts small and numerous, rods in masses
	7 weeks	Yeasts small and few, rods in masses
Ready for market	8 weeks	Yeasts small and few, rods in masses

(Reproduced with permission of A.A. Ernstrom, *Journal of Dairy Science.*)

Similar results were obtained by Macy & Erekson[11] in their studies on microbiological changes on the surface of Roquefort, Port du Salut, Tilsiter & Limburger, and by Morris *et al.*[12] who examined Minnesota Blue cheese. However, they[11,12] stated that the microorganisms which develop successively on the surface of the cheese following yeasts included not only *Br. linens* but also micrococci.

Yale[13] isolated 243 bacterial cultures from the surfaces of various smear cheeses. He stated that the dominant microorganisms in smear were Gram-positive bacteria, followed by *Br. linens*. The quality of experimental Limburger cheeses was lower when only *Br. linens* was present on the surface than that of commercial cheeses from dairy plants.

Similarly, Hartley & Jezeski[14] stated that *Br. linens* was not always the dominant microorganism in smear and its numbers depended on pH and the temperature of cheese ripening. From Blue cheese smear, they isolated 167 pure cultures, three of which were classified as *Micrococcus*, 41 as *Bacterium erythrogenes* (now *Br. erythrogenes*) and eight as *Br. linens*. At ripening temperatures of 46–49°F (7·7–9·4°C), *Br. erythrogenes* developed intensively on the surface of the cheeses, while at 55–58°F (12·7–14·4°C), *Br. linens* predominated.

By controlling the intensity of smear development on the surface of cheeses, it is possible to influence their flavour properties.

Langhus *et al.*,[5] when studying the microflora of Brick cheeses, observed that 1–2 days after salting, yeasts began to develop on the cheese surface and they grew intensively until the 3rd–4th day; micrococci then grew to very high numbers and *Br. linens* to a lesser extent. Lubert & Frazier[15] isolated 136 yeast cultures and 329 micrococcal cultures from the smear of Brick cheeses. In all cheeses examined, *Micrococcus varians* was present in high numbers, followed by *M. caseolyticus* and *M. freudenreichii*. According to the authors, these micrococci play a decisive role in developing the typical flavour of Brick cheese, because cheeses containing only *Br. linens* in smear did not develop a flavour typical of this cheese.

Mulder *et al.*[16] stated that 90% of a microflora of Limburger cheese was constituted of coryneform bacteria. The dominant group was grey–white bacteria; orange-coloured bacteria constituted 9–24% and micrococci amounted to 3–6% of the total bacterial count (Table II).

Accolas *et al.*[1] studied changes in the smear microflora of Gruyère and Beaufort cheeses. In the smear, yeasts, coryneforms (90–95% of the total count), micrococci and Gram-negative bacilli were present. They also observed the presence

TABLE II
Occurrence of Coryneform Bacteria on Limburger Cheese Surfaces (from Mulder *et al.*[16])

| Cheese | Total plate count/g of rind | Coryneform bacteria (%) | | | Micro-cocci (%) | Other types (%) |
		Orange	Yellow	Grey–white		
Limburger I	$1·9 \times 10^{10}$	9	2	80	3	6
Limburger II	$2·6 \times 10^{10}$	24	3	65	6	2

(Reproduced with permission of *Journal of Applied Bacteriology.*)

TABLE III

Total Counts and Groups of Microorganisms in the Surface Layer of Limburger Cheese During Ripening (from El-Erian[17])

Age of cheese[a]	Total count per gramme of scraped surface material	Number of strains examined	Type of organism (as % of the total count)						
			Lactic acid bacteria	Yeast	Arthrobacter	Br. linens	Other coryneforms	Sarcina	Mould
Fresh cheese	$4 \cdot 20 \times 10^{7}$	66	100·0	—	—	—	—	—	—
After salting, days									
5	$2 \cdot 41 \times 10^{7}$	72	19·4	20·8	30·6	15·3	11·1	2·8	—
9	$1 \cdot 48 \times 10^{8}$	84	—	100·0	—	—	—	—	—
14	$4 \cdot 41 \times 10^{8}$	76	—	52·6	47·4	—	—	—	—
20	$1 \cdot 81 \times 10^{8}$	70	—	—	91·4	8·6	—	—	—
27	$4 \cdot 70 \times 10^{9}$	68	—	—	80·9	19·1	—	—	—
35	$4 \cdot 60 \times 10^{9}$	80	—	—	73·8	26·2	—	—	—
44	$3 \cdot 17 \times 10^{9}$	75	—	—	62·7	37·3	—	—	—

[a] The first sample was taken just before salting, the second after salting the cheese in the brine overnight. (Reproduced with permission of J. Doorenbos, Mededelingen Landbouwhogeschool, Wageningen.)

TABLE IV

Effect of Light on Pigmentation of Coryneform Bacteria from Different Cheese (from Mulder et al.[16])

Organism	Source[a]	No. of strains examined	Colour of colonies grown in the	
			Dark	Light
Coryneform bacteria from cheese; grey–white strains	Ed, Go, Lei, Limb, Mesh	11	Grey–white	Grey–white; growth reduction in 6 strains
	Go, Mesh	6	Grey–white	Light yellow; growth reduction in 2 strains
	Go, Limb, Mesh	6	Grey–white with pink shade	Grey–white with pink shade; growth reduction in 4 strains
Br. linens[3,5,6,7,8,9]	Culture collections	6	Orange	Orange
Br. linens[1,2]		2	Light cream–yellow	Orange
Orange-pigmented strains	Ed, Go, Ke, Limb, Mam, Mesh, Pe, Rom, St.P	13	Orange	Orange
	Ed, Go, He, Ho, Mam, Marv, Mesh, Mu, Rom, St.P, Vach	16	White	Orange; growth reduction in 1 strain
	Go, Ke	3	Light cream–yellow	Orange

[a] Cheese of the type: Ed, Edam; Go, Gouda; He, Hervse; Ho, Hohenheim; Ke, Kernhemmer; Lei, Leidse kanter; Limb, Limburger; Mam, Mamirolle; Marv, Marville; Mesh, Meshanger; Mu, Munster; Pe, Pénitent; Rom, Romadour; St.P, St. Paulin; Vach, Vacherin Mont d'Or. (Reproduced with permission of *Journal of Applied Bacteriology.*)

of moulds in the smear of cheeses but the massaging of cheeses does not permit their development.

El-Erian[17] isolated 251 bacterial strains from Limburger cheese and classified them into the *Arthrobacter* group, *Br. linens* and other coryneforms. On the basis of detailed studies on changes in surface microflora during ripening of Limburger cheese, he concluded that fresh cheese contained only lactic bacteria (Table III).

After salting, the composition of surface microflora of the cheese was the same as that of the brine. After 5 days of ripening, only yeasts were present on the surface of the cheeses but from the 9th day, the number of yeasts decreased while the Arthrobacter-type bacteria became dominant. *Br. linens* was not observed until as late as the 14th day of ripening but its numbers then increased rapidly. He did not mention the presence of moulds on the surface.

The colour of smear on the surface of cheeses is dependent on its microbiological composition. The curing procedure, thickness of the smear layer and exposure to light influence the colour of the smear. The influence of light on pigment synthesis by bacterial strains isolated from many types of cheese is summarized in Table IV.

On the basis of studies conducted to date, one fact appears certain, i.e., that yeasts are the first microorganisms which develop on the surface of smear cheeses. Subsequently, other microorganisms, especially coryneforms of the *Arthrobacter* and *Brevibacterium* types, may develop on the surface. Therefore, the composition of the smear microflora is not known exactly. Furthermore, it is observed that the interest of research workers in smear cheeses is rather low at present, which is a reflection of the low level of interest in the production of these cheeses.

3 SIGNIFICANCE OF MICROORGANISMS PRESENT IN SMEAR

3.1 Yeasts

Little is known about the role of yeasts in the process of cheese ripening. Previously it was considered that it involved only the lowering of the acidity of the cheese surface which makes the development of other microorganisms, which tolerate a high salt concentration, possible.

Kelly & Marquardt[6] stated that yeasts isolated from the surface of Limburger cheese (two groups, e.g. cultures 65 and 67, which differed in their growth patterns in a liquid medium) may grow at pH 3·5 to 8·5 in the presence of 18–20% NaCl (Table V).

Similar results were obtained by Iya & Frazier[18] who stated that Mycoderma yeasts isolated from the surface of Brick cheeses could grow at pH 3 to 8, in the presence of 15% NaCl. They also observed that yeasts synthesized substances that stimulate the growth of *Br. linens*.

Purko et al.[19] confirmed the findings of Burkholder et al.[20] who reported that *Br. linens* isolated from Limburger cheese, requires pantothenic acid for growth.

TABLE V

Influence of Temperature and pH on the Growth of Yeasts Isolated from Limburger Cheese (from Kelly & Marquardt[6])

pH	Incubation period (days)	Culture 65 — Incubation temperature, °C					Culture 67 — Incubation temperature, °C				
		18	25	30	37	45	18	25	30	37	45
3·5	1	—	—	—	—	—	—	—	—	—	—
	2	+	+	—	—	—	—	—	—	—	—
	3	++	++	—	—	—	—	—	—	—	—
	7	++	++	—	—	—	+	+	—	—	—
4·5	1	—	—	—	—	—	—	—	—	—	—
	2	+	++	—	—	—	—	—	—	—	—
	3	++	++	—	—	—	—	—	—	—	—
	7	++	+++	—	—	—	++	++	++	—	—
5·5	1	—	—	—	—	—	—	+	—	—	—
	2	+	++	—	—	—	+	+	—	—	—
	3	++	+++	+	—	—	+	+	—	—	—
	7	++	+++	++	—	—	++	++	++	—	—
6·5	1	—	+	—	—	—	—	—	—	—	—
	2	++	++	++	—	—	+	+	—	—	—
	3	++	+++	++	—	—	++	++	—	—	—
	7	+++	+++	++	—	—	++	++	++	—	—

7·5	1	—	—	—	—	—
	2	+	+	—	—	—
	3	+	+	—	+	—
	7	++	++	++	—	—
8·5	1	—	—	—	—	—
	2	—	—	—	—	—
	3	—	—	—	+	—
	7	+	+	+	+	—
9·5	1	—	—	—	—	—
	2	—	—	—	—	—
	3	—	—	—	—	—
	7	—	—	—	—	—

7·5	1	—	—	—	—	—	—
	2	++	++	++	++	—	—
	3	++	++	++	++	—	—
	7	++	++	++	++	—	—
8·5	1	—	—	—	—	—	—
	2	—	—	+	+	—	—
	3	+	+	+	+	—	—
	7	++	++	++	++	—	—
9·5	1	—	—	—	—	—	—
	2	—	—	—	—	—	—
	3	—	—	—	—	—	—
	7	—	—	—	—	—	—

— No growth; + First sign of growth; ++ Good growth; +++ Heavy growth. (Reproduced with permission of A.A. Ernstrom, *Journal of Dairy Science.*)

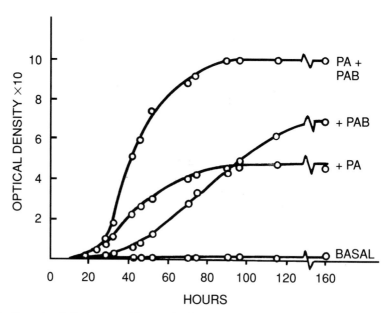

Fig. 1. Growth of *Br. linens* with p-aminobenzoic acid (PAB) or pantotheic acid (PA). (Reproduced from Ref. 19 with permission of A.A. Ernstrom, *Journal of Dairy Science.*)

A similar effect on the growth of *Br. linens* is exerted by the presence of p-aminobenzoic acid in the medium, and the presence of both these acids is particularly favourable (Fig. 1).

Purko *et al.*[21] also showed that several strains of yeasts isolated from Limburger cheese synthesize considerable amounts of pantothenic acid, niacin and riboflavin (Table VI).

Yeasts isolated from the smear of Brick cheese (two groups differing in proteolytic and lipolytic activities) also synthesize substances that stimulate the growth of *Micrococcus caseolyticus, M. freudenreichii* and *M. varians*[15] (Table VII).

However, yeasts also synthesize proteolytic enzymes. The endoproteinases of *Debaryomyces*, and especially of *Trichosporon* yeasts, isolated from the smear of Trappist cheese, hydrolyse casein at pH 5–7, optimum 5·8[22] and also degrade polypeptides released from casein by *Br. linens* enzymes.[23] According to Szumski & Cone,[22] the fact that the pH of the outer layer is close to the optimum for the activity of the endopeptidases of *Debaryomyces* and *Trichosporon* yeasts during the rapid decrease of yeast numbers on the surface of cheese, indicates that these yeasts participate in the ripening process. This is confirmed by the fact that the content of free amino acids in the surface layer of Trappist cheese (6·25 mm) was considerably higher when surface yeasts and *Br. linens* spp. were present than in cheese on which only *Br. linens* was found[24] (Table VIII).

The data present show that the development of yeasts on the surface of the cheese is indispensable for the growth of other smear-forming microorganisms and consequently affects the cheese ripening process.

TABLE VI
Synthesis of Vitamins by Yeasts (from Purko et al.[21])

Yeast culture	Vitamins present (µg/ml)			
	Pantothenate	Niacin	Riboflavin	Biotin
A	12	20	41	0
B	1	10	19	0
C	15	15	49	0
D	15	45	200	0
E	8	15	52	0
F	9	20	53	0
G	15	19	49	0
H	21	19	47	0
I	20	19	43	0
J	27	10	6	0
K	1	20	32	0
L	26	181	47	0
M	3	28	27	0
N	170	104	33	3
O	35	20	55	0
P	49	17	18	2
Q	15	43	16	0
R	15	13	20	0
S	14	40	12	0
T	13	33	25	0
U	44	36	47	0
V	54	40	53	0
W	10	20	50	0
X	190	104	32	4
Y	10	60	22	0
Z	14	16	20	0
AA	2	57	5	0
AB	5	17	16	0
AC	10	40	13	0

(Reproduced with permission of A.A. Ernstrom, *Journal of Dairy Science.*)

3.2 Moulds

The presence of moulds on the surface of smear cheeses seems to be accidental and the correct execution of cheese curing does not permit the development of moulds on the surface of the cheeses. The most common mould on the surface of smear cheeses is *Geotrichum candidum* at 15–25°C and it is capable of growth at up to 10% NaCl and at high/medium acidity, optimum about pH 4.[25]

Geotrichum candidum synthesizes lipolytic enzymes[25-27] with the activity dependent on the strain,[28] acidity and composition of the medium.[29] Synthesis of lipases by *G. candidum* depends on the presence of oxygen; the quantity of water-insoluble fatty acids released from the fat of cream during incubation of *G. candidum* was proportional to the surface area of the cream.[30] The optimum pH for the lipases of *G. candidum* is about 6.[31]

TABLE VII

Effect on Autolysates of Two Film Yeast on the Growth of Micrococci in Skim Milk at Room Temperature (from Lubert & Frazier[15])

Numbers of micrococci per ml. (×1000)

Micrococcus	*Control*		*B6 autolysate*		*B7 autolysate*	
	0 h	*48 h*	*0 h*	*48 h*	*0 h*	*48 h*
Mc 11	280	640 000	190	1 400 000	330	1 300 000
Mc 32	6900	1 100 000	6700	2 200 000	9600	1 700 000
Mf 43	10 000	3 800 000	17 000	7 800 000	14 000	6 400 000
Mv 22	220	490 000	220	1 200 000	190	800 000

(Reproduced with permission of A.A. Ernstrom, *Journal of Dairy Science*.)

TABLE VIII

Average Content of Free Amino Acids Formed in the Surface Layer of Trappist Cheese During Three Ripening Studies (modified from Ades & Cone[24])

	0 days		7 days		14 days		21 days		28 days		35 days	
	1[a]	2[b]	1	2	1	2	1	2	1	2	1	2
					μmoles/10 g cheese							
Lysine	4·288	7·489	9·643	9·422	18·550	13·240	25·300	14·320	24·810	17·280	32·880	25·670
Histidine	0·713	1·398	1·715	1·941	2·203	1·628	2·059	1·439	8·776	1·667	4·825	1·760
Arginine	0·644	2·114	2·152	1·631	11·899	1·598	1·585	1·537	1·790	1·552	2·664	1·254
Aspartic Acid	1·477	3·344	3·000	2·529	3·885	3·549	5·173	4·116	6·225	5·129	9·760	5·934
Threonine	Trace	2·001	Trace	Trace	1·033	0·773	1·134	1·035	2·248	1·198	3·736	1·764
Serine	1·257	4·148	5·890	3·950	5·784	3·690	5·253	4·236	7·362	4·917	13·630	6·452
Glutamic Acid	3·983	7·658	22·670	15·400	33·250	25·530	42·160	30·300	44·650	29·680	53·300	29·220
Proline	2·518	2·910	20·820	20·980	13·820	16·360	10·450	8·385	10·060	11·710	16·390	12·890
Glycine	0·412	2·523	1·805	1·506	4·133	3·478	5·624	3·719	5·757	3·805	7·884	4·158
Alanine	1·787	6·424	9·342	5·069	12·210	6·594	12·890	8·257	16·180	10·740	22·890	15·430
Cysteine	0·109	0·220	0·433	0·231	0·795	0·300	0·587	0·634	0·916	0·913	1·024	0·911
Valine	1·068	3·504	5·057	3·962	11·850	6·457	16·210	7·812	17·500	9·663	24·820	18·480
Methionine	0·072	0·777	0·652	0·591	1·765	1·382	1·808	1·286	1·820	1·255	2·163	1·728
Isoleucine	0·485	2·531	1·415	1·284	4·305	2·469	5·923	3·066	5·177	3·756	9·890	6·120
Leucine	2·107	5·421	11·280	9·450	21·140	12·700	24·530	15·520	21·190	18·600	37·410	24·510
Tyrosine	0·967	1·269	1·670	2·742	2·718	3·161	2·989	2·416	2·788	2·252	3·182	2·717
Phenylalanine	1·609	3·175	5·560	4·823	10·030	8·800	12·380	8·516	17·380	11·610	16·370	12·080
Total	23·496	56·906	103·104	85·511	159·370	111·709	176·055	116·594	188·629	135·727	262·818	171·078

[a] Control cheese with a mixed smear microflora.
[b] Pure cultured cheese containing only Br. linens on the surface.
(Reproduced with permission of A.A. Ernstrom, Journal of Dairy Science.)

Pulss[32] believed that the proteolytic enzymes of *G. candidum* do not participate in the ripening process, the principal influence of this organism on ripening being in lowering of acidity of the cheese surface. However, the studies of Dłuzewski & Bruderer[33] demonstrated that the presence of these moulds on the surface of cheeses affects the increase in the level of water-soluble nitrogenous compounds (Table IX).

TABLE IX

Hydrolysis of Protein in Cheese Ripened With and Without *Geotrichum candidum* **(from Dłuzewski & Bruderer[33])**

Samples	Soluble nitrogen	Amino nitrogen
	as % of total nitrogen	
Without *G. candidum*	2·4	0·77
With *G. candidum*	8·4	1·93

(Reproduced with permission of M. Dłuzewski, *Bull. Polon. Acad. Sci.*)

TABLE X

Proteolytic Activity of Enzymes from *Geotrichum candidum* **(from Gueguen & Lenoir[35])**

Strain	Extracellular activity (μg tyrosine/ml/h)	Intracellular activity (μg tyrosine/ml/h)
D.49	125	29
O.25	150	28
G.36	70	17
G.59	145	37
G.116	255	20
G.117	95	22
G.410	145	23
G.618	60	17
633	615	40
635	530	39
637	235	32
638	165	19
G.802	215	36
G.812	135	32
G.813	135	41
G.816	205	37
G.817	160	38
G.819	615	69
G.820	205	46
G.821	115	26
G.822	160	23
G.823	110	38
G.824	170	24
G.825	90	32
G.826	995	—
G.827	195	28
G.830	615	55
G.832	215	30

(Reproduced with permission of *Le Lait.*)

That the proteinases of *G. candidum* are capable of hydrolysing α- and β-caseins was confirmed by Chen & Ledford,[34] who studied the proteolytic activity of the enzymes from a culture of this mould isolated from the surface of Limburger cheese. Gueguen & Lenoir[35,36] studied the proteolytic activity of many strains of *G. candidum* isolated from the surface of many types of cheese. The optimum pH for casein hydrolysis was 5·5–6·0. The proteolytic activity of individual strains was very different; especially large differences between strains were observed during studies on the proteolytic activities of extracelluar enzymes (Table X).

The authors suggest that the proteolytic activity of a particular strain may be a criterion of their suitability for cheesemaking.

Greenberg & Ledford,[37] who examined the proteolytic activity of the intracellular enzymes of strains of *G. candidum* isolated from the smear of Limburger cheese and from raw milk, stated that these enzymes are capable of deaminating glutamic and aspartic acids; the resulting ammonia reduces the acidity of the medium.

Mourgues et al.[38] examined a suitability of many strains of *G. candidum* for modification of the flavour of Camembert cheese. They found that growth of *G. candidum* on the cheese surface led to a rapid increase in pH and this restrained growth of *Penicillum candidum*.

Vassal & Gripon[39] stated that the presence of *G. candidum* on the cheese surface inhibited growth of *P. candidum*, and this resulted in less intense proteolysis and hence decreased bitterness.

The data presented demonstrate that *G. candidum* cannot be classified as a component of the microflora of smear cheeses. It may, however, develop on the surface of cheeses simultaneously with yeasts and its presence may have an influence on the cheese ripening process.

3.3 Micrococci

Recent investigations indicate that micrococci have a significant influence on the ripening process of cheeses.[40] The majority of authors who have studied the microbiological composition of smear observed the presence of micrococci. Micrococci have been found on the surface and in the interior of many types of cheese but to date, most of these studies have concerned Cheddar[41] and mould-type cheeses.[42]

The addition of starter containing micrococci to a cheese milk affects positively the organoleptic properties of Cheddar cheese.[43] However, it is not the quantity but the type of micrococci that is of practical significance, since not all micrococci influence the organoleptic properties of cheese.[44] Microccoci isolated from Cheddar possess considerable lipolytic and proteolytic activities.[45,46] Moreno & Kosikowski[47,48] studied the activity of enzymes from *M. freudenreichii*, *M. caseolyticus* and *M. candidus* on β-casein. Maximum proteolytic activity was found at pH 5·5 and 22°C. They released large amounts of methionine, which is the precursor of flavouring substances, and therefore they may contribute to the formation of typical Cheddar cheese flavour.

Lubert & Frazier[15] suggest that certain strains of *M. freudenreichii*, *M. caseolyticus* and *M. varians* present in the smear of Brick cheese are decisive factors in the formation of the typical flavour of this cheese (Table XI).

Cheeses inoculated only with *Br. linens* or *Br. linens* and yeasts did not possess the characteristic flavour of Brick cheese. The best results were obtained when the surface of Brick cheese was inoculated with yeasts and a mixture of six strains of micrococci representing the three above species (Table XII).

The importance of micrococci for cheese ripening is well-known and has industrial implications. A French company, Roussel-Uclaf, produces a preparation, Rulactine, for accelerating cheese ripening. The preparation is obtained from cultures of *M. caseolyticus*. Addition of 450 Units of Rulactine per litre of milk intensified proteolysis and modified curd texture.[49] With 450 Units of Rulactine per litre of milk, bitterness occurred after 30 days of ripening. The yield of Saint Paulin type cheese was reduced. Electrophoretograms of curd showed a significant increase of β-casein breakdown (Table XIII).

TABLE XI

Effect of Adding Various Microccoci Plus a Pair of Film Yeasts to the Surface on the Flavour of Brick Cheese (from Lubert & Frazier[15])

Micrococcus added[a]	*Flavour grade*[b]	*Total grade*	*Brick cheese flavour*	*Description of flavour*
Mc 11	4·0	3·7	—	salty, acid, fermented
Mc 32	4·0	3·7	—	salty, sl.bitter, sl.acid, fermented
Control[c]	4·0	3·7	—	salty, acid, fermented
Mc 11	3·0	3·3	—	salty, bitter, acid, yeasty, pleasant
Control	2·3	2·0	—	yeasty, pleasant
Mc 11	3·8	4·0	+	salty, sl.acid, yeasty, sl.fermented
Mc 32	3·2	3·3	+	salty, sl.acid, fermented, glutamic
Mc 58	3·8	3·8	++	salty, bitter, acid, glutamic
Mf 15	3·2	3·5	—	salty, sl. acid, yeasty, sl.fermented
Mf 43	3·2	3·3	+++	salty, sl.acid, sl.unclean, yeasty
Mv 22	3·8	3·8	+	salty, bitter, sl.acid, yeasty, fermented
Control	3·2	3·5	—	salty, acid, sl.unclean, fermented
Mc 11	2·7	2·8	++	salty, sl.acid, sl.yeasty, pleasant
Mc 32	2·6	2·6	+	salty, sl.bitter, sl.yeasty, acetic
Mc 58	2·8	3·8	—	salty, sl.acid, sl.yeasty. sl.unclean
Mf 15	3·0	3·1	—	salty, sl.acid, yeasty, acetic
Mf 43	2·9	2·9	—	salty, sl.acid, yeasty, sl.unclean, fruity
Mv 22	2·7	2·7	—	salty, sl.acid, yeasty, acetic
Control	3·6	3·7	—	salty, acid, yeasty, Limburger, acetic

[a] Cheese previously inoculated with a saline suspension containing both B6 and B7 film yeasts.
[b] 1—excellent,   2—desirable,   3—satisfactory,   4—objectionable,   5—very objectionable, 6—not saleable as original cheese.
[c] Control without added micrococci.
Mc—*Micrococcus caseolyticus*;   Mf—*Micrococcus freudenreichii*; Mv—*Micrococcus varians*.
(Reproduced with permission of A.A. Ernstrom, *Journal of Dairy Science*.)

TABLE XII

Effect of Adding a Mixture of Six Micrococci and a Pair of Film Yeasts together to the Surface on Flavour of Brick Cheese (from Lubert & Frazier[15])

	Flavour grade[a]	Total grade	Brick cheese flavour	Description of flavour
Test	4·0	4·0	+	salty, acid, fermented, sl.Limburger
Control[b]	3·2	3·5	—	salty, acid, sl.unclean, yeasty
Test	2·8	3·8	+	salty, acid, yeasty, sl.fermented, sl.Cheddar
Control	3·6	3·7	—	salty, acid, yeasty, Limburger, acetic

[a] See Table XI for numerical grading.
[b] See Control without adding microccoci.
(Reproduced with permission of A.A. Ernstrom, *Journal of Dairy Science*.)

TABLE XIII

Percentage of Milk Nitrogen Measured in the Curds (from Alkhalaf et al.[49])

Dose of Rulactine (Units/litre)	Control curd	Rulactine	Difference
900	75·33	67·86	7·47
	73·86	67·13	6·73
	74·13	65·14	8·99
450	74·26	72·65	1·61
	74·63	71·93	2·70
	71·98	69·31	2·67
90	73·33	73·06	+0·27
	73·04	72·45	+0·59
	71·68	71·72	−0·04
	73·19	73·63	−0·44
	72·35	71·97	+0·38

(Reproduced by permision of J. Hommel, *Le Lait*.)

At the level of 90 Units of Rulactine per litre of milk no differences were observed between the experimental and control cheeses.

More favourable results were obtained by Pirard *et al.*[50] using Rulactine entrapped in multilamellar visicles (MLV) for the production of Saint Paulin-type cheese. Even when Rulactine was added at a concentration of 1050 Units/litre, no loss in yield was observed. Proteolysis of milk protein was enhanced by Rulactine as compared to that in the control cheeses, and only a faint bitter flavour was detected.

The results presented show that micrococci may be present in the smear. However, their influence on the ripening process in smear cheeses, especially on the development of their characteristic flavour, requires further study.

3.4 Coryneforms

Studies to date have revealed that the most numerous microorganisms in smear are coryneform bacteria[1,16,51] from the genus *Arthrobacter*, in which (a matter of discussion) *Brevibacterium* spp. are included.[52]

TABLE XIV
Effect of Salt on the Growth of Arthrobacter Isolated from Cheese (from El-Erian[17])

Colour of colonies	Number of tested strains	Added NaCl, %							
		0	2	4	6	9	12	16	20
		Number of strains growing							
Cream	62	62	62	62	62	62	22	—	—
Grey–white	42	42	42	42	42	42	8	—	—
Red	36	36	36	36	36	36	12	—	—
Greenish–yellow	33	33	33	33	33	33	33	—	—

(Reproduced with permission of J. Doorenbos, *Mededelingen Landbouwhogeschool, Wageningen.*)

3.4.1 Arthrobacter

Mulder *et al.*[16] isolated 22 strains of *Arthrobacter* from the surface of many types of cheese; they grew, forming grey-white colonies. In the presence of NaCl, they could develop at pH 5·5, i.e. considerably earlier than *Br. linens.*

Arthrobacter organisms appear in high numbers in smear on the surface of Limburger cheese. El-Erian[17] isolated 173 strains of *Arthrobacter* from the smear of this cheese and classified them into four groups according to the colour of the colonies formed during growth. He stated that they may grow at high concentrations of salt (Table XIV).

The presence of such high numbers of *Arthrobacter* spp., which are capable of hydrolysing casein,[16,17,42] in smear undoubtedly has an influence on cheese ripening. Nevertheless, this problem requires further study.

TABLE XV
Occurrence of *Brevibacterium linens* in Different Types of Cheese (from Toolens & Koning-Theune[54])

Cheese type		Peptone meat extract agar, 5·5 % NaCl		LGCS agar
		Total	Br. linens	Br. linens
Camembert du Castel	exterior			$3·4 \times 10^6$
de Beauval	interior			$1·2 \times 10^5$
Camembert 50	exterior			$3·4 \times 10^8$
	interior			$1·4 \times 10^5$
Edelweiss	exterior			$1·7 \times 10^8$
	interior			$1·9 \times 10^6$
Limburger	exterior	$14·4 \times 10^8$	$8·6 \times 10^7$	$7·2 \times 10^7$
	interior	$7·3 \times 10^5$	$3·7 \times 10^5$	$4·9 \times 10^5$
Pont l'Eveque	exterior	$8·6 \times 10^9$	$8·2 \times 10^9$	$7·6 \times 10^9$
	interior	$8·6 \times 10^6$	$8·6 \times 10^5$	$1·3 \times 10^6$
Port Salut	exterior	$3·7 \times 10^8$	$3·7 \times 10^7$	$3·6 \times 10^7$
	interior	$7·7 \times 10^6$	$3·9 \times 10^5$	$2·9 \times 10^5$
Kernhemmer		$1·4 \times 10^8$	$1·2 \times 10^8$	$1·0 \times 10^8$
Roquefort		$2·7 \times 10^8$	10^4	$1·0 \times 10^4$

(Reproduced with permission of H.W. Kay, *Milchwissenschaft.*)

3.4.2 Brevibacterium

The properties of *Br. linens* are well known.[53] This bacterium appears in high numbers in smear on the cheese surfaces. Mulder *et al.*[16] stated that 40% of cultures isolated from the surfaces of various types of cheese show resemblance to *Br. linens*. El-Erian[17] isolated 52 strains from the smear of Limburger cheese and included them in the *Br. linens* group. *Br. linens* was found in the microflora of the following types of cheeses: Beer, Beaufort, Bel Paese, Blue, Brick, Brie, Camembert, Danbo, Edelweiss, Fontina, Gouda, Gruyère de Comte, Harz, Herve, Hohenheimer, Kernheimer, Liederkranz, Limbourg, Limburger, Livarot, Mamirolle, Maroilles, Munster, Penitent, Pont l'Eveque, Port Salut, Romadour, St Paulin, Stilton, Tilsit, Trappist and Vacherin (Table XV).

Br. linens affects decisively the colour of smear, giving it an orange or orange–brown colour. Albert *et al.*[55] noticed that the colour of *Br. linens* colonies during growth depends on the composition of the medium, age of the culture and the presence of oxygen. Mulder *et al.*[16] (Table IV), El-Erian[17] and Crombach[51] observed that some of the strains of *Br. linens* examined were capable of synthesizing orange pigment only in the presence of light. El-Erian[17] reported that among 52 strains isolated from the smear of Limburger cheese, 24 strains of *Br. linens* did not synthesize orange pigment in darkness. In further studies, El-Erian & El-Gamal[56] found that among 150 strains of *Br. linens* examined, 72 synthesized orange pigment only when exposed to light.

In spite of the fact that *Br. linens* tolerates high NaCl concentrations, it may develop on the surface of cheese only when the pH of the cheese is > 5.85[6] (Tables XVI and XVII).

During growth, *Br. linens* synthesizes highly active proteolytic enzymes. Thomasow[57] observed two maxima in the synthesis of proteolytic enzymes

TABLE XVI

Effect of Salt on the Growth of Coryneform Bacteria Isolated from Cheese (from Mulder et al.[16])

Organism	Growth[a]					
	0	3	5	8	12	15
	(Added salt, %)					
Br. linens strain 7	+	+	+	±(+)	0(+)	0(±)
Br. linens strain 6	+	+	+	+	0(+)	0
Br. linens strain 1[b]	+	+	+	+	±	0
Br. linens strain 2[b]	+	+	+	±(+)	0(+)	0(±)
Br. linens strain 3	+	+	+	±(+)	±	0(±)
Br. linens strain 5	+	+	+	+	0(±)	0(±)
Strain 251 orange isolates	+	+	+	+	+	0(±)
Strain 252 from cheese[b]	+	+	+	+	+	0(±)

[a] Growth was recorded after 7 days, but marks in brackets indicate growth after 31 days: 0, no growth; ± slight growth.

[b] Orange in light only.

(Reproduced with permission of *Journal of Applied Bacteriology*.)

TABLE XVII

The Influence of Temperature and pH on the Growth of *Brevibacterium linens* (cultures 4 and 56a) (from Kelly & Marquardt[6])

pH	Incubation period (days)	Temperature, °C				
		18	25	30	37	45
3·5	2	—	—	—	—	—
	3	—	—	—	—	—
4·5	2	—	—	—	—	—
	3	—	—	—	—	—
5·5	2	—	—	—	—	—
	3	—	—	—	—	—
5·85	2		+			
	3		+			
6·0	2		+			
	3		+			
6·5	2	+	+	+	—	—
	3	++	++	++	—	—
7·5	2	+	+	+	—	—
	3	++	++	+	—	—
8·5	2	+	+	+	—	—
	3	++	++	+	—	—
9·5	2	+	+	+	—	—
	3	++	++	+	—	—

— No growth; + Slight growth; ++ Good growth.
(Reproduced with permission of A.A. Ernstrom, *Journal of Dairy Science*.)

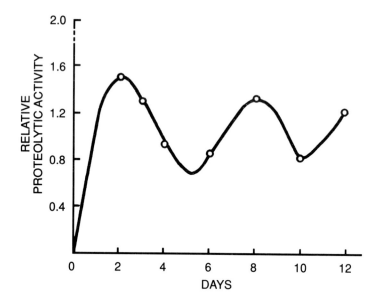

Fig. 2. Extracellular proteolytic activity during growth of *Br. linens* (from Friedman *et al.*[58]). (Reproduced from Ref. 58 with permission of A.A. Ernstrom, *Journal of Dairy Science*.)

during the growth of *Br. linens*. This was confirmed by the studies of Friedman *et al.*[58] who reported intense synthesis of intracellular enzymes up to the 2nd day and from the 6th to the 8th day of growth (Fig. 2).

The proteolytic activity of extracelluar enzymes was many times higher than that of intracellular enzymes (Table XVIII).

Tuckey & Sahasrabudhe[59] reported that the proteinases of *Br. linens* hydrolyse casein intensively, in contrast to yeasts and starter bacteria, and release high concentrations of amino acids (Table XIX).

Foissy[60] observed that *Br. linens* synthesizes from three to six proteinases. Sørhaug,[61] who examined six strains of *Br. linens*, observed the presence of 18 peptidases, three of which were common to all strains.

TABLE XVIII
Activity of Proteinase from *Brevibacterium linens* (from Friedman *et al.*[58])

Enzyme preparations	12% TCA-insoluble Casein	12% TCA-soluble	
		Polypeptides	Peptides, amino acids
Extracellular	3·50	1·99	0·44
Cell extract	0·80	0·07	0·03
Extracellular +	4·00	2·43	0·60

+ Cell extract.
(Reproduced with permission of A.A. Ernstrom, *Journal of Dairy Science.*)

TABLE XIX
Amino Acids (μg/ml) Liberated in Shaken Sterile Skim Milk Cultures Incubated at 60°F (from Tuckey & Sahasrabudhe[59])

Culture	Br. linens 9174	Yeast culture	Starter culture	Starter and rennet extract	Starter and rennet and 2% NaCl
Age (days)	12	39	7	23	52
pH	8·2	7·5	4·45	4·6	4·47
Alanine	184	+	20	37	108
Aspartic acid	765	+	+	—	+
Glutamic acid	298	38	55	48	+
Glycine	344	+	—	—	—
Leucine and Isoleucine	411	+	+	73	233
Methionine	285	—	—	+	119
Serine	94	50	—	—	+
Threonine	133	+	14	—	+
Tyrosine	1500	—	+	+	+
Valine	475	+	+	39	68

+ Present; — absent.
(Reproduced with permission of A.A. Ernstrom, *Journal of Dairy Science.*)

TABLE XX
Effect of Temperature on Five Serine Proteinases from *Br. linens* (Modified from Hayashi et al.[62])

Proteinases	Molecular weight	Specific activity (units mg⁻¹)	Effect of temperature optimum (°C)	Effect of temperature stability (°C)
A	37 000	2·69	40	35
B	37 000	2·70	40	35
C	44 000	3·23	55	45
D	127 000	3·19	55	45
E	325 000	2·66	55	45

(Reproduced with permission of *International Journal of Food Science and Technology*.)

Hayashi et al.[62] examined properties of five serine proteinases, which differed from the protease studied by Tokito & Hosono.[63] The serine proteinases, A, B, C, D and E, exhibited similar enzymatic properties but differed in molecular activity and molecular weight (Table XX).

The enzymes were classified into two groups based on the temperature stability and optimum temperature. The enzymatic activity of proteinases A and B was many fold lower than that of proteinases C, D and E. The optimum pH of the proteinases investigated was 11.

Partly purified proteinase C, and proteinases D and E were used for Cheddar cheese production.[64] More intensive proteolysis was observed in the cheeses containing the proteinases; β-casein was particularly strongly hydrolysed. After two months of ripening at 12°C, the organoleptic properties of the cheese, particularly the flavour, were better than those of control cheeses. The authors suggested that serine proteases can be used to accelerate cheese ripening (Table XXI).

Foissy[65] found that *Br. linens* synthesises one extracelluar aminopeptidase. On the other hand, Hayashi & Law[66] isolated two aminopeptidases from this bacterium.

TABLE XXI
Effect of Serine Proteinases of *Brevibacterium linens* on Flavour Development of Cheddar Cheese (from Hayashi et al.[64])

Amount of proteinase, units/kg	Flavour intensity (lowest 0; highest 8) Pro-C	Flavour intensity Pro-D-E	Bitter defect (lowest 0; highest 4) Pro-C	Bitter defect Pro-D-E	Proteolysis TCA N (% of control) Pro-C	Proteolysis Pro-D-E
0	2·94	2·94	0·1	0	100	100
26	3·88	3·76	0·4	0	125	152
8·6	3·24	3·59	0·4	0	110	148
2·9	3·35	2·76	0·2	0	112	146

(Reproduced with permission of R.L. Richter, *Journal of Dairy Science*.)

TABLE XXII

Deamination of Amino Acids by *Brevibacterium linens* (from Hemme *et al.*[67])

Amino acid	Distribution of strains according to intensity of deamination			
	NH_4 produced ($\mu g.$ 30 min^{-1} mg^{-1})			
	200	51–200	11–50	0–12
Serine	2	7	12	2
Glutamine	0	7	7	9
Asparagine	0	4	11	8
Threonine	0	0	10	13
Arginine	0	1	3	19
Alanine	0	1	7	15
Glutamic acid	0	0	5	18
Lysine	0	0	4	19
Glycine	0	0	2	21

(Reproduced with permission of J.L. Multon, *Sciences des Aliments*.)

Protein degradation, the formation of free amino acids and particularly amino acid metabolism are important factors influencing the organoleptic properties of cheeses. Among many compounds formed during amino acid metabolism are flavour components or their precursors.

Br. linens possesses an ability to decarboxylate lysine, leucine and glutamic acid.[67] The optimum conditions for this activity are 30°C and pH 5–7. As a result, CO_2 and amines are formed.

Tokita & Hasono[68] stated that there were six volatile amines, namely cadaverine, monoethyl amine, monomethyl amine, demethylamine, triethylamine and ammonia, and two nonvolatile amines, tyramine and histamine, in broth after cultivation of *Br. linens*. Tsugo *et al.*[69] reported that cadaverine and putrescine play an important role in cheese flavour.

As a result of amino acid deamination, ammonia is formed and this compound is present in the majority of ripening cheeses.

Hemme *et al.*[67] reported that strains of *Br. linens* isolated from the smear of Comte and Beaufort cheeses very strongly degraded serine, glutamine and asparagine (Table XXII).

The strains investigated also deaminated phenylalanine, tryptophan and histidine. The authors concluded that deaminating activity may be a function of the growth phase, depending on the strain.

Many surface-ripened cheeses exhibit a characteristic 'garlic' aroma which results from sulphur-containing compounds. Bacteria that synthesize degrading enzymes are of paramount importance in the development of the flavour of many cheeses because, as a result of action of methionine demethiolases, methanethiol is formed which is a component and a precursor of many sulphur-containing compounds present in Livarot, Munster, Pont l'Eveque, Comte and Beaufort.[70] *Br. linens* has a fundamental role in the formation of sulphur flavour compounds.

Tokina & Hosono[71] showed the presence of volatile sulphur compounds during cultivation of *Br. linens*. These compounds influence the characteristic flavour of

Limburger cheese. Hemme & Richard[72] isolated 80 different strains of bacteria from the surface of Camembert cheese, 51 of which were corynebacteria. All except two of the strains able to produce methanethiol from L-methionine were orange coryneform bacteria related to *Br. linens*. Ferchichi *et al.*[73,74,75] investigated in detail the degradation of L-methionine by *Br. linens* (Table XXIII).

Br. linens also produces lipolytic enzymes.[76,77] Brandl & Petutschnig[78] demonstrated that there is a relation between the proteolytic and lipolytic activity of *Br. linens*: those strains of *Br. linens* which exhibit strong proteolytic activity also have significant lipolytic activity and vice versa.

TABLE XXIII

Growth of 80 Bacteria Isolated from a Raw Milk Camembert on Minimal Medium Containing L-methionine or α-ketobutyrate, and Methanethiol Production from L-methionine (from Hemme & Richard[72])

Bacterial strains	Number of strains for each type	Number of strains		
		Growing from		Producing methanethiol (capacity higher than 10 nkat/g dry wt)
		L-methionine	α-ketobutyrate	
Coryneform bacteria	51			
Orange colonies	29	26[b]	26[b]	24
White colonies	5	0	0	0
Yellow colonies	17	0	0	0
Staphylococcus xylosus	12	1[c]	1	2[c]
Enterobacteria	7	0	0	0
Moraxella	10	0	0	1

[b] Only 28 strains tested;   [c] including the strain using L-methionine and α-ketobutyrate. (Reproduced with permission of J. Hommel, *Le Lait*.)

TABLE XXIV

Lipase and Esterase Activity of Selected Strains of *Brevibacterium linens* (from Sørhaug & Ordal[79])

Br. linens strain	Specific activity (nonoequivalents of fatty acid released per min per mg dry matter)					
	Substrates					
	Tributyrin		Triacetin		Methyl butyrate	
	Water	5% NaCl	Water	5% NaCl	Water	5% NaCl
ATCC	127	205	13	10	35	45
ATCC	127	226	38	103	35	52
ATCC	62	99	44	59	18	32
ATCC	72	148	50	52	18	41
Isolate from a Limburger cheese	30	69	18	30	17	18

(Reproduced with permission of *Applied Microbiology*.)

Sørhaug & Ordal,[79] who investigated five strains of *Br. linens*, found the presence of cell-bound glycerol ester hydrolase, lipase and carboxylesterase. The enzyme(s) was bound to the polysaccharides present in cell wall structures. Maximum lipase activity was at 35°C and pH 7·6 (H_2O) or pH 8·2 (5% NaCl). The inhibitor profile of the lipase indicated that the enzyme required a free sulphydryl group for activity (Table XXIV).

The above data indicate that particular strains of *Br. linens* or their mixtures present in the smear of cheeses may participate, to varying extents, in the cheese ripening process.

4   INFLUENCE OF SMEAR ON CHEESE RIPENING

Soft smear cheeses are characterized by rapid ripening and extensive proteolysis and lipolysis[17,80–83] which affect, in a decisive way, their consistency and flavour. Many research workers have stated that the development of smear on the surface of cheese has a major influence on the course of ripening, especially on the development of the typical rich aroma of these cheeses. The influence of smear on textural changes during ripening is also clearly visible, especially in the case of soft cheese. After cutting Romadour cheese, it may be seen that the consistency of the outer layer of the cheese, to a depth dependent on the duration of ripening, is softer than at the centre of the cheese.

The influence of the smear becomes especially apparent when changes in the concentration of particular components in different layers of cheese are analysed. As may be seen from Fig. 3, the pH of the smear that develops on Blue cheese increases to about 7·5 on about the 38th day of ripening and then, from the 60th to the 75th day, decreases to 7·0, at which it remains during the rest of the ripening period.

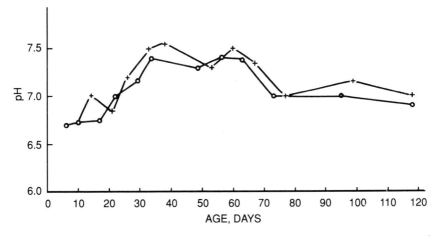

Fig. 3. Changes in the pH of smear from two lots of blue cheese. (Reproduced from Ref. 12 with permission of A.A. Ernstrom, *Journal of Dairy Science*.)

The pH of the smear is considerably higher than that of the surface and central layers of this cheese (Fig. 4).

Similarly, the pH of the smear of Limburger cheese increases to about 7·5 during the second week of ripening, and then remains at this level. As with Blue cheese, the pH of the smear is considerably higher than that of the surface layer of the cheese (Fig. 5).

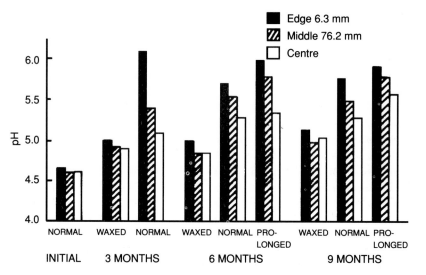

Fig. 4. Changes in pH of the edge, middle and central portions of blue cheese. (Reproduced from Ref. 12 with permission of A.A. Ernstrom, *Journal of Dairy Science.*)

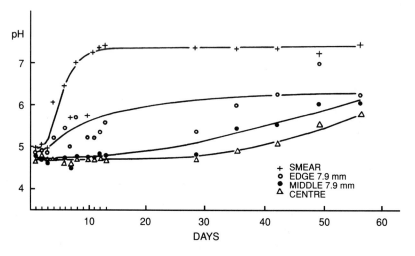

Fig. 5. Changes in the pH of Limburger cheese. (Reproduced from Ref. 6 with permission of A.A. Ernstrom, *Journal of Dairy Science.*)

The fact that the pH of the centre layer of smear cheeses is considerably lower than that of the surface layer points to the influence of smear on the cheese ripening process and its progress from the surface to the interior, i.e. ripening occurs faster in the surface layer of Limburger cheese than at the centre, e.g. the content of free amino acids in the surface layer of the cheese is several times higher than that at the centre (Table XXV). However, the same free amino acids were present in the surface and centre layers of the cheese.[55]

Ramanauskas[84] who studied the ripening process of semi-hard smear cheeses, also observed that smear effected a more intense degradation of proteins in the surface layer, which contained significantly higher concentrations of free amino acids and volatile fatty acids than the centre layer. The differences increased as the fat content of cheeses was increased and cheeses with a higher fast content had a fuller flavour profile.

After three and six months of ripening, the outer layer of Blue cheese, ripened for two months with some smear, was softer than that of dry-rind cheeses.[12] It had a red–orange colour and a nutty-bitter flavour. No differences were reported between the cheeses in respect of their middle and central layers. The cheese ripened for four months with the development of smear had a considerably softer consistency after six months of ripening. The centre of the cheese was red–orange as a result of the migration of pigment from the surface layer. The central layer was creamy, in contrast to the white appearance of the remaining cheeses. This cheese had many organoleptic defects.

After nine months of ripening, the best quality cheeses were those ripened for two months with the participation of smear. According to experts, the best cheeses, characterized by a full, rich aroma, can be obtained only when a smear is allowed to develop for two months on their surfaces.

In cheeses ripened with a smear, the pH of the outer layer was considerably higher than that of the middle and central layers. In cheeses ripened without smear (waxed), the pH of all layers was approximately equal and considerably lower than that of cheeses with smear (Fig. 4). The volatile acidity of the outer layer of smear-ripened cheeses was considerably higher than that of waxed cheeses. Nevertheless, the differences in the levels of amino nitrogen between the particular layers of the cheese were not large, which according to the authors, was a result of mould development inside the cheese (Fig. 6).

Similarly, Hartley & Jezeski[14] expressed the opinion that it is possible to affect the intensity of the aroma of Blue cheese by controlling smear development on its surface.

Brick cheese is a semi-hard cheese, which ripens uniformly throughout, i.e. after cutting, no layers differing in colour and consistency are observed. Langhus et al.,[5] who examined the influence of smear on the ripening of Brick cheese, stated that the increase in the pH of the outer layer from the 3rd day of ripening was a result of smear development. After removal of smear and paraffin-coating of cheeses, a decrease in pH was observed and later a considerably slower increase in pH occurred during further ripening at 4·4–7·2°C. The pH values of the centre and middle layers were similar. When smear was not removed from

TABLE XXV

Amino Acids Liberated During Ripening of Limburger Cheese (mg/g of Cheese) (from Tuckey & Sahasrabudhe[59])

Age (days)	1	11		18		32		46		80		101	
Portion	Whole	Rind	Int.	Rind	Int.	Rind	Int.	Rind	Int.	Rind	Int.	Rind	Int.
pH	4.95	5.25	4.90	5.65	5.0	5.95	5.00	6.2	5.45	7.0	—	7.2	—
Alanine	0.15	0.38	0.13	0.60	0.23	0.39	0.13	0.39	0.29	+	0.80	0.73	—
Aspartic Acid	+	+	Lost	+	+	+	+	0.32	+	0.80	+	+	
Glutamic Acid	0.24	0.47	Lost	1.11	0.21	0.81	0.08	0.62	0.44	1.26	0.96	1.96	
Glycine	—	+	Lost	0.08	+	+	+	0.15	+	0.28	0.14	0.20	
Leucines and Isoleucine	0.19	0.35	0.32	1.39	0.43	1.38	0.50	1.38	1.05	+	2.47	2.42	
Serine	—	+	Lost	0.29	+	0.19	+	0.12	+	0.84	0.20	0.15	
Tyrosine	+	0.52	0.12	0.14	0.19	0.32	0.92	0.58	1.32	+	0.36	0.99	
Valine	0.23	0.28	0.23	0.26	+	0.25	0.16	0.31	0.30	+	0.40	0.71	

Rind—6.3 mm cut from outer surface;   Interior—centre portion;   + present;   — absent.
(Reproduced with permission of A.A. Ernstrom, *Journal of Dairy Science*.)

the surface of the cheeses, i.e. those which were wrapped in foil (film) only, the differences between the pH of the surface and the inner layers were considerably higher. Cheeses wrapped in film were characterized by a richer aroma than paraffin-coated cheeses (Fig. 7).

During ripening, an increase in the level of water-soluble nitrogenous compounds was observed, being intense between the 10th and 17th day of ripening; the rapid increase in the concentrations of these compounds in the outer layer of the cheese was the result of smear development. If the smear was removed, the level of soluble nitrogen in the outer layer was similar to that in other layers.

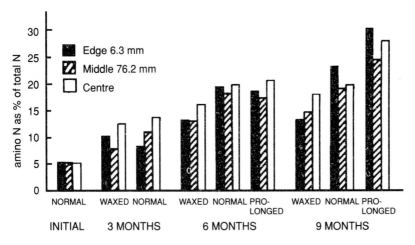

Fig. 6. Changes in amino nitrogen as a percentage of total N in edge, middle and central portions of blue cheese that had been waxed (no smear) or had undergone normal or prolonged smear development. (Reproduced from Ref. 12 with permission of A.A. Ernstrom, *Journal of Dairy Science.*)

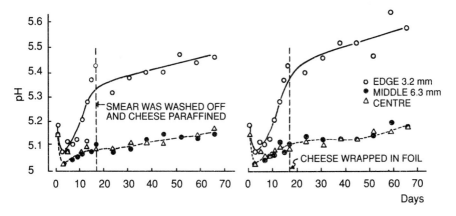

Fig. 7. Changes in the pH of zones of Brick cheese during ripening. (Reproduced from Ref. 5 with permission of A.A. Ernstrom, *Journal of Dairy Science.*)

However, the presence of smear did not influence the level of 80% ethanol-soluble nitrogenous compounds. In cheeses wrapped in film, no differences in the levels of soluble nitrogen between the various layers were observed. The more intense formation of soluble nitrogen was attributed by the authors to a higher temperature (14·4–15·5°C) during further ripening of cheeses after wrapping in film (Fig. 8).

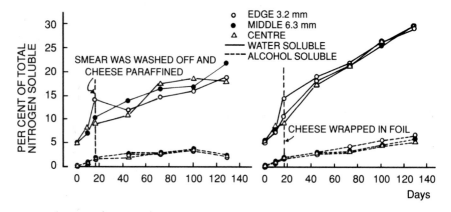

Fig. 8. Percentage of the total nitrogen in Brick blue cheese soluble in water or 80% alcohol. (Reproduced from Ref. 5 with permission from A.A. Ernstrom, *Journal of Dairy Science*.)

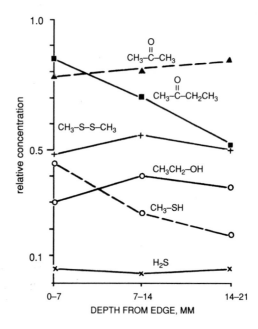

Fig. 9. Volatile compounds in Limburger cheese at various depths. (Modified from Ref. 87 with permission of B.F. Polansky, *Journal of Agricultural and Food Chemistry*. Copyright 1982, American Chemical Society.)

The above data show that the influence of smear proteolysis in Brick cheese is small. However, as stated by the authors, the presence of smear on the surface of the cheese has a decisive influence on the development of the specific flavour of the cheese. Flavouring substances contained in the smear are absorbed by the cheese. Simultaneously, the intensity of Brick cheese aroma can be regulated by controlling smear development on its surface.

Many studies have been conducted, especially on Limburger, with the aim of explaining the basis of the characteristic aroma of smear cheeses.[85-89] Many chemical compounds which participate in the formation of its flavour have been identified. Following examination of the influence of smear on the formation of cheese aroma, Parliment et al.[87] concluded that the concentrations of certain compounds which affect the aroma of Limburger cheese were lower at the surface than in the interior of the cheese; only the levels of methyl mercaptan and 2-butanone were higher at the surface (Fig. 9).

TABLE XXVI
Nitrogen-Containing Volatiles of Swiss Gruyère Cheese (from Linardan et al.[89])

Retention index	Component Name/MS[a]	Relative abundance[b]		
		Edge 15 mm	Middle 20 mm	Centre
1095	m/e[c] 112, 86, 83, 71, 58, 42	60	—	—
1122	m/e 113, 98, 72 (2 isomers)	190	—	—
1127		210	—	—
1318	2,5-Dimethylpyrazine	60	2	—
1327	2,6-Dimethylpyrazine	1000	10	—
1340	Ethylpyrazine (T)	15	—	—
1346	2,3-Dimemthylpyrazine	40	3	—
1385	Ethylmethylpyrazine	60	—	—
1405	Trimethylpyrazine	700	15	—
1446	n.a.	70	—	—
1450	n.a.	50	—	—
1456	m/e 112, 71, 42	45	2	—
1460	n.a.	45	—	—
1476	Tetramethylpyrazine	170	10	—
1515	Ethyltrimethylpyrazine	75	3	—
1591	m/e 112, 81, 54, 43	15	—	—
1600	m/e 112, 71, 42 (2 isomers)	160	—	—
1602	n.a.	85	—	—
1618	m/e 112, 98, 71, 43	30	—	—
1621	m/e 126, 43	80	—	—
1640–1720	m/e 112 or 126	100	—	—
1941	Benzothiazole	12	22	10
2420	Indole	100	7	—

[a] T: tentative identification; n.a.: mass spectrum not available.
[b] Normalized with respect to the abundance of 2,6-dimethylpyrazine in the outer zone sample.
[c] — m/e represents international making typical of the analytical method applied.
(Reproduced with permission of L. Lorenz, *Lebensmittel–Eissenschaft & Technologie*.)

Linardan *et al.*[89] studied the aroma profile of Gruyère cheese ripened with a smear. Among many alkaline volatile compounds present in the surface layer of the cheese only a few, in small quantities, were present in the middle layer, and only one was found in the central layer (Table XXVI).

Le Gret *et al.*[90] observed that during the ripening of Beaufort cheese, calcium, magnesium, zinc and iron moved from the interior to smear. The calcium, magnesium, iron and zinc contents in smear were 41·31, 1·17, 27·0 and 170·0 mg/kg, respectively, while in the middle part of cheese, they were 8·07, 0·24, 2·0 and 45·0 mg/kg, respectively. The authors suggested that the migration of these elements results from the formation of insoluble carbonates or phosphates favoured by high pH and by surface treatment.

The data presented show that microorganisms present in the smear influence the cheese ripening process. This influence is different for each particular type of cheese, the greatest effect being for soft cheeses but the presence of smear on the surface of semi-hard cheeses has a significant influence on the development of the typical flavour of these cheeses also.

5   CONCLUSIONS

During the ripening of cheese, red–orange, smear-forming microorganisms may develop on their surfaces. In the case of certain types of cheese, the so-called smear cheeses, this is a favourable phenomenon because it has a decisive influence on the ripening process and on the development of the specific flavour of these cheeses.

The composition of the smear microflora depends on the microflora of the brine and rooms in which cheeses are ripened and on the correctness of the curing procedure.

The first microorganisms which develop on the surface of cheese are yeasts which tolerate high acidity and high NaCl concentrations. Then, after the acidity has been reduced, coryneform bacteria develop. The red–orange colour of smear is a result of the development of *Br. linens*. Other microorganisms may also grow in the surface smear.

The influence of enzymes secreted by microorganisms in the smear on the cheese ripening process depends on the type of cheese, the composition of the microflora and on the length of time the smear is present on the surface of the cheese.

Soft smear cheeses ripen with the participation of bacterial enzymes from the smear, from the surface toward the interior of the cheese; the influence of smear enzymes is especially evident in a cross-section of these cheeses. In semi-hard cheeses ripening occurs more uniformly throughout but the presence of smear has a definite influence on the development of a specific flavour in such cheeses also.

REFERENCES

1. Accolas, J.P., Melcion, D. & Vassal, L., 1978. *Proc. 20th Intern. Dairy Congress, Paris*, p. 762.
2. Federation International Laitaire, Doc. 141, 1981.
3. Mocquet, G., 1979. *J. Dairy Sci.*, **46**, 133.
4. Foster, E.M., Garey, J.C. & Frazier, W.C., 1942. *J. Dairy Sci.*, **25**, 313.
5. Langhus, W.L., Price, W.V., Sommer, H.H. & Frazier, W.C., 1945. *J. Dairy Sci.*, **28**, 827.
6. Kelly, C.D. & Marquardt, J.C., 1939. *J. Dairy Sci.*, **22**, 309.
7. Yale, M.W., 1940. *New York Agr. Exp. Sta., Tech. Bull.* No. 253, p. 1.
8. Palo, V. & Baćiková, D., 1962. *Proc. 16th Intern. Dairy Congress, Copenhagen*, Vol. D, p. 497.
9. Laxa, O., 1899. *Cent. f. Bakt.*, II Abt., **5**, 755.
10. Kelly, C.D., 1937. *J. Dairy Sci.*, **20**, 239.
11. Macy, H. & Erekson, J.A., 1937. *J. Dairy Sci.*, **20**, 464.
12. Morris, H.A., Combs, W.B. & Coulter, S.T., 1951. *J. Dairy Sci.*, **34**, 209.
13. Yale, M.W., 1943. *New York Agr. Exp. Sta., Tech. Bull.* No. 268, 1.
14. Hartley, C.B. & Jezeski, J.J., 1954. *J. Dairy Sci.*, **37**, 436.
15. Lubert, D.J. & Frazier, W.C., 1955. *J. Dairy Sci.*, **38**, 981.
16. Mulder, E.G., Adamse, A.D., Antheunisse, J., Deinema, M.H., Woldendorf, J.W. & Zevenhuizen, L.P.T.M., 1966. *J. Appl. Bacteriol*, **29**, 44.
17. El-Erian, A.F.M., 1969. *Meded. Landb. Hogesch.*, Wageningen, **69**(12), 1.
18. Iya, K.K. & Frazier, W.C., 1949. *J. Dairy Sci.*, **32**, 475.
19. Purko, M., Nelson, W.O. & Wood, W.A., 1951. *J. Dairy Sci.*, **34**, 874.
20. Burkholder, P.R., Collier, J. & Moyer, D., 1943. *Food Res.*, **8**, 314.
21. Purko, M., Nelson, W.O. & Wood, W.A., 1951. *J. Dairy Sci.*, **34**, 699.
22. Szumski, S.A. & Cone, J., 1962. *J. Dairy Sci.*, **45**, 349.
23. Maginnis, R.L. & Cone, J.F., 1958. *J. Dairy Sci.*, **41**, 706.
24. Ades, G.L. & Cone, J.F., 1969. *J. Dairy Sci.*, **52**, 957.
25. Macy, H. & Gibson, D.L., 1937. *J. Dairy Sci.*, **20**, 447.
26. Tsujisaka, Y., Iwai, M. & Tominaga, Y., 1973. *Agr. Biol. Chem.*, **37**, 1457.
27. Wilcox, J.C., Nelson, W.O. & Wood, W.A., 1954. *J. Dairy Sci.*, **38**, 775.
28. Dłuzewski, M., 1963. *Bull. Polon. Acad. Sci.,* Serie des Sci. Biol., **11**, 227.
29. Nelson, W.O., 1953. *J. Dairy Sci.*, **26**, 143.
30. Purko, M. & Nelson, W.O., 1951. *J. Dairy Sci.*, **34**, 477.
31. Nelson, W.O., 1952. *J. Dairy Sci.*, **35**, 455.
32. Pulss, G., 1955. *Kieler Milch. Forsch. Ber.*, **7**, 385.
33. Dłuzewski, M. & Bruderer, G., 1963. *Bull. Polon. Acad. Sci.*, Serie des Sci. Biol., **11**, 221.
34. Chen, M.H. & Ledford, R.A., 1972. *J. Dairy Sci.*, **55**, 666.
35. Gueguen, M. & Lenoir, J., 1975. *Le Lait*, **55**, 145.
36. Gueguen, M. & Lenoir, J., 1976. *Le Lait*, **56**, 439.
37. Greenberg, R.S. & Ledford, R.A., 1979, *J. Dairy Sci.*, **62**, 368.
38. Mourgues, R., Bergere, J.L. & Vassal, L., 1983. *Tech. Lait.*, **978**, 11.
39. Vassal, L. & Gripon, J.C., 1984. *Le Lait*, **64**, 397.
40. Bhowmik, T. & Marth, E.H., 1990. *J. Dairy Sci.*, **73**, 859.
41. Feagan, J.T. & Dawson, D.J., 1959. *Aust. J. Dairy Technol.*, **14**, 59.
42. Lenoir, J., 1948. Federation International Laitaire, Doc. No. 171.
43. Alford, J.A. & Frazier, W.C., 1950. *J. Dairy Sci.*, **33**, 115.
44. Robertson, P.S. & Perry, K.D., 1961. *J. Dairy Res.*, **28**, 245.
45. Marth, E.H., 1963. *J. Dairy Sci.*, **48**, 869.
46. Nath, K.R. & Ledford, R.A., 1972. *J. Dairy Sci.*, **55**, 1424.
47. Moreno, V. & Kosikowski, F.V., 1973. *J. Dairy Sci.*, **56**, 33.

48. Moreno, V. & Kosikowski, F.V., 1973. *J. Dairy Sci.*, **56**, 39.
49. Alkhalaf, W., Vassal, L., Desmazeaud, M.J. & Gripon, J.C., 1987. *Le Lait*, **67**, 173.
50. Piard, J.C., Alkhalaf, W., El Soda, M., Desmazeaud, M.J., Gripon, J.C. & Vassal, L., 1986. *Microbiol. Alim. Nut.*, **4**, 111.
51. Crombach, W.H.J., 1974. *Antonie de Leeuwenhoek*, **40**, 361.
52. Holt, J.G., 1977. *The Shorter Bergey's Manual of Determinative Bacteriology.* Williams, Wilkins Co., Baltimore.
53. Boyaval, P. & Desmazeaud, M.J., 1983. *Le Lait*, **63**, 187.
54. Toolens, A.P. & Koning-Theune, W., 1970. *Milchwissenschaft*, **25**, 79.
55. Albert, J.O., Long, H.P. & Hammer, B.W., 1944. *Iowa Agr. Exp. Sta., Res. Bull.* No. 328, p. 235.
56. El-Erian, A.F.M. & El-Gamal, S., 1975. *Proc. 3rd Conf. Microbiol., Cairo*, p. 73
57. Thomasow, J.T., 1950. *Kieler Milch. Forsch. Ber.*, **1**, 35.
58. Friedman, M.E., Nelson, W.O. & Wood, W.A., 1953. *J. Dairy Sci.*, **36**, 1124.
59. Tuckey, S.L. & Sahasrabudhe, M.R., 1957. *J. Dairy Sci.*, **40**, 1329.
60. Foissy, H., 1974. *J. Gen. Microbiol.*, **80**, 197.
61. Sørhaug, T., 1981. *Milchwissenschaft*, **36**, 137.
62. Hayashi, K., Cliffe, A.J. & Law, B.A., 1990. *Int. J. Food Sci. Technol.*, **25**, 180.
63. Tokita, F. & Hosono, A., 1972. *Jap. J. Zootech. Sci.*, **43**, 39.
64. Hayashi, K., Revell, D.F. & Law, B.A., 1990. *J. Dairy Sci.*, **73**, 579.
65. Foissy, H., 1978. *Milchwissenschaft*, **33**, 221.
66. Hayashi, K. & Law, B.A., 1989. *J. Gen. Microbiol.*, **135**, 2027.
67. Hemme, D., Bouillanne, C., Metro, F. & Desmazeaud, M.J., 1982. *Sci. Aliments*, **2**, 113.
68. Tokita, F. & Hosono, A., 1968. *Milchwissenschaft*, **23**, 690.
69. Tsugo, T., Matsuoka, K. & Hirata, H., 1966. *Proc. 17th Intern. Dairy Congr., Munich*, Vol. D, p. 275.
70. Hemme, D., Ferchichi, M. & Desmazeaud, M.J., 1986. *Ind. Alim. Agri.*, **103**, 318.
71. Tokita, F. & Hosona, A., 1968. *Jap. J. Zootech. Sci.*, **39**, 127.
72. Hemme, D. & Richard, J., 1986. *Le Lait*, **66**, 135.
73. Ferchichi, M., Hemme, D., Nardi, M. & Pamboukdjian, N., 1985. *J. Gen. Microbiol.*, **131**, 715.
74. Ferchichi, M., Hemme, D. & Bouillanne, C., 1986. *Appl. Environ. Microbiol.*, **51**, 725.
75. Ferchichi, M., Hemme, D. & Nardi, M., 1987. *Appl. Environ. Microbiol.*, **53**, 2159.
76. San Clemente, C. L. & Wadhera, D.V., 1967. *Appl. Microbiol.*, **15**, 110.
77. Stadhouders, J. & Mulder, W., 1959. *Neth. Milk Dairy J.*, **13**, 291.
78. Brandl, E. & Petutschnig, K., 1972. *Öst. Milchwisschenschaft*, **17**, 3, 17.
79. Sørhaug, T. & Ordal, Z.J., 1974. *Appl. Microbiol.*, **27**, 607.
80. Schober, R., Christ, W. & Enkelzann, D., 1957. *Milchwissenschaft*, **14**, 206.
81. Singh, S. & Tuckey, S.L., 1968. *J. Dairy Sci.*, **51**, 942.
82. Tokita, F. & Hosono, A., 1968. *Milchwissenschaft*, **23**, 758.
83. Woo, A.H., Kollodge, S. & Lindsay, R.C., 1984. *J. Dairy Sci.*, **67**, 874.
84. Ramanauskas, R., 1978. *Proc. 20th Intern. Dairy Congr., Paris*, p. 777.
85. Dumont, J.P., Degas, Ch. & Adda, J., 1976. *Le Lait*, **56**, 177.
86. Grill, H., Patton, S. & Cone, J.F., 1966. *J. Dairy Sci.*, **49**, 409.
87. Parliment, T.H., Kolor, M.G. & Rizzo, D.J., 1982. *J. Agric. Food Chem.*, **30**, 1006.
88. Simonart, P. & Mayaudon, J., 1956. *Neth. Milk Dairy J.*, **10**, 261.
89. Liardan, R., Bosset, J.O. & Blanc, B., 1982. *Lebensm. Wiss. u. Technol.*, **15**, 143.
90. Le Gret, Y., Brule, G., Maubois, J.L. & Oeuvrad, G., 1986. *Le Lait*, **66**, 391.

6

Iberian Cheeses

A. Marcos and M.A. Esteban

Departamento de Ciencia y Tecnología de los Alimentos, Faculdad de Veterinaria, Universidad de Córdoba, E-14005 Córdoba, Spain

1 Introduction

In the Iberian Peninsula, as in other Mediterranean lands, herds of sheep are abundant and widely distributed. The most representative cheese varieties of Spain and Portugal are manufactured from ewes' milk. These are 'Queso Manchego', originally made by shepherds in the central plain of La Mancha, and 'Queijo Serra', produced in farmhouses in the Serra da Estrela mountains by curdling milk with extracts from thistle ('cardo') flowers.

Cheesemaking from cows' milk is concentrated in the North of the Peninsula, although insular production (on the Balearic Islands and Azores Archipelago) is also very important. Goats' milk cheese is made mainly in the South and Southeast of the Peninsula, as well as in the Canary Islands.

Autochthonous cheese varieties of major commercial importance have been subject to industrial manufacture along with other foreign varieties, processed cheeses and several fresh dairy products. Edam cheese, popularly known by the Portuguese as 'Flamengo' cheese due to its Dutch origin, and by Spaniards as 'Bola' cheese due to its spherical shape, is the major foreign variety produced in Iberian countries.

Both countries have strong local cheesemaking traditions. Many varieties are still made for consumption on the producing farm or for direct trade at local markets within their production area. Measures are being taken to save traditional cheesemaking practices and the uniqueness and authenticity of their produce.

2 Spanish Cheeses

Spain occupies an intermediate position in cheese production among European countries.[1] In 1989, total national production amounted to about 200 000 tonnes, almost 90% of which was produced industrially. Industrial cheese production

TABLE I
Industrial Production of Spanish Cheese in 1989[a]

Type of milk used	Production (tonnes)	Percent of total production
Cows'[b]	96 694	57
Sheep's[c]	42 428	25
Goats'	1836	1
Processed	29 749	17

[a] Excluding production on the Canary Archipelago.
[b] Including mixtures with sheep's and goats' milk.
[c] Including mixtures with goats' milk.
Source: MAPA,[2] Manual de Estadística Agraria 1990.

from various types of milk—excluding those of the Canary Islands—is listed in Table I.[2] Farm production of cheese, which is rather difficult to assess, is based mainly on sheep's and goats' milk, to which variable proportions of cows' milk are frequently added.[2]

Although there are over 100 autochthonous cheese varieties in Spain, 84 of which are officially recognized[3,4] and six of which are protected by Spain's Denomination of Origin (DO),[5–10] a number of cheese varieties are imported from different European countries.[2,11,12] The manufacture (Table II) and final features (Table III) of cheeses with DOs are controlled by their respective Regulator Councils.

The annual consumption of cheese in Spain *per capita* is moderately high, amounting to 6 kg in 1989,[13] of which 3 kg were hard cheese, 2 kg were fresh cheese and 1 kg was processed cheese.

Essential steps of traditional manufacturing methods for 36 cheese varieties, major features of the cheeses ready for consumption and other items of information of general interest were compiled in a national catalogue, the second edition of which was published in 1973.[14]

Data on a large number of chemical and two physical parameters of the principal varieties sold at market and obtained in a comprehensive systematic analytical study aimed at establishing physico-chemical correlations were compiled and reported by Marcos *et al.*[15] in 1985.

Most research papers dealing with cheese chemistry, physics and microbiology published up to 1986 were reviewed in the first edition of this chapter, which has been revised and updated to the end of 1990 by including new scientific and technological information.

2.1 Tables of Chemical and Physical Features

The following tables give average values and standard deviations of chemical composition (Table IV), salt and mineral contents (Table V), major esterified and total free fatty acid contents (Table VI), levels of insoluble and soluble

TABLE II

Essential Requirements for the Manufacture and Ripening of Cheese Protected by Spain's Denomination of Origin

Cheese name	Type of milk	Curdling	Cutting/heating	Moulding/pressing	Salting	Ripening[a]
Cantabria[8]	Cow, Friesian pasteurized; Cantabrian country	Lactic starter, animal rennet,[b] 30°C for 40 min	Grains 5 mm, heating at 34°C	Cylindrical or parallelepipedal moulds, pressing 24 h maximum	Brining for 24 h maximum	1 week minimum
Mahón[7]	Cow, Friesian, Brown Swiss or autochthonous breed, pasteurized or raw; Minorca Island	Lactic acid curd, rennet, 30–37°C for 30 min or longer	Cutting and draining at 30–37°C	Parallelepipedic moulds, 5–9 cm high, pressing	Brining (18–20% NaCl) for 48 h at 10–15°C	2 months minimum
Manchego[6]	Sheep, Manchega pasteurized or raw; La Mancha region	Lactic starter, natural rennet,[b] 28–32°C for 45–60 min	Grains 5–10 mm, stirring, gradual heating up to 40°C maximum	Cylindrical hoops, 8–12 cm high, 18–22 cm diameter, pressing	Brining for 48 h maximum, dry-salting or both	2 months minimum
Roncal[5]	Sheep, Rasa or Lacha; Roncal valley of Navarre	Lactic acid curd, rennet, 32–37°C for 1 h minimum	At 32–37°C	Cylindrical hoops, 8–10 cm high, variable diameter, pressing	Dry salting or brining for 48 h maximum	4 months minimum
Idiazábal[9]	Sheep, Lacha and Carranzana; Basque country	Lactic acid curd, animal rennet,[b] 25–35°C	Grains 5–10 mm, stirring and optional heating up to 40°C maximum	Cylindrical hoops, 8–12 cm high, 10–30 cm diameter, pressing	Brining for 48 h maximum or dry-salting	2 months minimum
Cabrales[10]	Binary and ternary mixtures of cow, sheep and/or goat, raw; Cabrales district	Lactic acid curd, rennet, 22–35°C for 1 h minimum	Grains 10–20 mm	Cylindrical hoops, 7–15 cm high, variable diameter, slight pressing	Outer (and inner) dry-salting (2–3 g NaCl per 100 g curd)	2 months minimum, mountain caves

[a] See some required important features of the finished products in Table III.
[b] Or other permitted coagulants.

TABLE III
Some Features of the Finished Cheeses from Table II

Cheese name	Type of consistency	Interior	Exterior	Weight (kg)	F/DM[a] (minimum %)	Water (maximum %)
Cantabria[8]	Fresh to semi-soft	Creamy, pale yellow, compact, no openings, characteristic flavour and taste	Soft-rind, ivory–yellow	0·4–2·8	45	—
Mahón[7]	Fresh to semi-hard	Firm, ivory–yellow, a few round eyes of variable size, characteristic flavour and taste	Dry-rind, greasy, compact, yellow to brown–yellow	1–4	38	50
Manchego[6]	Semi-hard to hard	Firm, compact, white to ivory–yellow, no holes or many small ones, characteristic aroma and taste	Hard-rind, pale yellow or dark green by surface moulds	2–3·5	50	45
Roncal[5]	Hard	Hard, white or pale yellow, pores, without eyes, slight piquant aroma and flavour	Natural rind, thick, rough feel, greasy, brown or straw yellow, with or without surface moulds	Variable	60	40
Idiazábal[9]	Hard	Firm, compact, white ivory yellowish, few small eyes, *sui generis* odour, delicate, smoky flavour	Hard-rind, smooth, shiny, pale yellow, dark brown when smoked	0·5–3·5	45	45
Cabrales[10]	Soft by internal moulds	Compact, but spreadable, without openings, white with blue-green veins, slightly piquant taste	Natural rind, soft, thin, sticky, grey with yellow-reddish patches	Variable	45	30 minimum

[a] F/DM = g fat/100 g dry matter.

TABLE IV

Chemical Composition (g/100 g), Means, Standard Deviations and Metabolizable Energy (per 100 g) of Some Major Spanish Cheeses Purchased from 1982 to 1984 at Large Supermarket Chains

Cheese variety	Moisture	Fat	Protein	Lactose	Lactic acid	Ash	kcal	M/FFC[a]	F/DM[b]
Burgos	54·0	24·0	16·0	1·6	0·3	2·7	286	71·0	52·1
	4·2	*3·0*	*1·8*	*0·1*	*0·1*	*0·5*	*31*	*3·0*	*2·9*
Arzúa	46·6	28·0	19·4	—	1·0	3·5	329	64·7	52·3
	2·6	*3·1*	*2·7*		*0·2*	*0·5*	*26*	*1·1*	*3·4*
Mahón	31·7	32·6	26·9	—	1·7	6·8	400	47·0	47·8
	2·9	*1·9*	*2·6*		*0·2*	*0·9*	*14*	*4·4*	*3·3*
Manchego	37·5	33·6	23·0	—	1·4	4·6	394	56·6	53·8
	1·6	*2·0*	*2·6*		*0·2*	*0·4*	*18*	*2·7*	*3·2*
Castellano	30·4	37·3	25·5	—	1·7	4·6	438	48·3	53·6
	6·1	*5·2*	*3·6*		*0·4*	*0·7*	*44*	*7·4*	*5·9*
Zamorano	32·0	36·3	25·3	—	1·6	4·7	428	50·2	53·4
	2·7	*1·8*	*2·1*		*0·3*	*0·6*	*15*	*4·0*	*3·0*
Roncal	29·4	38·8	24·7	—	1·6	4·8	447	48·1	54·9
	1·5	*1·5*	*2·9*		*0·1*	*0·3*	*10*	*1·9*	*1·6*
Idiazábal	33·2	37·8	23·3	—	1·7	4·0	433	53·2	56·6
	4·8	*3·2*	*1·3*		*0·2*	*0·4*	*32*	*4·9*	*1·0*
Majorero	45·3	23·4	25·2	—	1·8	5·0	311	59·1	42·6
	4·6	*4·2*	*1·7*		*0·3*	*0·8*	*38*	*3·8*	*4·9*
Cabrales	41·8	32·6	21·5	—	2·2	5·8	379	62·0	56·0
	1·8	*1·0*	*2·0*		*0·3*	*0·7*	*16*	*2·0*	*0·9*

[a] M/FFC = g moisture/100 g fat-free cheese.
[b] F/DM = g fat/100 g dry matter.
Source: Ref. 15.

TABLE V
Concentrations of NaCl, Macrominerals and Principal Microminerals (per 100 g) in the Cheeses from Table IV

Cheese variety	NaCl (g)	Ca (mg)	P (mg)	Na (mg)	K (mg)	Mg (mg)	Zn (µg)	Fe (µg)	Cu (µg)	Mn (µg)
Burgos	0·54	622	385	222	93	21	2417	613	75	167
	0·33	75	67	97	44	3	517	216	24	190
Arzúa	1·67	559	394	547	55	16	2316	630	72	65
	0·46	57	29	121	9	2	266	406	15	13
Mahón	4·29	559	478	1274	143	21	3078	388	71	56
	0·45	88	41	224	37	1	696	83	9	10
Manchego	2·39	685	544	670	80	59	2376	544	102	98
	0·38	131	160	82	15	89	250	67	18	24
Castellano	2·03	626	566	603	88	23	2667	712	94	107
	0·65	59	66	178	15	3	465	212	29	23
Zamorano	2·24	615	534	661	98	22	3111	602	94	150
	0·25	166	137	57	10	3	515	139	4	31
Roncal	2·37	753	534	658	89	22	2895	766	84	122
	0·41	29	31	87	12	1	336	196	10	9
Idiazábal	1·57	757	522	443	77	21	2491	466	84	107
	0·36	68	32	117	4	3	198	108	27	39
Majorero	2·14	727	501	878	153	29	1558	634	95	65
	0·63	85	60	222	37	7	365	167	8	30
Cabrales	3·70	358	379	1067	95	16	2324	500	66	77
	0·24	89	54	136	7	1	223	178	7	42

Source: Ref. 15.

TABLE VI
Concentrations of the Principal Esterified Fatty Acids (g/100 g of Total Fatty Acids) and Free Fatty Acids (FFA) in the Cheeses from Table IV

Cheese variety	4:0	6:0	8:0	10:0	12:0	14:0	14:1	15:0	16:0	16:1	17:0	18:0	18:1	18:2	18:3	FFA[a]
Burgos	4·3	3·4	2·7	7·2	4·4	11·2	1·1	1·0	26·7	3·1	0·7	8·5	21·1	2·7	1·0	—
	0·8	0·6	0·7	2·4	0·5	0·6	0·4	0·2	2·5	0·2	0·2	0·7	1·1	0·3	0·3	—
Arzúa	3·4	2·4	1·4	3·1	3·3	10·9	1·9	1·5	26·8	3·2	1·0	10·4	25·1	2·1	1·5	0·89
	0·3	0·3	0·2	0·5	0·5	0·8	0·2	0·3	1·4	0·5	0·5	1·7	2·0	0·7	0·3	0·24
Mahón	2·5	2·0	1·3	2·8	3·1	10·4	1·8	1·4	24·6	2·8	1·2	12·1	27·0	2·8	1·8	1·40
	0·4	0·4	0·1	0·3	0·4	0·5	0·2	0·4	1·4	0·5	0·3	1·2	0·9	0·8	0·3	0·43
Manchego	2·5	2·9	2·9	9·5	5·3	11·6	0·7	0·9	26·5	3·1	1·0	7·8	20·8	1·5	1·3	1·08
	0·8	0·5	0·5	1·4	0·8	0·9	0·3	0·2	1·9	0·5	0·2	1·5	2·9	0·4	0·2	0·35
Castellano	2·7	2·9	2·8	8·4	4·7	11·5	1·0	2·2	26·5	3·2	1·0	8·2	21·5	1·5	1·4	1·45
	0·3	0·7	0·6	1·7	0·8	0·9	0·4	2·9	2·1	0·2	0·1	1·2	2·6	0·8	0·4	0·70
Zamorano	2·6	3·3	3·1	8·9	5·4	12·0	0·8	0·9	26·3	3·6	1·1	7·6	20·4	1·1	1·2	1·21
	0·3	0·6	0·6	1·5	0·5	0·2	0·6	0·3	1·8	0·5	0·1	0·5	1·3	0·2	0·1	0·34
Roncal	3·0	3·6	3·3	9·5	4·9	10·9	1·1	1·5	21·5	2·9	1·8	8·7	20·7	1·8	2·2	0·82
	0·4	0·5	0·5	1·4	0·5	0·5	0·6	0·4	1·6	0·8	0·5	0·7	2·0	0·5	0·2	0·37
Idiazábal	3·3	3·5	2·8	7·5	4·5	11·5	1·1	1·7	26·6	2·7	1·7	7·3	20·7	1·4	1·2	1·06
	0·3	0·8	0·9	3·0	1·1	1·0	0·6	0·9	2·7	1·5	1·0	0·5	0·5	0·5	0·5	0·48
Majorero	2·3	3·6	4·2	14·1	5·9	10·7	0·2	1·1	24·1	2·2	0·5	5·9	20·0	2·7	0·8	1·25
	0·3	0·1	0·2	0·8	0·3	0·4	0·0	0·1	0·5	0·3	0·2	0·5	0·9	0·6	0·4	0·23
Cabrales	2·0	1·7	1·1	2·4	3·0	9·6	1·1	0·8	32·4	3·1	0·6	9·3	28·6	2·6	1·3	2·57
	0·4	0·2	0·1	0·2	0·2	0·6	0·1	0·2	2·0	0·7	0·2	0·6	2·6	0·3	0·4	0·90

[a]FFA = ml 0·05 N KOH/g cheese.
Source: Ref. 15.

TABLE VII
Nitrogen Distribution in the Cheeses from Table IV

Cheese variety	Quantitative PAGE[a] (Relative percentages)					Nitrogen fractions[b] (per 100 of total N)				UV measurements[c] (mg/100 g cheese)	
	α_s-I	α_s-cn	β-cn	γ-cn	Origin	SN	NPN	FN	NH_3N	SN-Tyr	SN-Trp
Burgos	2·0	38·4	29·1	24·2	6·2	13·6	4·4	3·1	0·5	87	67
1·2	*3·1*	*5·9*	*5·6*	*3·1*	*2·3*	*1·0*	*0·9*	*0·2*	*57*	*15*
Arzúa	12·0	20·9	32·8	27·1	7·2	24·5	12·4	7·1	0·8	124	130
3·5	*1·8*	*2·7*	*3·2*	*2·1*	*2·8*	*1·7*	*2·0*	*0·3*	*55*	*29*
Mahón	8·3	23·5	25·7	37·5	3·3	31·1	18·6	7·3	1·4	334	160
1·4	*4·1*	*3·5*	*3·8*	*0·3*	*7·9*	*4·4*	*2·2*	*0·5*	*223*	*61*
Manchego	9·0	21·7	38·6	23·6	7·1	25·9	14·4	5·2	1·1	161	97
3·3	*2·0*	*5·6*	*7·0*	*2·4*	*3·7*	*2·0*	*1·0*	*0·4*	*100*	*25*
Castellano	6·9	22·1	37·4	23·5	10·0	28·6	20·3	9·6	2·3	265	110
2·1	*3·4*	*5·4*	*6·2*	*2·3*	*5·8*	*5·3*	*3·1*	*1·0*	*114*	*9*
Zamorano	5·8	20·1	37·6	26·9	9·5	27·9	19·8	8·3	2·4	226	115
2·5	*2·8*	*5·6*	*5·9*	*1·5*	*6·2*	*4·8*	*1·8*	*0·8*	*61*	*25*
Roncal	9·9	28·2	34·0	22·7	5·4	26·2	19·6	8·9	2·7	322	89
0·8	*2·9*	*1·1*	*2·7*	*0·9*	*5·8*	*4·2*	*2·3*	*0·6*	*36*	*10*
Idiazábal	6·8	20·0	34·7	32·3	6·0	29·0	20·2	8·6	2·3	387	123
3·9	*2·7*	*6·6*	*3·6*	*1·2*	*4·9*	*5·3*	*2·8*	*0·8*	*99*	*26*
Majorero	15·5	14·2	33·0	33·8	3·3	27·4	15·9	5·5	1·0	247	88
3·5	*0·4*	*2·5*	*5·5*	*0·7*	*3·4*	*4·6*	*1·4*	*0·5*	*44*	*24*
Cabrales	8·1	13·2	8·0	66·4	4·2	58·1	55·4	29·6	10·4	440	226
4·9	*2·5*	*4·2*	*7·5*	*1·6*	*6·3*	*7·5*	*5·9*	*2·5*	*163*	*54*

[a] PAGE = polyacrylamide gel electrophoresis; cn = casein; α_s-I = insoluble fragments breakdown from α_s-cn with greater electrophoretic mobility (bovine α_{s1}-I and homologous ovine and caprine fragments); α_s-cn and β-cn represent peptides that migrate with mobilities similar to these proteins but do not necessarily represent the original proteins only. See Section 2.1.
[b] N = nitrogen; SN = soluble N; NPN = non-protein N; FN = formol N; NH_3N = ammonia N.
[c] SN-Tyr = tyrosine from SN; SN-Trp = tryptophan from SN.
Source: Ref. 15.

TABLE VIII
Water Activity (a$_w$) and pH of Cheeses from Table IV

Cheese variety	a$_w$		pH	
	$\bar{x}$	$\pm s$	$\bar{x}$	$\pm s$
Burgos	0·994	0·002	5·94	0·31
Arzúa	0·967	0·008	5·23	0·37
Mahón	0·881	0·028	4·94	0·10
Manchego	0·945	0·008	5·05	0·16
Castellano	0·918	0·031	5·23	0·23
Zamorano	0·923	0·011	5·19	0·23
Roncal	0·919	0·016	5·10	0·12
Idiazábal	0·944	0·016	5·03	0·17
Majorero	0·942	0·014	4·88	0·19
Cabrales	0·887	0·013	5·67	0·53

Source: Ref. 15.

nitrogen (Table VII) and two intrinsically physical parameters (Table VIII) for at least six samples of each of 10 of the most representative cheeses. Some data listed in the tables warrant special comments.

Insoluble nitrogen compounds (Table VII and other tables and figures in ref. 15) were resolved by polyacrylamide gel electrophoresis (PAGE) with 4 M urea. The Amido Black-stained gels were then scanned with a computerized densitometer in order to sort the relative percentages of groups of components found into five regions or zones named, in increasing order of anodic migration at a basic pH (8·4), origin (immobile material), γ-casein, β-casein, α_s-casein and α_s-I. Although the relative mobilities (Rm) of the principal bands were virtually identical

TABLE IX
Correlation Coefficients (and Significance) Between Soluble Tyrosine and Soluble Tryptophan Contents of Cheese (mg/100 g) and the Relative Amounts of Components in the Different Electrophoretic Regions[a]

Region	Soluble tyrosine		Soluble tryptophan	
	—(r)—	—(P)—	—(r)—	—(P)—
α_s-I Fragment	−0·19	(>0·1)	−0·23	(>0·1)
α_s-Casein	0·07	(>0·1)	0·08	(>0·1)
β-Casein	−0·40	(<0·05)	−0·58	(<0·001)
γ_1-Casein	−0·02	(>0·1)	0·06	(>0·1)
γ_2-Casein	0·40	(<0·05)	0·62	(<0·001)
γ_3-Casein	0·63	(<0·001)	0·74	(<0·001)

[a] The three major γ-casein bands are subscripted in decreasing order of electrophoretic mobility (together with each major band minor satellite bands were quantified). See Section 2.1.
Source: J. Dairy Sci.,[12] with permission of the American Dairy Science Association.

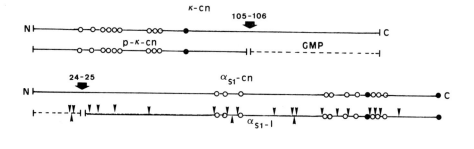

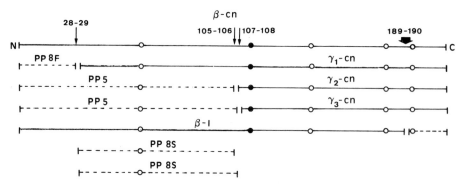

Fig. 1. Hydrolysis of major bovine caseins by chymosin (thick arrows) and by plasmin (thin arrows) to yield insoluble (solid lines) and soluble (dashed lines) polypeptides. The tyrosinyl (white circles) and tryptophanyl (black circles) residues in the primary structure of the casein molecules and their enzymatic cleavage fragments are also shown. Small arrows pointing to α_{s1}-cn fragments indicate other potentially but less sensitive bonds to chymosin, most of which probably are not hydrolysed in cheese—at least during the early stages of proteolysis. (For a more recent nomenclature of caseins and casein fragments released by plasmin see Eigel *et al.*[18])

to those of the α_s-, β- and γ-caseins present in the casein references from the different species, not only the residual caseins were quantified, but also breakdown peptides with Rm values similar to the original α_s-caseins, β-caseins, etc. Therefore, what is denoted by α_s-cn or β-cn may not be the original casein, but peptides with similar mobilities (cf. chapter 9, Volume 1).

Soluble tyrosine and tryptophan contents of cheeses (included also in Table VII) are each highly related to soluble nitrogen in Cheddar cheese[16] and have been used as rapid methods for monitoring cheese ripening. In assorted cheese varieties it was found[12] that neither soluble tyrosine nor tryptophan is related to the hydrolysis of α_s-casein but both are highly related to β-casein hydrolysis (cf. Table IX). Soluble tyrosine, and particularly soluble tryptophan concentrations were higher in cheese varieties where β-casein was extensively hydrolysed (Cheddar, Edam, Emmental, Gruyère, Parmesan, Provolone, Tilsit, Raclette, Mozzarella and blue cheeses).[12] Consequently, such values seem to be specific indexes of the type or stage of proteolysis.[17] The relationships for soluble tyrosine are explained on the basis of the primary structures of the casein molecules

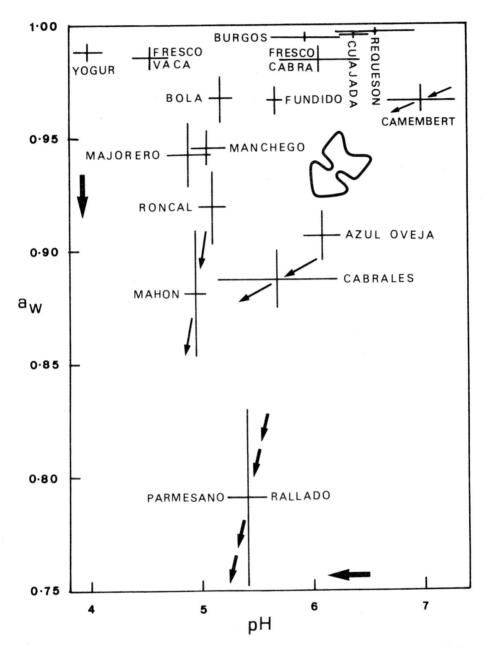

Fig. 2. Stability coordinates (pH vs a_w) of some Spanish cheese varieties and related dairy products. The intercepts of the crosses indicate average values; their branches denote standard deviations. The stability and safety of the products increase from higher values towards lower ones as shown by the arrows (cf. Chapter 11, Volume 1).

and their most sensitive bonds to chymosin and plasmin,[12] as shown in Fig. 1. In fact, the first soluble product released from α_{s1}-casein, namely the casein most precociously and extensively proteolysed by rennet, is an N-terminal polypeptide devoid of both tyrosinyl and tryptophanyl residues, so during the initial phase of proteolysis, the large amounts of soluble nitrogen cannot be measured spectrophotometrically because of the lack of aromatic amino acids (i.e. no relationship can be established). Later, when β-casein starts to be hydrolysed by alkaline milk protease (plasmin), soluble peptides containing tyrosinyl residues (but no tryptophan) start to be released from the N-terminus (PP5 and PP 8S). At this stage, therefore, the amount of soluble nitrogen should be directly proportional to that of soluble tyrosine but not to that of soluble tryptophan. Only in the later phase of proteolysis involving microbial action on primary breakdown peptides are soluble secondary nitrogen products containing both aromatic amino acids produced; however, the direct relationship between soluble nitrogen and tyrosine or tryptophan may be weakened by further enzymatic degradation of these amino acids.

The water activity (a_w) and pH of cheese (see Table VIII) are major physical parameters related to its stability and safety (cf. Chapter 11, Volume 1) and are also the only two that can be readily and commonly measured in commercial cheeses found on the market. The water activity was determined by an isopiestic technique with an accuracy of ± 0.005 a_w units between 0.75–0.99 and an underestimation above 0.99 of less than -0.003 a_w units.[19] The influence of suboptimal values of both pH and a_w on the growth of spoilage or pathogenic organisms and toxin production, tend to be enhanced when acting concomitantly so that the stability and safety of products mapped in Fig. 2 increases as both values decrease, until either a_w or pH reach a *per se* low-enough inhibitory 'upper limit'.

For cheese varieties italicized in the manuscript, data on chemical and physical features are given in Tables IV–VIII.

2.2 Fresh Cheeses and Related Products

Unripe cheeses are those that undergo only partial lactose fermentation yielding lactic acid. All are characterized by high moisture contents (and high a_w) and some residual lactose which contributes to their perishability. Curd cheese ('cuajada') and whey cheese ('requesón') are other fresh and perishable cheese-like dairy products made in dairy factories and distributed together with fresh cheese by refrigerated transport through the same commercial channels.

'Cuajada' is a traditional farmhouse product made from ewes' milk, consumption of which has increased steadily, so that currently it is produced in dairy plants, mainly from cows' milk. According to a quality standard,[20] 'cuajada' is the semi-solid product obtained from heat-treated (HTST or UHT) whole, skimmed or partly skimmed milk, curdled by rennet or other permitted coagulant, without added starter culture or whey drainage. The finished product must contain at least 15% total dry matter. Because the treated milk and the product are exposed to post-pasteurization contaminants, legal regulations allow the addition of sorbic acid, benzoic acid and their Na and K salts up to a total of

2000 mg/kg as preservatives. The presence of additives with aromatic structures (in this and other perishable products) interferes with the UV spectrophotometric measurement of soluble tyrosine and tryptophan.[15]

Fresh cheese manufactured from pasteurized cows' milk to which suitable starter cultures (*Lactococcus lactis* and *L. cremoris*) may be added, is industrially produced in large amounts. Miscellaneous commercial brands found on the market have been analysed.[15]

Burgos cheese and Villalón cheese are the major fresh varieties (traditionally made in old Castile) from raw sheep's milk; both are similar except for their shape. Since legal regulations do not permit consumption of any cheese type made from unpasteurized milk within two months of manufacture,[21] Burgos (low cylinder) and Villalón (loaf-shaped) type cheeses are currently factory-made outside their original loci from mixtures of sheep's and cows' milks. A new approach to detecting mixtures of milks applicable to fresh cheeses, in which proteolysis is rather limited, may be based on the almost non-overlapping ranges of the α_s/β-casein ratio found by Storry *et al.*[22] in cows' (1.78–0.95), sheep's (0.98–0.78) and goats' (0.64–0.28) milk; when applied to data (in Tables VI and VII) for Burgos-type cheese (made from mixtures of sheep's and cows' milks, according to their labels), this procedure worked better than that based on the ranges of the C_{12}/C_{10} ratios of fatty acids in the milk fats from these species.[23-27]

The industrial manufacture of these varieties[28] from pasteurized milk involves heating at 75–78°C for 16–22 s and addition of $CaCl_2$ (0.2 g per litre) followed by rennet with no added lactic starter. Flocculation (at $30 \pm 2°C$) begins within 15 s and after 30 min, the coagulum is cut into cubes (edge length, *ca.* 2 cm). After cutting, the temperature of Villalón curds may be raised to 36°C. On the other hand, the curds of Burgos cheese are scooped into cylindrical hoops and allowed to drain immediately for 4 h, with hourly turning; however, it is not salted, but perhaps slightly brined in 10% NaCl for 30 min at the most, while the curds of Villalón cheese are gently pressed for about 3 h and then brined for 2 h. The fresh cheeses are marketed at refrigeration temperature for consumption within 5 days of manufacture.

The chemical compositions, a_w and pH of Burgos and Villalón varieties have been reported, along with those of other fresh and ripe cheeses.[29] Although the extent of proteolysis undergone by fresh cheeses is obviously insignificant, the nitrogen distribution in Burgos cheese is schematized in Fig. 3 by way of reference for further comparison with bacterial and mould-ripened varieties. The high water activity and pH of these fresh varieties[29] (cf. Fig. 2) make them rather liable to microbial growth. At the retail level, the hygienic quality of cheese is affected by seasonal changes,[30] even though high microbial counts have been reported by García *et al.*[31] for Burgos cheese throughout the year. Thus, staphylococcal thermonuclease was found in one third of the samples at levels up to 2 ng/g with an average of about 300 picograms, although enterotoxins were not detected in any sample; also, coliform and psychrotroph counts were positively related to both the a_w and pH of the cheeses. No enterotoxin was detected even in Burgos cheese experimentally inoculated with enterotoxigenic strains of *Staphylo-*

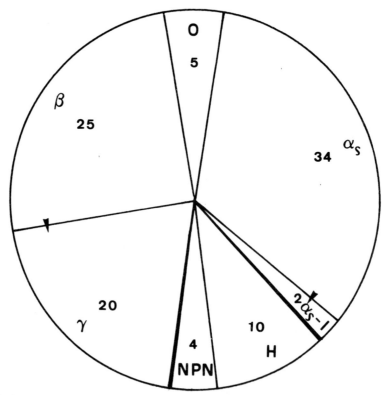

Fig. 3. Sectorial partitioning of nitrogen in Burgos cheese (as percentages of total nitrogen; see Section 2.1). O = Origin (immobile) peptides in PAGE; $\alpha_s = \alpha_s$-cn; $\beta = \beta$-cn; α_s-I = insoluble fragments of α_s-cns homologous to bovine α_{s1}-I (cf. Fig. 1); $\gamma = \gamma$-cns; remaining SN composed of H = soluble high M_r, peptides (cf. Fig. 1); NPN = non-protein N. (Recalculated from data in Table VII.)

coccus aureus and manufactured and stored under abusive temperature conditions.[32] Use of a lactic starter in the manufacture of this variety was suggested as a means of improving its hygienic quality.[33,34] Enterotoxigenic *S. aureus* may reach high numbers in raw sheep's milk,[35,36] but this pathogen is currently considered to be of low risk to cheese because its growth and toxin production can be readily inhibited by using modern lactic starter technology and acidity (pH) control in cheesemaking.[37]

Fresh cheeses from goats' milk[15] are common everywhere,[38] but mainly in Andalusia, where important varieties such as those from Málaga and Cádiz are eaten both fresh and slightly cured; these varieties have been the subject of some chemical and physical studies.[39-42] Andalusian cheese varieties were recently characterized by Montero.[43] Pasteurization of goats' milk is not commonly practised in rural cheesemaking and the fresh product may harbour *Brucella melitensis* which occasionally causes outbreaks of undulant fever when legal and sanitary measures by veterinarian inspection are evaded.

'Requesón' is produced by heat precipitation of whey proteins in acidic media to yield a soft paste.[21] The whey used as raw material in its manufacture is a by-product from the manufacture of pressed varieties. Whey cheese proteins are highly nutritive albumins and globulins from milk; only trace (unquantifiable) amounts of residual caseins were electrophoretically detectable in the samples analysed.[15] The average percentage of soluble nitrogen in whey cheese was inter-mediate between those found in cuajada and in fresh cheeses.[15] The a_w of whey cheese exceeds those of fresh cheese and cuajada[15] (cf. Fig. 2), and even that of fluid milk ($a_w \approx 0.995$). The higher a_w of whey cheese may be due, perhaps, to the thermal denaturation of the whey proteins, which renders them insoluble by re-ducing protein–water interactions and promoting protein-small ion and protein-protein interactions.

In fresh cheeses—and some other related dairy products—the maximum a_w can be calculated from the NaCl concentration in the aqueous phase of the product[44] or graphically estimated from moisture and salt percentages[45] (cf. Chapter 11, Volume 1). For more accurate calculation, the equation of López et al.,[46] which also takes account of pH, must be applied.

2.3 Bacterial Ripened Cheeses

Cheeses ripened by bacteria constitute the main group, both in quantity and in diversity. Varieties within this group comprise those based on cows', sheep's and goats' milk, as well as milk mixtures. Reference will be made to some varieties within each sub-group.

2.3.1 Varieties Made from Cows' Milk

The national production of Dutch-type cheese, particularly Edam and, to a lesser extent, Gouda, exceeds the production of all other cheese varieties made from cows' milk.[47] There is also limited production of other foreign varieties, primarily used as raw materials for processed cheese manufacture.

Indigenous cows' milk varieties, made in significant amounts, are produced in the Northwest of the Peninsula, along the Cantabrian mountain range and in the farthest corner of Galicia, as well as in the Balearic Islands.

Major steps involved in the manufacture and features of Cantabrian cheese with DOs[8] are included in Tables II and III. Whole raw milk with an acidity of not more than 20 Dornic degrees is pasteurized and then curdled with added cultures of L. lactis and L. cremoris. Cheeses from 2 to 2·5 kg may be sold in rectangular portions (0·3–0·7 kg) or radial sections (0·2–0·6 kg) wrapped in shrinkable film or vacuum packed. Four different cheese varieties, named Picón, Ahumado de Aliva, Quesuco and Pido, are also made in Cantabria from cows' milk and mixtures of cows' with sheep's and/or goats' milk, under the generic denomination 'Liébana cheese'.[3]

Descriptions of rural cheesemaking practices traditionally used in Galicia, to-gether with data on chemical composition and counts of some microbial groups in typical varieties (Ulloa, Tetilla, San Simón and Cebrero cheeses) gathered directly from producers and at local markets, were reported in 1961 by Compairé.[48]

The biochemical and microbiological aspects of Ulloa cheese, designed previously by the more generic name of Gallego cheese[14,15,29] and now called *Arzúa cheese*,[3] were investigated by Ordóñez & Burgos.[49-52] They followed proteolysis and the evolution of the major microbial groups during the ripening process of traditionally-made cheese from raw milk: lactic acid bacteria and enterococci increased steadily during the first few days and stabilized afterwards, while staphylococci and micrococci decreased progressively throughout maturation.[49] For cheese manufacture from pasteurized milk, they propose the use of a mixture of *L. lactis, L. cremoris, Lactobacillus plantarum* and *Str. faecalis* var. *liquefaciens* as starter, the last being assumed to play a significant role in the curing process.[50] Free amino acids[51] and a detailed analysis of the lipid composition of curds and cheese made from raw and pasteurized milk with the above starter have been reported.[52]

A similar soft variety, also with a washed rind, is Tetilla cheese, which takes its name from the similarity of its shape to a woman's breast. Its chemical and physico-chemical features[15,29] are virtually identical to those of Arzúa cheese (cf. Tables IV–VIII). The a_w of Ulloa and Tetilla cheeses have been calculated[53] from literature data.[48]

San Simón cheese is a smoked variety now produced industrially in two versions, one with the traditional pear shape and a softer one (about 60% moisture on a fat-free basis)[15] with a flat-disk shape. Since their moisture content is highly variable (ranging at the marketing stage from >40% to <20%), their a_w is closely related both to the moisture content and to the aqueous concentration of ash.[54] The latter relationship may be applied to most natural cheese varieties ripened by bacteria in order to estimate their a_w.[15] Both relationships were used[55] to calculate the a_w of San Simón cheese from previously published compositional data.[48] When applied to reported data,[15] discriminant analysis enabled 100% correct classification of both versions of San Simón cheese, Arzúa cheese and Tetilla cheese.[56]

Mahón cheese takes its name after the capital of Minorca (Balearic Islands), where it is produced. According to DO regulations,[7] its manufacture (Table II) —prohibited in July and August—and ripening is restricted to Minorca Island and cows' milk, although addition of indigenous sheep's milk in proportions up to 5% is currently also allowed. When made from pasteurized milk, the ripe cheese (Table III), of parallelepipedal shape with rounded edges and vertices, can be marketed fresh (after a minimum of 10 days' ripening), half-cured (2 months' ripening), cured (5 months' ripening) or aged (10 months' ripening). The rind is sometimes periodically treated with olive oil. Although sold in the natural form, mainly in the Archipelago, Catalonia and Madrid, most of the overall production is transformed into processed cheese ('queso fundido'), a fraction of which is marketed in powder form as grated cheese ('queso rallado'). Information of the same type as that included in Tables IV–VIII for natural Mahón cheese (with the exception of the fatty acid composition of lipids) was obtained earlier on natural, processed and grated preparations.[57–60] Natural cheese has some peculiar features (cf. Tables IV–VIII), namely: a low moisture content, high levels of salt, soluble nitrogen, non-protein nitrogen, reduced a_w and a C_{16}/C_{18} ratio lower than that found in any other cheeses analysed.[15] Ramos et al.[61] followed proteolysis and the microbial flora during ripening in order to identify a suitable specific starter culture for this variety; the results suggest that the starter should contain L. lactis, Lb. plantarum, Lb. casei and Str. durans in different ratios. Taxonomic studies on yeast[62] and bacterial[63] strains isolated from milk, curd and maturing cheese have also been carried out and selected by Suárez et al.[64] on the basis of acidifying and enzymatic activities to be used later in experimental cheese production. Also, lactic acid bacteria isolated from different batches have been identified to assess their potential for acidification and flavour development.[65] Peptides and free amino acids at several ripening stages have been analysed[66] on three farmhouse batches manufactured from raw milk and one industrial batch manufactured from pasteurized milk with starter added and ripened at 12°C and 97% RH for 4 months. The total content of free amino acids (FAAs) increased throughout ripening except in one batch in which it reached a maximum of 3·3 mg/g dry matter in the third month and then decreased to the typical level reached by the other batches. The prevailing FAAs, which accounted for 67–80% of total FAAs, were Phe, Val, Pro, Glu and Ile. No difference in the FAA pattern was found between traditionally and industrially manufactured cheeses.

Some compositional, physical and microbiological information has been collected on commercial Bola cheese (ball-shaped Edam).[67–69] Data obtained for Edam cheese and Nata cheese[15] were very similar to those found for Bola cheese.[15]

2.3.2 Varieties Made from Sheep's Milk

The main varieties in this group are hard, uncooked cheeses manufactured from whole milk. All have strong flavours and excellent keeping qualities. This type of cheese is produced extensively in both Castilian plateaux and, to a lesser extent, in Navarre and the Basque country. Even smaller amounts of a few semi-soft varieties made from sheep's milk curdled by vegetable rennet are made in farmhouses in the southwest of Spain.

Manchego cheese is named after the La Mancha region, where the original product was traditionally made from raw milk by shepherds. The product became popular and highly appreciated, so its manufacture was spread throughout Spain by transhumance and traditional cheesemaking practices involving sheep's milk were gradually superseded by factory production from pasteurized milk. The growing demand for the product throughout the year, together with seasonal shortages of sheep's milk, compelled the use of variable amounts of milk from various species other than sheep (cows' and, to a lesser extent, goats') and hence of binary or ternary mixtures of milk. All this resulted in diversification of the genuine product as a wide variety of Manchego-type cheeses.

By the end of 1984, the authorities decided on the definitive DO 'Queso Manchego' and established the corresponding legal regulations.[6] In addition to requirements for the manufacture of the cheeses (Table II) and a description of the ripe product (Table III), cheeses with this DO must have been produced in a given area (*viz.* the provinces of Albacete, Ciudad Real, Cuenca and Toledo), the minimum milk composition must include 6% fat, 4% lactose, 4·5% protein, 0·8% ash and 16·5% total dry matter, the methylene blue reduction time must be longer than 3 h and the maximum acidity must be 30 Dornic degrees. Milk for 'artesanal' (craft) varieties is not heat-treated, but for 'industrial' varieties it is pasteurized. Cheeses must be ripened for at least two months after moulding in curing rooms at 12–15°C and 75–85% RH, with turning and cleaning (washing) if needed, and stored below 10°C. They are packed with natural or washed rinds and may be covered with wax or an inert transparent substance, or impregnated with olive oil. The words 'Manchego cheese' and the number of the week of manufacture (1–52) must be printed on the cheese, the rind of which, in turn, must be printed with weave markings ('pleitas') on the sides and typically grooved or striated ('flor') on both flat surfaces. The physical, physico-chemical and microbiological features of the ripe cheese are as follows: pH 5·1 to 5·8; minimum protein in dry matter, 30%; maximum NaCl content, 2·3%. Ranges of indices for the cheese fat are: refractive index at 40°C, 1·4539–1·4557; Reichert-Meissl value, 26–32; Polenske value, 5–8; Kirschner value, 19–27. The maximum cholesterol level must be 98% of the unsaponifiable sterol fraction and the maximum microbiological limits are 1×10^1 cfu *Escherichia coli*/g, 1×10^2 cfu *Staphylococcus aureus*/g and no *Salmonella-Shigella* in 25 g.

The cheesemaking process used by a major industrial producer of Manchego cheese with a DO is basically as follows: sheep's milk, collected over two consecutive days from herds in the demarcated area and with an acidity of 20°D, is checked for the absence of milks of other species in amounts detectable by radial immunodiffusion,[70] cleaned by filtration and centrifugation, cooled to 6°C and stored in isothermal tanks with agitation for 16 h. The refrigerated milk is pasteurized (73 ± 2°C for 20 s), cooled to 30 ± 1°C and delivered to the vats (capacity 8000 litres); after filling (*ca.* 20 min), 1 litre of microbial rennet from *Mucor miehei* (strength 1 : 30 000) is added to the vat together with 80 litres of a mixed starter culture of mesophilic lactic acid streptococci (*L. lactis, L. cremoris* and *L. lactis* subsp. *diacetylactis*), to yield a soft curd, with no added $CaCl_2$, in 35 min, allowed to stand at 28°C. Then the coagulum is cut into small grains the size of a pea. The curd particles are heated to 37°C in 20 min and then stirred for another 30 min. After removal of the whey from the vat, the grains are transferred to a curd strainer and the beds of strained curd are cut into cubic blocks, each of which is placed in a cylindrical PVC hoop, lined with a smooth cloth, in which the curd is moulded and pressed pneumatically at 3 atm for 5 h. The cloths are then removed, and the curds pressed again at the same pressure for 17 h, after which they are immersed in a circulating brine tank (22° Brix) at 14°C for about 36 h. The fresh salted cheeses are placed in a drying room at 14°C and 85% RH, where they are stored, with periodic turning, for 10 days, after which they are transferred to a curing chamber at 9°C and 95% RH. After 2, 6 or 12 months, depending on the target market outlet, the cured cheeses are brushed and polymer-coated with a polyvinyl acetate emulsion containing an antimycotic agent.

Román[71] studied the physical and chemical events that occur during the ripening of Manchego cheese with DO and manufactured under the rigorously controlled conditions described above. He monitored throughout one year the dynamic changes in a 2-cm thick subcortical layer (mantle) and in the core (accounting for about 40% of the whole edible mass) of 17 cheeses, randomly selected from the same manufacturing batch in search for physico-chemical relationships.

During the first three weeks, both parts underwent a rapid constant loss of moisture from 40·5% to 35% and from 45 to 39% in the outer and inner parts, respectively (i.e. a drop in moisture content of about 6% in each). The outer layer also underwent a period of slow moisture loss from 35% (third week) to 27% after about 5–6 months of ripening, followed by a stationary phase at this moisture level (equilibrium moisture content) until the end of ripening (12 months), while the inner part lost moisture at a roughly constant rate from 39% (third week) to 32% at the end of the curing period.

The salt (NaCl) level remained constant (1·5%) in the outer layer for 2 months, while the core, initially containing 0% NaCl, reached the level in the outer layer, by diffusion, within one month of salting. After the second month, the salt content of both parts increased slightly and gradually to 2% at 12 months, as a result of drying.

The dynamics of water activity (a_w), which reflect approximately those of moisture loss, were roughly inverse to the variations in salt levels. The water activity of the mantle remained constant at 0·95 for the first 2 months, while that of the core decreased from an initial value of 0·99 to that of the mantle (0·95) between the second and third months, after which the a_w of both parts decreased at roughly identical rates to a final value of about 0·92 (0·91 and 0·93 in the mantle and core, respectively).

The water activity was positively correlated with the moisture content and inversely with the salt-in-cheese moisture ($P < 0.001$).

The sequential changes in the lactic acid content and pH induced by glycolysis were both qualitatively and chronologically similar throughout ripening in both the outer and inner parts. The lactic acid content increased initially from 0·7% in the mantle and 0·8% in the core to about 1·2 and 1·1%, respectively, during the first two weeks and then decreased during the next week to about 1·0 and 0·8%, respectively, after which it increased again to 1·2% in both parts and remained constant until the second month, subsequent to which it increased to 1·7 and 1·5% in the core and mantle, respectively, at the end of ripening. The initial pH (ca. 5·0 on the first day after salting) decreased by 0·1–0·2 units during the first few days and then increased to about 5·1–5·2 towards the third week, decreased again by 0·2–0·3 units by the end of the first month and increased once more, alternately for the inner and outer parts, up to the same final value of 5·75.

Proteolysis is the most significant biochemical process influencing the quality (flavour and texture) of this type of cheese. Inasmuch as the absolute amount of

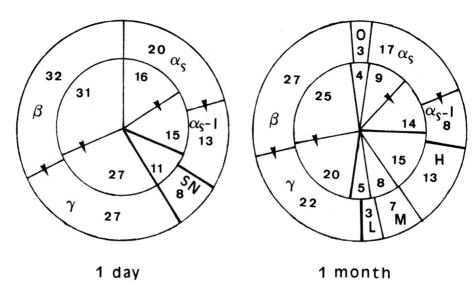

1 day   **1 month**

Fig. 4. Sharp changes in nitrogen partitioning occurring during the first month of ripening of Manchego cheese and proteolytic differences between its inner and outer parts, whose surface areas, bound by the concentric circles, are roughly proportional to the volume ratio (1:1·5) of the core and subcortical samples of the edible part. The meanings of letters assigned to sectors is as follows: O = Origin (immobile) peptides in PAGE; $\alpha_s = \alpha_s$-cn; $\beta = \beta$-cn; α_s-I = insoluble fragments of α_s-cns homologous to bovine α_{s1}-I (cf. Fig. 1); $\gamma = \gamma$-cns; SN = Soluble N composed of H = high M_r peptides (SN − NPN) plus M = intermediate M_r peptides (NPN − FN) plus L = low M_r peptides, amino acids and other compounds (determined by formol titration). The numbers denote percentages of total nitrogen in each sample. (Drawing based on data from Ref. 71.)

total nitrogen (TN) remains quantitatively constant (on a dry basis) throughout ripening, irrespective of other changing variables, this appears to be the most accurate means of expressing the intrinsic changes undergone by caseins for the appearance of various protein breakdown products. The extent of proteolysis increases rapidly during the first month of ripening (see Fig. 4) and then much more slowly in both the inner and outer parts; the soluble nitrogen (SN) level is significantly higher in the core than in the drier mantle throughout the ripening, as is the depth of proteolysis (i.e. the relative proportions of lower M_r nitrogen compounds). Like most other cheeses, cleavage of α_s-caseins by rennet to yield insoluble C-terminal fragments (homologous to bovine α_{s1}-I) and soluble N-terminal fragments (see Fig. 1) is much more rapid and extensive than the hydrolysis of β-casein by plasmin to γ-caseins and polar (soluble) N-terminal fragments (polypeptides PP 5, 8F and 8S). Both insoluble and soluble high M_r polypeptides are subsequently further hydrolysed by the action of bacterial peptidases to intermediate and low M_r peptides, and amino acids, some of which are deaminated or decarboxylated by bacterial enzymes with the production of ammonia, amines, CO_2, etc.

The SN content of the cheeses was found to be highly significantly correlated ($P < 0.001$) with the contents of α_s-casein (r = -0.8) and γ-casein (r = $+0.8$), showing that most of the SN results from the breakdown of α_s-casein and that as the SN content increases, the γ-caseins released from β-casein accumulate (positive correlation) as a result of being resistant to proteolysis.

Table X shows the nitrogen distribution in the outer and inner portions of the cheese at the three main stages of ripening, i.e. tender (two months), cured (six months) and one year old ('añejo').

Ripe cheese can occasionally be preserved for up to 2 years by immersing it in olive oil ('Manchego en aceite'). The ripening of two batches of Manchego cheese made industrially from raw milk and immersed in olive oil after 3 months'

TABLE X

Nitrogen Distribution in the Outer and Inner Zones of the Edible Part of Manchego Cheese[a] at Different Stages of Ripening

Industrial designation Ripening time (months)	Tender (2 months)		Cured (6 months)		Full-cured (12 months)	
			% of TN			
Caseins[b]	α_s-Cn	β-Cn	α_s-Cn	β-Cn	α_s-Cn	β-Cn
Outer zone (1·5 vol)	12	25	10	23	9	20
Inner zone (1·0 vol)	10	24	6	22	9	20
First insoluble C-terminal PP[b]	α_s-I	γ-Cn	α_s-I	γ-Cn	α_s-I	γ-Cn
Outer	12	24	9	28	9	25
Inner	14	27	13	28	7	22
Soluble nitrogen (SN at pH 4·6)						
Outer	22		27		37	
Inner	24		30		42	
Soluble nitrogen distribution	*% of TN*					
High M_r peptides (SN-NPN)						
Outer	11		12		16	
Inner	10		9		15	
Medium M_r peptides (NPN-amino nitrogen)						
Outer	6		7		10	
Inner	9		12		17	
Low M_r peptides, amino acids, etc. (Amino N by formol titration)						
Outer	5		8		11	
Inner	5		9		10	

[a] Recalculated from Ref. 71.
[b] See Section 2.1 and diagram in Fig. 1.

ripening was monitored by Ordóñez *et al.*[72] for 11 months. They found that proteolysis increased gradually both in extent and depth throughout the ripening period, at the end of which 80% of α_s-casein and 10% of β-casein had been hydrolysed and the resulting soluble, non-protein and amino nitrogen amounted to 40–50%, 25–35% and 10–18% of total nitrogen, respectively. Ordóñez & Burgos[73] studied the formation of free amino acids on one batch and found the concentrations of all amino acids increased during curing, except glutamic acid, which levelled off during the fourth month and arginine, which reached a maximum after 6 months and decreased afterwards. The amino acids Lys, Leu, Val, Phe and Ile were those present at the highest final concentrations in the cheese. The fatty acid profiles obtained after immersion in the olive oil for periods of different length were also determined.[74] Cheese preserved in olive oil typically has a very strong flavour and a somewhat piquant taste.

The physical and chemical data listed in Tables IV to VIII correspond to cheeses manufactured—according to the labels—from ewes' milk in La Mancha (*loc cit.* provinces). 'Manchego en aceite' purchased at market (of unknown origin) had a significantly lower moisture content (and a_w) and exhibited more extensive proteolysis and lipolysis and, obviously, had a higher proportion of oleic acid in the total fatty acid content.[15]

Many chemical, physical and technological studies of the so-called 'Manchego cheese' involve cheeses manufactured from milk mixtures outside the La Mancha region, so their results may be equally applicable to homologous hard cheese varieties obtained from ewes' milk or milk mixtures traditionally marketed as Manchego or Manchego-type cheeses. Extensive physical and chemical research has been conducted on the gross composition,[29,67] nitrogen compounds,[75–79] free amino acids[80–82] and other flavour substances,[82] tyramine,[83] histamine,[84] minerals,[85] water activity and pH,[86] chemical and physical parameters related to fat and lipolysis,[87] proteolysis[17,72,73,88–90] and its assessment,[91] the use of ultrafiltration in cheesemaking,[92] accelerated ripening,[93,94] and freezing of curds or 'green cheese' for delayed ripening.[95–98]

The above considerations are applicable to microbiological research,[99–123] which has so far been concerned chiefly with the evolution of the major microbial groups throughout ripening and with the screening of microbial strains isolated from milk, curds and cheeses at different curing stages. Laboratory culture collections of isolated strains have been gathered and used for the identification and/or characterization of their physiological activity and potential biotechnological suitability for development of appropriate starter cultures, some of which have been subjected to experimental trials. The survival of pathogenic organisms such as *Staphylococcus aureus, Salmonella* spp., *Listeria monocytogenes, Enterobacteriaceae,* etc. and the production of toxins (e.g. enterotoxins, aflatoxins) during cheese ripening have also been investigated, as have hygienic quality and ways to improve it.[31,35,36,67,84,124–133]

According to a review by Núñez *et al.,*[28] the microbial flora of Manchego cheese made from raw milk is quite well known and consists mainly of lactococci (particularly *L. lactis*), which prevails during the first month of ripening, after which they are outnumbered by lactobacilli (chiefly *Lb. plantarum* and *Lb. casei*); enterococci (especially *Ent. faecium*), leuconostocs (particularly *Leuc. mesenteroides* subsp. *dextranicum* and *Leuc. paramesenteroides*) and pediococci (*Ped. pentosaceus*) typically occur in smaller numbers, while micrococci and coagulase-positive staphylococci, which are usually detected in freshly manufactured cheese, gradually decrease in number throughout ripening. Finally, coliform counts vary considerably after a few months' curing.

The microbiology of the ripening of cheese made from pasteurized milk has also been investigated. Starter lactococci prevail during the first month of ripening, during which lactobacilli (particularly *Lb. casei* and *Lb. plantarum*) reached counts of up to 10^8 cfu/g, i.e. close to those obtained for cheese made from raw milk, in addition to low levels of leuconostocs, micrococci and yeasts.

Due mention should be made here of the analyses for aroma compounds made in Germany by Wirotama *et al.*[82] on six ripe Manchego-type cheeses of different origin. The electrophoretic (PAGE) patterns of different zones of the cheeses (core, mantle and rind) were qualitatively identical within each cheese but differed between cheeses, indicating that each was probably made from a different milk mixture. After precipitation of protein, they isolated 17 free amino acids, of which $Cys_{1/2}$ and Arg were present in trace amounts, and identified four primary and two secondary alkyl amines. They also detected and quantified 17 free fatty acids with two to 18 carbon atoms, both saturated and unsaturated, and identified methyl-branched fatty acids. As far as free carbonyl compounds are concerned, they identified eight methyl ketones, two unsaturated methyl ketones containing seven and nine carbon atoms, eight aldehydes, 10 alcohols (five 1-alkanols and five 2-alkanols) and four ethyl esters of fatty acids. The average concentrations of the major volatile constituents of Manchego-type cheese are listed in Table XI.[134]

Hard cheese of high quality is manufactured industrially in large amounts from ewes' milk (Churra, Castellana and Manchega breeds) in the Northern Castilian Plateau, particularly in the provinces of Palencia, Valladolid and Zamora. The products are homologous to those manufactured in the Southern counterparts.

Castellano cheese exhibits greater diversity (eight samples of cheese made from ewes' milk in the provinces of Valladolid and Palencia according to labels) in gross composition (cf. Table IV) than Manchego cheese (eight samples), but the average fat-in-dry matter and C_{12}/C_{10} ratio were identical to those of Manchego cheese. However, the average moisture-in-fat free substances was 14% lower than that of Manchego, so its metabolizable energy is higher. The average salt content and Ca/P ratio are 17% and 12% lower, respectively (cf. Table V), while the proportions of residual caseins and their insoluble first breakdown polypeptides (electrophoretic patterns) are virtually identical. Nevertheless, the extent and depth of proteolysis (SN and NPN, respectively) are 15% and 50% higher, respectively, in Castellano cheese, the amino nitrogen content of which is also 100% higher (cf. Table VII). The extent of lipolysis in Castellano cheese is 36% greater (cf. Table VI). As a result of the lower moisture content and greater

TABLE XI

Average Concentrations (μg/g Cheese)[a] of the Major Volatile Substances Found in Manchego-type Cheeses

Ripening time	2 months	5 months	10 months
Free fatty acids			
Butyric	0·2	0·9	1·9
Hexanoic	4·1	14·1	23·5
Octanoic	11·0	23·4	33·7
Decanoic	12·6	27·3	31·6
Dodecanoic	2·2	3·9	5·9
Tetradecanoic	1·1	1·7	2·9
Methyl ketones			
2-Pentanone	0·7	0·5	0·8
2-Heptanone	1·5	1·6	4·4
2-Nonanone	3·9	5·4	17·3
2-Undecanone	0·4	0·5	1·4
Ethyl esters			
Butyrate	1·0	1·3	2·2
Hexanoate	2·5	3·5	2·5
Octanoate	1·0	2·9	2·4
Decanoate	1·0	3·7	3·1
Methyl esters			
Butyrate	0·7	0·3	0·5
Hexanoate	1·1	1·2	1·7
Decanoate	0·2	1·1	2·3
2-Alkanols			
2-Heptanol	0·3	0·4	0·6
2-Nonanol	0·2	0·2	1·0

[a] Average values from four batches of cheeses made from sheep's milk containing 0, 5, 10 and 20% cows' milk (no significant differences were found between batches). Source: Adapted from Ref. 134.

extent and depth of proteolysis, the average a_w of Castellano cheese is lower (0·92 vs 0·95), but its pH is somewhat higher (5·23 vs 5·05), as shown in Table VIII.

Zamorano cheese (cf. Tables IV–VIII for average values and standard deviations of seven samples) is similar to Castellano cheese in the extent and depth of proteolysis, a_w and pH, while its moisture-in-fat free cheese, salt content, Ca/P ratio and extent of lipolysis (FFA level) are intermediate between those of Castellano cheese and Manchego cheese. Also, its fat-in-dry matter content is 1% lower and its C_{12}/C_{10} ratio (0·6) is higher than those of its two counterparts.

Smaller (about 2 kg) hard cheese varieties are also manufactured industrially or on farms from ewes' milk in the North of the Iberian Peninsula (Navarre and the Basque country).

Roncal cheese, the first variety with a definitive DO and subject to legal regulations since 1981,[5] takes its name from the Navarrese Valley where it is produced from December to July. The changes in pH, some chemical parameters, major nitrogen fractions and main microbial groups were monitored[135] during the first 3 months of ripening of two farmhouse-made batches; lactic acid bacteria (*L. lactis, Lb. casei, Lb. plantarum, Leuc. dextranicum* and *Leuc. lactis*) increased markedly during the first few days and stabilized thereafter; *Micrococcus* (identified species were *M. saprophyticus, M. lactis* and *M. roseus*) and *Staphylococcus* decreased slightly; counts of yeasts and moulds remained constant throughout maturation and coliform organisms decreased progressively until disappearing. In 1962, López *et al.*[80] followed the quantitative evolution of several free amino acids during the first weeks of ripening.

Another hard variety made from whole sheep's milk is *Idiazábal cheese*, traditionally produced in mountain caves in the Basque country, where it is cured (at 10°C and 85–90% RH) for at least two months. After the first month of ripening the cheese may be smoked by burning beech wood and the finished cheese can be stored for consumption for up to 1 year. Its DO was officially awarded in 1987.[9] Some physical parameters and gross composition have been determined at several stages of ripening of cheeses made in villages throughout the production area.[136]

In Western Andalusia and, particularly, in Extremadura, several semi-soft cheese varieties are traditionally made from ewes' milk curdled with vegetable rennet (a coagulant from *Cynara* spp. thistles). The three most important such varieties are those made in Los Pedroches Valley (Córdoba), La Serena Valley (Badajoz) and 'Torta del Casar' (province of Cáceres).

The earliest thorough research conducted on the virtually unknown proteolysis brought about by this peculiar coagulant (cf. Section 3.2) was reported by Fernández-Salguero[137–140] on the ripening of Los Pedroches cheese; he characterized cured cheeses gathered from different farmhouses, both physically and chemically.[141–143]

The manufacture and ripening of La Serena cheese was described and studied by Fernández del Pozo *et al.*[144,145] From January to May, both after morning and evening milkings of Merino ewes, the raw milk (8–10% fat) is curdled at 27–30°C for 45–75 min by adding a cloth-filtered aqueous extract of dry petals from the cardoon *C. cardunculus*, previously soaked for 24 h. The coagulum is cut to rice grain size, stirred gently for 2–3 min, allowed to stand for 10 min before removing the whey, scooped into woven matweed fibre moulds, pressed by hand and, after several hourly turnings, intended to expel further whey, the cheese is dry-salted for 16–24 h and allowed to ripen for 2 months at ambient temperature (10–18°C) and relative humidity (70–95%). The cured cheese has a flat cylindrical shape (*ca.* 18 × 4 cm). The biochemical (proteolysis and lipolysis), microbiological, textural and sensory changes undergone throughout ripening were monitored on 10 batches of cheese. Proteolysis, faster for α_s-casein than β-casein, was enhanced by high pH values and moisture contents (i.e. by higher water activities). As far as the rheological behaviour of the cheese is concerned, the elastic modulus, shearing strength and hardness reached minimum values during the first month of ripening and were dependent on the extent of hydrolysis of α_s- and β-caseins, pH, moisture content and NaCl-in-cheese moisture (i.e. on every major factor related to water activity). After 15 days' ripening, lactic acid bacteria prevailed inside the cheese mass, while these and lactic acid-utilizing yeast spp. and moulds were prevalent on the rind. Lactobacilli and yeasts influenced the breakdown of α_s-casein, while the hydrolysis of β-casein was effected only by yeasts. At the end of ripening, the cheese contained ca. 19% and 50% of residual α_s-and β-caseins, respectively, and 13 meq FFA/100 g fat, but no coagulase-positive staphylococci or faecal coliforms, although the death rate of these two was higher in spring than in winter. One attempt at improving on the traditional manufacturing procedure by using novel technology was clearly unsuccessful.[146] The compositional and physical features of the ripe cheese were also deter-mined,[147,148] as were the counts of the principal microbial groups.[149]

Similar to the varieties described above is Torta del Casar, which takes its name from its usually flat shape, resulting from the action of vegetable rennet. Physical and chemical data such as those listed in Tables IV–VIII are available for the mature cheese,[15,150,151] which was studied microbiologically by screening microbial strains (isolation and identification of lactic acid bacteria, enterococci and yeast spp.).[152] The most outstanding feature common to all these varieties is the more extensive and deeper proteolysis (SN and NPN were 30–60% and 15–30% of total TN, respectively) which takes place in only 2 months, clearly surpassing those of other bacterial ripened cheeses after 6–12 months and closer to those of mould ripened cheeses. This is a result of the strong proteolytic activity of the plant coagulant in relation to its clotting ability (cf. Section 3.2).

2.3.3 Varieties Made from Goats' Milk

Spain is the third largest European producer of goats' milk, after France and Hungary.[1] Roughly 40% of the overall Spanish production of this type of milk is used for cheesemaking.[153] Several hundred tonnes of semi-hard *Majorejo cheese* are produced each year on Fuerteventura Island, the cheese being marketed not only in the Canary Archipelago, but also on the mainland. The industrial manu-facturing process[153] involves adding about 0·1–0·2 g $CaCl_2$ per litre to pasteurized goats' milk and allowing it to coagulate at 30°C with 3 g of starter (Hansen DVS M-3; *L. lactis* and *L. cremoris*) and 6 g of animal rennet (strength 1:10 000) per hectolitre. The curd is cut, heated to 37°C at 0·2°C/min, scooped into cylindrical hoops and pressed for 2 h. The resulting cheeses (2–4 kg in weight) are salted for 40 h and allowed to ripen at 10–12°C and 85–87% RH for 3 months. Early studies of these cheeses[154,155] were concerned chiefly with the chemical charac-terization of commercial ripe cheeses purchased at peninsular markets. Later, samples collected after 0, 5, 30, 60 and 90 days ripening were subjected to physico-chemical (pH, SN, NPN), microbiological (total and proteolytic counts) and enzyme analyses (neutral protease, aminopeptidase and carboxypeptidase) by Gómez *et al.*[156] who found that neutral protease and aminopeptidase activi-ties increased throughout ripening but detected no carboxypeptidase activity. More recently, the lack of microbiological information on cheeses made from goats' milk[157] prompted research[153] on the growth patterns of the principal groups of microorganisms involved in the ripening of industrially manufactured cheese. The overall count was found to increase rapidly initially, chiefly as a result of the growth of mesophilic lactic streptococci that later levelled off or decreased in number as lactobacilli increased, until becoming the dominant microorganisms (particularly *Lb. casei* var. *casei*) at the end of ripening. Coliform, enterococcus, yeast and mould counts remained below 10^2–10^3 cfu/g, and the maximum numbers of micrococci and staphylococci were detected after ripening for 15–30 days and decreased gradually by the end of ripening. No leuconostocs were detected in the cheeses, nor were *Staph. aureus* found in the curd or cheese.

Ibores cheese, typically rubbed with pimento powder, and Badaya cheese, an acid curd variety with unique features and a remarkably high degree of proteolysis, seemingly related to surface green moulds and the presence or absence of inhibitory fine herbs on the rind, are two other farmhouse varieties made from goats' milk. Microbiological studies on the unusual Badaya cheese, the most extensively hydrolysed of all goats' cheeses examined so far,[15] seem to be quite pertinent and interesting.

Biochemical and microbiological features of craft hard cheese made from goats' milk were reported recently,[158] as were the effects of freezing and frozen storage on the physico-chemical and sensory properties of four varieties of goats' cheese.[159]

2.3.4 Varieties Made from Milk Mixtures

Cheeses made from binary or ternary mixtures of milks from different species, annual production of which exceeds 50 000 tonnes,[2,47] have recently been the subject of regulations intended to standardize their composition and specific features for internal trade.[4] These cheeses are made from uncooked, pressed curd which results in a semi-hard consistency, and resemble Manchego cheese in their manufacture and in the features of the final product. There are currently three denominations, namely: 'Hispánico' cheese, which is made from ewes' milk (minimum 30% v/v) and cows' milk (minimum 50%, v/v); 'Ibérico' cheese, made from cows' milk (minimum 50%), ewes' milk (minimum 10%) and goats' milk (minimum 30%); and 'De la Mesta' cheese, manufactured from ewes' and cows' milk (minimum 75% and 15% respectively), and—optionally—goats' milk (maximum 5%).

The compositional requirements legally established for the above three varieties are as follows: a minimum dry matter of 55% for all three denominations and a minimum fat-in-dry matter of 45% for 'Hispánico' and 'Ibérico' and 50% for 'De la Mesta' cheese. They are cylindrical cheeses, usually no larger than 13 cm in height and 26 cm in diameter, and weighing about 3.0 ± 0.5 kg, with a hard, yellowish, dry, smooth rind and a firm, white-yellowish body with holes the size of a pea.

All three varieties are made from milk pasteurised at 72–78°C for 15 s or at 63°C for 30 min, or from unpasteurized milk which is coagulated with rennet at 30°C for 15–20 min; milk for De la Mesta cheese is coagulated at 28–32°C for 45–60 min. The coagulum is cut and allowed to stand for 5 min, after which it is stirred while the temperature is gradually raised to 35–36°C—up to 40°C in 25–30 min for De la Mesta cheese—in order to obtain curds of the size of rice grains in about 30 min. The whey is removed after standing for 10 min and, once moulded, the curd is pressed at 16–18°C for 8–10 h under an increasing pressure of 2–8 kg—10 kg for De la Mesta cheese—per kg of cheese. The cheese is salted by brining below 13°C for 48 h and allowed to ripen by the action of lactic acid bacteria at 12–18°C and 80–85% RH for at least 1 month if pasteurized milk was used or for 2 months or longer otherwise.

The varieties described above have been the subject of no research as yet, nor are they usually commercially available; however, similar cheeses are widely available (cf. Section 2.3.2). Table XII lists compositional data providing information on the different types of cheese made from milk mixtures that were available on the market a few years ago. Data reported on Manchego, Castellano and Zamorano cheese, which are made from ewes' milk, and on similar cheeses manufactured from binary and ternary mixtures of milks, are currently being subjected to discriminant analysis (Cepeda; Rodero, pers. comm.).

TABLE XII

Some Chemical and Physical Features of Hard Cheeses Made from Binary and Ternary Milk Mixtures[a]

Milk mixture[b]	Sheep + Cow		Sheep + Goat	Sheep + Cow + Goat
Geographic region[b]	Old Castile	New Castile	Both Castilian plateaux	
	$\bar{x} \pm$ s.d.	$\bar{x} \pm$ s.d.	$\bar{x} \pm$ s.d.	$\bar{x} \pm$ s.d.
Moisture[c]	37 ± 4	36 ± 6	40 ± 1	37 ± 4
Fat[c]	34 ± 3	36 ± 4	34 ± 0	34 ± 3
Protein[c]	23 ± 2	22 ± 2	22 ± 1	24 ± 1
Lactic acid[c]	1·2 ± 0·2	1·3 ± 0·3	1·1 ± 0·2	1·4 ± 0·3
Ash[c]	4·7 ± 0·6	4·5 ± 0·4	3·6 ± 0·2	3·8 ± 0·4
kcal/100 g cheese	400 ± 40	410 ± 40	390 ± 10	400 ± 30
M/FFC[d]	57 ± 5	56 ± 6	60 ± 2	57 ± 3
F/DM[e]	55 ± 3	56 ± 2	56 ± 1	55 ± 2
NaCl[c]	2·3 ± 0·5	2·2 ± 0·4	1·7 ± 0·1	1·4 ± 0·3
Ca/P	1·39 —	1·37 —	1·48 —	1·22 —
C_{12}/C_{10}	0·70 —	0·59 —	0·47 —	0·63 —
FFA[f]	1·17 ± 0·34	1·23 ± 0·25	1·39 ± 0·73	1·11 ± 0·41
α_s-I frg.[g]	11 ± 3	9 ± 1	11 ± 3	7 ± 1
α_s-cn[g]	19 ± 2	22 ± 4	20 ± 1	16 ± 4
β-cn[g]	35 ± 6	34 ± 5	41 ± 2	42 ± 4
γ-cn[g]	30 ± 7	31 ± 5	23 ± 2	32 ± 3
SN[h] (pH 4·6)	23 ± 3	25 ± 6	19 ± 4	24 ± 5
NPN (sol. in 12% TCA)	15 ± 3	14 ± 3	11 ± 1	15 ± 4
FN (formol titration)	6 ± 2	6 ± 1	5 ± 1	6 ± 2
a_w	0·95 ± 0·02	0·94 ± 0·02	0·97 ± 0·01	0·96 ± 0·02
pH	5·19 ± 0·19	5·01 ± 0·14	5·03 ± 0·11	5·09 ± 0·14

[a] From Ref. 15 (rounded averages and standard deviations of six samples).
[b] According to labels.
[c] g/100g cheese; [d] g moisture/100 g fat-free cheese; [e] g fat/100 g dry matter;
[f] ml 0·05 N KOH/g cheese; [g] relative percentages (see Section 2.1); [h] g/100 g total nitrogen.

Many other varieties are traditionally made from milk mixtures, both on the Peninsula and in the Balearic and Canary Islands[3]—the last include Herreño cheese, which is made on the small Hierro Island, southwest of the Canary Archipelago, which was recently characterized.[160]

Knowledge of the type of milk used as a raw material in cheesemaking is of obvious interest to industrial manufacturers and the Administration alike. This issue was addressed over 20 years ago by Charro et al.,[161] followed by comprehensive research on this topic[162–175] that was reviewed in 1983 by Ramos & Juárez,[23] in 1986 by Juárez & Ramos[169] and in 1989 by Núñez et al.[28] The methods reported lately by Aranda et al.,[170] Amigo et al.,[171] García et al.[172–174] and Rodríguez et al.[175] appear to be virtually deterrent since they allow detection of economically insignificant amounts of extraneous milk (cows' or goats') added to sheep's milk, even in ripe cheeses.

2.4 Mould-Ripened Cheese

Within this group are several autochthonous blue-veined varieties from the Picos de Europa mountains (Asturias) such as Cabrales and Gamonedo cheeses, traditionally made from mixtures of cows', sheep's and goats' milk, which are ripened in natural limestone caves or abandoned mining galleries and are typically marketed wrapped in leaves of the maple tree or fern, as well as factory-developed commercial brands of blue cheese, which use sheep's milk (or milk mixtures) as raw material. Surface white mould-ripened varieties, mainly Camembert and Brie, are produced in smaller amounts from cows' milk; some varieties of this type are also manufactured from goats' milk.

The most genuine and representative of the national mould-ripened varieties is *Cabrales cheese*, whose definitive DO was officially awarded in 1990.[10] The raw milk mixture is coagulated by rennet with no starter culture or mould spores added to the milk or curd; the unpressed cheeses are salted and then ripened, unpierced, in caves, usually for about 3–4 months under natural air currents (at about 10°C and 90–95% RH), with periodic turning and cleaning; the ripe products may be sold wrapped in leaves of 'plágano' (*Acer pseudoplatanus*). The biochemistry and microbiology of the ripening process remained unknown until 1971, when Burgos et al.[176–179] reported that lactobacilli constituted the dominant flora[176] and *Penicillium roqueforti* was the main ripening agent among other isolated moulds;[177] some yeast species were also identified.[178] Sala & Burgos[179] monitored the release of individual free amino acids throughout curing. Further research by Núñez et al.[180–183] dealt with changes in internal and surface microflora during cheesemaking and maturation,[180] and with the isolation and identification in milk, curd and ripening cheese of strains of lactic acid bacteria,[181] micrococci and staphylococci[182] and yeasts and moulds.[183] Juárez et al.[184,185] analysed Cabrales cheese in relation to: composition, physical and chemical values of fat, fatty acid patterns, major nitrogen factions, soluble tyrosine and tryptophan, total free amino acids and casein breakdown in milk, curd and cheeses at three different curing stages. The chemical composition of ripened cheeses,[186,187] the genetic polymorphism of the proteins in the cows' milk (Brown Swiss) used in its manufacture,[188] the nitrogen compounds resulting from proteolysis,[189,190] the relationships between chemical composition and water activity,[191,192] the microbiology of ripening[193] and the features of wild strains of *L. lactis* isolated from cheese with a view to determining their potential use for the development of a specific starter for this type of cheese[194] were the subject of more recent studies. Alonso[195] studied the feasibility of freezing curds for subsequent ripening in any season,[196-198] and applied discriminant analysis to search for differences between Cabrales cheeses obtained by ripening of both unfrozen and thawed curds.[199]

Gamonedo cheese is similar to Cabrales but is smoked with ash or beech wood for two–three weeks prior to ripening in caves at 12–15°C and 85–90% RH. The ripe cheese has a brownish–red rind and some greenish–blue veins inside the body. Its physical, chemical and sensory characterization throughout ripening were reported by González.[200] Analysis of volatile flavour compounds formed during ripening was recently undertaken by González *et al.*[201] In the unripe cheese, volatile ($C \leq 14$) free fatty acids (FFAs) prevailed as result of the very low concentrations of other volatile compounds. However, the ripe cheese was found to contain high concentrations of about 40 volatile substances, particularly FFAs and carbonyl compounds such as methyl ketones, secondary alcohols and esters. Table XIII lists the ranges of the principal volatiles in ripe cheeses.

In relation to FFAs, the slightly water-soluble acids (C_8 and C_{10}) were present at the highest concentrations, while the water-insoluble (C_{12} and C_{14}) and water-soluble (C_4 and C_6) acids were at the lowest. About 90% of both volatile and non-volatile FFAs in the cheeses arose from lipolysis, but the FFA profile was rather different when triglycerides were hydrolysed by lipases in the rennet,

TABLE XIII

Concentrations (μg/g cheese) of the Principal Volatile Compounds in Mature (2–6 months) Gamonedo Cheese[a]

Main chain C-atoms	Free fatty acids	Methyl ketones	Secondary alcohols	Methyl esters	Ethyl esters
4	0·2–1·3	—	—	—	0·3–0·6
5	—	0·5–2·1	—	—	0·2–0·3
6	0·4–17·6	0·1–0·4	—	2·7–4·4	0·3–3·9
7	—	9·1–14·7	0·9–3·4	—	—
8	1·1–37·3	—	—	1·4–3·9	0·3–4·0
9	—	15·7–44·1	0·6–5·2	—	—
10	1·9–54·2	0·3–0·4	—	1·6–10·1	0·7–5·6
11	—	0·5–6·6	0·1–0·3	—	—
12	1·2–13·3	0·2–0·7	—	0·2–1·3	—
13	—	—	—	—	—
14	0·1–9·2	—	—	—	0·2–0·4
15	—	0·2–0·7	—	0·8–1·6	—

[a] Adapted from Ref. 201.

which act preferentially on the ester bond at the sn-3 carbon of the triglycerides to yield relatively high proportions of butyric acid and other short-chain FFAs than on hydrolysis with fungal lipases (chiefly alkaline lipase from *P. roqueforti*), which release small amounts of butyric acid. In blue cheeses, FFAs, are converted into methyl ketones via β-oxidation by *P. roqueforti* spores and mycelia, so the FFA content and the concentration of methyl ketones produced are positively correlated (cf. Chapter 4). Methyl ketones are highly aromatic and sapid and contribute rather significantly to the flavour of blue cheeses. Gamonedo cheese contains 2-nonanone and 2-heptanone as the principal methyl ketones, which reach their peak concentrations after 2 months' ripening (the concentrations at which the former are produced are equivalent to those reported for Roquefort and slightly lower than those reported for American blue cheese, while those of the latter are slightly lower than those found in the other two blue cheese varieties); the total concentration of methyl ketones is, however, similar in the three varieties (cf. Chapter 4). On reduction by *P. roqueforti*, methyl ketones are converted into secondary alcohols, particularly in 2-nonanol and 2-heptanol for Gamonedo cheese, where they also reach their maximum concentrations after two months' ripening—such concentrations, however, are only about 10% of the parent methyl ketones, but are still significant. Esters are believed to be formed in the cheese by microbial esterification of FFAs with alcohols. The methyl and ethyl esters detected in Gamonedo cheese include decanoates, hexanoates and octanoates in decreasing order of concentration. While methyl esters, like methyl ketones and alcohols, reach their maximum concentrations in 2-month cheese, ethyl esters are formed more slowly, but their production increases gradually over six months of ripening, after which it peaks at about 15 μg/g cheese, which suggests that the two types of esters are formed via different biochemical pathways.

Blue cheeses are also manufactured industrially from mixtures of cows' and ewes' milk and from pure ewes' milk. A comparative study on the ripening of Cabrales cheese and an industrially manufactured cheese made from 75% cows' milk and 25% ewes' milk showed that the former had a lower moisture content, a higher salt content, greater hardness, more extensive lipolysis, but a similar degree of proteolysis, and sensory quality.[202] Blue cheese from pure ewes' milk is made from HTST (72°C for 15 s) pasteurized ewes' milk to which a lactic starter (0·8–1·0%) and an inoculum of *P. roqueforti* spores are added before coagulation by rennet (9·25 ml Hansen's liquid rennet per hectolitre); the curd is cut into cubes (2·5 cm) and scalded (31°C for 190 min); after hooping, the unpressed cheese is dry-salted (for 5 days) and ripened for 4 months (at 7°C and 96–98% RH), being pierced after 2 and 4 weeks of ripening. The finished cheese (cylinder of 2·5 kg weight) is packed in aluminium foil and stored at 0°C. The composition of commercial blue cheese from ewes' milk is quite similar to that of Cabrales cheese, except for the fatty acid profile.[15]

Blue cheese made from ewes' milk is reportedly unsuitable for the synthesis of mycotoxins such as PR-toxin or roquefortin by *P. roqueforti*.[203]

As shown in Tables IV–VIII, distinctive features of blue-veined cheeses are, among others (cf. Chapter 4), their high salt content and degree of proteolysis

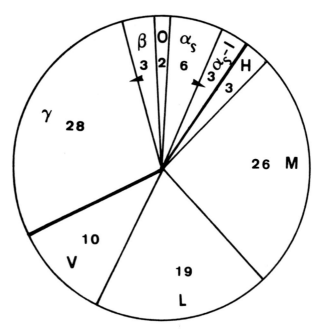

Fig. 5. Nitrogen distribution in highly proteolysed Cabrales cheese. For the meaning of symbols, see captions to Figs 3 and 4 (V denotes volatile nitrogen). Note the exceptional resistance of γ-cns to hydrolysis (also outstanding in the electrophoretograms) and the extreme hydrolysis of both β-cn and α_s-cn and even α_s-I fragments and soluble high M_r peptides to intermediate, low and very low M_r nitrogen compounds. (Constructed from data in Table VII).

(as well as lipolysis[15]) and, therefore, they have low a_w values due to the high aqueous concentration of NaCl and other low-molecular-weight compounds produced by the activity of fungal enzymes. The levels of water-soluble nitrogen (often 50 to 75% of total nitrogen) and non-protein nitrogen are high and usually rather similar (sometimes equal). This accounts for the small pool (or absence) of soluble polypeptides (proteoses-peptones), when expressed as a percentage of soluble nitrogen (cf. Fig. 5), compared to that found for most cheese varieties (cf. Fig. 4). Both α_s- and β-caseins are extensively proteolysed; hydrolysis of the latter yields higher levels of soluble tyrosine and tryptophan than in cheeses ripened by bacteria, in which α_s-casein breakdown is more extensive than that of β-casein (cf. Table VII). Consequently, electrophoretic patterns of mould-ripened varieties show very high relative levels of non-polar (insoluble) polypeptides released from β-casein by plasmin (cf. Fig. 1), which accumulate in the low mobility (γ-casein) zone (cf. Table VII and Fig. 5) because of their resistance to proteinases. Only in overripe samples do the electrophoretograms appear devoid of polypeptides stainable with Amido Black and which may be visualized with Coomassie Blue, a dye for which their affinity is much greater. As stated by Hewedi & Fox,[204] other techniques are consequently required for monitoring the advanced stages of proteolysis in mould-ripened varieties. Low M_r peptides of blue cheeses were recently characterized by González et al. (pers. comm.).

Comparative studies of the chemical composition, energy value and mineral content of commercially available samples of national versions of Brie and Camembert and their imported French counterparts were reported by Alcalá et al.[205,206] who found no significant differences. The measured water activity of the samples[207] was found to be highly significantly correlated to their ash and non-protein nitrogen contents, expressed on a moisture basis.[208] The microbiology of national Camembert has also been assessed.[209] Some surface white mould-ripened cheeses made from goats' milk have been the subject of physical, chemical and microbiological analyses.[156,157,210] One commercial brand of this type of cheese was found to share a number of physical features with its counterparts made from cows' milk.[15] Compared to internal mould-ripened cheeses, surface mould-ripened varieties have higher moisture contents and fat-in-dry matter levels, but lower concentrations of NaCl and non-protein nitrogen compounds, and hence higher water activities and pH (cf. Fig. 2).[15] The fatty acid profiles of blue and soft cheeses were also compared.[211]

An empirical equation for calculation of a_w from chemical composition data of all mould-ripened varieties, both soft and blue, was recently reported[212] (see Chapter 11, Volume 1).

3 PORTUGUESE CHEESES

In 1989, the overall Portuguese cheese production was about 50 000 tonnes.[1] That year, cheese imports were almost equal to exports and the overall national cheese consumption exceeded 43 000 tonnes. Some production data for the major cheese types and varieties made from milk from different species are summarized in Table XIV, which is included for orientation purposes only because the production figures are clearly underestimated as a result of substantially increased production in recent years (about 25% in the last decade) and of the difficulty involved in accurately assessing the overall farmhouse production, which is quite substantial in Portugal.

Cheese varieties currently protected by Portugal's Denomination of Origin (DO) are listed in Table XV.

Although, from a quantitative point of view, the industrial production of cheese from pasteurized cows' milk with starter cultures added and coagulated by animal rennet amount to 60% of total production, the most representative and genuine cheeses of Portugal are farmhouse varieties made from raw ewes' milk coagulated by plant rennet from 'cardo', the most famous of which, owing to its fine quality and almost uniqueness, is 'Queijo Serra da Estrela',[213] annual production of which exceeds 13 000 tonnes in its traditional production area.[214] Manufacture of goats' milk cheese is a wholly craft activity of low significance.[214-216] Home production of cheese from ewes' and goats' milk versus industrial production has been discussed by Vieira de Sá.[217]

TABLE XIV
Portuguese Cheese Production in 1984

Cheese made from[a]	Tonnes per annum	Major cheese types and varieties
Cows' milk	ca. 27 000	ca. 24 000 tonnes of Flamengo cheese (Dutch-type), ca. 3000 tonnes of Ilha (Açores) cheese (Cheddar-type), minor amounts of processed cheeses and others
Sheep's milk	16 000–17 000	mostly Serra cheese and smaller amounts of Serpa, Azeitão, Castelo Branco, Évora, Rabaçal, Tomar cheeses and others
Goats' milk	ca. 1600	mostly fresh cheeses and Castelo Branco (á Cabreira) cheese

[a] Many cheese varieties are made from mixtures of milk in variable proportions.
Source: Data from the Junta Nacional Dos Productos Pecuários supplied by courtesy of the Embaixada de Portugal em Madrid.

TABLE XV
Cheeses with Portugal's Denomination of Origin

Made in the Azores from cows' milk:
'Queijo São Jorge'

Made in the mainland from sheep's milk:
'Queijo Serra da Estrela'
'Queijo de Azeitão'
'Queijo Serpa'
'Queijo de Castelo Branco'

Made in the mainland from sheep's and/or goats' milk:
'Queijo Picante da Beira Baixa'
'Queijo Amarelo da Beira Baixa'

Source: Updated information (1991) from the Instituto de Apoio à Transformação e Comercialização dos Productos Agrários e Alimentares do Ministério da Agricultura de Portugal.

Research papers on Portuguese cheese varieties collected by international information services on food science and technology have been scanned to update this review. The sheep's milk cheeses, Serra and Serpa, together with the plant coagulant used in their manufacture, have been the subject of most studies, carried out mainly by the team for dairy research and development of the Departamento de Tecnologia de Indústrias Alimentares from the Laboratório Nacional de Engenharia e Tecnologia, at Lisbon.

3.1 Features and Manufacture of Major Cheese Varieties

One of the most comprehensive sources of general information on the domestic cheese varieties is the book by Carr[218] which includes features and other information on 19 Portuguese varieties. More detailed information and manufacturing methods for the principal varieties have been reported by Vieira de Sá & Barbosa.[219] These two sources were used as the main basis for compilation of Table XVI, which summarizes some important features of the major indigenous varieties.

3.1.1 Varieties Made from Cows' Milk

Industrial cheesemaking from cows' milk relies on the manufacture of foreign varieties, mainly of Dutch-type (Edam and Gouda). The production of 'Flamengo' cheese (Edam-type), improperly called 'Bola' cheese, accounts for almost 90% of the total production of cows' milk cheese.[219] Other foreign varieties manufactured in minor amounts are Camembert, Saint Paulin, Emmental, traditional fresh cheeses, Quark and Petit Suisse, as well as various processed cheeses, most of which use imported cheese as raw material.

In the Azores Islands not only is Flamengo cheese produced, but also significant amounts of a Cheddar-type cheese known on the mainland as 'Ilha' cheese, although this name applies more properly to the cheese produced in large creameries on São Miguel Island.

The essential principles of the manufacture of *Ilha cheese* are pasteurization of milk, addition of a starter culture and clotting by rennet at about 30°C for 30 min. The coagulum is cut into small cubes with a vertical harp and then with a horizontal harp; after a brief standing period, cutting is resumed until the curd particles are reduced to the size of a pea and these are then stirred for at least 30 min while the temperature is increased by about 5°C. After cooling to room temperature and removal of the whey, the curd is thoroughly mixed with dry salt (about 3%) and poured into large cylindrical hoops which are pressed for 24 h. The cheese is matured at 10–12°C and 75–85% RH for 2–3 months.

São Jorge cheese (DO) takes its name from the island on which it is produced from raw milk in small co-operative factories; this cheese is similar to Ilha cheese but has a stronger, more piquant flavour and aroma.

3.1.2 Varieties Made from Sheep's Milk

As stated above, the varieties in this group, farmhouse-made from raw milk and coagulated by a plant rennet, such as Serra, Serpa, Azeitão, among others (cf. Table XVI), are the most interesting and typical of the country, and are now protected by DOs. Traditional methods used in the farmhouse manufacture of Serra and Serpa cheeses were studied and described by Vieira de Sá et al.,[219,221] Vieira de Sá & Barbosa[222] and Barbosa.[223]

Internationally, *Serra da Estrela cheese* is the best known variety from Portugal. In 1982, a legal standard for the 'Queijo Serra da Estrela' was published in order to define it, establish its features and specify the packaging and preservation conditions, including a score for quality evaluation.[224]

This variety is farmhouse-made in the cold season from raw sheep's milk (Bordoleira breed) around the National Park of Serra da Estrela, mainly in villages in the Guarda and Viseu districts. Milk at 25–30°C is coagulated with an

TABLE XVI
Some Features of Major Portuguese Cheese Varieties[a]

Cheese name	Similar types	Type of milk	Type of consistency	Ripening (weeks)	Weight (kg)	Shape[b]
Mainland varieties						
Serra da Estrela	Unique	Sheep	Soft to semi-soft	4–6	1·0–1·7	LC
Serpa	Serra	Sheep	Soft to semi-soft	6–8	1·7–2·0	LC
Azeitão	Serra	Sheep	Soft	3–4	0·2–0·3	LC
Castelo Branco	Serra	Sheep	Semi-hard	3–4	1·0	LC
Évora	—	Sheep	Semi-hard to hard	26–52	0·1	LC
Rabaçal	—	Sheep	Fresh	—	1·0	TC
Tomar	Serra	Sheep	Semi-hard	2–3	0·03–0·04	LC
Varieties from the Azores Archipelago						
Ilha	Cheddar	Cow	Hard	8–12	5–10	TC
São Jorge	Cheddar	Cow	Hard	8–12	3·5–7	TC
Pico		Cow	Soft	3–4	0·5	LC

[a] Compiled from Carr,[218] Vieira de Sá & Barbosa[219] and Eekhof-Stork.[220]
[b] Abbreviations: LC = Low Cylinder; TC = Tall Cylinder.

aqueous extract from cardoon flowers (*Cynara cardunculus* L.). The coagulum is hand-cut and placed into plaited cheese-moulds, lined with cloths. After slight pressing for 6 to 12 h, the cheese is removed from the mould and dry-salted.

The ripening period is at least one month and comprises two phases. During the first 15 days the cheeses are turned daily in an atmosphere of a very high relative humidity (>95%) that promotes the development of a yeasty smear, called 'reima', on the surface. In the second phase, the cheeses are held in a dry environment to facilitate rind formation; during this period, the cheeses are turned every second day and the 'reima' is washed off periodically.

The finished product must have a minimum ripening index of 40% soluble N, a moisture content from 61 to 69% on a fat-free basis and a fat-in-dry matter of 45–60%. The mature cheese is a semi-soft type with a creamy body ('Serra amaintegado') and has the shape of a low cylinder (15–20 cm diam. × 4–5 cm

TABLE XVI—*contd.*
Some Features of Major Portuguese Cheese Varieties[a]—*contd.*

Interior	Exterior	Flavour and aroma	% Fat-in-dry matter
Creamy, white or pale yellow, no holes or a few small ones	Thin, smooth, straw yellow	Delicate, mildly lactic	45–60
Slightly creamy, some holes, white ivory	Smooth, straw yellow, sometimes with dry white or lemon moulds	Strong, piquant, sharp, peppery	45
Creamy, pale straw yellow	Thin, soft, smooth, straw yellow	Mild, slightly sourish	45
White, smooth, small holes	—	Strong, peppery	45
Pale yellow, crumbly	Dark yellow	Very strong, salty	45
White curd	No rind	—	45
Pale, crumbly, small holes	Tough, grey yellow	—	—
Firm-bodied, yellow	Hard, dark yellow	Mild to mellow and nutty	45
Crumbly	—	Strong, piquant	45
Creamy	Pale, smooth	Fairly piquant	45

high). It is stored at 0–5°C and has a maximum shelf-life of 4 months when held at refrigeration temperatures (below 10°C).

The average chemical composition of Serra cheeses, classed as excellent and good, is as follows:[219] 48·8% moisture, 28·8% fat, 19·9% protein, 0·7% lactic acid and 2·6% ash; the percentages of moisture in the fat-free cheese and fat-in-dry matter are 68·5 and 56·2, respectively, the salt content is 2·56%, the ripening index 49·2% soluble N, and the pH 5·1.

There is a semi-hard variant, called 'Serra velho', obtained by extending the ageing period by 6–12 months.

There are modern cheesemaking factories based on sheep's milk (to which cows' milk is often added) but the manufactured product, although more standardized, is not comparable in quality to the authentic product made by traditional methods.[214]

In order to protect the quality of this genuine national cheese, Vieira de Sá[225] has urged authorities to take management steps to improve the hygiene of ewes' milk production and establish quality standards, and to foster the development of appropriate cheesemaking technology to assure delivery of hygienically guaranteed produce, better delimitation of the demarcated area for cheese production and the establishment of the denomination of origin, currently in force (Table XV).

Serpa cheese, a variety similar to Serra cheese, also with a DO, is made in the Alenteijo region from raw sheep's milk (Merina breed). There are also two versions, one is a semi-soft type, ripened for 6–8 weeks ('Serpa amaintegado'), and the other a hard to very hard type, cured for over 6 months ('Serpa velho').

Fresh cheeses are made from ewes' milk and milk mixtures, both industrially and at farmhouses. Barbosa & Gonçalves[226] analysed 100 samples of labelled cheeses from 'fresh ewes' milk' by radial immunodiffusion and found that only 11% of the samples were made from pure ewes' milk; 72% were made from binary milk mixtures (52% and 20% containing cows' and goats' milk, respectively), and the remainder (17%) from ternary mixtures of the three types of milk. According to these authors, stricter control should be exercised in order to protect cheeses made from pure ewes' milk from adulteration.

3.1.3 Varieties Made from Goats' Milk

There is moderate production of several types of cheeses from goats' milk, generally for consumption in the fresh state. One of the most important is Castelo Branco ('á Cabreira') cheese with its fresh, cured ('amarelo') and piquant variants. Currently, there is a trend to increase the manufacture of pure goats' milk cheese and to this end, selected breeds of goat are being imported.[214]

Barbosa & Miranda[227] assessed the compositional and microbiological quality of 84 goats' milk samples collected on farms from different regions of the country and concluded that the hygienic quality of the milk obtained both by hand- and machine-milking procedures was unsatisfactory and should therefore be improved.

3.2 Thistle Rennet

The milk coagulant extracted with water from the flowers of cardoons from the genus *Cynara* (mainly *C. cardunculus* L., *C. scolymus* L. and *C. humilis* L.) is used traditionally in the manufacture of Serra and some other cheese varieties made from sheep's milk in the Iberian Peninsula. In some Latin American countries, the wild thistle *C. cardunculus*, known as 'cardo de Castilla' and profusely grown in Argentina and Chile, is also used in cheesemaking from goats' milk.[228]

Vieira de Sá & Barbosa[229] were the first to study scientifically the milk-clotting ability and proteolytic properties of *C. cardunculus* extract, and to confirm earlier observations.[230–232] The thistle extract is thermostable, like other proteolytic enzymes from vegetable sources; it exhibits a higher proteolytic activity than animal rennet, but is better than animal rennet for the production of Serra cheese from sheep's milk. However, its high proteolytic activity results in reduced yield and defective flavour and texture when used for the manufacture of Edam and Roquefort cheeses.

New trials on the potential use of the coagulant from *C. cardunculus* as a substitute for animal rennet in the manufacture of Camembert and Gruyère were made by Barbosa *et al.*;[233] both types of cheese developed a very bitter flavour when thistle rennet was used in their manufacture. Additional cheesemaking experiments were conducted by Barbosa *et al.*[234] on Italian-type cheeses.

Biochemical research on the enzyme system of the wild thistle, *C. humilis*, was undertaken for the first time in the authors' laboratory.[235-240] The petals of the flowers and their aqueous extracts show proteolytic, lipolytic and amylolytic activities; the crude fresh extract contains a complex proteinase system; the kinetics of thermal inactivation suggest the occurrence of three independent first-order reactions derived from the inactivation of proteinases with different thermostability ($D_{65°C}$ = 9·5, 23·5 and 37·0 min); the proteinases are localized in the petals, probably as zymogens since the milk-clotting activity of the extracts increases with time after extraction. The activated extracts show only two proteinases whose kinetic and thermodynamic parameters after activation at refrigeration temperature are shown in Table XVII.[237] Recent research on the Latin American wild thistle, *C. cardunculus*, confirmed the occurrence of three proteases as shown by the above-mentioned kinetics. Campos *et al.*[228] found that precipitation of the crude extract with $(NH_4)_2SO_4$ at 40, 50 and 80% saturation yielded

TABLE XVII

Kinetic and Thermodynamic Parameters of the Transition State During Thermal Denaturation of the Proteinases from the Thistle, *Cynara humilis* L., Extracted with Citric Acid (0·1 M)–Disodium Phosphate (0·2 M) (pH = 5·0), After 15 days of Activation at Refrigeration Temperature.[237]

Protease:	Thermolabile	Thermostable
% of total milk-clotting activity:	55	45
Kinetic constants		
Reaction rates at 65°C		
D (min)	17·00	43·65
k ($\times 10^2$ min^{-1})	13·55	5·28
$t_{1/2}$ (min)	5·12	13·13
Temperature dependence		
Z (°C)	5·5	4·7
E_a (kcal mol^{-1})	96	112
Q_{10}	66	134
Thermodynamic constants		
$\Delta F^{\ddagger}_+$ (cal mol^{-1})	23 978	24 605
$\Delta H^{\ddagger}_+$ (cal mol^{-1})	95 614	111 772
$\Delta S^{\ddagger}_+$ (e.u.)^a	212	258
Number of bonds broken^b	19	22

^{a} Entropy units in cal mol^{-1} K^{-1}.
^{b} Number of non-covalent bonds broken on denaturation = $\Delta H^{\ddagger}_+/5000$ where the average $\Delta H^{\ddagger}_+$ per bond is assumed to be 5000 cal mol^{-1}.

three proteases with increased clotting activity and decreased proteolytic activity —in a ratio close to that of commercially available rennet from *Mucor miehei*— compared to the crude extract. The 50% $(NH_4)_2SO_4$ fraction proved to have the highest milk-clotting activity. These results may open up new prospects for practical use, provided that petal collection becomes economically viable.

Studies on two ripened cheese varieties made from sheeps' milk coagulated by thistle rennet revealed[241-244] that some composition-related factors, such as moisture content, a_w, the ash/moisture and NaCl/moisture ratios, and others, were not related to the extent of hydrolysis of β-casein, but rather to the amounts of residual α_S-casein, which seems to indicate that, in this type of cheese, there is an inverse influence of solute concentration on the relative rates of hydrolysis of the major caseins, since Creamer[245] observed that the extent of hydrolysis of β-casein in Gouda cheese increases with increasing moisture content, while that of α_{s1}-casein is unaffected, thus suggesting[246] that a low a_w is responsible for the resistance of β-casein to proteolysis; Fox & Walley[247] and Phelan *et al.*[248] showed that the susceptibility of β-casein to proteolysis is affected more markedly by the NaCl concentration than that of α_{s1}-casein, thus suggesting[248] that the reduced relative susceptibility of β-casein to proteolysis is due to some concentration-dependent physical change in the β-casein molecule which renders it inaccessible to proteases, and that the salt concentration appears to influence this change.

ACKNOWLEDGEMENTS

The authors acknowledge financial support from the Planes Nacional y Andaluz de Investigación Científica y Desarrollo Tecnológico for the realization of research on cheeses. They are also grateful to the Ministerio de Agricultura, Pesca y Alimentación, and to the Ministério da Agricultura, Pescas e Alimentação for supply of updated technical information on Spanish and Portuguese cheeses, respectively.

REFERENCES

1. *FAO Production Yearbook*, 1989. FAO, Rome.
2. Ministerio de Agricultura, Pesca y Alimentación, 1991. *Manual de Estadística Agraria 1990*, Secretaría General Técnica, Madrid.
3. Ministerio de Agricultura, Pesca y Alimentación, 1989. *Quesos de España*, Dirección General de Política Alimentaria, Madrid.
4. Presidencia del Gobierno, 1987. *Boletín Oficial del Estado*, No. **170**.
5. Ministerio de Agricultura, 1981. *Boletín Oficial del Estado*, No. **85**.
6. Ministerio de Agricultura, Pesca y Alimentación, 1985. *Boletín Oficial del Estado*, No. **5**.
7. Ministerio de Agricultura, Pesca y Alimentación, 1985. *Boletín Oficial del Estado*, No. **160**.
8. Ministerio de Agricultura, Pesca y Alimentación, 1985. *Boletín Oficial del Estado*, No. **272**.
9. Ministerio de Agricultura, Pesca y Alimentación, 1987. *Boletín Oficial del Estado*, No. **241**.

10. Ministerio de Agricultura, Pesca y Alimentación, 1990. *Boletín Oficial del Estado*, No. **166**.
11. OECD, 1984. *Milk and Milk Products' Balances in OECD Countries*, 1974–1982, OECD, Paris.
12. Marcos, A., Esteban, M.A., León, F. & Fernández-Salguero, J, 1979. *J. Dairy Sci.*, **62**, 892.
13. Ministerio de Agricultura, Pesca y Alimentación, 1990. *El consumo Alimentario en España 1989*, Secretaría General Técnica, Madrid.
14. Ministerio de Agricultura, 1973.*Catálogo de Quesos Españoles*, 2nd edn, Servicio de Publicaciones Agrarias, Madrid.
15. Marcos, A., Fernández-Salguero, J., Esteban, M.A., León, F., Alcalá, M. & Beltrán de Heredia, F.H., 1985. *Quesos Españoles: Tablas de Composición, Valor Nutritivo y Estabilidad*, Servicio de Publicaciones de la Universidad, Córdoba.
16. Vakaleris, D.G. & Price, W.V., 1959. *J. Dairy Sci.*, **42**, 264.
17. Marcos, A. & Mora, M.T., 1981. *Arch. Zootecn.*, **30**, 253.
18. Eigel, W.N., Butler, J.E., Ernstrom, C.A., Farrell, H.M., Harwalkar, V.R., Jenness, R. & Whitney, R. McL., 1984. *J. Dairy Sci.*, **67**, 1599.
19. Marcos, A., Fernández-Salguero, J., Esteban, M.A. & Alcalá, M., 1985. *J. Food Technol.*, **20**, 523.
20. Presidencia del Gobierno, 1983. *Boletín Oficial del Estado*, **153**, 18015.
21. Ministerio de Agricultura, Pesca y Alimentación, 1982. *Recopilación Legislativa Alimentaria*, Capítulo 15.—Leches y derivados, Servicio de Publicaciones Agrarias, Madrid.
22. Storry, J.E., Grandison, A.S., Millard, D., Owen, A.J. & Ford, G.D., 1983. *J. Dairy Res.*, **50**, 215.
23. Ramos, M. & Juárez, M., 1983. *International Dairy Federation Bulletin*, A-Doc **74**.
24. García, R. & Coll, L., 1976. *Anal. Bromatol.*, **28**, 211.
25. Juárez, M., Martínez, I., Méndez, A. & Martín, P.J., 1978. *Estudio sobre la Composición de la Leche de Vaca en España*, Instituto de Productos Lácteos, Arganda, Madrid.
26. García, R., Carballido, A. & Arnáez, M., 1979. *Anal. Bromatol.*, **31**, 227.
27. García, R., Carballido, A. & Arnáez, M., 1980. *Anal. Bromatol.*, **32**, 169.
28. Núñez, M., Medina, M. & Gaya, P., 1989. *J. Dairy Res.*, **56**, 303.
29. Marcos, A., Millán, R., Esteban, M.A., Alcalá, M. & Fernández-Salguero, J., 1983. *J. Dairy Sci.*, **66**, 2488.
30. Chávarri, F.J., Núñez, J.A., Bautista, L. & Núñez, M. 1985. *J. Food Protect.*, **48**, 865.
31. García, M.C., Otero, A., García, M.L. & Moreno, B., 1987. *J. Dairy Res.*, **54**, 551.
32. Otero, A., García, M.C., García, M.L., Prieto, M. & Moreno, B., 1988. *J. Appl. Bacteriol.*, **64**, 117.
33. Chávarry, F.J., Núñez, J.A., De Paz, M. & Núñez, M., 1986. *Proc. XXII Intern. Dairy Congress*, The Hague, p. 98.
34. Núñez, M., Chávarry, F.J., García, B.E. & Gaytán, L.E., 1986, *Food Microbiol.*, **3**, 235.
35. Bautista, L., Bermejo, M.P. & Núñez, M., 1986. *J. Dairy Res.*, **53**, 1.
36. Bautista, L., Gaya, P., Medina, M. & Núñez, M., 1988. *Appl. Environ. Microbiol.*, **54**, 566.
37. Johnson, E.A., Nelson, J.H. & Johnson, M., 1990. *J. Food Protect.*, **53**, 441, 519, 610.
38. Martín, C., Juárez, M. & Ramos, M., 1984. *Alimentación*, **4**, 61.
39. Millán, R., Alcalá, M., Esteban, M.A. & Marcos, A., 1982. ITEA, **1**, 418.
40. Esteban, M.A., Millán, R., Alcalá, M. & Marcos, A., 1982. ITEA, **1**, 424.
41. Alcalá, M., Marcos, A., Esteban, M.A. & Millán, R., 1982. ITEA, **1**, 431.
42. Millán, R., Alcalá, M., Esteban, M.A. & Marcos, A., 1982. ITEA, **1**, 439.
43. Montero, E., 1990. PhD Thesis, University of Córdoba, Córdoba.

44. Marcos, A., Alcalá, M., León, F., Fernández-Salguero, J. & Esteban, M.A., 1981. *J. Dairy Sci.*, **64**, 622.
45. Marcos, A. & Esteban, M.A., 1982. *J. Dairy Sci*, **65**, 1795.
46. López, P., Marcos, A. & Esteban, M.A., 1990. *J. Dairy Res.*, **57**, 587.
47. Centre Français du Commerce Exterieur, 1978. *Le Marché des Fromages en Espagne*, Departement des Etudes de Marché Agricoles et Alimentaires, Paris.
48. Compairé, C., 1961. PhD Thesis, University of Oviedo, León.
49. Ordóñez, J.A. & Burgos, J., 1977. *Lait*, **57**, 150.
50. Burgos, J. & Ordóñez, J.A., 1977. *Lait.*, **57**, 278.
51. Ordóñez, J.A. & Burgos, J., 1977. *Lait*, **57**, 416.
52. Burgos, J. & Ordóñez, J.A., 1978. *Milchwissenschaft*, **33**, 555.
53. Esteban, M.A. & Alcalá, M., 1982. *Alimentaria*, **132**, 31.
54. Marcos, A., Esteban, M.A., Alcalá, M. & Millán, R., 1983. *J. Dairy Sci.*, **66**, 909.
55. Alcalá, M., Fernández-Salguero, J., Esteban, M.A. & Marcos, A., 1984. *Ind. Láct. Españolas*, **63**, 37.
56. Cepeda, A., Paseiro, P., López de Alda, M.L., & Rodríguez, J.L., 1990. *Alimentaria*, **213**, 25.
57. Alcalá, M., Beltrán de Heredia, F.H., Esteban, M.A. & Marcos, A., 1982. *Arch. Zootecn.*, **31**, 131.
58. Alcalá, M., Beltrán de Heredia, F.H. Esteban, M.A. & Marcos, A., 1982. *Arch. Zootecn.*, **31**, 257.
59. Esteban, M.A., Marcos, A., Alcalá, M. & Beltrán de Heredia, F.H., 1982. *Arch. Zootecn.*, **31**, 305.
60. Marcos, A., Esteban, M.A., Alcalá, M. & Beltrán de Heredia, F.H., 1983. *Arch. Zootecn.*, **32**, 17.
61. Ramos, M., Barneto, R., Suárez, J.A., & Iñigo, B., 1982. *Chem. Mikrobiol. Technol. Lebensm.*, **7**, 167.
62. Suárez, J.A. & Iñigo, B., 1982. *Chem. Mikrobiol. Technol. Lebensm.*, **7**, 173.
63. Suárez, J.A., Barneto, R. & Iñigo, B., 1983. *Chem. Mikrobiol. Technol. Lebensm.*, **8**, 52.
64. Suárez, J.A., Barneto, R. & Iñigo, B., 1984. *Chem. Mikrobiol. Technol. Lebensm.*, **8**, 147.
65. Ferrán, R., Carbó, R. & Sancho, J., 1990. *Alimentaria*, **156**, 31.
66. Polo, C., Ramos, M. & Sánchez, R., 1985. *Food Chem.*, **16**, 85.
67. Juárez, M., Román, M., Martínez, I. & Barros, C., 1972. *Alimentaria*, **9**, 43.
68. Madrid, I., 1980. *Ind. Láct. Españolas*, **21**, 15.
69. Esteban, M.A., Marcos, A., Fernández-Salguero, J. & León, F., 1979. *Arch. Zootecn.*, **28**, 301.
70. Levieux, D., 1978. *20ᵉᵐᵉ Congres International de Laitiere*, Paris. Conferences 15ST, p. 11.
71. Román, M.L., 1990. PhD Thesis, University of Córdoba, Córdoba.
72. Ordóñez, J.A., Barneto, R. & Ramos, M., 1978. *Milchwissenschaft*, **33**, 609.
73. Ordóñez, J.A. & Burgos, J., 1980. *Milchwissenschaft*, **35**, 69.
74. Juárez, M., Martinez, I. & Ramos, M., 1980. *III Congr. Nac. Quím. Agric. Aliment.*, Sevilla.
75. Marcos, A., Esteban, M.A., Fernández-Salguero, J., Mora, M.T. & Millán, R., 1976. *Anal. Bromatol.*, **28**, 57.
76. Marcos, A., Esteban, M.A., Fernández-Salguero, J., Mora, M.T. & Millán, R., 1976. *Anal. Bromatol.*, **28**, 69.
77. Marcos, A. & Esteban, M.A., 1976. *Anal. Bromatol.*, **28**, 401.
78. Esteban, M.A. & Marcos, A., 1977. *Anal. Bromatol.*, **29**, 35.
79. Marcos, A., Fernández-Salguero, J. & Esteban, M.A., 1978. *Arch. Zootecn.*, **27**, 341.
80. López, P., Sánz, B. & Burgos, J., 1962. *Anal. Bromatol.*, **14**, 221.
81. Marcos, A., Fernández-Salguero, J., Mora, M.T., Esteban, M.A. & León, F., 1979. *Arch. Zootecn*, **28**, 29.

82. Wirotama, I.P.G., Ney, K.H. & Freytag, W.G., 1973. *Z. Lebensm. Unters.-Forsch.*, **153**, 78.
83. Muñoz, M.H., Rivas, J.C., & Mariné, A., 1981. *Anal. Bromatol.*, **33**, 225.
84. Iñigo, B., Martín, D., Barneto, R., Quintana, M.A., Garrido, M.P., Burdaspal, P. & Bravo, F., 1986. *Alimentaria*, **177**, 33.
85. Juárez, M. & Martín, C., 1983. *Rev. Agroquím. Tecnol. Aliment.*, **23**, 417.
86. Marcos, A., Esteban, M.A. & Fernández-Salguero, J., 1979. *Anal. Bromatol.*, **31**, 91.
87. Ramos, M. & Martínez, I., 1976. *Lait*, **56**, 164.
88. Mora, M.T. & Marcos, A., 1981. *Arch. Zootecn.*, **30**, 139.
89. Mora, M.T. & Marcos, A., 1982. *Arch. Zootecn.*, **31**, 27.
90. Marcos, A. & Mora, M.T., 1982. *Arch. Zootecn.*, **31**, 115.
91. Santa-María, G., Ramos, M. & Ordóñez, J.A., 1986. *Food Chem.*, **19**, 225.
92. Jiménez, S. & Reuter, H., 1983. *Alimentaria*, **142**, 45.
93. Fernández, E., Ramos, M., Polo, C., Juárez, M. & Olano, A., 1988. *Food Chem.*, **28**, 63.
94. Fernández, E., López, R., Olano, A. & Ramos, M., 1990. *Milchwissenschaft*, **45**, 428.
95. Jiménez, F., 1978. *Alimentaria*, **98**, 35.
96. Peláez, C., 1980. *Ind. Láct. Españolas.*, **23**, 25.
97. Jiménez, F., Goicoechea, A., García-Matamoros, E. & Peláez, C., 1979. *XV Cong. Int. Froid*, C-2, 82.
98. Peláez, C., 1983. *Alimentaria*, **144**, 19.
99. Román, M., 1975. *Lait*, **55**, 401.
100. Núñez, M. & Martínez, J.L., 1976. *Anal. INIA*, **4**, 11.
101. Martínez, J.L. & Núñez, M., 1976. *Anal INIA*, **4**, 33.
102. Ordóñez, J.A., Barneto, R. & Mármol, M.P., 1978. *Anal. Bromatol.*, **30**, 361.
103. Martínez, J.L., 1976. *Anal. INIA*, **4**, 41.
104. Núñez, M., 1976. *Anal. INIA*, **4**, 57.
105. Núñez, M., 1976. *Anal. INIA*, **4**, 67.
106. Núñez, M., 1976. *Anal. INIA*, **4**, 75.
107. Martínez, J.L., 1976. *Anal. INIA*, **4**, 83.
108. Ortiz, M.J. & Ordóñez, J.A., 1979. *Anal. Bromatol.*, **31**, 11.
109. Martínez, J.L., 1976. *Anal. INIA*, **4**, 93.
110. Núñez, M., 1976. *Anal. INIA*, **4**, 103.
111. Núñez, M., 1976. *Anal. INIA*, **4**, 113.
112. Núñez, M., Núñez, J.A., Medina, A.L., García, C. & Rodríguez, M.A., 1981. *Anal. INIA*, **12**, 53.
113. Núñez, M., Martínez, J.L. & Medina, A.L., 1981. *Anal. INIA*, **12**, 65.
114. Marcos, A., Esteban, M.A., Espejo, J., Martínez, P. & Muñóz, M.T., 1977. *Arch. Zootecn.*, **26**, 189.
115. Ordóñez, J.A., Barneto, R. & Ramos, M., 1978. *Proc. XX Int. Dairy Congr.*, Paris, Vol. E, p. 573.
116. Barneto, R. & Ordóñez, J.A., 1979. *Alimentaria*, **107**, 39.
117. Ramos, M., Barneto, R. & Ordoñez, J.A., 1981. *Milchwissenschaft*, **36**, 528.
118. Núñez, M., Núñez, J.A. & Medina, A.L., 1982. *Milchwissenschaft*, **37**, 328.
119. Medina, M., Gaya, P. & Núñez, M., 1982. *J. Food Protect.*, **45**, 1091.
120. Gaya, P., Medina, M. & Núñez, M., 1983. *J. Food Protect.*, **46**, 305.
121. De Paz, M., Chávarri, F.J. & Núñez, M., 1988. *Biotechnol. Techn.*, **2**, 165.
122. Núñez, M., García-Aser, C., Rodríguez, M.A., Medina, M. & Gaya, P., 1986. *Food Chem.*, **21**, 115.
123. Núñez, M., Gaya, P., Medina, M., Rodríguez-Marín, M.A. & García-Aser, C., 1986. *J. Food Sci.*, **51**, 1451.
124. Núñez, M., Gaya, P. & Medina, M., 1985. *J. Dairy Sci.*, **68**, 794.
125. Gómez-Lucía, E., Blanco, J.L., Goyache, J., Fuente, R., Vázquez, J.A. & Súarez, G., 1986. *J. Appl. Bacteriol.*, **61**, 499.

126. Gaya, P., Medina, M. & Núñez, M., 1986. *Rev. Esp. Lechería*, **8**, 31.
127. Gómez-Lucía, E., Goyache, J., Blanco, J.L., Vadillo, S., Garayzábal, J.F.F. & Suárez, G., 1987. *Z. Lebensm.-Unter. Forsch.*, **184**, 304.
128. Domínguez, L., Garayzábal, J.F.F., Vázquez, J.A., Blanco, J.L. & Suárez, G., 1987. *Lett. Appl. Microbiol.*, **4**, 125.
129. Gaya, P., Medina, M., & Núñez, M., 1987. *J. Appl. Bacteriol.*, **62**, 321.
130. Blanco, J.L., Domínguez, L., Gómez-Lucía, E., Garayzábal, J.F.F., Goyache, J. & Suárez, G., 1988. *J. Appl. Bacteriol.*, **64**, 17.
131. Gaya, P., Medina, M., Bautista, L. & Núñez, M., 1988. *Inter. J. Food Microbiol.*, **6**, 249.
132. Núñez, M., Bautista, L., Medina, M. & Gaya, P., 1988. *J. Appl. Bacteriol.*, **65**, 29.
133. Blanco, J.L., Domínguez, L., Gómez-Lucía, E., Garayzábal, J.F.F., Goyache, J. & Suárez, G., 1988. *J. Food Sci.*, **53**, 1373.
134. Martínez, I., Sanz, J., Amigo, L., Ramos, M. & Martín, P., 1991. *J. Dairy Res.*, **58**, 239.
135. Ordóñez, J.A., Massó, J.A., Mármol, M.P. & Ramos, M., 1980. *Lait*, **60**, 283.
136. Rodríguez, L., Barcina, Y., Ros, G. & Rincón, F., 1988. *Alimentaria*, **189**, 57.
137. Fernández-Salguero, J., 1975. PhD Thesis, University of Córdoba, Córdoba.
138. Fernández-Salguero, J., 1978. *Anal. Bromatol.*, **30**, 123.
139. Fernández-Salguero, J., 1978. *Anal. Bromatol.*, **30**, 131.
140. Fernández-Salguero, J., 1978. *Anal. Bromatol.*, **30**, 136.
141. Fernández-Salguero, J., Esteban, M.A. & Marcos, A. 1977. *Trab. Cient. Univ. Córdoba*, **7**, 1.
142. Fernández-Salguero, J. & Marcos, A., 1977. *Trab. Cient. Univ. Córdoba*, **13**, 1.
143. Marcos, A., Esteban, M.A. & Fernández-Salguero, J., 1977. *Trab. Cient. Univ. Córdoba*, **15**, 1.
144. Fernández del Pozo, B., Gaya, P., Medina, M., Rodríguez, M.A. & Núñez, M., 1988. *J. Dairy Res.*, **55**, 449.
145. Fernández del Pozo, B., Gaya, P., Medina, M., Rodríguez, M.A. & Núñez, M., 1988. *J. Dairy Res.*, **55**, 457.
146. González, J., Mas, M., & López, F., 1990. *Rev. Agroquím. Tecnol. Aliment.*, **30**(3), 356.
147. Fernández-Salguero, J., Barreto, J.D. & Marsilla, B.A., 1978. *Arch. Zootecn.*, **27**, 365.
148. Marsilla, B.A., 1979. *Arch. Zootecn.*, **28**, 255.
149. Martínez, P. & Fernández-Salguero, J., 1978. *Arch. Zootecn.*, **27**, 93.
150. Fernández-Salguero, J., Ruíz, J. & Marcos, A. 1984. *Rev. Agroquím. Tecnol. Aliment.*, **24**, 383.
151. Ruíz, J., Fernández-Saguero, J., Esteban, M.A. & Marcos, A., 1984. *Arch. Zootecn.*, **33**, 301.
152. Suárez, J.A., Barneto, R. & Iñigo, B., 1984. *Ind. Láct. Españolas*, **67**, 25.
153. Gómez, R., Peláez, C. & De la Torre, E., 1989. *Intern. J. Food Sci. Technol.*, **24**, 147.
154. Barreto, J.D., 1979. *Arch. Zootecn.*, **28**, 287.
155. Fernández-Salguero, J., Barreto, J.D. & Marsilla, B.A., 1981. *Alimentaria*, **119**, 71.
156. Gómez, R., Peláez, C. & Martín, C., 1988. *Food Chem.*, **28**, 159.
157. De la Torre, E. & Peláez, C., 1986. *Proc. V Spanish Meeting of Food Microbiol.*, Zaragoza, p. 141.
158. Fontecha, J., Peláez, C., Juárez, M., Requena, T., Gómez, C. & Ramos, M., 1990. *J. Dairy Sci.*, **73**, 1150.
159. Martín, C., Juárez, M., Ramos, M. & Martín, P., 1990. *Z. Lebensm. Unters.-Forsch.*, **190**, 325.
160. Fernández-Salguero, J., Sanjuán, E., Gómez, R. & Alcalá, M., 1990. *Alimentación*, **9**, 103.

161. Charro, A., Simal, J., Creus, J.M. & Trigueros, J., 1969. *Anal. Bromatol.*, **21**, 7.
162. Palo, V., 1975. *International Dairy Federation Bulletin*, A7-Doc **3**.
163. Ramos, M., 1976. *Rev. Esp. Lechería*, **101**, 147.
164. Ramos, M., Martínez, I. & Juárez, M., 1977. *J. Dairy Sci.*, **60**, 870.
165. Krause, I., Belitz, H.D. & Kaiser, K.P., 1982. *Z. Lebensm. Unters. Forsch.*, **174**, 195.
166. Ramos, M. & Juárez, M., 1981. *Rev. Esp. Lechería*, **121**, 193.
167. Ramos, M. & Juárez, M., 1984. *International Dairy Federation Bulletin*, No. **181**, 3.
168. Ruíz, E. & Santillana, I., 1986. *Alimentaria*, **171**, 55.
169. Juárez, M. & Ramos, M., 1986. *International Dairy Federation Bulletin*, No. **202**, 175.
170. Aranda, P., Oria, R. & Calvo, M., 1988. *J. Dairy Res.*, **55**, 121.
171. Amigo, L., Ibáñez, I., Fernández, C., Santa-María, G. & Ramos, M., 1989. *Milchwissenschaft*, **44**, 215.
172. García, T., Rodríguez, M.E., Martín, R., Azcona, J.I., Hernández, P.E. & Sanz, B., 1988. *Proc. III Congr. Luso-Español Bioquímica*, Santiago de Compostela.
173. García, T., Martín, R., Rodríguez, E., Hernández, P.E. & Sanz, B., 1989. *J. Dairy Res.*, **56**, 691.
174. García, T., Martín, R., Rodríguez, E., Morales, P., Hernández, P.E. & Sanz, B., 1990. *J. Dairy Sci.*, **73**, 1489.
175. Rodríguez, E., Martín, R., García, T., Hernández, P.E. & Sanz, B., 1990. *J. Dairy Res.*, **57**, 197.
176. Burgos, J., López, A. & Sala, F.J., 1971. *Anal. Fac. Vet. León*, **17**, 109.
177. Sala, F.J., Burgos, J. & Ordóñez, J.A., 1971. *Anal. Fac. Vet. León.*, **17**, 115.
178. Sala, F.J. & Burgos, J., 1972. *Anal. Bromatol.*, **24**, 83.
179. Sala, F.J. & Burgos, J., 1972. *Anal. Bromatol.*, **24**, 61.
180. Núñez, M., 1978. *J. Dairy Res.*, **45**, 501.
181. Núñez, M. & Medina, M., 1979. *Lait*, **59**, 497.
182. Núñez, M. & Medina, M., 1980. *Lait*, **60**, 171.
183. Núñez, M., Medina, M., Gaya, P. & Dias-Amado, C., 1981. *Lait*, **61**, 62.
184. Juárez, M., Alonso, L. & Ramos, M., 1983. *Rev. Agroquím. Tecnol. Aliment.*, **23**, 541.
185. Alonso, L., Juárez, M. & Ramos, M., 1983. *Proc. Europ. Food Chem. II*, Rome, p. 437.
186. Marcos, I., Martín, D., Barneto, R. & Quintana, M.A., 1985. *Alimentaria*, **213**, 35.
187. Fernández-Salguero, J., Alcalá, M., Marcos, A. & Esteban, M.A., 1987. *Alimentación*, **6**, 279.
188. Ramos, M., Alonso, L., González, D. & Juárez, M., 1987. *Rev. Agroquím. Tecnol. Aliment.*, **27**, 575.
189. Fernández-Salguero, J., Marcos, A., Alcalá, M. & Esteban, M.A., 1989. *J. Dairy Res.*, **56**., 141.
190. González, D., Ramos, M. & Polo, C., 1987. *Chromatographia*, **23**, 764.
191. Fernández-Salguero, J., Alcalá, M., Marcos, A. & Esteban, M.A., 1986. *J. Dairy Res.*, **53**, 639.
192. Marcos, A., Esteban, M.A., Espejo, J. & Marcos, I., 1990. *Alimentación*, **9**, 97.
193. Marcos, I., Quintana, M., Iñigo, B., Martín, D. & Barneto, R., 1986. *Alimentaria*, **171**, 65.
194. Mayo, B., Hardisson, C. & Braña, A.F., 1990. *J. Dairy Res.*, **57**, 125.
195. Alonso, L., 1985. PhD Thesis, Complutense University, Madrid.
196. Alonso, L., Juárez, M., Ramos, M. & Martín, P.J., 1987. *Z. Lebensm. Unters. Forsch.*, **185**, 481.
197. Alonso, L., Juárez, M., Ramos, M. & Martín, P.J., 1987. *Intern. J. Food Sci. Technol.*, **22**, 525.
198. Ramos, M., Cáceres, I., Polo, C., Alonso, L. & Juárez, M., 1987. *Food Chem.*, **24**, 271.
199. Alonso, L., Ramos, M., Martín, P.J. & Juárez, M., 1987. *J. Dairy Sci.*, **70**, 905.
200. González, D., 1989. PhD Thesis, Complutense University, Madrid.
201. González, D., Ramos, M., Polo, C., Sanz, J. & Martínez, I., 1990. *J. Dairy Sci.*, **73**, 1676.

202. Alonso, L., Juárez, M. & Ramos, M., 1989. *Rev. Agroquím. Tecnol. Aliment.*, **29**, 77.
203. Medina, M., Gaya, P. & Núñez, M., 1985. *J. Food Protect.*, **48**, 118.
204. Hewedi, M.M. & Fox, P.F., 1984. *Milchwissenschaft*, **39**, 198.
205. Alcalá, M., Esteban, M.A., Marcos, A. & Gómez, R., 1990. *Alimentaria*, **213**, 29.
206. Alcalá, M., Marcos, A., Esteban, M.A., Gómez, R. & Mancha, R., 1990. *Alimentación*, **9**, 141.
207. Marcos, A., Esteban, M.A. & Alcalá, M., 1990. *Food Chem.*, **38**, 189.
208. Esteban, M.A., Marcos, A., Alcalá, M. & Gómez, R., 1991. *Food Chem.*, **40**, 147.
209. Caballero, M., Arroyo, J.A. & Yubero, A., 1986. *Rev. Esp. Lechería*, **10**, 31.
210. Martín, M.C., Juárez, M. & Ramos, M., 1988. *Food Chem.*, **30**, 191.
211. Fernández-Salguero, J., Florido, S., Alcalá, M., Marcos, A. & Esteban, M.A., 1986. *Grasas y Aceites*, **37**, 152.
212. Marcos, A. & Esteban, M.A., 1991. *Inter. Dairy J.*, **1**, 137.
213. Vieira de Sá, F., 1978. *O Leite e os seus Productos*, 4th edn, Livraria Clássica Editora, Lisboa.
214. *International Dairy Federation Bulletin*, 1983. A-Doc. 158.
215. *International Dairy Federation Bulletin*, 1980. A-Doc. 2779.
216. Vieira de Sá, F. & Barbosa, M., 1970. *Rev. Portug. Ciénc. Vet.*, **65**(413), 17.
217. Vieira de Sá, F., 1983. *International Dairy Federation Bulletin*, No. **158**, 66.
218. Carr, S., 1981. *The Mitchell Beazley Pocket Guide to Cheese*, Mitchell Beazley Publishers, London.
219. Vieira de Sá, F. & Barbosa, M., 1982. *Ind. Láct. Españolas*, **45**, 39.
220. Eekhof-Stork, N., 1977. *The World Atlas of Cheese*, Paddington Press Ltd, London.
221. Vieira de Sá, F., Reis, B., Rafael-Pinto, O.P., Vicente da Cruz, I.M., Dias, M.J., Antunes, M.M. & Costa, M.M., 1970. INII. *Química e Biologia* No. 6, Lisboa.
222. Vieira de Sá, F. & Barbosa, M., 1982. *Proc. XXI Intern. Dairy Congress*, Moscow, **1**, p. 462.
223. Barbosa, M., 1986. *International Dairy Federation Bulletin*, No. **202**, 133.
224. Norma Portuguesa, 1982. *Queijo Serra da Estrela: Definição, Características, Acondicionamento e Conservação* NP-1922.
225. Vieira de Sá, F., 1986. *International Dairy Federation Bulletin*, No. **202**, 201.
226. Barbosa, M. & Gonçalves, I., 1986. *International Dairy Federation Bulletin*, No. **202**, 188.
227. Barbosa, M. & Miranda, R., 1986. *International Dairy Federation Bulletin*, No. **202**, 84.
228. Campos, R., Guerra, R., Aguilar, M., Ventura, O. & Camacho, L., 1990. *Food Chem.*, **35**, 89.
229. Vieira de Sá, F. & Barbosa, M., 1972. *J. Dairy Res.*, **39**, 335.
230. Christen, C. & Virasoro, E., 1935. *Lait*, **15**, 354.
231. Christen, C. & Virasoro, E., 1935. *Lait*, **15**, 496.
232. Pereira de Matos, A.A. & Vieira de Sá, F., 1948. *Boletim Pecuário*, **16**, 6.
233. Barbosa, M., Valles, E., Vassal, L. & Mocquot, G., 1976. *Lait*, **54**, 1.
234. Barbosa, M., Corradini, C. & Battistotti, B., 1981. *Sci. Técn. Latt.-Cas.*, **34**, 203.
235. Serrano, E. & Marcos, A., 1980. *Arch. Zootecn.*, **29**, 11.
236. Martínez, E. & Esteban, M.A., 1980. *Arch. Zootecn.*, **29**, 107.
237. Marcos, A., Esteban, M.A., Martínez, E., Alcalá, M. & Fernández-Salguero, J., 1980. *Arch. Zootecn.*, **29**, 283.
238. Cabezas, L., Esteban, M.A. & Marcos, A., 1981. *Alimentaria*, **128**, 17.
239. Marcos, A., Cabezas, L. & Esteban, M.A., 1982. *Alimentaria*, **129**, 33.
240. Esteban, M.A., Marcos, A. & Cabezas, L., 1982. *Alimentaria*, **130**, 19.
241. Marcos, A., Esteban, M.A. & Fernández-Salguero, J., 1976. *Arch. Zootecn.*, **25**, 73.
242. Marcos, A., Fernández-Salguero, J. & Esteban, M.A., 1978. *Anal. Bromatol.*, **30**, 314.

243. Marcos, A., Esteban, M.A. & Fernández-Salguero, J., 1978. *Arch. Zootecn.*, **27**, 285.
244. Marcos, A., Fernández-Salguero, J, Esteban, M.A. & León, F., 1979. *J. Dairy Sci.*, **62**, 392.
245. Creamer, L.K., 1970. *N.Z. J. Dairy Sci. Technol.*, **5**, 152.
246. Creamer, L.K., 1971. *N.Z. J. Dairy Sci. Technol.*, **6**, 91.
247. Fox, P.F. & Walley, B.F., 1971. *J. Dairy Res.*, **38**, 165.
248. Phelan, J.A., Guiney, J. & Fox, P.F., 1973. *J. Dairy Res.*, **40**, 105.

7

Italian Cheese

B. Battistotti

*Istituto di Microbiologia, Università Cattolica, Via Emilia Parmense,
84-Piacenza, Italy*

&

C. Corradini

*Istituto di Tecnologie Alimentari, Università di Udine, Via Marangoni,
97-Udine, Italy*

1 Introduction

In the previous edition of this book, Fox and Guinee wrote that while Italy may
not be ranked among the leading dairying countries of the world, at least relative to its size, its cheese industry is of the highest order. It is in many respects
unique, with a history of more than 2500 years. To cite Reinhold[1] 'Like Italian
art, architecture, music and literature, Italian cheese is a product of an ancient
culture. Cheese graced the banquet tables of the Caesars, served as rations for
the conquering Roman armies, and, today, is part of traditional dishes.'

Today, Italy has a greater range of cheese varieties than any other country, with
the exception of France, and the same Italian cheeses, Gorgonzola and Grana
(Parmigiano Reggiano and Grana Padano), rank among the famous international
cheese varieties. While Roquefort and Stilton may challenge Gorgonzola as the
prime blue-mould cheese, Grana cheeses are at the forefront of grating cheeses.
Italy is also the principal producer of that rather unique family of cheeses—the
Pasta filata or stretched curd cheeses—of which Provolone, Caciocavallo and
Mozzarella are the best known members. Italy is probably unique in using milk
commercially from four species (cow, sheep, goat and buffalo); goat cheeses
('caprini' or 'di capra'), sheep cheeses ('pecorini' or 'di pecora'), buffalo cheeses
('bufalini' or 'di bufala') and cheeses from mixtures of such milks (i.e., with cow
and either goat or sheep milk; with cow and buffalo milk; with goat and sheep
milk) are commercially produced.

For the level of its production, 760 426 tonnes in 1989 (Table I),[2] Italy is the
third largest cheesemaking country in Europe after France and the former West

TABLE I
Cheese Production in Italy, 1989

Cheese	Milk species	Milk quantity hl	Cheese yield	Butter yield	Cheese production, tonnes
				kg/hl	
Parmigiano-Reggiano	cow	15 784 160	6·25	1·75	98 651
Grana Padano	cow	14 986 451	6·03	1·80	90 368
Other Grana cheeses	cow	880 399	6·02	1·80	5300
Asiago ('d'allevo' and pressato)	cow	1 604 746	9·86	1·50	15 823
Montasio	cow	1 588 235	8·50	1·50	15 300
Fontina 'Valle d'Aosta'	cow	335 752	10·50	0·90	3525
Other semi-hard cheeses	cow	1 322 355	9·00	0·90	11 901
Provolone and similar cheeses	cow	5 312 500	8·00	1·00	42 500
Gorgonzola	cow	3 661 402	11·20	0·30	41 008
Italico, Quartirolo, Crescenza	cow	8 437 092	12·00	0·50	101 245
Taleggio	cow	1 291 667	12·00	0·50	15 500
Other fresh 'pasta	cow	18 750 000	12·00	0·50	225 000
filata' or generally	sheep	1 827 675	12·00	—	21 932
fresh cheeses	goat	237 125	12·00	—	2846
	buffalo	816 692	12·00	—	9800
Pecorino Romano (Sardinia)	sheep	1 232 788	17·00	—	20 958
Pecorino Romano (Rome)	sheep	346 306	18·00	—	6234
Fiore Sardo and other sheep cheeses	sheep	1 518 325	17·85	—	27 102
Caprini	cow	723 330	10·00	—	7233

Source: Associazione Italiana Lattiero-Casearia (1990) (Ref. 2).

Germany.[3] A higher proportion (about 45%) of milk production is converted into cheese in Italy than in any other major dairying country but even at this level, local supply is inadequate to meet demand and Italy is currently a large cheese importer (300 000 tonnes in 1989), mainly from the former West Germany, France and Switzerland. The home trade absorbs most of the Italian cheese production; in 1989 only 70 093 tonnes (about 10% of production) were exported to European countries (58 000 tonnes) or to North America (10 330 tonnes); Pecorino Romano, Fiore Sardo, Parmigiano Reggiano, Grana Padano and Gorgonzola represent most of the exported cheese.

As a result of the large number of Italian emigrants to the US and a few European countries, and to a lesser extent, Australia, Argentina and Brazil, some of the principal Italian cheese varieties, especially 'Parmesan', Romano and Mozzarella, have become international varieties, even if in most cases their characteristics are clearly different from cheeses with the same name made in Italy. Other factors which have contributed to the spread of Italian cheeses are the increasing popularity of Italian dishes, notably pizza and spaghetti, which has created an international demand for Mozzarella-type and Parmesan-type cheeses, respectively, and the large imports of cheese into Italy, some of which are of Italian-types produced in neighbouring European countries.

Cheesemaking in Italy is concentrated in the North of the country (Po valley and Prealpine districts) where the principal varieties, Parmigiano Reggiano, Grana Padano, Asiago, Montasio, Gorgonzola, Taleggio, Fontina and Italico (i.e., Bel Paese) are manufactured on a large scale. The Pasta filata types were, traditionally, produced principally in southern Italy and Sicily, frequently from buffalo milk; at present, however, this production is concentrated in the Po valley, often on big dairy farms. Traditionally, Pecorino Romano, and other sheep and goat cheeses of smaller size, are produced principally in the regions around Rome and in Sardinia.

2 CLASSIFICATION OF ITALIAN CHEESES

From remote antiquity, localized cheese production, which was previously produced domestically and later by little firms, but always for private use or local trade, has resulted in the differentiation of Italian cheeses into many varieties. Nevertheless, some cheeses have similar technology of production, similar sensorial characteristics, and similar chemical composition, but others have particular features related to the milk employed, to the particular technology and to the geographical environment.

In Italy, therefore, the following cheeses are legally designated by 'DOC' ('Denominazione d'origine controllata' i.e., controlled denomination of origin): Parmigiano Reggiano, Grana Padano, Gorgonzola, Fontina della Valle d'Aosta, Asiago, Robiola di Roccaverano, Formai de Mut, Castelmagno, Raschera, Montasio, Bra, Casciotta d'Urbino and Taleggio, all made from cows' milk, and Pecorino Romano, Pecorino Siciliano, Pecorino Toscano, Fiore Sardo, Canestrato Pugliese and Murazzano made from sheep's milk; these cheeses can be produced only in the traditional area. On the contrary, Mozzarella di Bufala, made from buffaloes' milk, and Provolone, Caciocavallo and Ragusano made with cows' milk, are always by law, 'typical', i.e. they can be produced throughout Italy.

The classification of Italian cheeses is complicated by the use of different names for the same or very similar cheeses in different regions, and by qualifications based on the species from which the milk is obtained and even the season when the cheese is produced. Many varieties are consumed after different degrees of ripening, e.g. as table cheeses after 2–4 months but as grating (low moisture) cheeses if ripened for a longer period, e.g. 12 months; it is therefore impossible to classify many cheeses on the basis of moisture content.

The previous edition of this book reported a modification of the scheme employed by Reinbold[1] to classify the 77 varieties listed by Walter & Hargrave.[4]

However, that list included varieties that are no longer produced, several repeated and unusual denominations, and cheeses made by different technologies and with different sensorial characteristics were placed in the same group. Moreover, more correct classification can be based on ripening factors.

TABLE II
Characteristics of the More

Cheese	Type of milk	Type of cheese	Starter	Type of rennet	Cooking temperature °C
Parmigiano Reggiano	cow raw partly skimmed	very hard	natural whey culture (thermophilic, rod shaped, lactic acid bacteria)	calf powder	54–55
Grana Padano	cow raw partly skimmed	very hard	natural whey culture (thermophilic lactic acid bacteria)	calf powder	53–54
Provolone piccante (piquant)	cow raw whole	semi-hard or hard	natural whey culture (thermophilic lactic acid bacteria; *S. thermophilus* <10%)	kid or lamb paste	50–52
Provolone dolce	cow raw whole	semi-hard	natural whey culture (thermophilic lactic acid bacteria; *S. thermophilus* <10%)	calf paste	48–50
Fontina della Valle d'Aosta	cow raw whole	semi-hard	none or thermophilic and mesophilic bacteria in milk	calf	48–50
Crescenza	cow pasteurized	soft	*S. thermophilus* or natural whey culture	calf liquid	37–39
Montasio	cow raw, whole or partly skimmed	semi-hard or hard	natural culture in whey or milk[a]	calf powder or liquid	43–48
Asiago d'Allevo	cow raw partly skimmed	semi-hard or hard	nothing or natural culture in whey or milk[a]	calf powder or liquid	46–48
Asiago pressato	cow pasteurized whole	soft	natural culture in whey or milk[a]	calf or bovine liquid or powder	46–48
Gorgonzola	cow pasteurized whole	soft	*L. delbrueckii* subsp. *bulgaricus* *S. thermophilus* *Penicillium roqueforti*	calf powder or liquid	32–34
Taleggio	cow pasteurized whole	soft	*L. delbrueckii* subsp. *bulgaricus* *S. thermophilus*	calf liquid	33–35
Italico (e.g. Bel Paese)	cow pasteurized whole	soft	natural milk culture (*S. thermophilus*)	calf liquid	42–44
Mozzarella di bufala	buffalo raw or thermalized whole	fresh	none or natural whey culture (*S. thermophilus*)	calf liquid	33–36
Pecorino Romano	sheep raw or thermalized whole	hard	*S. thermophilus* *L. delbrueckii* subsp. *helveticus* subsp. *lactis*	lamb paste	45–48

[a] *L. helveticus* and *L. delbrueckii* subsp. *lactis*; *S. thermophilus*.

Representative Italian Cheeses

Days and temperature °C of ripening	Enzymatic processes during ripening: proteolysis	lipolysis	In ripened cheeses amount of lactic acid bacteria with vital functions	pH	carbonyls content
540–730, 18–20	deep, slow	weak	very poor	5·5–5·6	high
360–470, 18–20	deep, slow	weak	very poor	5·4–5·5	high
120–360, 15–16	fairly deep	strong	poor	5·1–5·3	poor
120–180, 10–15	weak	weak	fairly good	5·3–5·4	poor
90–120, 8	weak	weak	high	5·6	fairly good
6–7, 6–7	fairly deep	weak	good	5·6	poor
120–360, 14–16	fairly deep	weak	fairly good	5·6	fairly good
120–360, 15–16	fairly deep	weak	fairly good	5·5	fairly good
25–30, 15–16	fairly deep	weak	fairly good	5·5	fairly good
50–80, 5–8	deep	strong	high	5·8–6·0	good
40–50, 6–8	fairly deep	weak	high	5·4–5·5	good
20–30, 6–8	weak	weak	high	5·4	fairly poor
— —	weak	weak	high	5·3	fairly poor
240–360, 10–14	deep	strong	very poor	5·3–5·4	no data

According to such concepts, the particular characteristics of the more representative Italian cheeses are summarized in Table II.

The principal cheeses made in Italy are as follows:

Besides Parmigiano Reggiano and Grana Padano, Bagozzo and Lodigiano are also Grana cheeses, but, because of very limited production, they have practically disappeared from the market. The 'Grana' are hard or very hard cheeses with a granular texture, made from partially skimmed milk and with a very long ripening period, characterized by slow proteolysis and limited lipolysis.

'Pasta filata' cheeses are: Mozzarella, made from both buffaloes' and cows' milk (the latter is also named 'fior di latte'), Scamorza, Scamorzina, Provola, Burrino, Manteca, Provolone with several shapes (Topolino, Pancetta, Pancettone, Mandarino), Caciocavallo, Ragusano and Silano which differ from one another in the duration of ripening and consistency. The most important are Provolone, a hard or semihard cheese, and Mozzarella, a fresh cheese.

Semihard cheeses, traditionally produced in Alpine and Prealpine Valleys, are: Asiago, Montasio, Fontina della Valle d'Aosta, Bitto, Sbrinz, Monte Veronese, Formai de mut, Bra, Castelmagno and similar cheeses; all these are eaten as table cheeses or used in traditional dishes.

Hard and semihard cheeses obtained from cooked or lightly cooked curds, whether stretched or not, are the most classical and representative of Italian production; they are characterized by a long ripening period of at least 6 months, according to the shape and the weight of the cheese. The taste is generally marked and sometimes, for the cheeses made with rennet paste containing lipases, also 'piccante'.

Soft cheeses are: Crescenza, Raschera, Robiola di Roccaverano, Caciotta (cows'), Toma, Tomino, Gorgonzola, Pannerone, Formagella, Robiola, Robiolino, Taleggio, Quartirolo and Italico (e.g. Bel Paese), all of which are table cheeses. They are made from whole milk, pasteurized or 'termizzato' (i.e. heated to 63°C without a holding period at this temperature), inoculated with starter or natural culture. The texture, depending on the acidity at coagulation, is elastic or fribable. The most representative cheeses are Gorgonzola and Italico.

Apart from Mozzarella, very few fresh cheeses are produced in Italy; this group includes varieties that have recently been in good demand, such as caprino cheeses, made by acid coagulation, and presamic curds (Casatella) which because of their acidic or very soft taste can be used in combined dishes, mostly as sources of protein and lipid.

Among the traditional Italian cheeses, there are varieties made from fully skimmed milk while some, such as Grana and Montasio cheeses, are made from partially skimmed milk, with fat contents of about 40% of the dry matter.

Generally, salting is in brine; only Montasio and Fontina are sometimes dry salted. For Asiago pressato, the salt can be partially added to the milk or to the curd before pressing.

Ricotta and sometimes also Mascarpone are not classified as cheeses, because they are obtained from whey and cream, respectively; both are coagulated without rennet but by the combined effects of high temperature and acidity. They are fresh, unsalted dairy products.

Apart from 'Mozzarella di Bufala', all the above mentioned cheeses are made from cows' milk.

In Italy, cheeses made from goats' milk have limited significance, but sheep's milk cheeses have great economic significance, particularly in central and southern Italy and in Sardinia. The best known sheep's milk cheeses are: Caciotta, Canestrato, Pecorino Romano, Pecorino Toscano, Pecorino Siciliano, Fiore Sardo, Murazzano, Calcagno; Ricotta made from sheep's milk whey can be salted and ripened ('Ricotta salata', i.e. salted Ricotta).

Normally, all these are table cheeses, but after prolonged ripening they can be used also as grating cheeses (e.g. Pecorino Romano).

3   PARMIGIANO REGGIANO AND GRANA PADANO CHEESES

The 'grana' cheeses (Parmigiano Reggiano and Grana Padano) are the most important 'characteristic' group of Italian cheeses from a technological point of view (the other 'characteristic' groups are Pasta filata and Pecorino Romano types').

Particularly, Parmigiano Reggiano is the best known (in the world as 'Parmesan') and is produced according to a traditional technology from raw milk in a restricted region of the Po valley. In this area, the feed of the cows is regulated by careful instructions, e.g. silage fodder is not allowed to control the content of gas-producing bacteria.

The raw milk for Grana cheesemaking is partially skimmed by natural creaming. In Parmigiano Reggiano cheesemaking, a mixture of milks from two consecutive milkings, in which only the milk from the evening milking is skimmed after overnight creaming, is used. Grana Padano is made from milk from a single milking, skimmed by creaming for about eight hours. In this way, the fat content of the milk in the vat is about 2·4–2·5% for Parmigiano Reggiano and about 2·1–2·2% for Grana Padano. During creaming, the microbial content of the milk is reduced.

The vats used in Grana cheesemaking have a capacity of 10–12 hl; therefore, from each vat, two cheeses, each weighing about 38–40 kg 24 h after manufacture (i.e. 33–35 kg after ripening), are produced. Grana cheeses have a cylindrical shape with a diameter of 33–45 cm and a height of 18–25 cm.

The manufacturing protocol of Grana cheeses is summarized in Table III and their composition in Table IV.

Ripened Grana cheeses are classified as extra hard because of their low moisture content, decreasing during ripening to about 30% (Fig. 1).

These cheeses have a typical texture ('di grana', i.e. 'grained'); they are compact with or without many very small eyes. Grana cheeses melt in the mouth with a sweet flavour, which is the result of very slow ripening, in which the proteins are the principal components involved.

TABLE III
Manufacturing Protocol for Grana Cheeses

Acidity of milk in vat	0·13–0·16%
Starter	Natural whey culture[a]
Acidity of whey culture	1·25–1·35%
Acidity of milk after starter addition	0·19–0·20%
Coagulation temperature	33–35°C
Rennet	calf rennet
Coagulation time	9–10 min
Setting time	2–3 min
Cooking temperature and time	to 54–55°C in 10–12 min
Time from rennet addition to end of cooking	22–23 min
pH of whey at draining	6·35
Salting	in saturated brine
Salting time	20–23 days for Parmigiano Reggiano
	25–26 days for Grana Padano
Ripening time	about 2 years for Parmigiano Reggiano
	14–16 months for Grana Padano
Ripening temperature (if controlled)	18–20°C

[a] The natural whey culture is obtained from whey from a previous cheesemaking, held for 24 h at decreasing temperature (from 50°C to about 35°C). Its microbiological composition is complex; the main microorganisms present are thermophilic lactic acid bacteria ($0·8–1 \times 10^9$ cfu/ml): *Lactobacillus helveticus, L. delbrueckii* subsp. *lactis* and, in a ratio of about 1 to 10 with the above mentioned species, *L. fermentum* (Ref. 5).

TABLE IV
Composition of Ripened Parmigiano Reggiano and Grana Padano (Average Data)

	Parmigiano Reggiano	Grana Padano
	g/100 g of cheese	
Moisture	30·8	32·0
Total protein (N × 6·38)	33·0	33·0
Fat	28·4	27·0
Ash	4·6	4·9
Calcium	1·15	1·15
Phosphorus	0·7	0·7
Magnesium	0·04	0·043
Salt (NaCl)	1·4	1·7
$\dfrac{\text{Soluble N}}{\text{Total N}}$ %	32	34
In the ripened cheeses there are also:	Vit. A	300 μg
	Vit. B$_1$	22 μg
	Vit. B$_2$	370 μg
	Vit. B$_6$	106 μg
	Colin	42 μg
	Biotin	22 μg
	Vit. B$_{12}$	4 μg

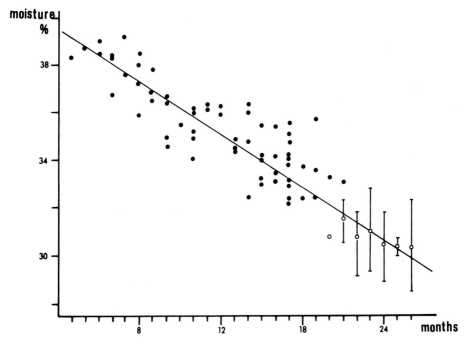

Fig. 1. Moisture (g/100 g of cheese) in Parmigiano Reggiano during ripening. Black points are for single cheeses of different age; white points are the average of few cheeses, of the same age, at the end of ripening (Ref. 6).

The principal proteolytic agents in these cheeses are microbial enzymes (mainly from thermophilic lactic acid bacteria of whey cultures which develop during the first hours after manufacture) and indigenous milk proteinases. The enzymes of calf rennet are practically inert during the cooking.[7]

Lactic acid fermentation in Parmigiano Reggiano cheese during the first 48 h after manufacture was studied by Mora et al.[8] The results showed that the growth of thermophilic lactic acid bacteria and lactose hydrolysis depended mainly on the rate at which the curd cooled after removal from the cheese vat. The temperature at the centre of the cheeses remained > 50°C for about 10 h (the curd was at 55°C at hooping) while that of the exterior of the cheeses decreased to about 42°C in 2 h. Consequently, bacterial growth commenced earlier and was more intense in the exterior than in the interior of the cheeses and this affected sugar metabolism and acidity development (Figs 2 and 3). After 48 h the number of bacteria, pH and lactate concentration had not attained equal values throughout the cheese and the authors suggest that these differences may affect subsequent ripening.

Proteolysis in Grana cheeses has been studied by many authors using many different analytical methods. Starch gel electrophoresis (see Fig. 4) of samples during the first months of ripening shows the formation of two bands with faster mobility than α_{s1}-casein, i.e. the primary degradation products of α_{s1}-casein:

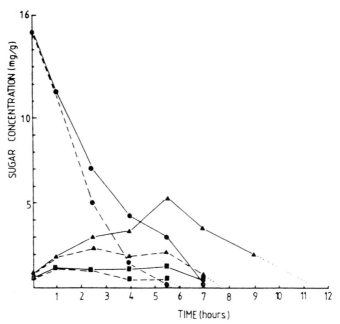

Fig. 2. Changes of lactose (●), glucose (■) and galactose (▲) concentrations in the external (----) and internal (——) parts of Parmigiano cheese (Ref. 8).

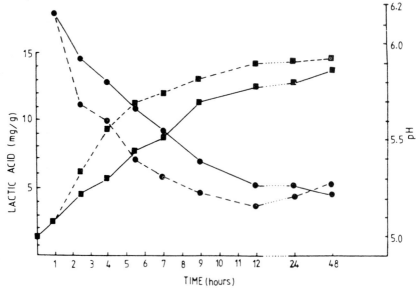

Fig. 3. Changes of pH (●) and lactic acid concentration (■) in the external (----) and internal (——) parts of Parmigiano cheese (Ref. 8).

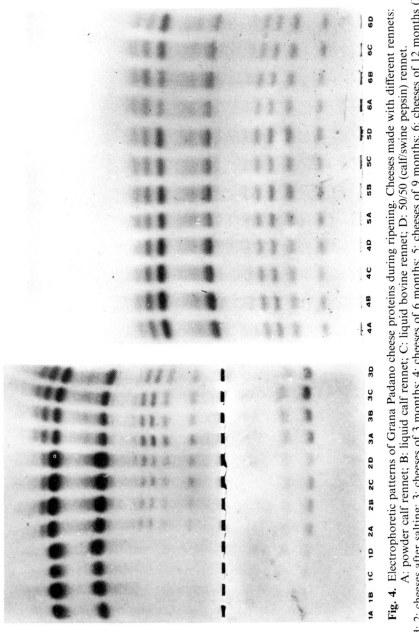

Fig. 4. Electrophoretic patterns of Grana Padano cheese proteins during ripening. Cheeses made with different rennets: A: powder calf rennet; B: liquid calf rennet; C: liquid bovine rennet; D: 50/50 (calf/swine pepsin) rennet. 1: curd; 2: cheeses after salting; 3: cheeses of 3 months; 4: cheeses of 6 months; 5: cheeses of 9 months; 6: cheeses of 12 months (Ref. 7).

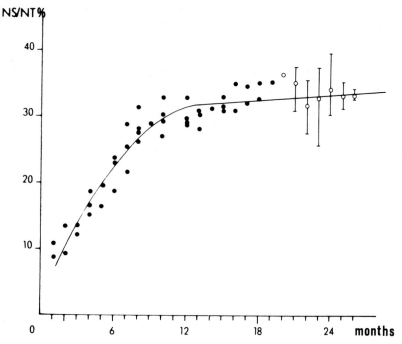

Fig. 5. Increases in the level of water soluble nitrogen in Parmigiano Reggiano cheese during ripening. Open circles are the average of a few cheeses, of the same age, at the end of ripening (Ref. 10).

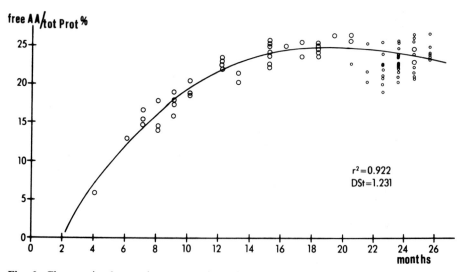

Fig. 6. Changes in the total concentration of free amino acids in Parmigiano Reggiano cheese during ripening (Ref. 11).

α_{s1}-I and α_{s1}-II casein; in parallel, the intensity of the bands corresponding to the γ-caseins, from the degradation of β-casein, increased. These results were recently confirmed by Addeo et al.[9] by PAGE (polyacrylamide-agarose gel electrophoresis) and PAGIF (isoelectric focusing on polyacrylamide gel).

The degradation of the principal caseins is reflected in the increasing proportion of water soluble nitrogen (see Fig. 5)[10]; in this manner, the ratio soluble N/ total N%, can be adopted as a ripening coefficient for Grana cheeses.

As the final step in proteolysis, during the slow ripening, there is a very large increase in the concentration of free amino acids (Fig. 6), which in ripened Parmigiano Reggiano reach the distribution and concentration reported in Table V.[11] This high concentration of free amino acids is important from a nutritional point of view.

TABLE V
Concentration of Free Amino Acids (g/100 g) in Parmiagiano Reggiano
Cheese at the End Ripening (Average of 60 Cheeses) (Ref. 11)

	Average	Minimum	Maximum
ASP	0·99	0·73	1·32
GLU	4·41	0·73	5·18
ASN	0·87	0·59	1·18
SER	1·36	0·96	1·71
GLN	0·14	0·07	0·24
HIS	0·82	0·17	1·10
GLY	0·64	0·52	0·83
THR	1·23	0·86	1·66
ARG	0·25	0·00	1·22
ALA	0·69	0·52	0·81
GABA[a]	0·02	0·00	0·17
TYR	0·63	0·34	0·88
AABA[b]	0·02	0·00	0·14
MET	0·72	0·57	0·90
VAL	1·84	1·49	2·15
PHE	1·32	1·08	1·57
ILE	1·59	1·24	1·89
LEU	2·22	1·85	2·67
ORN	0·38	0·12	0·67
LYS	3·08	2·43	3·91
$\dfrac{\text{Total free AA}}{\text{Total prot.}}$ %	23·21	18·91	27·33
$\dfrac{\text{Total free AA}}{\text{Cheese}}$ %	7·61	6·09	8·85

[a] γ-amino butyric acid.
[b] α-amino butyric acid.

4 PASTE FILATE

4.1 Provolone

The archetypes of 'pasta filata' cheeses were originally produced in the south of Italy, domestically on small dairy farms, but the name 'provolone' was adopted only at the end of the last century when industrial dairies in the north of Italy commenced the production of a cheese similar to the 'caciocavallo'.

Provolone is made in many shapes and with different weights. At present, there are Provolone cheeses, with different names according to the weight, of the following shapes: semi-conical (calabrese, silano, gigantino), cylindrical (topolino, pancetta, pancettone, salame, gigante), spherical (mandarino, provoletta, mellone), parallelepipeded (ragusano). The weights vary from 0·5 kg (minimum) to about 100 kg.

Provolone is a semihard/hard cheese, with a compact but rather flaky texture, without eyes but with a few small slits in the more ripened types. The rind is thin with a yellow-gold colour: in contrast, the colour of the cheese is pale yellow or white, tending to straw-coloured cream. The flavour is quite mild in the sweet types (made with calf rennet) or typical 'piccante' in the cheeses made by kid or lamb rennet pastes. Today, most people prefer the sweet cheeses.

The manufacturing protocol for Provolone cheese with a piquant taste after medium ripening (25 kg of weight and 6–8 months of ripening) is summarized in the following and in Table VI.

The raw milk, generally not kept cooled at the farm, is put into vats, warmed to 35–38°C and inoculated at 1–2% with natural whey culture. This culture is obtained from whey from previous cheesemaking, held at 40–42°C so as to reach an acidity of 38–40° SH; the microorganisms in such a culture are principally thermophilic bacteria.[12]

For coagulation, kid or lamb rennet paste (calf rennet paste for sweeter types) is normally used; very rarely, liquid calf rennet and pregastric lipase extracts or microbial lipases are used.

TABLE VI
Manufacturing Protocol for Provolone Cheese ('Piccante' Type)

Acidity of raw milk in vat	0·16–0·18%	pH: 6·5–6·6
Starter	Natural whey culture	
Acidity of whey culture	1·0–1·1%	pH: 3·6
Acidity of milk after starter addition	0·18–0·19%	pH: 6·4
Coagulating temperature	35–38°C	
Rennet	Kid or lamb rennet paste	
Coagulation time	10–15 min	
Setting time	10–15 min	
Cooking temperature and time	to 48–52°C in 15–20 min	
Acidification-ripening of curd	3–5 h	pH: from 6·2–4·9
Curd after stretching process		pH: 5·1
Duration of salting in brine	Depends on weight	
Ripening time and temperature	3–12 months at 12–15°C	pH: from 5·2 to 5·3–5·5

The coagulation time is about 20–22 min; at this point, the curd is cut and, after a short mixing period, the curd particles are allowed to settle to the bottom. After 5–10 min, the whey is drawn off and the curd dispersed in a portion of the whey which had been warmed to 63–65°C; the suspension reaches the cooking temperature of 48–52°C, at which point it is stirred for 10–12 min.

At this point, the curd is removed from the vat and spread on tables as large blocks, 30–40 cm high, for 3–5 h. During this period, the temperature is held at about 30°C, permitting a rapid increase in acidity and the consequent transfer of calcium from the casein complexes to soluble forms. When the pH reaches 4·9–5·1, about 50% of the calcium is solubilized and the curd can be stretched.

At this point, the curd, cut in little strips, is kneaded in hot water (85–90°C) until it can be stretched. Traditionally, this work was done manually in wooden vats, but today the curd is stretched in a steel mechanical apparatus, often using continuous systems.

In this way, the curd is stretched, formed (and when of greater weight, put into steel moulds) and refrigerated in cool water.

After salting in brine (at 10–15°C for different periods depending on their weight) the cheeses are dried, sometimes covered with a paraffin film, and ripened in rooms at 15–16°C for 1–12 months, depending on the weight and shape.

The chemical composition of ripened Provolone cheese is summarized in Table VII.

The sensorial characteristics of the different Provolone cheeses depend on the extent of lipolysis, due mainly to the action of pregastric esterase in rennet pastes which are employed as the source of both coagulant and lipolytic agents in manufacture of these cheeses.

Rennet pastes are prepared by macerating the engorged stomachs, including contents (curdled milk), of young calves, kid goats or lambs, which are slaughtered immediately after suckling or pail-feeding. The stomachs and contents are held for about 60 days before macerating. Pregastric esterase, the physiological role of which is to aid in digestion of fat by young animals with limited pancreatic lipase, is secreted during suckling and is carried into the stomach with ingested milk.

TABLE VII
Chemical Composition of Provolone Cheese ('Piccante' Type)
(Average Data)

	g/100 g of cheese
Moisture	38·0
Dry matter	62·0
Fat	28·5
Total protein (N × 6·38)	24·0
Ash	4·8
Non-casein nitrogen	0·9
Non-protein nitrogen	0·7

Harper & Gould,[13] who analyzed rennet pastes and commercial pregastric esterase preparations from calves, kids and lambs for lipolytic activity on milk fat, identified multiple lipase systems in each. Connoisseurs of Italian cheese claim that rennet pastes give cheeses a flavour superior to those made using pregastric esterase preparations. The superiority of rennet pastes may be due to the presence of enzymes in addition to pregastric esterase and to the normal gastric

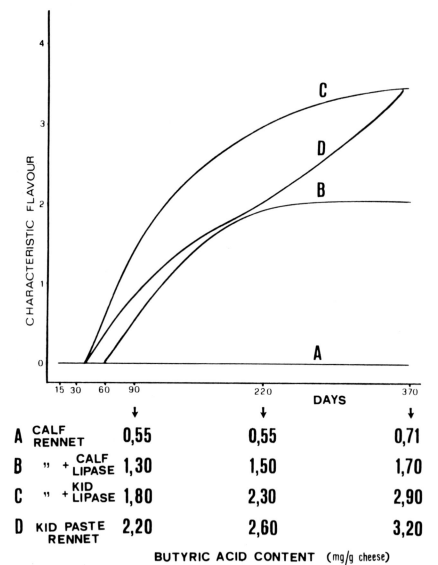

Fig. 7. Characteristic flavour development and free butyric acid content in Provolone cheeses produced with different rennets and lipase preparations (from data of Long & Harper — Ref. 16).

TABLE VIII
Free Fatty Acids in 71 Ripened Provolone Cheeses (Ref. 17)

	Average	Maximum	Minimum
		mg/100 g of cheese	
Total FFA	6350	25 819	2095
$C_{4:0}$	932	2760	100
$C_{6:0}$	363	942	49
$C_{8:0}$	171	458	43
$C_{10:0}$	316	993	89
$C_{12:0}$	329	1095	98
$C_{14:0}$	678	3208	229
$C_{16:0}$	1413	8284	394
$C_{18:0}$	368	1273	80
$C_{18:1}$	1181	5999	343
$C_{18:2}$	186	713	42

proteinases (and possibly gastric lipase). Such enzymes could be of microbial origin since considerable bacterial growth probably occurs during the ageing period (up to 60 days) prior to maceration of the stomachs.

The superiority of rennet pastes was confirmed by Harper & Gould,[14] who investigated flavour development in commercial Romano and Provolone cheeses, selected to include those made with either rennet extract, rennet pastes from calves, kids or lambs, or pregastric esterase preparations from kids or calves. Cheeses made with either commercial rennet extracts or purified rennet pastes did not develop satisfactory flavour, whereas those made with either unpurified pastes or pregastric esterase preparations did.

According to Harper & Long,[15] the FFA and FAA levels exhibited wide variations in cheeses of the same age. Butyric acid was the only FFA definitely related to the type of enzyme preparations used and to the desirable 'piccante' flavour of Provolone cheese (Fig. 7).

Recently, De Felice et al.,[17] who studied free fatty acid formation in Provolone cheeses made in Italy, have confirmed such findings; their results for the free fatty acid concentration in ripened cheeses are reported in Table VIII. However, in the sensorial evaluation of Provolone cheese, the quality must be related to the ratios of a few free fatty acids: e.g. in ripened Provolone with good flavour, the ratio of C_4/C_6 is normally 2·5 to 4·0.[18]

On the other hand, wide variations in the concentrations of free amino acids in Provolone cheeses was shown by Resmini et al.[19] (see Table IX). According to these authors, the variability both in free amino acid content and in their relative ratios can be due to different processing and ripening conditions.

4.2 Mozzarella

The other 'pasta filata' cheeses are produced according to the traditional method of manufacture; Mozzarella, either from buffaloes' or cows' milk, is produced

TABLE IX

Free Amino Acid Content (as % of Total Protein) in 12 Commercial Provolone Cheese Samples After 5-6 Months of Ripening (Ref. 19)

Sample	ASP	GLU	ASN	SER	GLN	HIS	GLY	THR	ARG	ALA	GABA	TYR	AABA	MET	VAL	PHE	ILE	LEU	ORN	LYS	Σ
I	0·48	2·53	0·80	0·59	0·68	0·32	0·29	0·77	0·07	0·35	0·39	0·43	0·03	0·50	1·26	1·07	0·91	1·66	0·51	1·60	15·33
II	0·45	1·05	0·44	0·40	0·51	0·39	0·20	0·45	0·68	0·25	0·73	0·39	0·02	0·29	0·83	0·75	0·61	1·21	0·04	1·24	10·93
III	0·50	2·65	0·82	0·69	0·74	0·52	0·30	0·85	0·51	0·40	0·16	0·44	0·02	0·52	1·28	1·09	0·99	1·90	0·28	1·93	16·59
IV	0·34	1·83	0·69	0·44	0·73	0·59	0·26	0·41	0·89	0·34	0·29	0·51	0·02	0·37	0·82	0·85	0·57	1·50	0·00	1·15	12·60
V	0·24	1·45	0·46	0·28	0·32	0·27	0·19	0·30	0·73	0·23	0·00	0·33	0·02	0·29	0·67	0·64	0·43	1·23	0·00	1·41	9·49
VIa	0·31	1·88	0·55	0·35	0·72	0·00	0·21	0·35	0·06	0·29	0·00	0·32	0·03	0·35	0·83	0·76	0·65	1·52	0·57	1·30	11·05
VIb	0·29	1·10	0·47	0·34	0·46	0·47	0·19	0·43	0·26	0·27	0·40	0·33	0·02	0·30	0·71	0·59	0·55	1·05	0·22	1·09	9·54
VII	0·18	0·37	0·50	0·24	0·37	0·33	0·14	0·32	0·62	0·17	0·87	0·30	0·01	0·27	0·60	0·56	0·38	1·06	0·04	0·90	8·23
VIIIa	0·31	1·94	0·59	0·42	0·52	0·38	0·16	0·53	0·30	0·25	0·00	0·28	0·01	0·28	0·78	0·70	0·66	1·18	0·29	1·17	10·75
VIIIb	0·35	1·94	0·58	0·48	0·54	0·28	0·17	0·56	0·17	0·26	0·04	0·38	0·00	0·33	0·87	0·74	0·61	1·10	0·37	1·22	10·99
IX	0·42	1·47	0·72	0·56	0·61	0·68	0·19	0·66	0·74	0·35	0·89	0·52	0·00	0·49	1·09	0·94	0·89	1·54	0·06	1·60	14·42
X	0·44	1·52	0·22	0·32	0·33	0·16	0·17	0·31	0·71	0·22	0·00	0·31	0·00	0·28	0·68	0·60	0·40	1·15	0·04	0·89	8·75

GABA = γ amino butyric acid; AABA = α amino butyric acid.
The roman numbers indicate the producer; a and b are cheeses of the same producer.

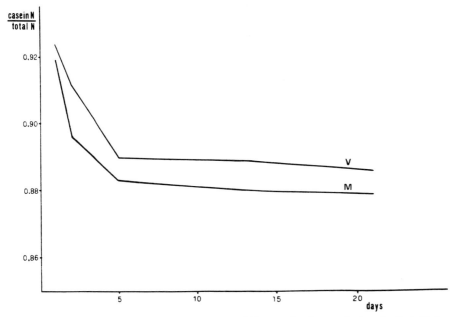

Fig. 8. Changes in the casein nitrogen content of Mozzarella cheeses during the first 20 days after manufacture. The Mozzarella cheeses were made from cows' milk using calf rennet (V) or *M. pusillus* proteinases (M) (Ref. 21). The casein nitrogen was determined by the method of Schulz & Mrowetz (Ref. 22).

through biological acidification of curd by a culture in milk of *Str. thermophilus*. The milk is coagulated at 33–37°C with liquid or powdered rennet. The limited cutting (0·5–1 cm) and syneresis of the curd produce a stretched cheese containing 48–52% of moisture which is used directly as a table cheese with salads or on pizza.

The extent of salting during stretching or in 10% NaCl brine, is limited (about 0·1–0·5% NaCl in cheese).

The weight of the spherical cheeses ranges from 7 g to 2000 g.

The little cheeses are packed with water (containing 0·5% NaCl, at the most) and commercialized for 10–15 days. Mozzarella for pizza containing 48% moisture is packed without water and is commercialized for 20–25 days.

During the very short period before sale, the level of proteolysis in Mozzarella cheese is quite low. However, Di Matteo *et al.*,[20] who studied proteolysis in Mozzarella by electrophoresis, showed the almost complete breakdown of α_{s1}-casein to α_{s1}-I and α_{s1}-II, indicating that some rennet activity survives the kneading and stretching process; compared to other cheese varieties, however, the extent of proteolysis in Mozzarella is very limited. The changes in casein nitrogen reported in Fig. 8 confirm such results and show that the extent of casein proteolysis in cows' Mozzarella during the first 20 days after manufacture is about 4%.[21]

Recently, proteolysis both in bovine and buffalo Mozzarella was studied by Chianese *et al.*,[23] using a two-dimensional electrophoretic procedure, which gave a better resolution of α_{s1} and α_{s1}-I caseins in such cheeses (Fig. 9). This method permits the detection of bovine milk in buffalo Mozzarella in amounts as low as 2·5%, even when a comparatively extensive degradation of the caseins has taken place in the cheese.

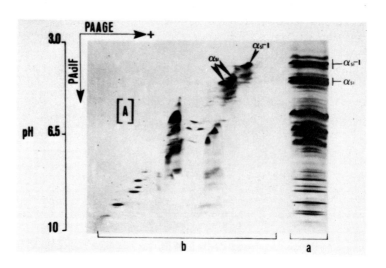

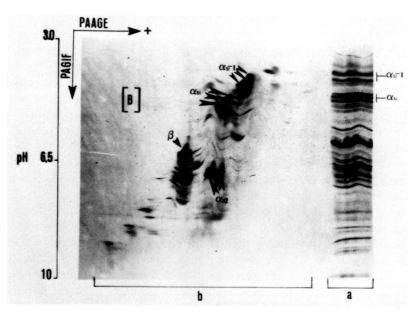

Fig. 9. Two-dimensional electrophoresis of the total proteins of bovine (A) and water buffalo (B) Mozzarella cheese samples showing limited hydrolysis of α_{s1}-casein. Slot a: PAGIF patterns; slot b: two-dimensional electrophoresis (Ref. 23).

5 PECORINO ROMANO

Pecorino Romano cheese is a hard cheese made exclusively from ewes' milk and its manufacture is, by law, restricted to regions around Rome and in Sardinia.

Pecorino Toscano (previously called Toscanello), Pecorino Sardo, Fiore Sardo, Pecorino Moliterno, Canestrato Pugliese, Crotonese, Pecorino Siciliano (also called Calcagno, Canestrato or Incanestrato Siciliano) and Pepato Siciliano (flavoured with peppers) are also hard cheeses made from ewes' milk, but Pecorino Romano is the best known.

The manufacturing protocol of Pecorino Romano and its composition are summarized in Tables X and XI, respectively.

The milk for this cheese is usually thermalized and inoculated with 'scotta-fermento', a natural culture obtained by acidifying the 'scotta', i.e. the residual whey from Ricotta manufacture.

TABLE X
Manufacturing Protocol for Pecorino Romano Cheese

Milk	Ewes' milk, raw or, usually, thermalized
Starter	Natural culture obtained from Ricotta whey (scotta)
Starter inoculum	400–500 ml per 100 litre of milk
Rennet	Lamb rennet pastes (30–50 g per 100 litre of milk)
Coagulation temperature	37–39°C
Coagulation time	14–16 min
Curd setting time	2–3 min
Cooking temperature	45–46°C
Time from beginning of cooking to deposition of curd particles on vat bottom	28–30 min
Salting	Dry salting for 30–60 days
Ripening temperature	10–14°C
Ripening time	8–12 months

TABLE XI
Chemical Composition of Ripened Pecorino Romano Cheese
(Average Data)

	g/100 g of cheese
Moisture	30–32
Dry matter	68–70
Fat	28·0–30·0
Total protein (N × 6·38)	28·0–29·5
Ash	8·5–8·6
NaCl (salt)	3·2–4·5
Calcium	0·36–0·41
Phosphorus	0·31–0·32
$\dfrac{\text{Soluble N}}{\text{Total N}}\ \%$	20–25

The milk is coagulated at 37–39°C using lamb rennet pastes and the cut curds are cooked to 45–46°C.

After removal from the vat, the curds are pressed into moulds and, in order to facilitate whey drainage, drilled with fingers or sticks (the name of this operation is 'frugatura'). The curd is dry salted for 30–60 days.

The cheeses have a cylindrical shape, 25–32 cm high and 25–30 cm in diameter, and weigh 22–32 kg. They are ripened at 10–14°C for not less than 8 months and are salted periodically during the first 2 months.

The sensorial characteristics of ripened Pecorino Romano depend mainly on the lipolysis due to the use of lamb rennet pastes. According to Arnold et al.,[24] in Pecorino Romano type cheese, there is a direct relationship between the flavour intensity of the ripened cheese and the butyric acid content, but, at present, there are no precise indications on the relative proportions of the various free fatty acids and standard flavour of cheese made in Italy.

On the other hand, the level of proteolysis in ripened Pecorino Romano shows wide variations; the influence of the processing and ripening conditions and the season of manufacture was pointed out by Pettinau and Bottazzi,[25] who determined the soluble nitrogen content in 183 cheeses made in 19 dairies in Sardinia. However, the ratio of soluble nitrogen/total nitrogen in ripened cheese (not less than 8 months old) is always less than 30.

REFERENCES

1. Reinbold, G.W., 1963. *Italian Cheese Varieties*, Pfizer Cheese Monographs, Vol. I New York.
2. Associazione Italiana Lattiero-Casearia, *Relazione anno 1989*, Milano 8 June 1990.
3. FIL-IDF, 1990. *The World Dairy Situation 1990*—C-Doc. 137.
4. Walter, H.E. & Hargrove, R.C., 1972. *Cheese of the World*, U.S. Department of Agriculture, Dover Publications Inc., New York.
5. Bottazzi, V., 1972. *Annali Microbiologia* **12**, 59.
6. Panari, G., Pecorari, M., Fossa, E. & Menozzi, P., 1988. *Atti Giornata di Studio*, Consorzio del Formaggio Parmigiano Reggiano p. 73, Reggio Emilia.
7. Bottazzi, V., Corradini, C. & Battistotti, B., 1974. *Scienza e Tecnica Lattiero-Casearia*, **24**, 43.
8. Mora, R., Nanni, M. & Panari, G., 1984. *Scienza e Tecnica Lattiero-Casearia*, **35**, 20.
9. Addeo, F., Moio, L. & Stingo, C., 1988. *Atti Giornata di Studio*, Consorzio del Formaggio Parmigiano Reggiano p. 21, Reggio Emilia.
10. Panari, G., Mongardi, M. & Nanni, M., 1988. *Atti Giornata di Studio*, Consorzio del Formaggio Parmigiano Reggiano p. 85, Reggio Emilia.
11. Resmini, P., Pellegrino, L., Hogenboom, J. & Bertuccioli, M., 1988. *Atti Giornata di Studio*, Consorzio del Formaggio Parmigiano Reggiano p. 41, Reggio Emilia.
12. Bottazzi, V. & Battistotti, B., 1980. *Industria del Latte*, **16**, 19.
13. Harper, W.J. & Gould, I.A., 1955. *J. Dairy Sci.*, **38**, 19.
14. Harper, W.J. & Gould, I.A., 1952. *Butter, Cheese, Milk Prod. J.*, **43**, 22, 44.
15. Harper, W.J. & Long, J.E., 1956. *J. Dairy Sci.*, **39**, 46.
16. Long, J.E. & Harper, W.J., 1956. *J. Dairy Sci.* 1956, **39**, 245.
17. De Felice, M., Gomes, T. & De Leonardis, T., 1989. *Industria del Latte*, **25**, 27.

18. Battistotti, B., Scolari, G.L., Corradini, C., Resmini, P., Lodi, R. & Battelli, G., 1986. *Atti Convegno 'Le fermentazioni gasogene del Provlone'* p. 15, Camera di Commercio I.A.A., Cremona.
19. Resmini, P., Pellegrino, L., Hogenboom, J.A. & Bertuccioli, M., 1988. *Scienza e Tecnica Lattiero-casearia*, **39**, 81.
20. Di Matteo, M., Chiovitti, G. & Addeo, F., 1982. *Scienza e Tecnica Lattiero-Casearia*, **32**, 197.
21. Corradini, C., 1973. *Scienza e Tecnica Lattiero-Casearia*, **24**, 240.
22. Schulz, M.E. & Mrowetz, G., 1952. *Deutsche Molkerei Zeit.*, Kempten (Alyan), **73**, 495.
23. Chianese, L., Laezza, P., Smaldoni, L.A., Del Giovane, L. & Addeo, F., 1990. *Scienza e Tecnica Lattiero-Casearia*, **41**, 315.
24. Arnold, R.G., Shahani, K.M. & Dwivedi, B.K., 1975. *J. Dairy Sci.*, **58**, 1127.
25. Pettinau, M. & Bottazzi, V., 1971. *Scienza e Tecnica Lattiero-Casearia*, **22**, 1.

8

North European Varieties of Cheese

I   INTRODUCTION

1.1 History

Little is known about the characteristics of the cheeses produced until the 16th century in the Nordic countries, Finland, Sweden, Norway, Iceland and Denmark. However, reports exist from the 17th and 18th centuries on considerable variations in the quality of cheese produced in different regions of Scandinavia.[1] Cheese was produced primarily from cultured milk, but renneted milk was also used.

In the 19th and 20th centuries, information on cheesemaking practises was exchanged increasingly between the Nordic countries. In addition, new cheesemaking practices were imported, especially from England, Holland, Germany and Switzerland.

During the last decades of the 20th century, cheesemaking practices from all over the world have been taken up by Nordic cheese plants, so that today the cheeses produced in the Nordic countries represent cheesemaking traditions in many countries throughout the world.

However, a number of the varieties have such a long tradition in the Nordic countries that they can be regarded as characteristic of this region along with the cheeses that were originally developed here.

1.2 Production and Consumption

Most nordic cheeses are produced in co-operative cheese plants from cows' milk. There are also a number of privately owned cheese factories, some of medium size, others quite small. Goats' milk is used for cheesemaking in limited quantities, while no ewes' milk is used.

In 1989, the production of cheese in the Nordic countries amounted to 573 000 tonnes, which is about 8% of the total European cheese production.[2] About 280 000 tonnes (49%) were exported, while only 31 000 tonnes of cheese were imported from other countries.

In Nordic households, cheese is usually eaten in slices on buttered bread. For this purpose, semi-hard and semi-soft cheeses with a sliceable texture are preferred. However, cheese is being used increasingly as a dessert and as an ingredient in cold salads and hot dishes. For these applications, all types of cheeses are used. In 1988, the consumption of cheese was 12–16 kg per person[3] in the Nordic countries.

1.3 Characteristics of Nordic Cheeses

In the following, the chemical, physical and microbiological characteristics of the cheeses that are typical for each Nordic country are described:

Denmark: Danbo group: Danbo
                       Samsoe
                       Fynbo
                       Molbo
         Havarti and Maribo
         Danablu and Mycella
         UF-Feta

Norway: Brunost: Gudbrandsdalsost
                  Ekte Geitost
                  Fløtemyseost
                  Prim

       Gamalost
       Pultost
       Rosendal and Balsfjord
       Ridder
       Jarlsberg

Sweden: Herrgårdsost
       Drabantost
       Wästerbottensost
       Prästost
       Svecia

Finland: Turunmaa
       Karelia
       Juustoleipä
       Ilves

Iceland: Skyr

REFERENCES

1. Jensen, H.M., 1974. *Ost*. Danske Mejeriers Fællesorganisation, Aarhus, Denmark.
2. Sliter, J., 1990. *IDF Bulletin* No. 249, p. 11.
3. Milk Marketing Board of England and Wales, 1990. *IDF Bulletin* No. 246, p. 1.

II DANISH CHEESE VARIETIES

The traditional Danish cheese seems to have been a semi-hard, kneaded type, much like the present Maribo variety. However, today, as has been the case for many decades, the majority of Danish cheese production is based on varieties originating abroad. The production methods are, however, adapted and further developed according to the demands of customers in Denmark and abroad, and according to the prevailing conditions of production, transportation and distribution.

Table I gives the chemical characteristics of the major types of Danish cheese.

2.1 Danbo Cheese Group

This group includes the cheeses named Danbo, Samsoe, Elbo, Tybo, Fynbo and Molbo, which differ from one another mainly with respect to size and shape.

2.1.1 Physical Characteristics

Danbo cheeses are produced with weights from 1 to 14 kg. The shape of the cheeses may be square, cylindrical (Samsoe, Fynbo) or spherical (Molbo).

TABLE I
Chemical Characteristics of Danish Cheeses[1,2]

Cheese	Water in fat-free cheese %	NaCl %	pH of young cheese (2 days)	pH of 3 months' old cheese	Indices of ripening at 3 months			
					% of total N			Acidity of fat, meq per 100 g fat
					Soluble N	Amino N	NH₃ N	
Danbo group	58–60	1·8	5·2	5·6	30	8	2	1–2
Havarti + Maribo	58–60	2·0	5·2	5·6	30	8	2	1–2
Danablu	61–63	3·3	4·7	5·5	50	15	6	30–50
Mycella	61–63	3·3	4·7	6·0	50	15	10	15–20
UF-Feta	66–68	4·0	4·6	4·2	15	2	0·3	1–5

The structure of Danbo cheeses is characterized by round holes with a diameter of 4–8 mm in an otherwise 'closed' body; typically there are five to 10 holes in a cross section. The 'closed' cheese body is obtained by pressing the cheese curd under whey, with exclusion of air, before cutting the curd block into cheese loaves. The round holes are formed during the first weeks of ripening as a consequence of the formation of CO_2 from the fermentation of citrate by the aroma-producing bacteria in the starter.

Danbo and the other cheeses of the Danbo group are semi-hard varieties with typically 58–60% water in fat-free cheese. Samsoe contains 56–58% water.

The consistency permits slicing of the cheeses. The young cheese may be somewhat rubber-like, while a very old Danbo, 8–12 months of age, may develop quite a brittle consistency.

2.1.2 Chemical Characteristics
The legal requirements as to the chemical composition of the Danbo group of cheeses are as follows:

> with 45% fat in dry matter: max. 46% moisture (Samsoe 44%)
> with 30% fat in dry matter: max. 54% moisture (Samsoe 52%)
> with 20% fat in dry matter: max. 57% moisture.

The figures for maximum moisture correspond to approximately 61–62% water in fat-free cheese. However, optimum quality is generally considered to be achieved with 58–60% water in fat-free cheese; for Samsoe, the corresponding figures are 56–58%.

The above-mentioned figures apply to Danbo-type cheeses for export. For the home market, the maximum moisture content is 2% higher (i.e. must not exceed 48, 56 and 59%, respectively). The corresponding figures of 63–64% water in fat-free cheese places Danbo for the home market more as a semi-soft cheese.

2.1.3 Microbiological and Biochemical Characteristics

Milk for the production of Danbo-type cheese is always pasteurized for 15 s, although there is no legal requirement for pasteurization.

Starter culture grown in skim milk is added to the milk at a level of about 1% prior to renneting. The starter is a mesophilic DL-type, containing *Lactococcus lactis* subsp. *lactis, L. lactis* subsp. *cremoris, L. lactis* subsp. *lactis biovar. diacetylactis* and *Leuconostoc mesenteroides* subsp. *cremoris*.

At the temperatures used in the production of Danbo-type cheese—renneting at about 30°C, scalding at 35–39°C (depending on the fat content and the desired moisture content), cooling to about 15°C after pressing, brining at 12°C and ripening at 16–18°C for 2–4 weeks and then at 10–12°C—all the lactose in the cheese is fermented to lactic acid in less than 24 h. Meanwhile, the pH decreases from about 6·4 at moulding to a minimum of about 5·2 in less than 24 h, before the cheese is brined.

It is of very great importance for the quality of Danbo-type cheese that the correct minimum pH of about 5·2 is actually attained. If the pH minimum value is too low, the cheese will become too acid, the consistency will become more brittle and not sliceable, and the holes will have cracks and crevices. If the minimum pH is too high the consistency may become rubber-like and there is a risk for the development of an unclean flavour due to the growth of non-starter bacteria, e.g. clostridia. The value of the minimum pH can be adjusted via the lactose content, in practice by addition of water to the whey and curd during stirring and scalding.

During ripening, the pH of the cheese increases slowly, reaching pH 5·6–5·8 after 3–6 months, due to proteolysis.

The citrate of the cheese, 1·5–2·0 g/kg of the fresh curd at moulding, is fermented by the aroma-producing bacteria in the starter, *L. lactis* subsp. *lactis biovar. diacetylactis* and *Leuconostoc mesenteroides* subsp. *cremoris*, during the first 7 to 10 days. The diacetyl thereby produced imparts an aromatic flavour to the cheese while the CO_2 from this fermentation is responsible for the formation of the round eyes which develop during the first 10 to 20 days, depending on the ripening temperature and the velocity of the citrate fermentation. If the citrate fermentation is too fast, numerous small, irregular holes and crevices will be formed, while too slow a fermentation may result in a 'blind' cheese without eyes. The velocity of the citrate fermentation can be regulated by replacing part of the DL-starter by a starter containing only *L. lactis* subsp. *cremoris* and *L. lactis* subsp. *lactis* (a so-called O-starter) which does not ferment citrate. The rate of the citrate fermentation can also be regulated by adjusting the temperature overnight after moulding and pressing.

For Danbo-type cheese there is always a risk for growth of detrimental anaerobic, lactate-fermenting clostridia, because pH and moisture are relatively high, salting by diffusion from brine is slow and the ripening temperature is rather high, e.g. compared to Cheddar. As the spores of these bacteria always seem to be present in milk, although in small numbers (about one spore per ml), and as they cannot be killed by pasteurization, nitrate or lysozyme must be added to the milk for Danbo-type cheeses. It is permitted to add up to 15 g of KNO_3 per 100 litres of cheesemilk. In the cheese, nitrate is slowly reduced to nitrite and further by the milk enzyme, xanthine oxidase, and it is the nitrite which inhibits the development of the spores of butyric acid bacteria. After 2–3 months of ripening, only very small amounts of nitrate and nitrite can be detected in the cheese.[3,4]

2.1.4 Ripening of Danbo Cheeses

During ripening, degradation of the proteins by the action of rennet enzymes and proteolytic enzymes from the starter bacteria follows the same pattern as that for other semi-hard cheeses made with mesophilic starters. Proteolysis causes a change in consistency, which is rubber-like initially, to a more sliceable form, a process which after more than 6 months may produce a brittle, crumbly texture.

The amino acids liberated during proteolysis, and compounds produced by the degradation and transformation of the amino acids, together with diacetyl, lactic acid and other compounds from the fermentation of lactose and citrate, are considered to be responsible for the flavour of the ripened cheese. For Danbo-type cheese, fat hydrolysis is normally only slight, because of the inactivation of about 95% of the indigenous milk lipase by pasteurization. However, lipolytic products may play a significant role in the flavour of old Danbo.

For ordinary Danbo-type cheese, surface ripening, i.e. development of a smear of *Brevibacterium linens* and other bacteria and yeasts on the surface, and the diffusion of breakdown products therefrom into the cheese, plays an important role in the development of the characteristic Danbo flavour. The components of this flavour have not yet been identified for Danbo-type cheeses. However, based on investigations on other types of cheese with a similar surface smear, it is known that the contribution of the smear to cheese flavour comes from low molecular breakdown products and the transformation products of amino acids and fatty acids, among which a number of low molecular sulphur compounds at very low concentrations may play a significant role.[5]

The intensity of this flavour becomes more pronounced for small cheeses, for cheeses with a high moisture content, and for cheese which is kept for a long time before cleaning and waxing.

2.2 Havarti and Maribo

Havarti and Maribo cheeses are made in rectangular or cylindrical shapes with weights ranging from less than 1 kg to 14 kg, and either with rind and surface smear, or rindless.

Compared to the Danbo group, Havarti and Maribo are characterized by having numerous small, irregular openings throughout the cheese body. This texture originates from the treatment prior to moulding and pressing. For these cheeses, the whey is drained from the curd grains before filling into moulds and pressing; thereby, air is included between the grains and hence a complete fusion of the grains into a closed cheese body, as for Danbo cheese, is prevented.

The numerous small openings between the grains expand somewhat during the first weeks of ripening, owing to the pressure of CO_2 formed during the fermentation of citrate.

Apart from this, the description given above for Danbo-type cheeses also applies to Havarti and Maribo cheese. The moisture content is typically 58–60% in fat-free Havarti (as for Danbo) and 56–58% in fat-free Maribo (as for Samsoe).

The more open structure of Havarti and Maribo may facilitate diffusion of flavour compounds from the surface smear, when present, thereby giving Havarti and Maribo a more spicy flavour compared to Danbo-type cheeses, provided other conditions are equal.

2.3 Danish Blue Cheeses: Danablu and Mycella

2.3.1 Chemical and Physical Characteristics
Danablu and Mycella are cylindrical or rectangular cheeses with a weight of 3–6 kg. The body of the cheese has irregular openings and crevices, and perforations, overgrown with the blue mould, *Penicillium roqueforti*. The consistency is at first brittle and crumbly, but becomes spreadable as the ripening processes advance.

Danablu is made with a minimum of 50 or 60% fat in dry matter (Mycella with only 50%) and with a maximum moisture content of 47%, corresponding to 64–69% water in fat-free cheese. Optimum quality is obtained at 61–63% moisture in fat-free cheese, corresponding to a semi-soft type of cheese.

The salt content is high compared to other types of cheese; optimum quality is obtained with 3·0–3·5% NaCl.

2.3.2 Microbiological and Biochemical Characteristics
Traditionally, the milk for Danablu and Mycella is not pasteurized, but heated to 60–65°C for 15 s. The reason for this mild treatment is that it has not been possible to produce these cheeses with the traditional quality from pasteurized milk. There may be several explanations for this, one being the inactivation of more than 95% of the indigenous milk lipase by pasteurization at 72°C for 15 s.

A unique feature of the Danablu technology is that the milk is homogenized. Apparently this treatment was implemented—some 80 years ago—in order to make the cheese appear whiter in an attempt to imitate the colour of Roquefort sheep's milk cheese. However, homogenization has other important effects which at first went unnoticed: it facilitates lipolysis and increases the water binding capacity of the cheese.

Lipolysis proceeds at a high rate during the first few hours after homogenization of the unpasteurized milk. At moulding, the acidity of the cheese fat, determined by a modified BDI-method, typically exceeds 10 meq/100 g fat. It has been shown that in Danablu without growth of *P. roqueforti* the acidity of the fat may reach values as high as 20–30 meq/100 g fat during the ripening period, while the acidity of the fat of a normal Danablu with growth of *P. roqueforti* typically reaches 30–50 meq/100 g fat.[6]

In Mycella, for which milk is not homogenized, the acidity of the fat increases at a significantly lower rate than for Danablu, and consequently Mycella has a milder flavour than Danablu after the same period of ripening.

The homogenization of the milk for Danablu and the resulting high rate of lipolysis may be one of the very few examples of a commercially successful accelerated ripening process, apparently invented by chance.

The milk for Danablu is inoculated with *P. roqueforti* and a lactic starter culture of the same type as for Danbo (a DL-starter). The CO_2 produced by the fermentation of citrate expands the cavities of the young cheese so that the subsequent development of the *P. roqueforti* is facilitated.

In the production of Danablu, no water is added to the whey, and therefore the lactose content of the fresh cheese is high compared to that of Danbo. Consequently, the minimum pH of Danablu is low, about pH 4·7, and this is responsible for the crumbly texture of young Danablu. After 1–2 weeks of ripening, the pH increases due to the proteolytic activity of the mould, and after 2 months, the pH is normally at 5·5 or higher.

The growth of *P. roqueforti* is facilitated by the low pH of the young cheese and by its rather open texture. Furthermore, the cheese, after cooling and salting, is perforated in order to give atmospheric oxygen access to the interior of the cheese. Compared to other types of mould, *P. roqueforti* can grow at low oxygen concentrations (2–5% O_2).

P. roqueforti produces powerful proteolytic and lipolytic enzymes which cause very rapid ripening of Danablu and Mycella compared to Danbo-type cheeses. They result in the formation of a number of compounds, including degradation and transformation products of amino acids and fatty acids, e.g. ammonia, esters and thioesters, other low molecular sulphur compounds, methyl ketones, etc., that contribute to the characteristic flavour and aroma of these cheeses.

2.4 UF-Feta

2.4.1 Physical and Chemical Characteristics

Danish UF-Feta cheese, made by ultrafiltration, is normally produced in rectangular pieces with a weight of 1 kg, placed in brine in tins with a total weight of 1–20 kg. The cheese has a homogeneous structure without openings, a brittle, crumbly and rather soft consistency, and a white colour.

Danish UF-Feta may be made with 45, 40 or 30% fat in dry matter and with not more than 53, 58 and 61% moisture, respectively, corresponding to 67–70% water in fat-free cheese; therefore, it is a soft type of cheese. The minimum NaCl content is 2% , but is typically around 4%. The pH is as low as 4·2–4·5.

UF-Feta is made from homogenized retentate from ultrafiltered milk, at a total solids content of 39·5% (for Feta with 40% fat in dry matter). Rennet and starter are added to the concentrate and the cheese can be 'cast' directly into the tins. After acidification, during which the pH decreases to <4·6, about 4·0–4·5% salt is added: slight syneresis occurs. Feta made in this way contains most of the whey proteins of the milk which may constitute as much as about 15% of total protein of the cheese, while in traditionally-made cheese the whey proteins account for only 1·0–1·5% of the protein.

2.4.2 Microbiological and Biochemical Characteristics

For Danish UF-Feta cheese, only O-starters, i.e. starters with only *L. lactis* subsp. *cremoris* and *L. lactis* subsp. *lactis*, are used because it is important that no CO_2 is produced in the cheese, in order to prevent blowing of the tins.

The low pH of UF-Feta, about 4·6 for the fresh cheese, is due to its high lactose content (high moisture content and no addition of water) and relatively low buffering capacity—compared to Danbo-type cheeses. During ripening/storage the residual lactose is fermented and the pH decreases to about 4·2. In contrast to Danbo and Danablu, there is no increase in pH during ripening, because proteolysis is very limited.

The low level of proteolysis (Table I) is due primarily to the low pH and the high salt content, which inhibit the action of proteases of the starter bacteria especially. The high concentration of whey proteins in UF-Feta may also contribute to the slow proteolysis.[7,8]

Because the milk or concentrate used in the production of UF-Feta cheese is pasteurized to about 72–77°C for 15–60 s, there is practically no lipase activity; hence the extent of lipolysis during ripening is limited unless lipase is added to the concentrate to induce some lipolysis.

The taste and flavour of the ripened/stored Feta cheese is dominated by salt and lactic acid, and—if lipase has been added—fatty acids. The production of a low level of free amino acids seems to sweeten the taste but little is known about other compounds which might contribute to the slight but distinct changes that occur in the flavour of UF-Feta during ripening.

REFERENCES

1. Steen, K., 1981. *Mælkeritidende*, **94**, 52.
2. Nielsen, E. Waagner, 1990. *Omdannelser i ost under syrning og modning*, Mejeribrugsinstituttet, Den kgl. Veterinær- og Landbohøskole, København, p. 167.
3. Galesloot, T. E., 1961. *Neth. Milk Dairy J.*, **14**, 395.
4. Jensen, F. & Werner, H., 1977. *Annex 226th Report Statens Forsøgsmejeri*, p. 18.
5. Parliment, T.H., Kolor, M.G. & Rizzo, D.J., 1982. *J. Agric. Food Chem.*, **30**, 1006.
6. Nielsen, E. Waagner, & Edelsten, D., 1974. *Proc. 19th Intern. Dairy Congr., Delhi*, **IE**, p. 252.
7. Koning, P.J. de, Boer, R. de, Both, P. & Nooy, P.F.C., 1981. *Neth. Milk Dairy J.*, **34**, 35.
8. Qvist, K.B., Thomsen, D. & Kjærgaard Jensen, G., 1986. *268th Report Statens Forsøgsmejeri*.

III   Swedish Cheese Varieties

About 80% of all cheeses made in Sweden are pressed, semi-hard or hard cheese varieties of national origin. They are made of pasteurized cows' milk using mesophilic starter cultures of the DL type and calf/bovine rennet mixtures.

The flavour and consistency differ markedly between the cheese varieties. However, the maturation process described in chemical and microbiological terms follows the same basic principles. This is described below, as are the characteristics of each cheese variety.

3.1 Physical Characteristics

The cheese varieties could be classified into two groups according to their structure. If air has been excluded from the curd, the eyes will be round, regular and large, as in Herrgårdsost and Drabantost. Otherwise, the eyes will be irregular and very small, as in Wästerbottensost, Prästost and Svecia.

3.2 Chemical Characteristics

Most of the lactose (>90%) is fermented during the first 24 h. At the end of the lactose fermentation, the citrate is metabolized and almost no citrate remains after a week. The CO_2 produced from the citrate effects eye formation. The cheeses are formed, pressed and then salted in brine for two to five days when the salt content of the cheese will be 0·7 to 1·6%, depending on the cheese variety. Sometimes, salt is also added to the curd before draining all the whey from the vat.

Proteolysis is the principal activity in these kinds of cheeses during maturation. Lipolysis occurs, but to a minor extent, and the products of lipolysis, even at low concentrations, cause off-flavours. The first steps of proteolysis are effected by proteolytic enzymes from different sources, i.e. rennet, starter bacteria and cheesemilk (mainly plasmin and heat stable proteases from psychrotrophic bacteria). The environment is favourable for many of these enzymes and their activity is controlled by maintaining optimal temperature, water activity, salt concentration and pH. Different types of bacterial peptidases are important for attaining thorough proteolysis during cheese maturation and also for degrading bitter peptides produced from casein.

3.3 Microbiological Characteristics

The mesophilic starters used are mixtures of several strains of *L. lactis* subsp. *lactis, L. lactis* subsp. *cremoris, L. lactis* subsp. *lactis* biovar. *diacetylactis* and *Leuconostoc mesenteroides* subsp. *cremoris*. The lactococci use lactose as an energy source and grow to about 10^9 bacteria per gramme of cheese during the first day. The number of starter bacteria then decreases and only about 10% are viable after two to three weeks. However, starter bacteria are found in the cheeses at 10^6–10^7 cfu/g throughout a ripening period of six months.[1] Figure 1

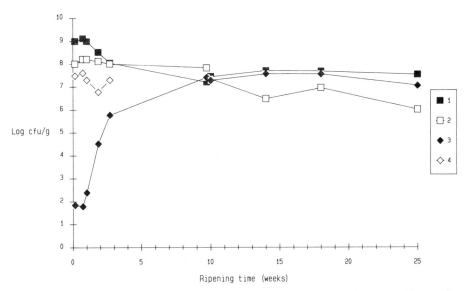

Fig. 1. Development of the microbial flora in Swedish round-eyed cheese. The results before and after 10 weeks are from two different trials.[1,2]

1 = Mainly starter lactococci, that do not utilize citrate (KcA)
2 = Mainly *Lactococcus lactis* subsp. *lactis* biovar. *diacetylactis*. After
   10 weeks, graph 2 represents all citrate utilizing bateria growing on KcA.
3 = Mainly lactobacilli; anaerobically growing bacteria on AcA before
   10 weeks and on MRS after 10 weeks.
4 = Mainly *Leuconostoc*; citrate-utilizing β-galactosidase-positive bacteria on
   KcA.
cfu/g = colony forming units per gramme of cheese

shows the typical changes in the bacterial flora of round-eyed cheeses, like Drabantost and Herrgårdsost. The growth of wild lactobacilli is typical for these cheeses and probably contributes to cheese maturation. However, excessive numbers of lactobacilli or certain lactobacilli strains, i.e. gas-forming or off-flavour-producing strains, cause defects.

Starter bacteria develop in a similar way in the cheeses with small and irregular eyes, e.g. Prästost and Svecia. In Prästost, however, wild bacterial strains reach higher numbers. The number of viable lactococci (i.e. bacteria forming colonies on streptoselagar (BBL)) after three to six months of ripening has been shown to be rather stable at 10^6 to 10^7 cfu/g in Prästost, as well as in the round-eyed cheese, Herrgårdsost.[1]

3.4 Characteristics of Individual Cheeses

The data given below on the extent of proteolysis are estimated from results of investigations at our laboratory. The other characteristics for the individual cheese varieties are described in Ref. 3.

3.4.1 Drabantost

This is a semi-hard cheese with a slightly elastic consistency and round eyes (3–12 mm). It is ripened for at least 6 weeks and has a mild, clean, slightly acidic taste. The salt concentration is 1·2–1·6%, fat in dry matter is 20, 30 or 45%, and water in fat-free cheese is 55–59%. Water-soluble N is about 24–28% of total N. The PTA (phosphotungstic acid)-soluble N is only 2·5–3·0% of total N.

3.4.2 Herrgårdsost

The most common cheese in Sweden, Herrgårdsost has round eyes (5–15 mm) and a mild, rich flavour (nutty). It has a fat in dry matter of at least 45% and a water in fat-free cheese of 53–57%. It is most often consumed after 4–6 months of maturation when about 26–30% of its casein has been hydrolysed into water soluble N compounds. It contains 0·9–1·3% salt. Acetaldehyde, acetic acid and ethanol are important metabolites for flavour development.[4] Herrgårdsost is a semi-hard pressed cheese. Its consistency is firm and somewhat elastic, and it melts evenly.

3.4.3 Wästerbottensost

This is a long-established and highly-valued hard cheese with irregular small eyes (2–6 mm). The aroma is strong, rich, aromatic and slightly bitter. The fat in dry matter is at least 50% and the water content of the fat-free cheese is only 50–53%. The consistency is firm, tender, somewhat brittle and granular. During production, the curd is heated and stirred for several hours. By this treatment the cheese will become dry and the activity of the starter bacteria will be very limited, giving the cheese a slow and very special maturation. The ripening time is about one year. Little is known about its microbiology and its chemical composition.

3.4.4 Prästost

This cheese has an open texture with small irregular eyes (2–6 mm). Its flavour is strong, aromatic and slightly bitter. The minimum fat content in dry matter is 50% and the water content of the fat-free cheese is 55–57%. A low-fat variety of Prästost is also made with a fat in dry matter of 30%. The maturation time is 4–12 months. About 30% of the casein is hydrolysed to water soluble N compounds after 4–5 months and about 42% after a year. The cheese is semi-hard and has a smooth body that can easily be melted or sliced.

3.4.5 Svecia

This cheese has been made at Swedish farms and chalets from time immemorial. It is a pressed, semi-hard cheese with an open texture and small (2–6 mm), irregular eyes. It is made with fat in dry matter levels from 30 to 60%. The ripening time is 2–12 months. The flavour ranges from mild, slightly acidic and rich to strong and aromatic. In a typical young Svecia, about 30% of the total N is water-soluble and about 7% is PTA-soluble compounds, i.e. amino acids and very small peptides.

REFERENCES

1. Pettersson, H.E., 1982. *Internal Report SMRC-0101/A-07-82*, SMR, Malmö.
2. Ardö, Y. & Christiansson, A., 1990. Unpublished data, SMR (Swedish Dairies Association) R&D, Lund.
3. *Svenska Kontrollanstalten för Mejeriprodukter och Ägg, Ostsorter i Sverige*, 1984, KMÄ, Malmö.
4. Lööf, N.G. & Persson, H., 1983. *Internal Report SMRC-0101/A-19-83*, SMR, Malmö.

IV NORWEGIAN CHEESE VARIETIES

A large number of cheese varieties are produced in Norway. Most of them are variants of other European types, but there are also varieties that are typical for Norway.

The Norwegian cheese varieties may be divided into four categories:

1. traditional cheeses,
2. cheeses from goats' milk,
3. whey cheeses,
4. cheeses developed in recent years (e.g. Jarlsberg).

4.1 Brunost

Brunost is Norwegian for brown cheese. It is most often referred to as Whey Cheese in English. Brunost is the second most important cheese group with a per capita consumption of 3·5 kg per year. As the English name indicates, this is a cheese based on whey from cheese or casein production. Milk or cream is added to the whey and the mixture is concentrated by evaporation to ~80% dry matter. The heat treatment gives the cheese a characteristic brown colour due to Maillard reactions between lactose and protein components.

Concentrating whey to make cheese was a method for utilizing all the components of milk on farms. About 1885 to 1890 in the valley of Gudbrandsdalen, a milkmaid got the idea of adding cream to the whey. This improved the quality of the whey cheese and the farm was able to obtain a better price than could be achieved by the sale of butter. More and more farms took up this practice, and in 1911 industry-scale production commenced. Production has expanded since and is now an important part of the dairy industry.

Brunost is produced in a number of varieties. Gudbrandsdalsost (Ski Queen) is made from whey from cows' milk with goats' milk added equivalent to a minimum of 1 litre per kg of cheese. Ekte Geitost is made solely from goats' milk components and Fløtemyseost is made solely from cows' milk components.

The typical chemical composition is a dry matter content of 78–82% with fat in dry matter varying from 20 to 35%, depending on cheese type. Lactose is the main component (36–46%), with a protein content as high as 12%. Brunost is an important iron source in the Norwegian diet. Earlier, this was the result of the process itself through the use of iron pans. Today, with modern stainless steel equipment, iron is added during processing.

After final concentration and browning, the cheese is cooled in such a way that the lactose crystallises as small crystals. Most of the lactose is present as crystals of lactose monohydrate, and combined with fat, denatured whey protein and casein, the crystals build up the structure of the cheese. It is cast while still hot into rectangular blocks weighing from 0·25 to 4 kg, with 0·5 and 1 kg being the normal sizes.

The cheese is ready for consumption immediately after manufacture. The taste is mild and sweet with a flavour of goats' milk (except Fløtemyseost). The consistency is semi-hard and it is suitable for slicing.

Finally, a spreadable variety, Prim, should be mentioned. The water content of this is higher, being around 30%, and the fat in dry matter is only 10%.

4.2 Gamalost

Gamalost literally means old cheese. The cheese is so-called because it is a long-established and very traditional cheese in Norway; it can be traced back to the Viking age, a thousand years ago.

Gamalost is an acid-curd cheese made from skim milk. No salt is added and the cheese consists mainly of protein (and water). The cheese is, therefore, regarded by many as a very healthy nutrient. Also, quite a number of myths are connected to its healing effects. Still, it is not everybody's type of cheese, but many Norwegians regard Gamalost as a real delicacy.

Gamalost has a rich, characteristic flavour caused by the growth of *Mucor mucedo* or *M. racemosus* on the surface and throughout the interior.

The cheese is yellow-brown with a light yellow core, and is sliceable. According to Codex standards, it should be classified as a hard, internal mould-ripened cheese. It has a cylindrical shape, weighing approximately 1·6 kg.

Gamalost is a skim cheese, with a fat content of only 0·5%. The protein content is 50% (dry matter is 52%) and carbohydrates, 0·5%.

In spite of its name, Gamalost is ready for consumption after only five weeks of ripening. During ripening, the cheese is penetrated by the Mucor mycelium, which secretes extracellular proteases and lipases. The decomposition of the caseins is wide and deep (soluble N around 90% of total N), and the small amounts of fat are more or less hydrolysed.

4.3 Pultost

Pultost is another traditional Norwegian cheese. Different varieties were made on farms under many different names. The meaning of the name 'pult' is not clear but it may be derived from the Latin word 'pulta', which means porridge.

Pultost is, like Gamalost, an acid-curd cheese, made from skim milk. It is a granular, unshaped, matured cheese. Caraway seeds are added to the cheese curd. The colour of Pultost is light yellow and the taste is sharp and rich.

The cheese contains 40% protein, 0·9% fat, 0·4% carbohydrates and 3% salt, giving a total of 44% dry matter. The pH is around 5·5.

Traditionally, ripening was due to a natural flora of yeasts, bacteria and sometimes moulds. Fresh cheese was inoculated with matured cheese.

Today, a mixed culture of *Lb. lactis* and *Candida rugosa* is added to get a more controlled fermentation. This process is interrupted after a few days by adding salt (5%), and is followed by a maturing period of two months at a much lower temperature (8–10°C). Manufacture of pultost demands a very high hygienic standard because the cheese composition and the fermentation temperature are ideal for undesirable microorganisms.

4.4 Cheeses Made From Goats' Milk

Although the amount of goats' milk produced in Norway is very small, only 25 million litres per year, it plays an important role on farms in some regions. Rennet cheeses made from goats' milk have some tradition, even if most of the milk is used in whey cheese production (see above).

Two novelties from recent years deserve mention:

Rosendal (Hardanger) is a semi-hard full-fat interior-ripened cheese with a clean and slightly sour taste. It contains 61% dry matter and 45% fat in dry matter. The cheese has no holes. It is cylindrical, weighs 1·6 kg and is covered with a black plastic coating. The ripening organisms are cultures of lactic acid bacteria.

Balsfjord is a semi-soft, full-fat, interior-ripened cheese with a mild flavour of goats' milk. It is made from a mixture of goats' and cows' milk and contains 53% dry matter and 45% fat in dry matter. It is cylindrical, rindless and weighs 300 g. The ripening organisms are cultures of *L. lactis* subsp. *lactis* and *L. lactis* subsp. *cremoris*. It has a smooth texture, irregular holes and is suitable for cutting. The cheese was developed as an alternative for those consumers who prefer mild-tasting goats' milk cheeses.

4.5 Ridder

Ridder (knight) cheese was developed in 1969 to fill a need for cheeses with more taste than the large-volume cheeses on the market. It soon became a success on the domestic market and it is also exported in small quantities. Ridder cheese can be classified as a semi-soft, full-fat surface-ripened cheese.

The cheeses weigh 1·5 kg, are cylindrical, have no holes, and are suitable for slicing. They have the characteristic taste from surface-ripening with *Brevibacterium linens*. The chemical characteristics are a dry matter content of 61% and a fat in dry matter content of 60–65%.

4.6 Jarlsberg

Jarlsberg is a widely exported cheese that can be categorized as a Swiss-type cheese, but with its own special characteristics. Jarlsberg was redeveloped in the early 1950's from old formulae used by Swiss cheesemakers who settled in Norway during the first half of the 19th century.

Jarlsberg is a cylindrical cheese with a weight of approximately 10 kg. The diameter is ~330 mm and the height is 95–105 mm. It has a smooth texture with many large, regular-shaped eyes. The taste is clean and rich, with a slightly sweet and nutty flavour. Jarlsberg contains 60% dry matter with 45% fat in dry matter and 1·25% salt.

The ripening organisms are ordinarily lactic acid bacteria and *Propionibacterium freudreichii shermanii*. In fully-matured cheeses, the level of water soluble N is typically 21–23%, and the amino N around 4–6% of the total N.

The propionic acid fermentation results in concentrations of 35–45 mM propionic acid and ~25 mM of acetic acid. The pH is 5·6–5·7.

Although wheels are the usual shape, Jarlsberg is also produced as rectangular, rindless blocks. In 1990, a rindless Jarlsberg Lite variety was introduced onto the market.

V FINNISH CHEESE VARIETIES

5.1 Turunmaa

Turunmaa cheese (also called cream cheese) is a semi-hard, full-fat cheese which originated in South Finland.

The cheese is made from pasteurized milk and ripened using starters containing mesophilic lactococci and leuconostocs. The coagulation time is approximately 30 min. The curd is cut and stirred for 10–15 min, after which part of the whey is removed. The curd-whey mixture is scalded at 37–39°C for 40–60 min. The total cooking time is 120–140 min. The curd is pre-pressed for about 20 min, after which the curd is milled, put into the moulds and pressed for about one hour. The cheeses are salted in 20% brine at 10–12°C for 48 h and vacuum packed in Cryovac pouches. The ripening time is at least 7 weeks at 12–13°C.

Turunmaa cheese is normally moulded as cylinders 20 cm high and weighing about 7 kg. Block cheeses weighing about 3 kg are also made.

The flavour and aroma of the cheese are mild and slightly acid. The texture is firm but pliable. The structure shows irregular gas holes and some spaces between curd particles.

The cheese contains 30% fat, a maximum of 44% moisture and a minimum of 50% fat in dry matter.

5.2 Karelia

Karelia is a quite recently developed semi-hard, medium or full-fat cheese. It is made from pasteurized milk using starters containing mesophilic lactococci, leuconostocs, lactobacilli and suitable propionic acid bacteria. The cheese is manufactured by the Casomatic method. The scalding temperature is about

36–38°C and the total cooking time is about 150 min. Whey is removed before moulding. The pressing time is about 3 h and the cheeses (blocks weighing about 13 kg) are salted in brine at 10–12°C for 48 h. The cheeses are ripened for about 2 weeks at 20–24°C and then for about six weeks at 10°C.

The flavour and aroma of the cheese is slightly acidic and similar to Jarlsberg cheese, but milder. The structure of the cheese is open, with irregular holes.

The fat content of the full-fat cheese is about 23%. The maximum moisture content is 40% and the minimum value for fat in dry matter is 45%.

5.3 Juustoleipä

Juustoleipä ('cheese bread') is a traditional fresh farmhouse cheese which originated in North Finland where it was originally made from colostrum. Nowadays, the cheese is made from pasteurized, full-fat milk which is coagulated using rennet (150–250 ml per 100 litre milk). Salt is sometimes added to the milk. The renneting temperature is about 30°C and the renneting time about 30 min. The curd is cut, stirred and heated to 40–45°C. The heating time and temperature vary with manufacturing practices. After heating, the whey is drained off and the curd particles placed in moulds. The cheese may be pressed slightly if it is too soft. After moulding, the cheeses are about 5 cm high but during grilling (roasting) on both sides for about 7 min, the thickness of the cheese is reduced to 1·5–2·0 cm. The surface of the cheese has brown patches from the grilling.

The cheese has a mild, neutral taste and a soft, slightly tough texture. Its fat content is about 17%, the maximum moisture content is 55% and the minimum value for fat in dry matter is 38%. The keeping quality of the cheese is about 2–3 weeks.

5.4 Ilves

Ilves cheese is a traditional fresh farmhouse cheese which originated in South Finland. It is made from pasteurized full-fat milk. The milk is heated to 100°C after which egg (5%), mesophilic starters (15%) and lactic acid (0·15%) are added. Heating is continued until the curd rises to the surface and the whey can be drained off. Salt (2·5%) and sugar (3·8%) are added and mixed into the curd for about 15 min. The cheeses are kept in the moulds overnight without pressing, after which they are baked for about 30 min in an oven at 190°C.

The taste of Ilves cheese is mild. The fat content is 16%, the maximum moisture content is 56% and the minimum value for fat in dry matter is 30%.

VI ICELANDIC CHEESE VARIETIES

The cheeses produced in Iceland are the same as those produced in the other Nordic countries except for one product that is specific to Iceland: Skyr.

6.1 Skyr

Skyr is a fresh cheese made from cows' milk that has been made on farms in Iceland for hundreds of years, but is now also produced on a larger scale at dairy plants. Skyr is traditionally eaten as a dessert mixed with cream or milk and sprinkled with sugar or berries.

It is made from skim or whole milk that is pasteurized at 85–90°C. Starter culture (traditionally, some good Skyr from a previous production is used) and rennet are added to the milk at 40°C. The coagulum is coarsely cut and whey is drained off. Skyr is then ready for consumption.

Skyr has a paste-like consistency (like quarg) and quite a sharp acidic taste. The flavour is mild and aromatic, like most cultured milk products, but has, in addition, a faint yeast aroma.

Skyr contains about 15% protein and, when made from skim milk, about 20% dry matter; the pH is about 4·3. The concentration of citric acid is the same as in fresh milk, but it contains up to 15 mg acetaldehyde per litre.[1] This shows a difference between Skyr and cultured milks made with normal starter cultures.

REFERENCE

1. Nielsen, E. Waagner & Finnson, J., 1968. *Ladelund Elevforenings Årsskrift 1968*, p. 18.

9

Ripened Cheese Varieties Native to the Balkan Countries

Marijana Carić

Faculty of Technology, University of Novi Sad, Yugoslavia

1 Kashkaval

1.1 General Characteristics

One of the most popular hard cheeses in Balkan countries, dating back to the 11th and 12th centuries, is Kashkaval, which was first brought to Bulgaria by nomadic tribes from the East. Nowadays, it is produced in an area stretching from the former southern USSR (Crimea, South Ukraine, the Caucasus) and Turkey, through Greece, Bulgaria, Romania, Yugoslavia, Albania and Hungary to Italy, Algeria, Tunisia, Egypt and Morocco. Although there are many varieties of Kashkaval due to certain differences in some operations in their production, there are three essentially distinct technological processes for Kashkaval production today: Balkan (native to Balkan countries), Russian, which is very similar to the first, and Italian. According to differences in language and production, the following variants of Kashkaval are produced in Balkan countries and in the former southern USSR: Kaškaval Balkan, Kaškaval Preslav, Kaškaval Vitoša (Bulgaria), Kačkavalj (Yugoslavia), Kačkaval, Kačekavalo (former USSR), Τυροζ καοελον (Greece), Košer (Turkey, Albania) and Cascaval Dobrogen (Romania).[1,2] It is also given different names according to the production district, e.g. Pirdop in Bulgaria, Epir in Greece, or Šarplaninski and Pirotski Kačkaval in Yugoslavia.[3] Its Italian relative is Caciocavallo; in Greece it is also known as Kasseri, and in Egypt the name Romy is commonly used.[4]

Kashkaval is manufactured from cows', sheep's, goats' or mixed milk, raw or pasteurized. For example, in Bulgaria, Kaškaval Preslav is produced from mixed milk, Kaškaval Balkan is produced from sheep's milk, while Kaškaval Vitoša is produced from cows' milk; in Yugoslavia, Kačkavalj is produced from sheep's, cows' or mixed milks, and in Romania, Cascaval Dobrogen is produced from sheep's milk only.

The typical form of Kashkaval is flat cylindrical, with a smooth, amber-coloured rind; the typical size is: diameter 30 cm; height 10–13 cm; and weight 7–8 kg.[5] Average composition is shown in Table I.[2,6] The European Economic Community

TABLE I
Average composition of Kashkaval Cheese

Type of Kashkaval	Fat (%)	Total solids (%)	Total protein (%)	Salt (%)	Ash (%)	pH	Author
Bulgaria	30·0	60·14	19·60	4·01	5·69	5·0	Kosikowski[6]
Greece	33·88	66·36	25·14	2·17	4·38	5·13	Kosikowski[6]
Yugoslavia	27–32	60–65	—	2–3·5	—	4·9–5	Pejić[2]

recently specified the characteristics of Kashkaval as: 'Kashkaval cheese from sheep's milk, matured for at least 2 months, with a minimum fat content of 45%, by weight, in the dry matter, and a dry matter of at least 58%, in whole cheeses of a net maximum weight of 10 kg, whether or not wrapped in plastic'.[7]

1.2 Unique Features of the Manufacturing Procedure

Different parameters in Kashkaval technology have been investigated by numerous authors.[3-40] The main characteristics of Kashkaval cheese technology are that it consists of two independent stages: (1) production of the curd, and (2) heat treatment of ripe curd by soaking it in hot water. The operations in the process are as follows: renneting → cutting → fresh curd (with low heat treatment: 38–40°C) → curd ripening (cheddaring) → slicing → texturing (heat treatment) → forming → salting → ripening.

The unique feature in the manufacture of Kashkaval varieties is texturing (soaking the curd in salt water, 12–18% NaCl, at 72–75°C for 35–50 s),[1] which has a pasteurizing effect that encourages correct fermentation and ripening,[2,8] thus resulting in high quality cheese with good keeping qualities. Šutić[8] found no Escherichia coli or other coliform bacteria in textured curd, although these were present in the curd after cheddaring. The inclusion of the above operation enables production in hot climates and has the further advantages that milk with high acidity can be used as raw material for cheese production.[9] Antonova et al.[10] produced high quality Kaškaval Balkan and Kaškaval Vitoša using two bacterial enzyme preparations from Bacillus mesentericus instead of calf rennet; the experimental cheese had the same physico-chemical properties, taste and texture as the controls, even after 12 months' storage.

Specific technology, together with texturing, yields Kashkaval cheese with characteristic structure, i.e. pliable, elastic, laminar, very close with visible layers and occasional slits, but no gas holes. Although Kaskaval was originally hand-formed, nowadays the second stage of manufacture of this cheese (slicing and texturing with heat treatment, in particular), is generally mechanized, the equipment used having a significant influence on cheese quality and structure. One of the very convenient machines for continuous mixing/cooking is used nowadays not only for Kashkaval, but other pasta filata cheeses also. This is featured in

Fig. 1. Continuous mixing/cooking machine used in Kashkaval production, shown together with automatic moulding device (equipment by MM, Modena, located in Mlekoprodukt Dairy, Zrenjanin, Yugoslavia).

Fig. 1, together with an automatic moulding device. The development of cheese structure by the different types of mechanized cheddaring equipment used in Cheddar production was thoroughly investigated by Kaláb *et al.*[11] and Lowrie *et al.*,[12] while Omar *et al.*[13] investigated the structure of Kashkaval. The structure of Kashkaval at different stages of production is shown in Fig. 2a–d. SEM micrographs in Fig. 2 show that the character of the curd (paracasein aggregates) is altered, changing from a loose structure (Fig. 2a) to a fibrous material (Fig. 2b) during cheddaring, similar to changes in Cheddar cheese.[11] With further processing, casein fibres disintegrate to form a compact, homogeneous structure (Fig. 2c,d). SEM micrographs of Kashkaval and other cheese varieties obtained by Hassan,[14] showed distinct differences in the internal structure of the cheeses, related to their textural properties.

Texturing has physico-chemical as well as microbiological and structural consequences: partial protein denaturation, removal of a significant quantity of water-soluble substances and fat, and a decrease in the water content of ripe curd have been observed during this operation.[15-18] On the basis of its textural

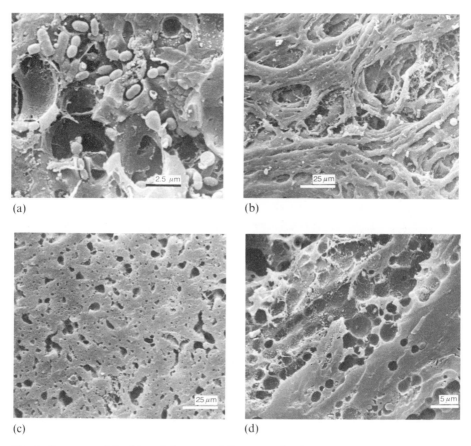

(a) (b)

(c) (d)

Fig. 2. Structure of Kashkaval cheese at different production steps:
a) ripe curd showing bacteria aggregated in nests;
b) textured curd with protein matrix having uniform orientation;
c) final product;
d) final product from another plant (courtesy M. Kalab).

and compositional characteristics, Kashkaval can be frozen successfully, in the form of ripe curd or salted cheese.[19-22]

Possible differences between numerous Kashkaval variants arise from the fact that the technology used in its manufacture may be subjected to many variations in respect to curd composition, added cultures, degree of curd ripening and heat treatment temperatures.[8,17]

1.3 Ripening Changes

Kashkaval is ripened at a temperature of 12–16°C and a humidity of 85% RH, for 2 months.[1,2] Its shelf-life is 10–18 months at 2–4°C. In the production of Kashkaval cheese, 0·1–0·5% of a culture containing *Str. thermophilus, L. lactis* subsp. *diacetylactis, Leuc. mesenteroides* subsp. *dextranicus, Lb. delbrueckii* subsp. *bulgaricus, Lb. helveticus* and *Lb. casei* is added.[1,15,23] The following combinations are usual: *Lb. helveticus* and *Str. thermophilus* or *L. lactis* subsp. *diacetylactis, Str. thermophilus* and *Lb. casei* or *L. lactis* subsp. *diacetylactis, Str. thermophilus, Lb. casei* and *Lb. delbrueckii* subsp. *bulgaricus.*[1]

The changes in cheese components, particularly proteolysis, start during curd fermentation (cheddaring), which Pejić[24] and Djordjević[25] characterize as the first stage of the ripening process. According to Pejić,[24] the extent of ripening in this stage is 25% of the total ripening in the Balkan procedure, 33% in the Russian and even 46% in the Italian procedure, leaving 75%, 67% and 54% of ripening, respectively, after forming and salting of the cheese. During cheddaring, lactic acid fermentation occurs with an increase of acidity to pH 5·4–5·5 (or 150–170°T) when cows' milk is used, and pH 5·2–5·3 (or 170–190°T) when sheep's milk is used.[1] This leads to an increase in the concentration of soluble calcium (the curd after cheddaring contains 53% more soluble calcium than fresh curd[16] and the formation of monocalcium paracaseinate according to the following reaction:[26]

$$2NH_2R(COO)_6Ca_3 \quad + \quad 10C_3H_6O_3 \rightarrow$$
Calcium paracaseinate $\qquad$ Lactic acid

$$(NH_2R(COOH)_5COO)_2Ca \quad + \quad 5\ Ca\ (C_3H_5O_3)_2$$
$\qquad$ Monocalcium paracaseinate $\qquad$ Calcium lactate

Monocalcium paracaseinate nitrogen in the ripe curd amounts to 1·73%, which is 48·7% of the total nitrogen and 3·6 times higher than at the beginning of cheddaring, or 1·6 times higher than in Cheddar cheese at the beginning of pressing. This is due to the difference in manufacturing procedure between these two cheeses: in Kashkaval manufacture, the curd is subjected to a heat treatment and its plastic properties must be developed without pressing.[25]

Monocalcium paracaseinate is responsible for the ability of the ripe curd to be stretched as fine, long, ductile threads when heated, thus influencing the rheological properties and providing the characteristic structure of the finished product.[15,25]

Both lactic acid producing bacteria and the acid produced by them, inhibit the growth of many species of microorganisms that can cause defects in the finished cheese (gas-forming, proteolytic, lipolytic, etc.).[27]

Ripening processes continue in the formed and salted cheese. Since this stage of ripening occurs after the curd has been textured at high temperatures, which inactivate the added culture, the ripening process is slow and not very intensive.

Proteinases in dairy technology have been reviewed by Fox[28] and proteolysis of cheese proteins during ripening by Grappin *et al.*[29] and in Chapter 10, Volume 1 of this text. Marcos *et al.*[30] investigated the electrophoretic patterns of proteins in 34 different varieties of European cheeses after ripening. A detailed electrophoretic study of proteins during Kashkaval ripening (90 days) was undertaken by Alrubai.[31] The polyacrylamide gel electrophoretic patterns obtained

TABLE II
Changes in the Relative Proportions of Casein Fractions in Kashkaval Cheese during Ripening[31]

Ripening days	α_S-Casein				β-Casein	κ-Casein[a]			para-κ-Casein		
	α_{S1}	α_{S1-I}	α_{S1-II}	Total α_S		I	II	Total κ	I	II	Total para-κ
1	39·10	15·79	—	54·89	22·81	—	—	11·03	—	—	11·28
15	23·78	29·26	—	53·04	21·96	—	—	—	18·07	6·93	25·00
30	15·28	11·36	15·12	41·76	17·59	11·68	16·13	27·81	—	—	12·83
45	13·26	17·36	15·91	46·53	25·08	3·86	14·47	18·33	6·75	3·32	10·07
60	23·19	17·04	—	40·23	26·50	8·78	15·50	24·29	6·87	2·12	8·99
75	24·38	14·63	—	39·01	25·52	5·91	16·01	21·92	8·67	4·88	13·55
90	17·26	10·81	—	28·07	18·79	13·21	21·38	34·59	12·23	6·23	18·55

[a] This protein is referred to as κ-casein in Ref. 31 but it appears likely that the protein in question is γ- or γ-like casein.

were quantitatively evaluated and the results are presented in Table II. The number and quantities of protein fractions evidently differ from those for similar cheese types (e.g. Cheddar).[31] Comparative investigations of the extent of ripening of nine cheese varieties on the Egyptian market were carried out by quantitative gel electrophoresis, thin layer chromatography and other methods. Polyacrylamide gel electrophoretograms showed that the proteins of Kashkaval and Provolone were degraded more extensively than those of other cheese samples (Ras, Gouda, Roquefort, Processed, Domiati, Kariesh and Mish).[14] However, another group of Egyptian scientists[32] obtained electrophoretic patterns of proteins from Kashkaval cheeses produced from several types of raw and heated milk with very little variation during manufacture. β-Casein appeared to be more resistant to hydrolysis than α_s-casein, which was rapidly degraded during ripening. Cheese made from raw cows' milk or heated buffaloes' milk revealed the presence of an increased number of degradation products compared to those from raw buffaloes' milk.

Law[33] and Law & Wigmore[34] accelerated cheese ripening by the addition of a bacterial neutral proteinase to the curd. Working with Cheddar cheese, the authors[33,34] obtained about a 20% increase in the level of proteolysis in 2 months with the typical flavour intensity of a 4 months-old untreated cheese.

There are controversial data in literature about the effect of moisture and salt content on the hydrolysis of casein, while the extent of proteolysis in cheese increases with increasing pH.[14]

In Kashkaval cheese production, the concentration of water-soluble N compounds increases slowly during ripening (Table III) and a high content of monocalcium paracaseinate, which is specific to this type of cheese, is retained even after advanced ripening.[25,26] According to Romanian authors,[35] the content and proportion of soluble N increases during ripening in the smoked Kashkaval also. Teama et al.[36] found, by gel filtration, that different heat treatments of milk

TABLE III
Changes in Water Soluble N as Percentage of Total
N in Kashkaval Cheese during Ripening[26]

Ripening (days)	Water soluble N (% of total N)
0	4·13
30	10·64
60	12·61
90	13·74
120	15·35
150	16·76
180	17·53
270	20·65
550	30·64

prior to the manufacture of Kashkaval cheese did not markedly affect molecular weight of soluble N compounds formed during ripening.

Djordjević[25] has shown (Table IV) a significant increase in monocalcium para-caseinate content during the first month of ripening, reaching a maximum at the beginning of the second month (40 days). At this stage, monocalcium para-caseinate nitrogen represents ~81% of the total and ~92% of the insoluble nitrogen in this cheese. Later, the absolute amount remains nearly constant up to six months, with a slight decrease afterwards. The author has discussed the advantages of this in detail elsewhere.[26] Proteolysis during the ripening of Kashkaval cheese is shown in Fig. 3.[26]

The free amino acid profile in Kashkaval cheese has been investigated by Alrubai[31] at the end of ripening (Table V); glutamic acid was present in the highest concentration, followed by leucine and lysine.

TABLE IV
Dynamics of Monocalcium Paracaseinate during Kashkaval Cheese Ripening[25]

Monocalcium paracaseinate	Ripening (months)								
	0	1	2	3	4	5	6	9	18
Monocalcium paracaseinate N (%)	1·56	2·99	3·23	3·14	3·19	3·24	3·28	2·97	2·60
Monocalcium paracaseinate N as % of the total N	41·48	76·71	78·67	74·89	74·46	74·90	75·00	64·79	50·76
Monocalcium paracaseinate N as % of the insoluble N	43·27	83·81	90·50	86·53	87·75	89·54	90·64	81·22	73·47

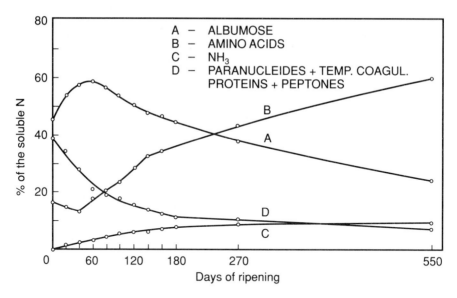

Fig. 3. Changes in the different water-soluble N fractions during ripening; albumose-fraction coagulating on addition of $ZnSO_4$; paranucleids-fraction coagulating on addition of 1% HCl (not entirely defined by Van Slyke).[9,26]

TABLE V
Free Amino Acid Composition in Ripened Kashkaval Cheese[31]

Amino acid	Free amino acid (as % of total amino acids)
Lys	17·75
His	0·33
Try	0·42
Arg	0·08
Asp	1·57
Thr + Ser	8·87
Glu	21·22
Pro	6·03
Gly	1·73
Ala	2·53
Cys	0·69
Val	7·26
Met	1·86
Ile	5·05
Leu	18·21
Tyr	0·56
Phe	5·84

TABLE VI
Free Fatty Acids Composition of Kashkaval Cheese (mg/kg Cheese)[13]

Fatty acid	Age of cheese					
	Young		2 months		4 months	
	(mg/kg cheese)	(% of total)	(mg/kg cheese)	(% of total)	(mg/kg cheese)	(% of total)
C_2	0·04	0·02	0·74	0·09	0·31	0·02
C_3	0·02	0·01	0·66	0·08	0·43	0·03
C_4	0·39	0·16	4·85	0·62	5·61	0·35
iso-C_5	0·01	—	0·09	0·01	0·79	0·05
C_5	0·42	0·17	0·07	0·01	0·17	0·01
C_6	0·73	0·30	2·37	0·30	5·27	0·33
C_8	0·23	0·09	1·69	0·22	3·63	0·23
C_{10}	6·16	2·51	6·62	0·84	5·48	0·34
$C_{10:1}$	0·99	0·40	0·81	0·10	1·53	0·10
C_{12}	8·59	3·50	12·70	1·62	17·50	1·09
$C_{12:1}$	0·69	0·28	1·89	0·24	4·18	0·26
iso-C_{14}	0·48	0·20	1·68	0·21	3·98	0·25
C_{14}	24·10	9·83	81·90	10·40	168·00	10·40
$C_{14:1}$	4·86	1·98	14·40	1·84	23·60	1·47
C_{15}	3·48	1·42	13·62	1·73	26·74	1·66
iso-C_{16}	0·60	0·24	4·48	0·57	46·19	2·87
C_{16}	79·70	32·50	275·00	35·10	554·00	34·40
$C_{16:1}$	5·38	2·19	21·50	2·74	28·00	1·74
C_{17}	1·79	0·73	9·97	1·27	13·70	0·85
$C_{17:1}$	1·88	0·78	3·35	0·47	2·68	0·17
C_{18}	24·00	9·78	80·60	10·30	180·00	11·20
$C_{18:1}$	71·70	29·10	226·00	28·80	456·00	28·30
$C_{18:2}$	5·81	2·37	2·97	1·14	45·10	2·80
$C_{18:3}$	3·64	1·48	11·40	1·45	16·90	1·05
Total	245·00		785·00		1609·00	

Only a low level of lipolysis occurs in Kashkaval during ripening, due mainly to the destruction of lipolytic microorganisms during milk pasteurization, leaving only lipases from thermoresistant bacteria.[37] However, volatile lower fatty acids, including butyric, caproic, caprylic and capric are detectable.[27] An increase in the free fatty acid level in the cheese mass has been observed during ageing, which gives a certain piquancy to the taste of the cheese.[8] The free fatty acid (FFA) compositions of young and aged Kashkaval cheese, are shown in Table VI. Evidently, the presence of high and increasing concentrations of butyric, caproic, caprylic and capric acids indicates their importance in the flavour of Kashkaval cheese. At the same time, the limited quantities of C_2, C_3, C_5, and iso-C_5, indicate their minor role in the characteristic flavour of Kashkaval cheese, unlike Cheddar, where acetic acid dominates.[13] Based on the results of an investigation into the lipid fractions of Kashkaval by thin layer chromatography

TABLE VII
Data of Chemical Analysis and Sensorial Evaluation of Kashkaval Vitosha, Produced with Different Starter Cultures after Prolonged Storage[38]

Indexes	Starter culture							
	Sofia		Stara Zagora		Vidin		Plovdiv	
	Storage [months]							
	2·5	8	2·5	8	2·5	8	2·5	8
VAC [mg/kg]								
Acetic aldehyde	0·07	—	0·94	—	0·04	0·01	0·12	0·04
Acetone	0·12	0·03	0·30	0·02	0·24	0·02	0·19	0·04
Ethylacetate	—	0·01	—	—	—	0·01	0·01	—
2-Butanone	—	0·01	0·10	—	0·03	—	0·05	—
Diacetyl	1·40	1·81	0·99	1·39	1·78	0·25	1·56	0·66
Ethanol	3·80	16·80	1·40	19·10	0·80	19·70	5·30	16·00
Acidity [°T]	280·00	254·00	213·00	197·00	217·00	182·00	200·00	232·00
pH	5·50	5·75	5·85	5·90	5·85	5·80	5·80	5·70
Sodium chloride [%]	2·19	2·87	2·58	2·94	2·64	2·82	2·22	2·57
Sensorial evaluation (position I–IV)	II	I	III	III	I	IV	IV	II

(TLC), Hassan & El-Deeb[14] reported the following composition: triglycerides 51·7%, free fatty acids 17·5%, 1,3-diglycerides 16·2% and 1,2-diglycerides, 14·6%.

As early as 1905, Van Slyke (cf. Ref. 24) stated the importance of the degree of lactose hydrolysis and noticed the correlation between pH (or acidity) and general cheese quality, including structure. Hydrolysis of lactose (lactic acid fermentation) occurs through different stages of Kashkaval production, reaching its maximum at the end of ripening: pH 4·8, compared to pH 4·9 for cheese made by the Russian, or pH 5·2 by the Italian methods.[24] A group of Bulgarian authors[38] investigated the changes in some volatile aromatic compounds (VAC) during storage of Kashkaval Balkan and Kashkaval Vitosha produced with four new high-quality commercial cheese starters. In Vitosha cheese, the concentrations of acetaldehyde and acetone decreased, whilst that of ethanol increased; the changes of diacetyl concentration were dependent on the starter used. Balkan cheeses contained higher concentrations of ethanol than Vitosha cheeses, with no diacetyl and little or no acetaldehyde after 10 months' storage. As regards Kashkaval Vitosha, the concentrations of diacetyl, ethanol and acetaldehyde were associated with the quality of the product (Table VII).

1.4 Quality Defects

The following important quality defects in Kashkaval cheese are encountered: (a) red or yellowish rind, due to the growth of the moulds *Oidium crustacea* (red) or *Oidium sulfurea* (yellow); (b) dirty white rind, caused by salt-resistant moulds, yeasts and bacteria, which also produce off-flavours (strong, bitter); (c) development of mites, *Tyroglyphus longior* or *Tyroglyphus siro*, on the rind due to low humidity; (d) late blown curd with gas holes can arise, caused by spore-forming anaerobic bacteria and followed by a sweet and butyric acid taste (early blown curd, caused by coliform bacteria, is prevented by heat treatment of the curd); (e) oxidative degradation of fatty acids to ketones, which arises from mould growth on the rind; (f) hard and dry curd, caused by incorrect, prolonged cheddaring (too much calcium removed).[1,5,23]

2 WHITE BRINED (PICKLED) CHEESE

White brined (pickled) cheese is a widespread cheese group produced in many varieties in hot countries: Feta (Greece), Bjalo salamureno sirene (Bulgaria), Beli sir u kriškama (Yugoslavia), Teleme (Romania, Greece), Lori, Imperetinskii, Limanskii, Osetinskii (former USSR),[41-45] Domiati (Egypt), Brinza (Czechoslovakia, Israel), Queso Blanco (South and Central America). White brined cheese is suitable for hot climates as it is actually stored in a concentrated brine (4 to 10–16% NaCl).[23,46] At temperatures lower than 8°C, it can be preserved for more than 3 months,[23,47,48] even up to 15 months, depending on storage temperature.[39]

Originally, it was manufactured by shepherds from sheep's, goats' or buffaloes' milk, but nowadays cows' and mixed milks are used also, and the cheeses are produced in modern mechanized or automated plants. However, certain differences in the technological processes used for the varieties mentioned are evident.

A common feature of white brined cheese technology is that ripening occurs in brine and lasts from a few weeks up to two months. Some of the varieties are produced from milk salted prior to renneting, and some from unsalted milk.[46] Specific parameters in the production schedules of white brined cheese varieties have been studied extensively.[41-44]

Different mixed cultures, including thermophilic and/or mesophilic microflora, are used in the production of white brined cheeses. The main flavour characteristics are acidity and saltiness. Obretenov *et al.*[49] showed that the total volatile fatty acid content of white brined cheese was 1226 mg/kg, of which 89% was acetic acid, while Dilanyan & Magak'yan[50] found 1181 mg/kg, of which 88% was acetic acid, showing a degree of fat lipolysis. A group of authors from the Australian CSIRO Division of Food Research[51] carried out a comprehensive investigation of the fat, headspace volatiles and free fatty acid content of Feta cheeses of different origin, using gas chromatography and mass spectrometry. The results (Table VIII) show that acetic acid (C_2) was the principal acid present: the levels of butyric (C_4), caproic (C_6) and caprylic (C_8) acids were low. The differences in the free fatty acid concentrations found by various authors may be related either to the techniques used for the distillation and extraction, or connected with the formation of acetic acid due to lactate fermentation (through pyruvic acid).[52]

TABLE VIII
Free Fatty Acid Concentrations in Feta Cheese Samples (mg/kg)[51]

Acid	Bulgarian	Romanian	Greek	Australian	
				Sample A	Sample B
C_2	604	673	639	411	41
C_3	12	—	19	—	—
C_4	50	35	521	17	470
C_6	26	19	292	12	219
C_8	14	6	146	10	76

Proteolysis in white brined cheeses has not been studied much, Magak'yan (cf. Ref. 53) found that the free amino acid content of a white brined cheese variety (Chanakh) increased to a maximum during ripening and then decreased, since the amino acids diffused into the brine. The author concluded that the main difference compared to other cheese groups was the predominance of di-n-butylamine, being about 20% of the total amine content, which could influence the characteristic flavour of white brined cheese. Polychroniadou & Vlachos[54] investigated the free amino acid, as well as amino acid composition, of another white brined cheese variety (Teleme) at different stages of ripening by ion-exchange column chromatography. The accumulation of free amino acids (leucine in the highest amounts followed by phenylalanine, lysine and valine) was accompanied by the development of a characteristic taste. Manolkidis et al.[55] estimated the rate and extent of proteolysis by the level of soluble N formed during the ripening of Teleme cheese, and Alichanidis et al.[56] studied proteolysis in Feta cheese.

The size and shape of this white, rindless cheese group varies, but the typical form is square, 10 cm³ with a weight of about 1 kg.[46] It has a white, firm texture, with no gas holes (only mechanical openings in the curd). The average chemical composition is presented in Table IX.

TABLE IX
Average Composition of White Brined Cheese

Type of white brined cheese	Fat-in-dry matter (%)	Dry matter (%)	Total protein (%)	NaCl (%)	pH or acidity	Reference
Feta (Greece)	48·52	46·37	—	5·05	4·3–4·4	Veinoglou et al.[48]
Bjalo Salamureno	27–31	40–42	10–12	3–4	290–320(°T)	Dimov et al.[1]
Sirene (Bulgaria)	43–45	42–44	16–18	2–3	280–300(°T)	
Beli sir u kriškama (Yugoslavia)	45–50	52–56	—	3–5	—	Šipka and Zonji (cf. Ref. 5)
	47·78	45·56	19·19	2·22	252(°T)	Carić et al.[57]

Most defects in white brined cheese are caused by coliform bacteria, in spite of the high acidity of the curd.[23] Yankov & Denkov[58] stated that slime in the brine caused by *Lb. plantarum* and/or *Lb. casei* could be prevented by maintaining the pH of the brine below 4·0 and the salt concentration above 8%.

2.1 Feta

Feta is one of the most popular, international, white brined cheeses made in many southern European and Middle Eastern countries: Greece, Yugoslavia, Bulgaria, Turkey, Egypt, Israel, but recently also in Denmark, United States, Canada and some other Western countries.[6,46,56,58,59]

The technology and characteristics of Feta cheese manufacture have been studied extensively,[51,53,56,58–63] partly due to the introduction of the new technology based on ultrafiltration, and are reviewed in detail in Chapter 11, Volume 2.

2.2 Bjalo Salamureno Sirene (White Brined Cheese)

Bjalo Salamureno Sirene (Belo Salamureno Sirene or Bjalo Sirene or, simply Sirene), is a very popular variety of white brined cheese in Bulgaria. Bjalo Salamureno Sirene has all the characteristics of the group, with a consistency that is neither too hard nor too soft and a slightly salty and acidic flavour. A culture composed of *L. lactis* subsp. *diacetylactis* and *Lb. casei* is used in its manufacture.[1,5,23] Continuous attempts are being made to improve the technology, which is partly mechanized, and the quality of this widespread cheese of Bulgaria.[1,64–66]

Contamination with micrococci, yeasts or moulds can cause colour changes. A reddish colour may also be the consequence of dosing animals with phenolthiazine. An atypical, sticky consistency is a defect due to *B. viscosum*, and an excessively hard consistency is caused by over-salting (Jotov, cf. Ref. 5).

2.3 Beli Sir U Kriškama (White Cheese in Pieces)

Beli Sir u kriškama is produced in Yugoslavia, under the names Srpski, Travnički, Sjenički, Sremski cheese, etc. It is manufactured from sheep's, cows' or mixed milks, and has all the typical characteristics of white brined cheese. A mixed culture, composed of *L. lactis* subsp. *diacetylactis* and *Lb. casei* is usually used.[24,46] Šutić *et al.*[67] have produced the white brined cheese, Sremski, using a mixed culture and ripened in vacuum-packed foil containers. The same cheese variety was successfully produced on an industrial scale using chymosin and pepsin as coagulants.[57] Other authors have also investigated the technology and quality of Travnički-type cheese.[68,69] Some additional varieties of white brined cheese have been investigated as well.[70–74]

3   OTHER BALKAN CHEESE VARIETIES

3.1   Kefalotyri (Greece)

Kefalotyri is a hard, salty Greek cheese, made exclusively from sheep's or goats' milk.[2,5,23,75] It is known as Pindos cheese, Skyros cheese, Vouscos cheese or Xynotiro[5] according to the district of manufacture. Kefalotyri has a flat cylindrical shape, diameter 30 cm; height 12 cm; and weight 5–10 kg. It has a high dry matter content, a firm texture with small gas holes and bigger slit holes. Its flavour is strong, piquant and salty. No starter is used in its manufacture. Average composition is: fat 28·25%; total solids 65·81%; total protein 24·76%; salt 4·77%; ash 4·60%; and pH 5·15.[6]

3.2   Paški (Yugoslavia)

Paški cheese means 'cheese from Pag', which is the name of the Adriatic island where it is produced. It is a hard cheese made from sheep's milk with characteristics typical of Parmesan, but the technology used in its manufacture is more similar to that used for Dutch cheeses. The ripening period is about two to three months, and possibly longer. Paški cheese has a flat cylindrical shape, diameter 18–22 cm; height 7–8 cm; and weight 2–4 kg. It has a high dry matter content and a firm, compact texture, with no holes. Its flavour is surprisingly piquant. No starter is used in its manufacture. Average composition is: fat 36%; total solids 66·2%; salt 2·9%; acidity 1·5%.[76]

3.3   Somborski (Yugoslavia)

Somborski cheese is a type of soft cheese with characteristics between those of white cheese and Trappist. It originates from Sombor, a city in Vojvodina, and was originally made only from sheep's milk. It is very popular in Yugoslavia, and now is also produced from mixed or cows' milk. The specific feature of its manufacture is that all operations after curd forming, i.e. salting, pressing, ripening, take place in wooden buckets. It has a soft, pasty body, with small gas holes and a slightly piquant flavour.[2]

The starter used in Somborski cheese manufacture consists of *Str. thermophilus* and *Lb. delbrueckii* subsp. *bulgaricus*. Average composition is: fat 26%; total solids 45·87%; total protein 17·32%; salt 2·32%; titratable acidity 152·10°.[77]

3.4   Chavdar (Bulgaria)

Chavdar is a soft cheese, produced in Bulgaria from cows' milk. It has a flat, cylindrical shape, diameter 20 cm; height 8 cm; and weight 2 kg. Chavdar has a homogeneous, creamy consistency, with small irregular holes, or without holes. Its flavour is specific, pleasant and acidic. A culture containing *L. lactis* subsp. *diacetylactis* and *L. lactis* subsp. *cremoris* is used in Chavdar manufacture. Average composition is: fat 25%; total solids 50%; salts 2·5–3·5%.[1]

4 MISCELLANEOUS CHEESE VARIETIES

Besides the cheese varieties described, which are native to all Balkan countries, many other cheeses are produced in the same area,[2,5,78,79] which have either local importance, or are counterparts of well-known western cheese groups. Since 80% of total cheese production in Bulgaria is white brined cheese (Bjalo Salamureno Sirene)[1] and in Greece, 75% is Feta,[5] and bearing in mind that there is also a significant production of Kashkaval (especially in Bulgaria), the quantities of other cheese varieties produced in those countries are rather small.

Some of the ripened cheeses which originated in, are produced in and are popular only in certain Balkan countries, besides those described, are:

1. Bulgaria:[1]
   Bezsalamureno Bjalo Sirene (means: unbrined white cheese);
2. Greece:[5]
   Kasseri (hard, Kashkaval type);
   Manur (hard cheese made of whey protein);
3. Yugoslavia:[2,46]
   Selam (hard);
   Njeguški (hard);
   Krčki (hard).

The most widely produced and popular counterparts of well-known western cheese varieties are: Parmesan (Greece, Yugoslavia); Cheddar (Bulgaria, Yugoslavia); Gruyère (Greece); Emmental (Bulgaria, Yugoslavia); Dutch types: Gouda, Edam (Bulgaria, Yugoslavia); Trappist (Bulgaria, Yugoslavia); Roquefort, Gorgonzola (Bulgaria, Yugoslavia); Camembert (Yugoslavia).

ACKNOWLEDGEMENTS

The author wishes to thank Prof. J. Djordjević for providing most of the references on Kashkaval and useful comments. Thanks are due to Prof. Z. Puhan for kind help in supplying important references, to Dr A. Kožev for providing helpful references on Bulgarian cheese varieties, and to Dr E. Alichanidis for supplying references on Greek cheeses varieties. Appreciation is expressed to Dr M. Kaláb for his unpublished electron micrographs of Kashkaval. Thanks are extended to Prof. Lj. Kršev for useful suggestions and to the author's co-workers: Dr D. Gavarić, Mr. sci. S. Milanović, Mr. sci. J. Kulić and Mrs. Z. Kosovac, Dairy Department, Faculty of Technology, University of Novi Sad, for their kind assistance in elaboration of the manuscript.

REFERENCES

1. Dimov, N., Kirov, N., Čomakov, H., Georgiev, I., Peičevski, I., Kožev, A., Denkov, C., Mineva, P., Petrova, N. & Konfortob, A., 1984. *Handbook for Dairy Technology*, Zemizdat, Sofia, pp. 175, 176, 203.

2. Pejić, O.M., 1956. *Dairy Technology*, Vol. 2. Technology of Milk Products, Naučna knjiga, Beograd, pp. 155, 175, 217.
3. Pejić, O., 1951. *Godišnjak Polj. fak.*, Zemun, **3**, 141.
4. El-Erian, A.F., Nour, M.A. & Shalaby, M., 1976. *Egyptian J. Dairy Sci.*, **4**, 91.
5. Mair-Waldburg, H., 1974. *Cheese Handbook, Cheeses of the World from A-Z.*, Volkswirtschaftlicher Verlag GmbH, Kempten, pp. 303, 544, 549.
6. Kosikowski, F.V., 1982. *Cheese and Fermented Milk Foods*, 2nd edn, F.V. Kosikowski and Associates, Brooktondale, New York, p. 690.
7. European Economic Community, 1990. *Official Journal of the European Communities*, **33**(L120), 56.
8. Šutić, M., 1966. PhD Thesis, Faculty of Agriculture, Beograd University, Beograd.
9. Djordjević, J., personal communication.
10. Antonova, T., Stefanova-Kondratenko, M., Bodurska, T., Manafova, N., Daov, T. & Nikoevska, T., 1984. *Acta Microbiologica Bulgaria*, **14**, 85.
11. Kaláb, M., Lowrie, J.R. & Nichols, D., 1982. *J. Dairy Sci.*, **65**, 1117.
12. Lowrie, J.R., Kaláb, M. & Nichols, D., 1982. *J. Dairy Sci.*, **65**, 1122.
13. Omar, M.M., El-Zayat & Ali I., 1986. *Food Chemistry*, **22**, 83.
14. Hassan, H.N. & El Deeb, S.A., 1988. *Food Chemistry*, **30**, 245.
15. Djordjević, J., 1960. PhD Thesis, Faculty of Agriculture, Beograd University, Beograd.
16. Djordjević, J., 1971. *Arh. Polj. Nauka*, **24**, 105.
17. Djordjević, J., 1974. *Mlekarstvo*, **24**, 54.
18. Stefanović, R., 1961. PhD Thesis, Faculty of Agriculture, Beograd University, Beograd.
19. Pejić, O., Djordjević, J., Stefanović, R. & Živković, Ž., 1964. *Zbornik radova Polj. fak.*, Beograd, **12**, 374.
20. Pejić, O., Djordjević, J. & Stefanović, R., 1965. *Hrana i ishrana*, **6**, 361.
21. Pejić, O., Djordjević, J. & Stefanović, R., 1966. *J. Sci. Agric. Res.*, **19**, 64.
22. Stefanovič, R. & Djordjevič, J., 1965. *Zbornik radova Polj. fak.*, Beograd, **13**, 407.
23. Scott, R., 1981. *Cheesemaking Practise*. Elsevier Applied Science Publishers, London, pp. 272, 437, 447.
24. Pejić, O., 1954. *Arh. Polj. nauka*, **7**, 17.
25. Djordjević, J., 1962. *Proc. 16th Intern. Dairy Congress*, Copenhagen, **B**, p. 490.
26. Djordjević, J., 1972. Kashkaval, Authorized postgraduate lectures, Faculty of Agriculture, Sarajevo.
27. Robinson, R.K., 1981. *Dairy Microbiology, Vol. 2. The Microbiology of Milk Products.* Elsevier Applied Science Publishers, London, pp. 179, 183.
28. Fox, P.F., 1981, *Neth. Milk Dairy J.*, **35**, 233.
29. Grappin, R., Rank, R.C. & Olson, N.F., 1985. *J. Dairy Sci.*, **68**, 531.
30. Marcos, A., Esteban, M.A., Leon, F. & Fernandez-Salguero, J., 1979. *J. Dairy Sci.* **62**, 892.
31. Alrubai, A., 1979. PhD Thesis, Faculty of Agriculture, Beograd University, Beograd.
32. Hofi, A.A., Teama, Z.Y., Abd-El-Salam, M.H. & El-Shabrawy, S.A., 1978. *Egyptian J. Dairy Sci.*, **6**, 177.
33. Law, B.A., 1980. *Dairy Industries Int.*, **45**(5), 15.
34. Law, B.A. & Wigmore, A., 1982. *J. Dairy Res.*, **49**, 137.
35. Buruiana, L.M. & Zeidan, I., 1982. *Egyptian J. Dairy Sci.*, **10**, 215.
36. Teama, Z.Y., Hofi, A.A., Abd-El-Salam, M.H. & El-Shabrawy, S.A., 1980. *Egyptian J. Dairy Sci.*, **8**, 71.
37. Klimovskii, I.I., 1966. *Biochemical and Microbiological Fundamentals of Cheese Technology*, Pishchevaja promishlenost, Moscow, p. 162.
38. Gyosheva, B., Stefanova, M. & Bankova, N., 1988. *Nahrung*, **32**, 121.
39. Kožev, A., 1972. *Molochnaja promishlenost*, **33**, 40.
40. Kožev, A., 1970. *Izv. Nauchnoizsled. Inst. Mlechna Promsht.*, Vidin, **4**, 137.
41. Dzobadze, K., 1965. *Molochnaja promishlenost*, **4**, 34.

42. Lomsadze, R.N., Mamatelashvili, G.S., Purceladze, N.G., Demurishvili, L.I. & Mandzavidze, N.P., 1976. *Molochnaja promishlenost*, **12**, 13.
43. Martirojan, A.A., Magak'jan, A.T. & Karasheninin, P.F., 1975. *Molochnaja promishlenost*, **1**, 21.
44. Hikalo, L. & Tadulev, B., 1964. *Molochnaja promishlenost*, **9**, 26.
45. Ramazanov, I.U., 1979. *Molochnaja promishlenost*, **9**, 15.
46. Slanovec, T., 1982. *Cheese Technology*, ČPZ Kmečki glas, Ljubljana, p. 141.
47. Veinoglou, B., Voyatzoglou, E. & Anifantakis, E., 1978. *Mljekarstvo*, **28**, 30.
48. Veinoglou, B.C., Boyazoglu, E.S. & Kotouza, E.D., 1979. *Dairy Ind. Intern.*, **44**, 29.
49. Obretenov, T., Dimitroff, D. & Obretenova, M., 1978. *Milchwissenschaft*, **33**, 545.
50. Dilanyan, Z.K. & Magak'yan, D.T., 1978. *Proc. 20th Intern. Dairy Congr.*, Paris, p. 295.
51. Horwood, J.F., Lloyd, G.T. & Stark, W., 1981. *Aust. J. Dairy Technol.*, **36**, 34.
52. Webb, B.H. & Johnson, A.H., 1965. *Fundamentals of Dairy Chemistry*. The AVI Publishing Company, Inc., Westport, p. 687.
53. Lloyd, G.T. & Ramshaw, E.H., 1979. *Aust. J. Dairy Technol.*, **34**, 180.
54. Polychroniadou, A. & Vlachos, J., 1979. *Le Lait*, **59**, 234.
55. Manolkidis, C., Polychroniadou, A. & Alichanidis, E., 1970. *Le Lait*, **50**, 128.
56. Alichanidis, E., Anifantakis, E.M., Polychroniadou, A. & Nanou, M., 1984. *J. Dairy Res.*, **51**, 141.
57. Carić, M., Milanović, S., Gavarić, D. & Španović, A., 1979. *Zbornik radova*, Tehn. fak., Novi Sad, **10**, 29.
58. Yankov, Y. & Denkov, T., 1972. *Izv. Nauchnoizsled. Inst.*, Mlechna Promsht., Vidin, **6**, 103.
59. Hansen, R., 1980. *Nordeuropaeik mejeri-tidsskrift*, **6**, 149.
60. Hansen, R., 1977. *Nordeuropaeik mejeri-tidsskrift*, **9**, 304.
61. Smietana, Z., Zuraw, J., Kaoka, E. & Poznanski, S., 1983. *Technologia Zywnoshci*, **18**, 55.
62. Carić, M., Gavarić, D., Milanović, S., Kulić, Lj., Popović, B., Cvetkov, D. & Kosovac, Z., 1985. *Investigations of the Usage of Ultrafiltration in Dairy Technology*, Faculty of Technology, Dairy Department, University of Novi Sad, p. 17.
63. Goncharov, A.I., Konanihin, A.V. & Tabachnikov, V.P., 1977. *Molochnaja promishlenost*, **12**, 14.
64. Baltadzhieva, M., Kyurkchiev, I. & Denkov, T.S., 1984. *Khranitelna promishlenost*, **33**, 14.
65. Baltadzhieva, M.A., Andreev, A.F., Sanechev, I.H., Todorov, T.L. & Minkov, T.H., 1983. Swiss Patent, CH 635 985 A5.
66. Stefanova-Kondratenko, M., Antonova, T., Bodurska, I., Manafova, N., Daov, T. & Nikoevska, T., 1984. *Acta Microbiologica Bulgarica*, **14**, 92.
67. Šutić, M., Obradovič, D., Pavlović, Ž., Marinković, L. & Birovljev, V., 1985. *Mljekarstvo*, **35**, 99.
68. Dozet, N., Stanišić, M., Bijeljac, S. & Perović, M., 1983. *Mljekarstvo*, **33**, 132.
69. Dozet, N., Stanišić, M., 1978. *Mljekarstvo*, **28**, 78.
70. Renner, E. & Ömeroglu, S., 1981. *Milchwissenschaft*, **36**, 334.
71. Tekinsen, O.C., 1983. *Ankara Üniversitesi Veteriner Fakültesi Dergisi*, **30**, 449.
72. Tekinsen, O.C. & Celik, C., 1983. *Ankara Üniversitesi Veteriner Fakültesi Dergisi*, **30**, 54.
73. Rakshy, S.E. & Attia, I., 1979. *Alexandria J. Agric. Res.*, **2**, 355.
74. Rakshy, S.E. & Attia, I., 1979. *Alexandria J. Agric. Res.*, **2**, 359.
75. Kristensen, J.M.B., 1983. *Maelkeritidende*, **96**, 7.
76. Markeš, M., Rubeša, M., Bašić, V., Koludrović, B., Stojak, Lj. & Lešić, Lj., 1979. *Mljekarstvo*, **29**, 26.
77. Carić, M., Milanović, S., Gavarić, D. & Litvai, V., 1983. *Zbornik radova*, Tehn. fak., Novi Sad, **14**, 7.
78. *IDF Catalogue of Cheeses*, 1981. Int. Dairy Federation, Doc. 141, Brussels.
79. Courtine, T.J., 1973. *Larousse des Fromages*, Librarie Larousse, Paris.

10

Cheeses of the former USSR

A.V. Gudkov

*Department of Microbiology, All-Union Research Institute for
Butter and Cheese-making Industry, Uglich, former USSR*

1 General Description of Cheese Production in the former USSR

The production and range of rennet cheeses made in the former USSR before 1984 have been described by Nebert *et al.*[1] Cheese production in the former USSR in 1989 is shown in Table I. Over the past 8 years, production has increased at an average annual rate of 2·67%. Farmhouse cheese products are not included in these data. Export and import of cheese are very limited.

Cheese is manufactured mainly from cows' milk. Sheep, goat and buffalo milks are used only for the production of some pickled and farmhouse cheeses.

Only Swiss-type cheese and some pickled varieties are permitted to be made from raw milk. Milk is usually pasteurized at 72°C for 20–25 s. At some dairies in the Baltic Republics, milk is pasteurized at 78–82°C. Usually, high temperature pasteurization is applied if the initial bacterial count in the raw milk is more than 10^6 cfu/ml. Milk pasteurized at high temperature is ripened by addition of 0·05–0·2% starter at 8–12°C for 12–20 h prior to processing into cheese. Cheeses made from milk pasteurized at a high temperature and subsequently ripened received higher total scores than cheeses made from the same milk pasteurized at 71–72°C.[2]

TABLE I
Cheese Production in the former USSR, 1989

Type of cheeses	Tonnes $\times 10^3$
Rennet cheeses	632·6
Processed cheeses	263·4
Whole milk tworog (9 or 18% fat)	720·0
Skim milk tworog	235·0
Total	1 851·0

Nitrate is used in the production of semi-hard cheeses. Hydrogen peroxide-catalase treatment of the cheesemilk is permitted but is very seldom used on a commercial scale.

Silage is used for feeding dairy cows everywhere. Special starters or organic acids are used for ensiling in some regions. Bulk starters for ensiling are prepared by direct inoculation of whey, enriched with growth factors, with concentrated freeze-dried cultures which comprise strains of *Lactococcus* and *Lactobacillus* which possess specific antagonistic action against butyric acid bacteria.[3]

As a rule, milk is delivered from farms to dairies every day (often twice a day). Therefore, psychrotrophs in milk do not pose a very serious problem for cheesemaking. However, the bacterial quality of raw milk is often poor, as cooling facilities are absent at some dairy farms.

Chymosin (only for hard cheese production), mixtures of bovine pepsin and chymosin (50:50 or 75:25), bovine and chicken pepsins (50:50) and bovine

TABLE II
Some Types of Cheese Starter Used in the former USSR

Commercial name of culture	Composition of freeze-dried culture	Counts of viable cells (cfu × 10⁹/g)
Bacterial preparate-U-4 (BP-U-4) and BCS[a]	*Lc. lactis, Lc. lactis* subsp. *cremoris, Lc. lactis* subsp. *diacetylactis* (6–8 strains)	250–700
Bioantibut	Mixed strains of lactococci and mesophilic lactobacilli possessing antagonistic action against clostridia	> 150, including 1–2 of lactobacilli
Thermophilic lactic bacteria (TMB)	Mixed strains of *Str. thermophilus, Lb. helveticus* or *Lc. lactis* (1–2 strains of each species)	100–200
Concentrate of mesophilic lactobacilli (BC-U-P)	Selected strains of mesophilic lactobacilli possessing antagonistic action against clostridia and coliforms	> 200
Concentrate of *Leuconostoc* spp. (BC-U-L)	Selected strains of *Leuconostoc cremoris*	> 40
ST TL LB LC	Selected strains of: *Str. thermophilus* *Lb. helveticus* or *Lb. lactis* *Lb. bulgaricus* *Lb. casei*	0·5–1·0 0·5–1·0 0·5–1·0 0·5–1·0

[a] BCS differs from BP-U-4 in the number of cells.

and sheep pepsins (25 : 75) are used to coagulate milk. Microbial proteinases are not used.

Freeze-dried concentrated starter cultures are usually used for cheese manufacture. Liquid starters are used only in the Baltic Republics. Deep-frozen cultures are not used because of the difficulties encountered in transporting them over long distances.

Only starter cultures with defined composition are used. Strain selection and starter culture production are performed at special laboratories attached to research institutes. The Scientific and Production Amalgamation of Butter and Cheese-making Industry 'Uglich' (SPA 'Uglich') is the main supplier of concentrated starter cultures for cheesemaking. The principal types of starter cultures supplied by SPA 'Uglich' are listed in Table II.

Strains of lactic acid bacteria for cheesemaking are selected on the basis of acid-producing activity in milk, phage resistance, proteolytic activity, especially failure to produce bitter flavour in a model milk medium, lipolytic activity, compatibility and antagonistic action against butyric acid bacteria and coliforms. A selection of 160 bacteriophages, isolated from cheeses and whey at cheese factories located in different regions of the former USSR, are used to type *Lactococcus* strains.

Selected strains of lactic acid bacteria are usually used as multiple-strain starters. Every month, SPA 'Uglich' provides cheese factories with three to four sets of multiple-strain starters composed of phage-unrelated strains. *Lactococcus* strains are used only in rotation. Strains of species that are more resistant to bacteriophages (e.g. lactobacilli) also may be used singly or in pairs without rotation.

Concentrated freeze-dried cultures prepared by SPA 'Uglich' are suitable for direct-to-vat inoculation of cheesemilk, but this method has not yet found commercial use for economic reasons. They are used for bulk starter preparation, without sub-culturing.

2 HARD RENNET CHEESES WITH HIGH SCALDING TEMPERATURES

Varieties and output of high-scald cheeses produced in the former USSR are shown in Table III. Their production was 3·7% of total rennet cheese output in 1987. Cheeses of this type are manufactured in the Altaï region (Sovetskiǐ, Altaiskiǐ, Biiskiǐ), Armenia (Swiss-type), Georgia (Altaïskiǐ), Estonia (Swiss-type in blocks) and the Ukraine (Ukrainian, Karpatskiǐ, Swiss-type in blocks).

Hard cheeses are made from milk containing not more than one lactate-iermenting clostridial per ml. Milk containing from one to 2·5 spores of these microorganisms per ml may be processed into hard cheeses if a special starter, possessing an antagonistic effect against clostridia, is used (Table II).

The use of *Streptococcus salivaricus* subsp. *thermophilus*, thermophilic lactobacilli and propionibacteria is a *sina qua non* for all varieties of high-scalded cheeses. Some years ago *Str. thermophilus* and thermophilic lactobacilli bulk starters were prepared individually from freeze-dried selected strains. Now,

TABLE III
Production and Some Characteristics of High-Scald Cheeses[a]

Feature	Sovet-skiĭ	Swiss-type	Swiss-type in blocks	Altaĭ-skiĭ	Biĭ-skiĭ	Gorny	Ukrai-nian	Karpat-skiĭ
Production (tonnes × 10³), 1987	5·9	2·1	2·9	1·3	1·5	3·0	4·6	0·8
Shape[b]	R	LC	R	LC	R	R	HC	LC
Size, cm: length	48–50	—	60–65	—	26–28	36–39	—	—
width	18–20	—	35–45	—	26–28	16–17	—	—
height	12–17	12–18	12–18	12–16	12–15	12–13	40–50	10–13
diameter	—	65–80	—	32–36	—	—	15–18	32–35
Weight, kg	11–18	40–90	30–45	12–18	8–10	7·5–9	8–10	12–15
Fat in dry matter, %	50	50	45	50	45	50	50	50
Moisture, %: opt.	36–38	36–38	38–39	36–37	38–40	37–39	39–41	39–41
max.	42	42	40	40	42	40	42	42
NaCl, %	1·5–2·5	1·5–2·5	0·5–1·8	1·5–2·0	1·2–1·8	1·3–1·6	1·2–1·6	1·2–1·5
pH (mature)	5·5–5·7	5·6–5·7	5·4–5·8	5·4–5·6	5·5–5·7	5·5–5·7	5·5–5·7	5·5–5·7

[a] There are state standards for Sovetskiĭ, Swiss-type, Swiss-type in blocks and Altaiskiĭ cheeses; for the other varieties, there are regional standards.
[b] R — Rectangle; LC — low cylinder; HC — high cylinder.

mixed bulk starters of these species are prepared with the use of TMB (Table II) as inoculum.[4] About 10^2–10^3 cfu/ml of propionibacteria, as a freeze-dried concentrate or liquid culture, are added to the pasteurized milk in the cheese vat.

The use of lactococci is not essential in the manufacture of hard cheese as they participate in the lactose fermentation only prior to scalding.[5] Lactose fermentation may be performed by Str. thermophilus alone. However, lactococci and Str. thermophilus are attacked by different phages and the use of both reduces the risk of starter failure. The size of the Str. thermophilus inoculum is increased if it is used without lactococci (Table IV).

The most popular variety of hard cheese in the former USSR is Sovetskiĭ, the manufacturing process for which was developed at the beginning of the 1930s.[6]

Sovetskiĭ cheese differs from Emmental by its smaller size and its shape (a loaf) (Table III). The smaller the cheese, the faster it expels moisture and the shorter the time required for pressing. Sovetskiĭ cheese is pressed for 4–6 h instead of 10–20 h for Swiss-type cheese. After pressing, Sovetskiĭ cheese is placed immediately in brine at 8–12°C. The short period of pressing and the low temperature of salting cause rapid cooling of Sovetskiĭ cheese to a temperature

TABLE IV
Features of Manufacturing Procedures for High-Scald Cheeses

Feature		Sovetskiĭ	Swiss-type	Swiss-type in blocks	Altaĭskii	Biĭskii	Gorny	Ukrainian	Karpatskiĭ
Type and level (%) of starters:	MS[a]	0·2–0·3	0·2–0·5	0·2–0·4	0·2–0·3	—	—	—	—
	TMB[b]	0·3–0·6	0·3–0·6	0·3–0·6	0·3–0·6	0·3–0·6			
	ST[b]	0·2–0·3	0·2–0·4	0·3–0·6	0·2–0·3	0·2–0·5	0·2–0·5	0·3–0·8	0·3–1·0
	TL[b]	0·05–0·15	0·05–0·20	0·05–0·20	0·1–0·2	0·05–0·20	0·15–0·35	0·03–0·07	0·05–0·07
Scalding temperature, °C		52–55	55–58	52–58	50–54	50–52	48–50	48–50	47–48
Duration of pressing, h		4–6	10–14	18–22	6–8	3·5–4·5	3–4	4–6	5–6
Moisture content after pressing, %		38–40	38–40	38–39	38–40	41–43	41–43	42–44	44–45
pH after pressing		5·4–5·7	5·5–5·6	5·4–5·6	5·5–5·6	5·5–5·7	5·3–5·6	5·7–5·9	5·6–5·9
Ripening: first period	°C	10–12	10–12	10–14	9–13	9–13	10–12	15–18	15–18
	days	15–25	15–25	15–25	10–15	10–20	10–15	20	15–20
second period	°C	20–24	16–18	20–24	21–25	19–23	20–22	10–12	12–14
	days	25–30	5–15	20–40	25–35	20–30	20–30	till ripe	till ripe
third period	°C	10–12	21–25	6–10	9–13	9–13	10–12		
	days	till ripe	20–40		till ripe				
fourth period	°C		10–12						
	days		till ripe						
Min. time of ripening, months		3	6	3	4	2	2	2	2

[a] MS — mesophilic streptococci; TMB — mixed culture of *Str. thermophilus* and thermophilic lactobacilli; ST — monoculture of *Str. thermophilus*; TL — monoculture thermophilic lactobacilli.
[b] TMB or ST + TL are used.

below the minimum for the growth of thermophilic bacteria. During brining, the temperature at the centre of Sovetskiĭ becomes equal to that of the brine bath after 10–15 h. Thermophilic starter does not have sufficient time to utilize all the lactose in Sovetskiĭ cheese, which contains lactose for up to 5–10 days while the population of thermophilic lactic acid bacteria does not increase after the end of pressing.[7,8] Also, lactococci do not multiply in Sovetskiĭ cheese after pressing as they are killed or inactivated by the high scalding temperature.

Residual lactose in Sovetskiĭ cheese is fermented mainly by mesophilic lacto-bacilli. Their counts increase from 10^2–10^3 of cfu g^{-1} in the cheese after pressing to 10^7–10^8 cfu g^{-1} at the end of ripening in the warm store, after which their number declines gradually.

Mesophilic lactobacilli are also one of the most numerous groups of micro-organisms in other hard cheeses (e.g. Emmental).[9] Their development in cheese depends on the amount of lactose which has not been utilized by the thermo-philic starter microorganisms. It is obvious that the numbers of mesophilic lacto-bacilli will be higher in hard cheeses which are cooled too rapidly.

Thermophilic lactobacilli usually promote the growth of propionibacteria in mixed cultures[10] but mesophilic lactobacilli may create favourable or un-favourable conditions for the growth of propionic acid bacteria in cheese. The type of interaction depends on the species and strains of lactobacilli. Some cells of mesophilic lactobacilli survive milk pasteurization and scalding and are, therefore, always present in cheeses. It is not possible to guarantee the absence of wild strains of mesophilic lactobacilli which may gain entry to the cheese and inhibit the growth of propionibacteria.

The pasteurized milk now used for the manufacture of Sovetskiĭ and some other varieties of hard cheese is direct-to-vat inoculated with 10^3–10^4 cfu/g of concentrated freeze-dried strains of mesophilic lactobacilli.[11] The strains are selected on the basis of compatibility with the propionibacteria strains used for the production of these cheeses and their antagonistic effect against butyric acid bacteria. Thus, mesophilic lactobacilli perform the functions of stimulating the growth of propionibacteria and inhibiting the development of butyric acid bac-teria. They do not effect the manufacturing process in the vat since only small numbers of them are added to the milk. Cheese made with selected strains of mesophilic lactobacilli achieves a higher flavour score than cheese made from the same milk to which only thermophilic starter is added. However, it is neces-sary to bear in mind that many *Lactobacillus* strains, although they inhibit the growth of butyric acid bacteria, produce off-flavours in cheese. Therefore, lacto-bacilli strains for cheesemaking must be selected very carefully.[12]

The optimum pH of Sovetskiĭ cheese at the end of pressing is 5·5–5·7 and it decreases to 5·3–5·4 within 5–10 days. The pH value is regulated by washing the curd with warm water (50–60°C) during scalding. The volume of water used for washing the curd is 0–20% that of the cheesemilk. The pH value of cheese exerts a strong influence on the growth of propionibacteria. Maximum numbers of propionibacteria in Sovetskiĭ cheese are about 280×10^6 per g at pH 5·31–5·36, 600×10^6 per g at pH 5·37–5·47 and 850×10^6 per g at pH 5·44–5·52.[13] Cheese made with additional wash water develops more pronounced flavour, a better consistency and better eye formation.

A high pH value in the cheese is also favourable for the growth of butyric acid bacteria. Therefore, the curd is washed with water when cheesemilk is free from spores of butyric acid bacteria, particularly in July–August.

The NaCl content of hard cheeses made in the former USSR is higher than that of similar cheeses made in other countries. Usually, Sovetskiĭ cheese is salted to 1·5–1·8% NaCl in the mature cheese. If cheesemilk is free of clostridial spores, the NaCl content in Sovetskiĭ cheese is reduced to 0·8–1·2% to create better conditions for the growth of propionibacteria. Increasing the salt content from 1·2 to 2·0% reduces the maximum propionibacteria count from 90×10^6 to 10×10^6 g.[14]

Sovetskiĭ cheese is salted in brine at 8–12°C. The minimum values for pH, water activity and temperature are higher for clostridial spore germination than for the subsequent vegetative growth of these bacteria. Most of the spores of butyric acid bacteria germinate during pressing and at the beginning of salting when conditions are more favourable for germination.[15] If cheesemilk is not free of clostridial spores it is advantageous to maintain the brine temperature at the lower end of the above range (i.e. 8°C) to retard spore germination. Later, their germination can be retarded by mesophilic lactobacilli and the reduction of water activity.

The following ripening procedure is recommended for Sovetskiĭ cheese: holding of brined cheese for 15–21 days at 10–12°C, at 20–25°C for 20–25 days and then at 10–12°C until final ripeness is achieved at 90 days.[16] Prolongation of the initial low-temperature period enables maximal development of lactic acid bacteria, including mesophilic lactobacilli, and the adaptation of propionibacteria to conditions in the cheese. By the end of this period the activity of enzymes produced by lactic acid bacteria in cheese reaches the maximum level; some of the intracellular enzymes are released into the cheese matrix which ensures a high rate of ripening during the later holding at 20–25°C. The quality of cheese ripened for 90 days by this procedure is similar to that ripened for 120 days by the procedure used previously (at 10–12°C for 8–10 days, then at 20–25°C for 20–25 days and further at 10–12°C to the end of ripening).

Propionibacteria begin to multiply in Sovetskiĭ cheese during the initial low-temperature period of ripening. At the end of this period their counts usually reach 10^4–10^5 cfu/g. However, the first eyes, 2·0–2·5 mm in diameter, appear after holding for 10–12 days in the warm store and their size increases progressively to 10–12 mm at the end of ripening.[17] Salt-tolerant strains of propionibacteria are now used for Sovetskiĭ cheese to prevent inhibition of their growth by high NaCl levels. Nevertheless, commercial cheeses sometimes do not develop good flavour due to poor growth of propionibacteria which is experienced more frequently in winter than in summer.[18] However, growth of *Clostridium tyrobutyricum* presents the greatest potential danger to Sovetskiĭ cheese.

The manufacture of Swiss-type cheese in blocks was commenced during the second half of the 1980s.[19] Four complete plants for the manufacture of this cheese were supplied by the Finnish firm, MKT. Cheese is made from pasteurized milk (72–74°C). According to the Finnish method, the mixture of curd grains and whey is pumped from the cheese vat to the cheese block-forming and pressing system. The curd grains are allowed to settle and form a layer under the whey. The layer of curd is levelled, covered with a perforated plate, compressed with a special device and after removing the whey, is pressed for 18–22 h at an external pressure of 3·15–4·73 kPa. The pressed layer of curd is cut into blocks that are transferred to the brine bath. Characteristics of this cheese and the method for its manufacture are summarized in Tables III and IV.

Altaiskiĭ cheese contains a little less moisture and salt than Sovetskiĭ (Table III). This cheese is ripened for not less than 4 months and is held in the warm store for 25–35 days instead of 20–25 days for Sovetskiĭ cheese. It has a more pronounced Emmental-type flavour than Sovetskiĭ. Usually, Altaiskiĭ cheese is made in mountainous regions in summer.

Biĭskiĭ cheese is smaller and has lower moisture and NaCl contents than Sovetskiĭ cheese (Table III).[20,21] The process for Biĭskiĭ cheesemaking differs from the procedure for Sovetskiĭ cheese by the addition of 200–300 g of NaCl/100 kg of cheesemilk to the mixture of curd grains and whey after scalding, and the use of a lower scalding temperature and shorter pressing and ripening times (Table IV). Biĭskiĭ cheese has a moderately pronounced Emmental-type flavour.

Addition of mineral salts to cheesemilk is the main distinctive feature of Gorny cheese manufacture.[22,23] The mixture of mineral salts consists of 2·3 g of $MnCl_24H_2O$, 0·1 g of $CoCl_26H_2O$, 1·34 g of $ZnCl_2H_2O$ and 0·57 g of $CuCl_22H_2O$ per tonne of milk. Minerals stimulate the growth of propionic acid bacteria and accelerate cheese ripening. Gorny cheese is ripened for 2 months. The favourable effect of these minerals on cheesemaking has been observed particularly with milks which have reduced concentrations of them but detailed information on the usefulness of these supplements for milks of normal mineral composition is lacking.

Ukranian and Karpatskiĭ cheeses were developed as intermediate varieties between hard and semi-hard cheeses. The curds for these cheeses are scalded at 48–50°C, the moisture content is 42–44 and 44–45% (after pressing), respectively; the temperature of the warm ripening store is 15–18°C (Table IV). It has been recommended to reduce the scalding temperature for these cheeses to 44–46°C and to increase the clotting milk temperature to 36°C.[24]

Ukrainian and Karpatskiĭ cheese have specific flavours resulting from the metabolic products of thermophilic lactic acid bacteria; eyes of small or moderate size and Karpatskiĭ may be devoid of eyes.

3 CHEESES WITH LOW OR MEDIUM SCALD TEMPERATURE

Full-fat rennet cheeses ripened by lactic acid bacteria only and scalded at 33–43°C are included in this group. The varieties and production of these cheeses are shown in Tables V and VI.

The scalding temperature for most of these cheeses is 38–42°C, i.e. it is a little higher than that usual for this type of cheese. The higher scald temperature increases the syneresis rate and has only a slight detrimental effect on the rate of lactose fermentation. It should be emphasized that the bacterial species used most commonly in mesophilic cheese starters in the former USSR is *Lc. lactis* that have a slightly higher maximum temperature for growth than *Lc. cremoris* which dominate in western mesophilic cheese starters. In addition, the inoculum volume, varies greatly and may be increased to 1·5% to accelerate lactose fermentation if the curd is scalded at 40–43°C. Starter containing only *Lc. cremoris* and *Lc. lactis* subsp. *diacetylactis* prepared at Kiev is used for the manufacture of Bucovina cheese,[25] which is scalded at a very low temperature.

In the manufacture of these cheeses, 200–300 g of NaCl per 100 kg of cheesemilk is added to the curd-whey mixture after scalding. This partial salting of the curd helps to control the water content of the cheese and is used only when the curd possesses normal syneresis properties.

All cheese varieties in this group are moulded from a curd layer under whey. Bucovina and Uglichskiĭ cheese also may be formed by draining the curd-whey mixture into moulds. Cheeses moulded in this manner have irregularly shaped eyes.

TABLE V
Production of Cheeses with Low or Medium Scalding Temperature

Cheese variety	Tonnes × 10^3
Kostromskoĭ (Kostroma)[a] and Poshechonskiĭ[b]	142·5
Hollandskiĭ brushkovyi (Dutch brick)	80·5
Bukovinskiĭ (Bukovina)[a]	25·1
Hollandskiĭ kruglyĭ (Dutch spherical)[a]	6·2
Estonskiĭ (Estonia)[a]	5·2
Stepnoĭ (Steppe)[a]	3·7
Jaroslavskiĭ[a]	1·0
Dnestrovskiĭ	0·9
Juznyĭ	0·9
Valmierskiĭ	0·4
Severnyĭ	0·3
Tchuiskiĭ	0·2
Novosibirskiĭ	0·2
Pjarnuskiĭ	0·2
Rossiiskiĭ[a,c]	105·9
Atlet[c]	6·9
Susaninskiĭ[a,c]	4·9
Cheddar[a,c]	2·5
Baltija[c]	5·1

[a] There are state standards for these varieties; for the other varieties there are regional standards.
[b] Kostroma and Poshechonskiĭ cheeses are now considered to be the same variety because the procedures for their manufacture are very similar.
[c] Cheeses with high levels of lactic acid fermentation.

TABLE VI

Characteristics of the Major Varieties of Cheeses with Low and Medium Scalding Temperature

Variety	Shape[a]	Size, cm				Weight, kg	Fat in DM, %	Moisture, %		NaCl, %		pH	Shape of eyes
		L^a	W^a	H^a	D^a			max.	opt	max.	opt		
Kostroma	LC	—	—	8–11	24–28	3·5–7·5	45	44	40–42	2·5	1·5–2·0	5·25–5·35	Round openings
Hollandskiĭ bruskovyĭ	R	24–30	12–15	9–12	—	2·5–6·0	45	44	40–42	3·0	2·0–2·5	5·25–5·35	Round openings
Bukovina	HC	—	—	40–45	12–14	4–6	45	44	40–43	2·5	1·5–2·0	5·5–5·7	Round or irregular openings
Hollandskiĭ kruglyĭ	S	—	—	—	12–16	1·8–2·5	50	43	39–41	3·0	2·0–2·5	5·2–5·3	Round openings
Estonia	HC	—	—	30–35	8–10	2–3	45	44	40–42	2·5	1·5–2·0	5·25–5·4	Round openings
Steppe	R	26–28	26–28	9–11	—	6·5–9·5	45	44	40–41	3·0	2·0–2·5	5·3–5·4	Round openings
Jaroslavskiĭ	HC	—	—	25–35	8–10	2–3	45	44	40–42	2·5	1·5–2·0	5·3–5·4	Round openings
Uglichskiĭ	R	24–30	12–15	9–12	—	2·5–6·0	45	45	41–43	2·5	1·5–2·0	5·3–5·4	Round or irregular openings

[a] R — rectangular; LC — low cylinder; S — spherical; HC — high cylinder; L — length; W — width; H — height; D — diameter.

Cheeses in moulds are held without external pressure for 20–50 min (self-pressing) or with a low pressure (1–2 kPa) for 20–30 min and are then pressed under gradually increasing pressure from 10–30 to 25–130 kPa for 1·5–2·5 h (Table VII). The pH of low-scalded cheeses is 5·6–5·8 after pressing and reaches the ultimate value within 1–3 days.

Cheeses are salted in brine, usually at 10–14°C for two to three days, but a lower brine temperature (sometimes 6–8°C) is used when cheesemilk is contaminated with *Cl. tyrobutyricum* spores. The salt content of these cheeses is higher than those in cheeses of this group made in most other countries, but customers object to lowering the salt content.

Recommended temperatures and duration of ripening for these cheeses are shown in Table VII. In practice, cheeses in this group are usually ripened at 10–14°C throughout, but a lower ripening temperature (9–10°C) is used if the cheesemilk is contaminated with *Cl. tyrobutyricum* spores.

The high concentration of salt, rapid cooling, ripening at comparatively low temperatures and the use of starters possessing specific antagonistic action against clostridia (Table II) increase the resistance of these cheeses to butyric acid fermentation.

Most of the cheeses in this group are ripened for 45–75 days; Bukovina and Estonia cheeses are ripened for about 30 days. The high moisture content of Bukovina cheese accelerates its ripening[26] and it is also ripened at a higher temperature (Table VII). From 0·1 to 1·0% of a lactic acid bacterial culture hydrolysed by pepsin is added to milk used for Estonia cheese manufacture to accelerate ripening. The flavour of this cheese is quite distinctive. Addition of the hydrolysate shortens the ripening period and improves the quality of other low-scald cheeses, e.g. Dutch-type and Rossiĭskiĭ cheese.[27]

4   CHEESES WITH A HIGH LEVEL OF LACTIC ACID FERMENTATION

Officially, Rossiĭskiĭ, Cheddar, Atlet and Baltija cheeses are included in this group. From our point of view, Susaninskiĭ cheese may be also classified in this group.[28] Production of these cheeses in the former USSR in 1987 is shown in Table V.

Cheeses of this group are scalded at low or medium temperatures. The salient point in the manufacture of these cheeses is the higher rate of acidification during the manufacturing stage, which is achieved by certain process modifications. First, a large inoculum volume is used: 0·7–1·5% for Rossiĭskiĭ cheese, 1·0–2·5% for Cheddar, 1·0–4·5% for Atlet, 1·5–3·0% for Baltija and 3·0–5·0% for Susaninskiĭ (Table VIII).

Lactobacilli cultures are also added to the cheesemilk: up to 0·2% of *Lb. casei* or *Lb. plantarum* cultures for Rossiĭskiĭ cheese, 0·04–0·06% of *Lb. bulgaricus* for Susaninskiĭ and up to 0·6% of mesophilic or thermophilic lactobacilli for Cheddar. Lactobacilli themselves contribute to lactose fermentation, stimulate the growth of lactococci and increase the resistance of the starter to bacteriophage.[29] It is not necessary to use lactobacilli if lactococci produce acid at a rate appropriate for the particular cheese being made.

TABLE VII

Characteristics of Procedures for the Manufacture of Low-Scald Temperature

Feature		Kostroma	Holland bruskov	Bukovina	Holland kruglyĭ	Estonia	Jaroslav-skiĭ	Uglichs-skiĭ	Steppe
Scalding temperature, °C		38–42	38–42	33–35	38–41	38–42	40–42	37–39	40–42
Inoculum volume, %		0·5–1·0	0·5–1·0	0·7–1·5	0·5–1·0	0·5–1·2	0·5–1·0	0·5–1·5	0·7–1·2
Volume of water for curd washing, % of whey		10–15	5–15	8–15	5–15	5–15	—	—	—
NaCl added to curd-whey mixture, g/100 kg of milk		200–300	200–300	—	200–300	—	200–300	200–300	200–300
Cheese after pressing	pH	5·6–5·8	5·5–5·8	5·4–5·7	5·5–5·8	5·2–5·4	5·6–5·9	5·4–5·6	5·6–5·8
	moisture, %	44–46	43–45	44–48	43–45	42–44	42–44	46–48	44–46
Ripening: first period	°C	10–12	10–12	12–14	10–12	10–16	10–12	12–14	13–14
	days	15–20	12–16	30	15–20	30	15–20	20–25	15–20
second period	°C	14–16	14–16	—	14–16	—	14–16	10–12	10–12
	days	25–30	30	—	30	—	20–30	35–40	55–60
third period	°C	—	12–14	—	10–12	—	12–14	—	—
	days	—	till ripe	—	till ripe	—	till ripe	—	—
Total ripening time, days		45	60	30	75	30	60	60	75

TABLE VIII

Some Characteristics of Cheeses with a High Level of Lactic Acid Fermentation

Feature		Cheese variety				
		Rossiĭskiĭ	*Susaninskiĭ*	*Cheddar*	*Atlet*	*Baltija*
Shape		LC	R	R	R	R
Size, cm:	length	—	24–30	27–29	35–37	27–31
	width	—	11–15	11–13	35–37	11–14
	height	10–16	6–10	8–10	10–13	7–10
	diameter	24–28	—	—	—	—
Weight, kg		4·7–11·0	2·0–2·5	2·5–4·0	13–19	2·5–4·0
Fat in dry matter, %		50	45	50	45	45
Moisture content, %	opt.	40–42	46–48	38–39	38–39	38–39
	max.	43	48	42	42	40
NaCl in mature	opt.	1·3–1·6	1·2–1·5	1·5–1·8	1·4–1·6	1·5–1·8
cheese, %	max.	1·8	1·8	2·0	2·0	2·5
pH		5·25–5·35	5·2–5·4	5·1–5·2	5·2–5·4	5·2–5·3
Shape of eyes		irregular	round	absent	irregular	absent
Volume of inoculum, %		0·7–1·5	3·0–5·0	1·0–2·5	1·0–4·5	1·5–3·0
Scalding temperature, °C		41–43	36–38	38–40	39–43	38–40
NaCl added to curd-whey mixture, g/100 litre milk		300–700	500–1000	—	500–700	—
Duration of pressing, h		6–18	1·5–2·0	12–14	2	12–18
Cheese after	pH	5·2–5·3	5·3–5·6	5·2–5·3	5·4–5·6	5·1–5·2
pressing:	moisture, %	43–45	50–53	39–42	40–43	—
Ripening:						
first period	°C	10–12	12–14	10–13	8–12	8–10
	days	10–14	15	30–45	45	45
second period	°C	13–15	—	6–8	—	6–8
	days	16–20	—	45–60	—	30
third period	°C	10–12	—	—	—	—
	days	30–60	—	—	—	—
Total time of ripening, days		60	15	90	45	75

Rossiĭskiĭ and Cheddar cheeses are pressed for long periods (Table VIII). Pressing decreases the rate of cooling of the cheese and as a result promotes the growth of starter microorganisms. These cheeses contain almost no lactose at the end of pressing.

Rapid acidification during the manufacturing stage enhances syneresis. Partial salting of Rossiĭskiĭ and Susaninskiĭ curd (300–1,000 g of NaCl per 100 kg of cheesemilk are added to curd-whey mixture) is used to control the moisture content at the optimal level. Addition of this amount of salt to the curd does not inhibit the growth of starter microorganisms.

If acidification during manufacture is too fast it may result in the development of a strong acid taste in the cheese. To prevent the impairment of cheese quality, curd is washed with pasteurized water at the beginning of scalding. The level of whey replacement with water is 5–10% for Rossiĭskiĭ cheese and 80–100% for Susaninskiĭ cheese. The lactose content of Susaninskiĭ is reduced sharply as a result of the extensive washing of the curd. Therefore, the ultimate pH value of Susaninskiĭ cheese is higher than that of Rossiĭskiĭ and Cheddar cheese.

Susaninskiĭ cheeses are moulded from a layer of curd under whey. The curd-whey mixture for Rossiĭskiĭ cheese is pumped to a rotary whey separator and is then poured into moulds. Eyes in Rossiĭskiĭ cheese are irregularly shaped because they are formed as a result of the incorporation of air bubbles into the cheese mass during moulding and eye formation does not depend on the gas-forming activity of the starter.

Susaninskiĭ cheese ripens very quickly due to its high moisture content and comparatively high pH. Rossiĭskiĭ cheese is ripened for less than 60 days. Both cheeses have distinctive flavour and consistency.

The faster the lactose fermentation, the less the opportunity for the growth of harmful microorganisms in cheese. This is especially true for Susaninskiĭ cheese. Rossiĭskiĭ cheese is also resistant to microbial spoilage, particularly to the butyric acid fermentation, but only if the acidification rate is high enough. If starter activity is low due to bacteriophage infection or for any other reason, the slow cooling of Rossiĭskiĭ cheese results in the rapid growth of harmful micro-organisms.

In the 1980s, a method for the production of a new variety of cheese with a high level of lactic acid fermentation was developed in Estonia.[30] This cheese variety was named Atlet, which is analogous to Rossiĭskiĭ cheese (Table VIII). Milk is processed into Atlet cheese immediately after pasteurization (72–75°C, 20–25 s) or after ripening for 10–24 h at 8–12°C. Pasteurized milk, which has been ripened, is inoculated with 0·2–0·3% of mesophilic starter. From 3 to 6% of the required amount of rennet may be added to the cheesemilk before ripening to save rennet as the amount that must be added to the milk at beginning of cheese manufacture in the vat is reduced by more than 3–6%.

The high acidification rate during the manufacturing stage of Atlet cheese results from a high inoculum of starter (up to 4·5%) and milk ripening. The pH of inoculated cheesemilk before renneting is 6·5–6·35.

Atlet curd is scalded at 39–43°C; the scalding temperature actually used is chosen taking into account the syneretic properties of the curd. From 100 to 150 g of NaCl per 100 kg of cheesemilk are added before renneting to reduce grain treatment time and to accelerate the growth of starter microorganisms.[30] After scalding, 400–550 g of NaCl/100 g are added to the curd-whey mixture. The salted curd-whey mixture is held for about 20 min and is then pumped to the rotary whey separator. After whey separation, curd grains are pumped to the top part of the column of a Caso-Matic curd-forming system. In the column, cheese grains fuse slowly into a continuous mass under its own weight. A special device located at the bottom of the column cuts off portions of cheese that correspond to the weight of a cheese loaf. These are put into moulds moving on a conveyor.

Atlet cheese is pressed in the Press-Matic system for 2 h. Then, the cheese is held in a room until the pH decreases to 5·45–5·65, usually for about 1 h. The pH of Atlet cheese reaches the ultimate value (5·1–5·2) after 1–2 days. The cheeses are salted in brine at 8–12°C for 2–3 days and ripened for not less than 45 days.

In the 1970s, a new variety of cheese with a high level of lactic acid fermentation, named 'Baltija', was developed in Latvia.[31] Baltija is produced only by a factory in Prejly, according to the local standard. Baltija is similar to Cheddar but it contains a little less fat (45% in dry matter) and a little more salt (up to 2·5%) than Russian Cheddar (Table IV).

According to the manufacturing process for Baltija cheese, the massed curd is Cheddared continuously at 32–36°C for 2–3 h in order to reduce the pH to 5·1–5·3. The curd is then cut into pieces and salted, also continuously. Salted curd is pressed in special chambers under vacuum for 12–18 h. Baltija is ripened in blocks, weighing 18–19 kg, for the first 45 days at 8–10°C and then for 30 days at 6–8°C. Blocks of mature cheese are cut into six loaves which are packaged in a plastic film. The complete plant for the production of Baltija was supplied by Pasilac.

5 PICKLED CHEESES

This group includes high-salted cheeses ripened only by lactic acid bacteria. These cheeses represented about 17% of total rennet cheese output in 1987 (Table IX). They are the most popular varieties in Armenia (Lory, Osetinskiĭ, Aikavanskiĭ, etc), Azerbaĭjan (Brynza, Tushinskiĭ, Chanach), Georgia (Suluguni Gruzinskiĭ, Svezhiĭ, Imeretinskiĭ, Stolovyĭ), Northern Caucasus (Brynza, Osetinskiĭ, Kobiĭskiĭ, Stolovyĭ), Moldavia (Moldavskiĭ) and a few regions of the Ukraine (Limanskiĭ, Brynza, Moldavskiĭ). The composition and some other characteristics of pickled cheeses are shown in Table X.

TABLE IX
Production of Pickled Cheese in the former USSR, 1987

Varieties	Tonnes $\times 10^3$
Brynza	27·2
Osetinskiĭ	24·2
Suluguni	16·5
Svezhiĭ	7·1
Gruzinskiĭ	4·3
Stolovyĭ	4·1
Moldavskiĭ	3·2
Limanskiĭ	3·0
Kobiĭskiĭ	1·2
Tushinskiĭ	1·3
Imeretinskiĭ	1·3
Aikavanskii	0·9
Chanach	0·8
Others (14 varieties)	8·8
Total	103·9

TABLE X

Characteristics of Some Pickled Cheese Varieties

Variety	Shape[a]	L[c]	W[c]	H[c]	D[c]	Weight, kg	Fat in DM, %	H_2O %	NaCl, %	pH	Duration of ripening, days
Brynza	R	10–12	10–12	7–9	—	1·0–1·5	45	53	3–5	5·0–5·1	20
Osetinskii[b] fresh	LC	—	—	10–14	24–28	4·5–8·0	45	54	2–4	4·9–5·0	5
ripe	LC	—	—	10–14	24–28	4·5–8·0	45	51	4–5	5·0–5·15	30
Suluguni	LC	—	—	2·5–3·5	15–20	0·5–1·5	45	50	1–5	4·9–5·1	1
Lory	R	28–30	14–15	10–12	—	4–6	50	44	3–4	5·2–5·35	45
Stolovyĭ fresh	R	24–30	12–15	10–14	—	3·0–6·5	40	53	1–3	5·2–5·4	5
ripe	R	24–30	12–15	10–14	—	3·0–6·5	40	50	2–4	5·2–5·35	15
Moldavskiĭ	R	26–30	13–15	10–12	—	3·4–5·5	40	60	< 4	—	5
Limanskiĭ	R	10–12	10–12	8–10	—	0·8–1·8	50	52	1·5–4	—	5
Kobiiskii	TC	—	—	17–19	21–25 13–16	4–6	45	51	4–5	5·1–5·2	30
Imeretinskii	R	18–20	8–10	6–7	—	0·5–1·2	45	52	2–4	5·0–5·1	1
	LC	—	—	3–5	14–17	0·5–1·2	45	52	2–5	5·0–5·1	1
Chanach	R	18–20	18–20	11–15	15–16 22–24	4–6	45	50	4–7	5·1–5·2	60
	DTC	—	—	18–20		4–7	45	50	4–7	5·1–5·2	60

[a] R — rectangular; LC — low cylinder; TC — truncated cone; DTC — two truncated cones joined at bases.
[b] Osetinskii and Stolovyĭ cheeses may be consumed fresh or ripened.
[c] L — length; W — width; H — height; D — diameter.

The salient points of the traditional manufacturing procedure for pickled cheeses are as follows. Cheesemilk (cows' milk or mixtures of cows', sheep and buffalo milk) is pasteurized at 72°C for 20–25 s. Pickled cheeses may be made from raw milk when cows are grazed on distant pastures. Cheeses made from raw milk must be ripened for not less than 60 days.

Ripened milk (cows' milk with a titratable acidity of 19–21°T or mixtures of cows', sheep and buffalo milks with a titratable acidity of 23–26°T) is used for the manufacture of pickled cheeses. Milk is inoculated with 0·5–2·0% of mesophilic starter and renneted within 30–50 min. Most pickled cheese varieties are scalded at 34–38°C.

Different methods of moulding are used, e.g. from the curd layer under whey, by draining the curd-whey mixture into moulds or by pouring the curd grains into the moulds after whey separation.

Typically, pickled cheeses contain 52–56% moisture and the pH is 5·0–5·2 at brining. The pH usually attains the ultimate value (4·9–5·0) within 2–3 days, but sometimes later. To hasten lactose fermentation in the cheese during salting in brine, salt-resistant lactococci strains, able to grow at NaCl concentrations of 6·5–7·4%, have been selected and used for the manufacture of pickled cheeses.[32,33] Osetinskiĭ cheese made using salt-resistant starters has increased contents of soluble protein and non-protein nitrogen compounds, a more pronounced flavour and better consistency than cheese made with the usual starter. Mesophilic lactobacilli, particularly strains of Lb. casei, which are more resistant to salt and acid than lactococci, are also used for the manufacture of pickled cheeses.

Traditionally, pickled cheeses are salted in 18–22% whey or water brine for 3–6 days at 8–12°C and ripened in 16–18% brine at 6–8°C.[34] Limanskiĭ and Moldavskiĭ cheeses are ripened in 14–19% and 13–15% brine, respectively. Fully brined cheeses (Brynza, Chanach, Limanskiĭ, Tushinskiĭ, Kobiĭskiĭ, Osetinskiĭ) are stored and sold in barrels filled with brine. After storage, the salt content of the cheeses may increase to 8%. These cheeses have a strong salty taste and usually a coarse consistency.

Noticeable progress has been made on the manufacturing process for pickled cheeses during the past two decades. A new procedure has been developed for Lory cheese.[35,36] Partial salting of the curd-whey mixture (300–500 g NaCl per 100 kg of cheesemilk) enables the period of salting in brine to be reduced from 6–8 days, compared to 14–15 days previously. After salting in brine, Lory is packed in plastic film under vacuum and ripened for 45 days (previously 60 days). Ripening out of brine improves the quality and increases the yield of cheese. The salt content of Lory, which is the most widespread pickled cheese in Armenia, does not exceed 4·5%, regardless of the storage period.

A newly developed pickled cheese is 'Stolovyĭ'.[37] Milk for the manufacture of this cheese is pasteurized at 85–90°C, which increases cheese yield by co-precipitation of the whey proteins with the caseins. Partial salting of the curd-whey mixture after removing 60–70% of the whey (200–300 g NaCl per 100 kg of milk) is used in the manufacture of Stolovyĭ cheese, which is pressed under an external pressure of 8–20 kPa for 25–30 min and salted in 18–20% brine at 10–20°C for one to four days. It is sold either fresh or mature (15 days old).

It has been shown that homogenization of cheesemilk, involving separation, homogenization of the cream only and its recombination with the skim milk, in the manufacture of pickled cheese increases cheese yield by an average of 5·8% and the greater retention of milk fat, by 6%, decreases hardness and improves consistency.[38]

Suluguni cheese is very popular in Georgia where its output represents about 27% of total cheese production.[39,40] Suluguni is scalded at 36–38°C but may be made without scalding if renneting is performed at 34–38°C. Cheddaring, one of the most important steps in the manufacture of Suluguni cheese, is carried out in whey at 34–35°C for 2–5 h to attain a pH of 5·0–5·2 or a titratable acidity of 140–160°T.

The next important step of Suluguni cheese manufacture is plasticizing: the curd is plasticized at 70–80°C in water, whey, 8–10°C NaCl brine or by direct heating for 5–7 min. Cheeses plasticized by the dry method have the lowest content of moisture, retain most fat, have the highest content of free amino acids and receive the highest score for taste and consistency.

Suluguni is salted in 16–20% water brine or 16–18% whey brine. Titratable acidity of the water and whey brines must not be higher than 25°T and 60°T, respectively.

Suluguni is used for direct consumption and for the manufacture of a wide range of cooked foods.

REFERENCES

1. Nebert, V.K., Silaeva, V.M., Vinogradova, R.P., Mirin, V.G. & Baryshev, G.A., 1984. *Moloch. Prom.*, **N3**, 15.
2. Ramanauskas, R., 1980. *Trudy Lit. Fil. vses. nauchno-issled.* Inst. masl. i syr. Prom., **14**, 71.
3. Gudkov, A.V., Poljanin, A.N., Michlin, E.D. & Krasheninin, P.F., 1976. In *Nauchn. osnovy konserv. rastitel'nych kormov*, Academy of Science of former USSR, Moskow, p 149.
4. Belova, L.P., Guseva, N.M. & Ostroumov, L.A., 1986. *Trudy vses. nauchno-issled.* Inst. masl. i syr. Prom., **46**, 58.
5. Kagan, Ya.R., Chistjakova, N.Ya., 1981. In *Biologia Microorganismov i ich Ispol'zovanie v Narodnom Khozyaistve*, ed. A.G. Grinevich, Irkutskii Gos. Univ., Irkutsk, 32, former USSR.
6. Granikov, D.A., 1972. *Sovetskii Cheese*, Pishchevaja Promyslennost', Moskow, p. 276.
7. Gudkov, A.V., Anischenko, I.P., Ostroumov, L.A. & Alekseeva, M.A., 1980. *Moloch. Prom.*, **N2**, 13.
8. Kagan, Ya.R. & Sergeeva, I.Ja., 1985. *Trudy vses. nauchno-issled.* Inst. masl. i syr. Prom., **42**, 50.
9. El-Tobqui, N.A.-L., 1979. *Milchwissenschaft*, **34**, 505.
10. Perez Chaiva, A., Pesce de ruiz Holgado, A. & Oliver, G., 1987. *Microbiologia-Aliments-Nutrition*, **5**, 325.

11. Alekseeva, M.A., Anischenko, I.P., Stepanova, E.A. & Ott, E.F., 1986. *Moloch. Prom.*, **N1**, 41.
12. Gudkov, A.V. & Perfil'ev, G.D., 1979. *Trudy vses. nauchno-issled.* Inst. masl. i syr. Prom., **30**, 75.
13. Babushkina, V.A., Ostroumov, L.A. & Govorytkina, S.A., 1977. In *Biologia Micro-organismov i ich Ispolizovanie v Narodnom. Khozaistve*, ed. A.G. Grinevich, Irkutskii Gos. Univ., Irkutsk, 64, former USSR.
14. Ostroumov, L.A., Alekseeva, M.A. & Babushkina, V.A., 1977. *Moloch. Prom.*, **N12**, 11.
15. Perfil'ev, G.D., Gudkov, A.V. & Matevosyan, L.C., 1979. *Trudy vses. nauchno-issled.* Inst. masl. i syr. Prom., **30**, 78.
16. Ostroumov, L.A., Babushkina, V.A. & Gudkov, A.V., 1977. *Trudy vses. nauchno-issled.* Inst. masl. i syr. Prom., **21**, 50.
17. Ostroumov, L.A., Maiorov, A.A. & Tabachnikov, V.P., 1979. *Trudy vses. nauchno-issled.* Inst. masl. i syr. Prom., **29**, 75.
18. Laht, T.I., Vili, P.O. & Kanger, K.O., 1989. In *Sovremennaj Technologij Syrodelija i Besothodnaja Pererabotka Moloka*, Aiastan, Ervan, former USSR, p. 184.
19. Babushkina, V.A., Anishchenko, I.P., Alekseeva, M.A. & Laht, T.I., 1988. *Trudy vses. nauchno-issled.* Inst. masl. i syr. Prom., **51**, 33.
20. Avdanina, E.A., 1979. *Trudy vses. nauchno-issled.* Inst. masl. i syr. Prom., **29**, 79.
21. Drobyshev, A.A. & Zinov'ev, I.I., 1982. *Moloch. Prom.*, **N3**, 35.
22. Saakyan, R.V., 1982. *Moloch Prom.*, **N3**, 27.
23. Dilanyan, Z.Kh., Makaryan, K.V. & Kochryan, N.A., 1985. *Biologich. Zhurnal Armenii*, **38**, 356.
24. Shalygina, A.M., Koziachenko, T.N. & Lozitskaya, T.N., 1989. In *Sovremennaja Technology syrodelija i Besotchodnaja Pererabotka Moloka*, Aiastan, Erevan, former USSR, p. 182.
25. Jankovskii, D.C., 1989. In *Sovremennaja Technology syrodelija i Besotchodnaja Pererabotka Moloka*, Aiastan, former USSR, p. 272.
26. Bryzgin, M.I., Shalygina, A.M., Popova, T.B. & Yaremenko, N.F., 1977. *Moloch. Prom.*, **N12**, 24.
27. Ul'janov, S.D. & Nedostoev, P.D., 1980. *Moloch. Prom.*, **N1**, 17.
28. Telegina, T.D., Vinogradova, P.P., Sakharov, S.D. & Nebert, V.K., 1979. *Moloch. Prom.*, **N8**, 34.
29. Gudkov, A.V., Ostroumova, T.L. & Alekseev, V.N., 1986. *Trudy vses. nauchno-issled.* Inst. masl. i syr. Prom., **46**, 12.
30. Puldmaa, Kh.Ya., Reshetnikov, A.F., Paema, A.Kh., Sarand, R.-Ya. & Toots, V.E., 1975. Former USSR Patent 477 715.
31. Ramanauskas, R., Treimane, V. & Golde, A., 1979. *Trudy Lit. Fil. vses. nauchno-issled.* Instl masl. i syr. Prom., **29**, 64.
32. Karlikanova, S.N. & Ramazanov, I.U., 1979. *Trudy vses. nauchno-issled.* Inst. masl. i syr. Prom., **30**, 71.
33. Ramazanov, I.U., Ramazanova, O.P., Vdovichenko, O.V. & Suyunichev, O.A., 1980. Former USSR Patent 784 852.
34. Pankova, M.S., Ramazanov, I.U., Suyunichev, O.A., Alferova, L.S. & D'jachenko, E.I., 1979. *Trudy vses. nauchno-issled.* Inst. masl. i syr. Prom., **27**, 50.
35. Martirosyan, A.A., Magakyan, A.T. & Krasheninin, P.F., 1975. *Moloch. Prom.*, **N1**, 21.
36. Krasheninin, P.F., Martirosyan, A.A., Nebert, V.K. & Magakyan, A.T., 1976. Former former USSR Patent 507 298.
37. Ramazanov, I.U., 1984. *Moloch. Prom.*, **N2**, 32.
38. Vaitkus, V.V., Lyubinskas, V.P. & Sauts, T.V., 1979. *Moloch. Prom.*, **N6**, 30.
39. Piranishvili, A.V., 1974. In *Sovershenstvovanie Technology Protsessov v Molochnoi Prom. Tom I. Chast' II.* Leningrad, Techn. Inst. Cholod. Prom., **64**.
40. Piranishvili, A.V., 1975. *Moloch. Prom.*, **N2**, 28.

11

Domiati and Feta Type Cheeses

M.H. Abd El-Salam

*Department of Food Technology & Dairying, National Research Centre, Dokki,
Cairo, Egypt*

&

E. Alichanidis and G.K. Zerfiridis

*Laboratory of Dairy Technology, Department of Agriculture, Atistotelian
University of Thessaloniki, 540 06 Thessaloniki, Greece*

1 INTRODUCTION

Manufacture of pickled cheeses in Egypt has been dated to the First Dynasty
(3200 BC). Earthenware cheese pots were found in the tomb of 'Hor Aha' at
Saqqara.[1]

Traditionally, the manufacture of pickled cheese varieties was limited to the
Mediterranean basin and the Balkans. Possibly, these cheese varieties share the
same origin, with various modifications to suit local conditions and needs. Pickled cheeses are of great importance in warm climates—under these conditions,
the shelf-life of milk is short and cheese deteriorates before it ripens. Storage in
pickle (usually salted whey) becomes one of the inevitable practices necessary for
cheese preservation. The pickling practice constitutes the main difference between this group and those cheese varieties produced in temperate zones. Production of pickled cheeses was limited for centuries to small-scale production
which made standardization of the technological properties and composition of
these cheese varieties difficult. Evolution of large-scale production of pickled
cheeses in their native countries greatly improved and defined their characteristics. Nowadays, pickled cheeses are gaining popularity, international recognition
and new markets all over the world. Their production has been extended to new
countries, e.g. Denmark, UK, USA, Australia, New Zealand and Ireland. Also,
standardized and advanced technologies have been adopted in their production,
including mechanization and ultrafiltration techniques.[2] Of the different categories of UF-cheese marketed in the world, UF-Feta cheese constitutes 56%.[3]

Many studies in the literature describe the manufacture, composition, quality
and ripening of pickled cheeses. Although pickled cheeses share the practice of

storage in pickle for extended periods, they are quite different in several aspects, including the type of milk used and the manufacturing and storage conditions. Most varieties within this group of cheeses are stored in sealed containers under almost anaerobic conditions, but some are stored in gas-permeable containers, e.g. barrels, which affects biochemical changes during ripening.

A few efforts[4-6] have been made to review work done on specific pickled cheese varieties but no attempt has been made to compile data on these cheeses in a systematic and comparative manner. This chapter deals with this topic.

2 CLASSIFICATION OF PICKLED CHEESES

There is no well-defined, clear classification of pickled cheeses. Confusion usually arises from the lack of such classification as can be observed from the published work on pickled cheeses. Sometimes, the cheese nomenclature does not fit the described procedure for cheese manufacture[7,8] and in many cases cheeses are described as white pickled cheeses, a general term which can apply to all the pickled cheeses. Pickled cheese varieties can be classified as follows (country of origin is given in parentheses).

2.1 Soft Cheese (Moisture Content 55–65%)

2.1.1 Acid Coagulation
Mish (Egypt). It is made from naturally fermented, partly skimmed milk remaining after gravity separation of sour cream. Salt is sprinkled on the coagulum and the curd is ladled onto cheese mats, cut into suitable pieces and pickled in earthenware containers for more than one year.[9]

2.1.2 Rennet Coagulation
Rennet-coagulated cheeses may be sub-classified based on the method of salting:

Salting of cheese curd (Feta type)
— Feta (Greece)
— Teleme(a) (Greece, Romania)
— Brinza (former USSR)
— Bli-Sir U-Kriskama (Yugoslavia)
— Bjalo (Belo Salamureno Sirene) (Bulgaria)
— Chanakh (former USSR)
— Beyaz peynir (Edirne peyniri, Salamura beyaz peynir, Teneka peyniri and Istambol peyiniri) (Turkey)
— Akaawi (Syria)

Salting of cheese milk (Domiati type)
— Domiati (Egypt)
— Danie (Egypt), a variant of Domiati cheese made from sheep's milk.[10]

2.2. Semi-Hard Cheeses (Moisture Content 45–55%)

— Halloumi (Cyprus)[11]
— Medafara, Magdola, Shinkalish (Syria, Sudan)[12,13]
— Arab (Iraq)[14]
— Baladi, Montanian (Lebanon).[14]

Data on cheese varieties other than those in class 2.1.2 are scarce; they are produced on a small scale and enjoy a limited market. Therefore, this chapter will deal mainly with cheese varieties falling in class 2.1.2, particularly Domiati and Feta cheeses.

3 ROLE OF SODIUM CHLORIDE IN PICKLED CHEESES

One of the characteristic features of pickled cheeses is their high salt content and storage for long periods in brine (salted water or whey). Therefore, NaCl has a definite role in determining the chemical, physical and biochemical changes in these cheeses. The chemical changes which occur in the colloidal system of milk and in cheese proteins on addition of NaCl can be summarized as follows.

3.1 Exchange of Colloidal Calcium for Sodium in Milk

Addition of NaCl to milk solubilizes part of the colloidal calcium.[15,16] The amount of calcium released increases with the amount of NaCl added up to 4 g/100 ml and there is no noticeable change thereafter.[16] In buffalo's and cows' milk, about 23–25% of the colloidal calcium can be solubilized by addition of NaCl but only 10% is solubilized in the case of goats' milk, indicating species differences in the exchangeability of the colloidal calcium.

3.2 Disaggregation and Dispersion of the Colloidal Phase of Milk

Addition of NaCl (to 1M) to milk or casein micelles of variable size in simulated milk ultrafiltrate[17–19] reduces their turbidity and the average micellar size, in addition to non-preferentially solubilizing some of the individual caseins. According to the model proposed by Schmidt,[20] the colloidal calcium phosphate linking micelle sub-units is unevenly distributed and the outer layer of the sub-units is not firmly attached to the core aggregates. Treatment with NaCl seems to remove the surface layer of the micelle sub-units, rendering these sub-micelles soluble and releasing their colloidal calcium.

Treatment of casein micelles with NaCl increases particle dispersion.[21] This leads to a decrease in the rate of aggregation of casein particles during rennet coagulation and a decrease in the stability of the structural elements of the coagulum,[22] reducing thixotropic characteristics and the intensity of syneresis. Therefore, in Domiati cheesemaking, curd is ladled into frames without cutting due to the fragility of the curd from salted milk. The rennet coagulation times of cows', buffalo's, goats' and sheep's milks increase with the amount of NaCl added up to 7·5–10% and decrease slightly with further increases in added salt.[23]

Besides its effect on the colloidal state of milk, NaCl affects the action of the coagulant; calf rennet is less affected by the addition of NaCl to milk than *Mucor miehei* protease.[24]

Addition of NaCl to milk increases its titratable acidity up to 5% NaCl, beyond which the acidity remains constant.[23] This has been attributed to a base exchange reaction of Na^+ for the NH_3^+ groups in casein micelles with the liberation of H^+.[25]

3.3 Interaction between Sodium Chloride with Milk and Cheese Proteins

The interaction between NaCl and the proteins of milk and cheese is evident from several studies using various conditions and techniques.[26–28] Absorption and desorption isotherms for NaCl, paracasein and their mixtures show that the components do not behave independently but more like paracasein, indicating interaction between NaCl and proteins.[28] The amount of NaCl which interacts with milk proteins increases with the concentration of added salt and decreases as the water activity (a_w) increases.[28] The binding of NaCl to paracasein reaches a maximum when the water binding capacity of the protein is maximal.[21] Salt seems to play a major role in regulating cheese consistency.[29] As a result of the interaction of NaCl with paracasein, the amount of strongly-bound moisture decreases.[29] Increasing the amount of weakly-bound moisture causes an increase in firmness but decreases the elasticity and plasticity of cheese.[29] Geurts *et al.*[30] hypothesized that the degree of NaCl interaction with paracasein depends on pH, NaCl and Ca^{2+}.

3.4 Solubilization of the Paracaseinate-Phosphate Complex

The effect of the NaCl or NaCl and lactic acid on a simulated soft cheese model[31] or on the paracaseinate-phosphate complex[32] shows the release of significant quantities of calcium phosphate and relatively small quantities of calcium paracaseinate when shaken with dilute NaCl solution. Aqueous solutions containing large quantities of NaCl attack both phosphate and paracaseinate with equal severity and a more extensive release of Ca and inorganic P occurs when lactic acid is added to extracting solutions. Moneib[31] reported that maximum peptidization occurred when using NaCl concentrations between 3 and 5% within pH range 5·3–5·6. At pH 5 or less, NaCl has only a very small effect on the amount of protein dissolved.

4 DOMIATI CHEESE

Domiati cheese is unique among cheese varieties in the addition of large quantities (8–15%) of NaCl to milk before renneting. This results in the need of more

Table I
Summary of Changes that Occur in Domiati Cheese During Pickling

Constituent	Trend of changes during storage	Responsible factor(s)
Moisture	decrease (about 2–5%)	exudation of cheese serum.
Fat in dry matter	increase (3–6%)	decrease in solids not fat.
Acidity	increase (1·0–1·5%)	lactic acid fermentation.
pH	decrease (2–2·5)	lactic acid fermentation.
Lactose	decrease (1·5–2%)	lactic acid fermentation.

rennet and longer coagulation time than other cheeses. Also, the formed curd is usually weak, is ladled without cutting and whey drainage is allowed for a long time (about 24 h). The cheese is either consumed fresh or pickled in soldered tins with salted whey from the same cheese. Ripening usually occurs at room temperature. Details of Domiati cheese manufacturers have been described.[33,34,35]

4.1 Changes in the General Composition During Pickling

Extensive data are found in the literature concerning the moisture, fat, salt, pH and acidity of Domiati cheese during storage in pickle.

Fresh Domiati cheese is characterized by relatively high pH (6·0–6·5), high moisture (60–65%) and high NaCl content (5–8%). The composition changes continuously during pickling, as summarized in Table I.

The role of developed acidity in determining the changes in gross composition of pickled Domiati cheese has been realized from the analysis of cheese with added preservative that inhibits acid development.[36] Acidity developed brings the pH of the cheese close to the isoelectric point of caseinate and partially solubilizes the colloidal calcium which causes shrinkage of the cheese matrix and exudation of cheese serum into the pickle.[37] The pH of pickled Domiati cheese reaches as low as 3·3.[38,39] This is due to two factors: firstly the high moisture content of fresh Domiati cheese which retains a high level of lactose; secondly, the whey used as a pickle is a rich source of lactose for bacterial fermentation within the cheese through diffusion. El-Abd et al.[40] used different mixtures of salted whey and brine of the same salt content as pickle for Domiati; reduction of the amount of whey in the pickle reduced acid development in the cheese.

Carbohydrates are found in Domiati even after six months of storage; lactose and galactose were identified but glucose was not detected in six month-old cheese.[41] It seems that in Domiati, lactose fermentation proceeds until the developed acidity inhibits the growth of the cheese microflora. However, the available lactose (lactose in cheese and pickle) is more than the cheese microflora can utilize, which explains the high residual lactose in the cheese (Table II) throughout the storage period. In the first step of fermentation, lactose is hydrolysed to glucose and galactose but the cheese microflora selectively utilize glucose with the accumulation of galactose in a similar manner to yoghurt[42] and Swiss cheese.

Table II
Changes in the Carbohydrate Content of Domiati Cheese During Storage (as Lactose)

Fresh	Storage period (days)				Reference
	15	*30*	*120*	*180*	
3·5	3·40	2·85	1·65	—	38
—	2·09	1·84	—	0·54	39

However, quantitative studies are needed to verify this. The identified microflora in Domiati[43,44] are mainly homofermentative, indicating that lactose fermentation yields primarily lactic acid.

Several interacting variables affect the general composition and acid development in Domiati. No attempt has been made to cover all these variables, but the most important ones are considered.

4.1.1 Type of Milk

Domiati from buffalo's milk contains more fat in DM (FDM), less moisture and lower developed acidity than cheese made from cows' milk (Table III). The use of reconstituted or recombined milks for pickled soft cheeses reduces the moisture content of the cheese.[45–47] However, raising the reconstitution ratio, i.e. total solids content of cheese milk, increases the moisture content in pickled soft cheese.[48–50]

4.1.2 Storage Temperature
Cold storage usually reduces the rate of biochemical changes in Domiati cheese as apparent from the lower acidity and higher pH than cheese stored at room temperature.[51,52] Also, the moisture content of the cheese increases during the early storage period at low temperatures due to increased swelling of the cheese proteins at the relatively high pH of the fresh cheese. Differences in the fat content of cheeses stored at different temperatures have been reported also.

4.1.3 Storage Period
The composition of Domiati changes continuously during storage.[38,52,53] The maximum rate of change occurs during the first month of storage, which coincides with the maximum growth of cheese microflora;[43,44] changes occur at slower rates thereafter.

Table III
General Composition of Domiati Cheese from Cows' and Buffalo's Milk[16]

	Cow milk cheese		Buffalo milk cheese	
	Fresh	*4 months*	*Fresh*	*4 months*
Moisture, %	59·0	55·0	55·0	52·0
FDM, %	43·9	44·4	53·3	52·1

Table IV
Gross Composition of Domiati Cheese as affected by the Level of NaCl Added to the Milk[57]

	8% NaCl		10% NaCl		12% NaCl		15% NaCl	
	Fresh	3 months	Fresh	3 months	Fresh	3 months	Fresh	3 months
Moisture, %	58·60	51·40	59·50	52·20	60·90	54·50	61·70	55·80
FDM, %	34·60	49·70	35·00	48·20	32·80	48·70	31·80	45·50
Acidity, %	0·27	2·24	0·24	2·02	0·21	1·42	0·11	1·00

4.1.4 Method of Storage

Storage in pouches or cans without pickle has been suggested for Domiati cheese; development of acidity in Domiati stored in pouches or cans is faster than that stored in pickle.[54-56]

4.1.5 Salt Content

Variable levels of salt are usually added to the milk for Domiati cheesemaking, depending on the season and quality of the milk. The higher the percentage of salt added to milk, the higher the moisture content of the cheese, either fresh or pickled[57] (Table IV). A high salt content weakens the cheese curd and it retains more moisture. Also, a high salt content retards acid development in the cheese during pickling.

4.1.6 Heat Treatment of Milk

Pasteurization of cheese milk has little effect on the gross composition of Domiati;[44,52,58] a slight increase in moisture content and a slight decrease in acid development are apparent in cheese made from pasteurized milk.

4.1.7 Ultrafiltration

The manufacture of Domiati by ultrafiltration has been described.[54-56,59-60] The moisture content of these cheeses is usually higher and the fat content lower than those of cheese made by the traditional technique[54,55,59] due to the high water holding capacity of whey proteins retained in UF cheeses (Table V).

Table V
Gross Composition of Domiati Cheese Made From Cows' or Buffalo's milk by Conventional and Ultrafiltration Techniques after 30 days of storage[56]

	Conventional		Ultrafiltration	
	Cow	Buffalo	Cow	Buffalo
Moisture, %	55·77	54·82	58·72	57·29
FDM, %	45·35	49·71	42·88	46·82
pH	4·55	4·70	5·15	4·91

4.2 Proteolysis

Domiati cheese undergoes continuous proteolysis during storage in pickle. Studies on factors that affect proteolysis, as indicated by determining nitrogen fractions, show that:

(1) Proteolysis is usually slowed down by low-temperature storage,[51,52] heat treatment of cheese milk,[44,52,53,58] H_2O_2-catalase treatment of milk[61,62] and the use of dried milks in cheese manufacture.[47]

(2) Cheese from cows' milk undergoes more rapid proteolysis than that from buffalo milk,[51] which reflects differences in the susceptibility to proteolysis of casein fractions from these two species.[63-65]

(3) Increasing the salt content in cheese slightly decreases proteolysis in Domiati.[47,52]

(4) Homogenization slightly increases proteolysis.[38]

(5) Addition of whey proteins,[66] phosphate and citrate,[67,68] and capsicum tincture (ethanol extract of paprika)[69-71] enhance proteolysis in Domiati.

(6) Manufacture of cheese by ultrafiltration techniques[54,72] or by direct acidification[73] increase proteolysis.

(7) The use of milk clotting enzymes other than calf rennet[74-80] modifies proteolysis.

(8) The use of different types and concentrations of cheese starters has only a limited influence on proteolysis[79] due to the high salt content of the cheese. However, the use of salt-tolerant strains[80] enhances proteolysis.

(9) Storage in containers (pouches/cans) without brine enhances proteolysis.[54,55]

Generally, the total nitrogen content of cheese gradually decreases while the levels of the soluble nitrogen fractions increase continuously during storage, indicating continuous proteolysis. Transfer of degradation products to the pickling solution by diffusion explains the decrease in total N during storage.

The use of electrophoretic and chromatographic techniques in cheese analysis gives a better insight than nitrogen solubility into changes in individual protein fractions of Domiati during storage.[81-83] The rennet contributes much to the proteolysis in Domiati. This is due to the high enzyme concentration used in milk coagulation (compared to most cheese varieties), the retention of a high proportion of the milk clotting enzymes in cheese curd owing to its high moisture content and to storage of the cheese in salted whey which contains residual rennet used in milk coagulation. In Domiati, α_s-casein is hydrolysed rapidly, while β-casein resists hydrolysis.[79-82] This pattern of change arises from the action of rennet on cheese proteins as affected by salt content.[84] The high salt content and the high storage temperature practised in the manufacture and ripening of Domiati cheese enhance the polymerization of β-casein and render it less susceptible to rennet action.[84] The use of milk clotting enzymes other than calf rennet alters the degradation pattern of the cheese proteins.[74]

A number of fast and slow-moving degradation products are apparent in the electrophoretograms of Domiati cheese. α_{s1} I peptide, which originates from the action of chymosin on α_{s1}-casein, was identified in the electrophoretograms of Domiati[85] while the slow-moving peptides are comparable to γ-caseins produced from β-casein by the action of indigenous milk proteinase (plasmin).[86] The β/α_{s1}–casein ratio in Domiati increases continuously during storage[79] and, after extended storage, the water-insoluble proteins are made up mainly of β-casein, which may explain the soft body and texture of ripened Domiati.[87] The use of different starters in Domiati cheese manufacture has only a slight effect on the electrophoretic pattern of the cheese proteins,[79] which indicates that bacterial proteinases are of limited significance to proteolysis as measured by these methods. Analysis of the soluble nitrogenous constituents of Domiati by gel permeation chromatography[81] shows that they are mainly low molecular weight compounds (amino acids and small peptides). The HPLC patterns of cheese proteins from fresh and 4 months-old Domiati (Fig. 1)[88] show that the proteins of fresh cheese consist of two major fractions with a small number of peptides at low concentrations. On the other hand, the number and concentration of peptides increase during storage with a noticeable decrease in the concentration of one of the original protein fractions.

Comparison of the free amino acid profile in Domiati[89] with those of cow and buffalo caseins reveals a marked reduction in the concentration of glutamic acid which is accompanied by the formation of γ-amino butyric acid through a deamination reaction. Also, arginine is almost absent with the appearance of

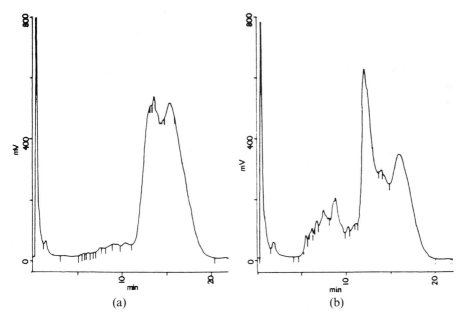

Fig. 1. HPLC profiles of Domiati cheese proteins: (a) fresh cheese, (b) 4 months-old pickled cheese.[88]

ornithine. The stored cheese has a high concentration of ammonia, which indicates the significance of deamination reactions occurring in Domiati and which contribute to flavour development in this type of cheese.

The concentration of biogenic amines in Domiati is very low.[90] Tyramine is the principal biogenic amine formed in Domiati together with low concen-

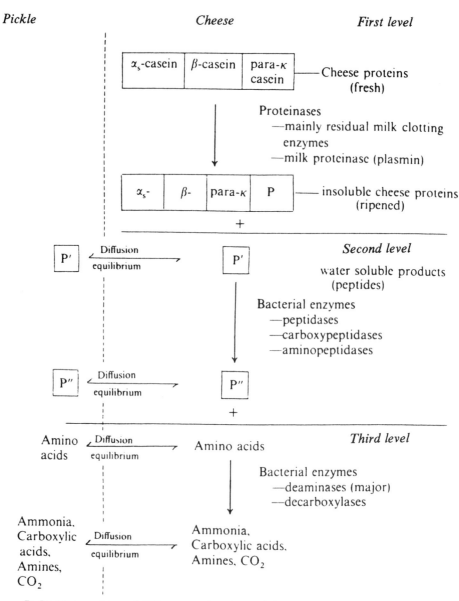

P, P', P" (peptides of different molecular weights)

Fig. 2. Proteolysis in pickled cheeses.

Table VI
Development of Volatile Fatty Acids in Domiati Cheese made from Buffalo's Milk During Storage (Expressed as Percentage of Acetic Acid)

Coagulant	Ripening period (days)				
	Fresh	15	30	60	90
Rennet[a]	0·073	0·150	0·158	0·173	0·179
M. pusillus protease[a]	0·073	0·121	0·142	0·184	0·219
Calf rennet[b]	0·034	0·089	0·107	0·109	0·111
Bovine pepsin[b]	0·044	0·092	0·097	0·107	0·109
M. miehei protease[b]	0·052	0·084	0·105	0·108	0·111
E. parasitica protease[b]	0·029	0·078	0·107	0·118	0·120

[a] Ref. 76.   [b] Ref. 74.

trations of histamine, tryptamine, phenylethylamine and putrescine. This suggests that deamination and not decarboxylation is the major pathway for amino acid catabolism in Domiati.

To conclude, proteins of Domiati cheese undergo three levels of proteolysis, as illustrated in Fig. 2. The soluble products from the different levels of proteolysis diffuse into the pickling solution which is in equilibrium with the cheese. Removal of the degradation products from the cheese allows the enzymatic reactions to continue throughout the storage period.

4.3 Lipolysis and Development of Free Fatty Acids

The literature on Domiati cheese contains voluminous data on the total volatile acidity as a measure of the formation of volatile fatty acids during pickling as affected by different processing conditions. Most of these data are expressed as millilitres of alkali used in their titration. Acetic acid constitutes the major part of the volatile fatty acids in this type of cheese.[91] The reported values for the total volatile fatty acids in Domiati were recalculated as percentages of acetic acid in cheese (Table VI).

Most of the changes occur during the first 15–30 days of storage which coincide with maximum bacterial growth.[44,71] The high concentration of total volatile fatty acids in pickled Domiati would contribute significantly to the total acidity of this type of cheese. Data (Table VII) on free fatty acids in Domiati[85,92] suggest lipolysis.

Table VII
Average Free Fatty Acid Contents for Domiati Cheese (% of Total Free Fatty Acids; Total Free Fatty Acids mg/kg)[85]

C_4	C_6	C_8	C_{10}	C_{12}	C_{14}	C_{16}	C_{18}	$C_{18:1}$	Total free fatty acids (mg/kg)
3·80	8·72	5·83	3·32	4·03	12·55	31·53	8·34	21·66	2308

Bacterial lipases or lipases in commercial preparations of milk clotting enzymes are responsible for fat hydrolysis in Domiati. The contribution of free fatty acids to flavour development in Domiati cheese has been realized from the analysis of cheeses of different fat contents.[91] Analysis of glycerides from pickled Domiati after different storage[93] periods clearly indicates lipolysis during storage. However, the origin of lipases responsible for fat hydrolysis in Domiati cheese is not clear.

Measurements of peroxide values indicate that fat oxidation occurs in Domiati cheese during storage.[92]

4.4 Volatile Flavour Compounds

Apart from the volatile fatty acids, the concentrations of total acidic and neutral carbonyls increase in Domiati during storage.[94] However, the role of these compounds in the cheese flavour is not evident.

4.5 Vitamin Content

Data on this subject are scarce. Almost all the vitamin A content of milk is retained in Domiati cheese and remains stable during storage.[95] On the other hand, variable percentages of thiamine, riboflavin and niacin[95] and biotin, B_{12} and folic acid[96] are retained in the fresh cheese. The levels of biotin, B_{12} and folic acid remain almost unchanged during storage[96] while the changes in riboflavin and niacin depend on the storage conditions (aerobic/anaerobic).[95]

4.6 Changes in the Composition of Pickle

It is unlikely that significant fermentation occurs in the pickle during cheese storage due to its high salt content. However, the composition of pickle changes continuously during storage as a result of the chemical and biochemical changes that occur in cheese and diffusion of soluble constituents between cheese and pickle. The changes in the composition of pickle are controlled by the following:

(1) Composition of fresh cheese and pickle.
(2) Rate and extent of biochemical changes in the cheese; these in turn are controlled primarily by the salt contents of the cheese and pickle, temperature of storage and heat treatment of the milk used for cheesemaking.
(3) Ratio of cheese to pickle. Practically, the cheese/pickle ratio is in order of 5–6:1. However, reported studies did not mention this point. Therefore, care must be taken in the interpretation of the reported results.

In Domiati cheese stored at room temperature, the volume of pickle increases significantly during the first month of storage (12·5%) but changes much less thereafter.[37] Losses of 17·4% of the cheese solids occur during the first month and about 70% of these appear in the pickle;[37] a further 5·9% of cheeses solids are lost during a further two months of storage. This is attributed to the partial exudation of cheese serum into the pickle due to acid development and shrink-

Table VIII

Changes in the Total N Content (%) of Domiati Cheese Pickle as Affected by Cheese Salt Content, Storage Temperature and Homogenization of the Cheese Milk

Storage period (days)	NaCl, %[a]			Storage[a]		Homogenization[b]	
	7	10	13	20–25°C	8–10°C	used	not used
Fresh	0·172	0·173	0·147	0·173	0·173	0·145	0·146
15	—	—	—	—	—	0·164	0·156
30	0·419	0·411	0·407	0·411	0·416	0·217	0·201
120	0·512	0·555	0·461	0·555	0·472	0·311	0·303

[a] Ref. 52.   [b] Ref. 38.

age of the cheese matrix. The reverse occurs during storage of Domiati at low temperature,[37,52] i.e. a decrease in the volume of pickle occurs during the early storage period through swelling of the cheese protein and an increase in its moisture content, but further storage is accompanied by an increase in pickle volume and changes in its composition.

The concentration of nitrogenous compounds in cheese pickle increases continuously during storage. The rate and extent of this increase depend on the storage temperature (Table VIII).

The Ca and P contents of the pickle also increase to a steady state after two months of storage and this coincides with changes in pH and acidity[38] (Table IX). Addition of disodium phosphate to milk for Domiati production[97] reduces the amount of Ca released during pickling. The Ca:P ratio in the pickle is in the order of 2·3–2·5:1, indicating that both calcium phosphate and paracaseinate-bound Ca are released by developed acidity. Increasing the concentration of NaCl in the pickle and the storage temperature increase the release of Ca and P into the pickle. It appears that the release of Ca and P into pickle is related to

Table IX

Changes in the Concentrations of Calcium and Inorganic Phosphorus in Domiati Cheese Pickle (g/litre)

Storage period (days)	Homogenized[a]		Unhomogenized[a]		Added disodium phosphate[b]			
					0·0%	0·2%	0·4%	0·8%
	Ca	P	Ca	P	Ca	Ca	Ca	Ca
Fresh	0·87	0·32	0·70	0·30	0·64	0·63	0·55	0·48
15	1·80	0·72	1·91	0·83	1·94	1·68	1·53	1·26
30	1·99	0·80	2·24	0·80	2·33	2·09	1·82	1·46
60	2·46	0·94	2·55	0·99	2·79	2·42	2·18	1·84
90	2·65	1·03	2·65	0·97	2·46	2·18	2·04	1·55
120	2·47	1·03	2·69	1·07	—	—	—	—

[a] Ref. 38.   [b] Ref. 97.

the solubility constants of calcium phosphate and calcium lactate in pickle as affected by ionic strength, temperature and pH. Therefore, Ca and P are removed from cheese into the pickle until the latter becomes saturated with respect to salts of these two elements. It is of interest to note that 25–30% of the Ca and P in the cheese is released into the brine.[38]

The whey used as a pickle for Domiati contains a small amount of fat that increases slightly as the amount of salt added to milk prior to cheesemaking and the severity of the heat treatment of the milk are increased. Both of these factors weaken the cheese matrix and increase fat losses in whey.[52] However, the fat content of pickle changes very little during storage.

Changes in the NaCl content of pickle depend on the ratio of its content in the pickle and cheese serum. An equilibrium in the distribution of NaCl between pickle and cheese serum occurs rapidly during the early days of storage. Equilibria also occur in the distributions of lactose and lactate as measured by developed acidity. The pickle of Domiati usually contains significant amounts of lactose after four months of storage.[38]

4.7 Texture and Structure

The texture profile and microstructure of Domiati reflect the physical changes in this type of cheese as affected by different factors.

The texture characteristics of fresh UF and traditional Domiati cheese are significantly different.[55] UF Domiati is harder and more adhesive than the traditional cheese, whereas the latter is characterized by more pronounced chewiness and gumminess. Both types of cheese increase in hardness, adhesiveness and gumminess during early storage, followed by a decrease after three months of storage in pickle.[55] Traditional Domiati is more elastic than the UF cheese throughout ripening. It seems that the increase in textural parameters during early ripening is related to the decrease in moisture and pH, leading to a firmer texture. During the latter stages, changes in texture are related more to changes in the protein matrix arising from proteolysis and loss of Ca.

The textural parameters of Domiati are also related to the method of storage, i.e. in pouches or in brine. Cheese stored in pouches is significantly harder, more cohesive and gummy than cheese stored in brine.[55] Also, the hardness of Domiati made from milk supplemented with whey protein concentrate decreases as the level of WPC is increased. The texture of fresh UF Domiati can be controlled by changing the homogenization pressure, heat treatment and pH of the pre-cheese concentrate.[56]

Electron microscopy of ultra-thin sections of Domiati[98,99] indicate that the internal structure of fresh cheese is composed of a framework of spherical casein aggregates held by bridges and enclosing fat. On storage in pickle, the casein aggregates dissociate into smaller spherical particles, forming loose structures. Additional proof that changes occur in the microstructure of Domiati during storage has been obtained by scanning electron microscopy.[100] It has been pointed out that the high salt content of this type of cheese has little effect on its morphological characteristics.[100] It is evident from these studies that fat globules are unlikely to change during storage. Most of the changes occur in the protein matrix. In fresh Domiati with a relatively high pH (> 5·8), the casein molecules have a net negative charge and while hydrophobic interactions persist, the ionic interactions between the molecules change from attractive to repulsive. Thus, the tight protein aggregates absorb water, partly to solvate the unneutralized ionic charges. Besides, the partial exchange of Na^+ for Ca^{2+} in fresh Domiati weakens the tight interactions in casein aggregates. With advanced storage, the pH of Domiati (< 4·0) is on the acid side of the isoelectric point of the caseins, thus retaining repulsive ionic forces in the cheese matrix, a factor responsible for the loose structure of the cheese. Another factor that may be responsible for the loose structure of Domiati is the fast disintegration of the casein aggregates as a result of the continuous proteolysis of α_s-casein, which can interact strongly with two or possibly other casein molecules[87] (either α_s- or β-casein) and can thus be linked into the protein network. Consequently, if α_s-casein molecules are cleaved so that they lose their ability to act as links in the protein network, then the network disintegrates. A third factor is the continuous loss of Ca from the cheese matrix into pickle;[38] this may be responsible for differences in the texture of cheese stored in pouches or in pickle.[55]

4.8 Microbiology

4.8.1 Microorganisms Present

Lactic acid bacteria are predominant in Domiati cheese—lactococci during early stages of pickling and later lactobacilli.[44,101,102] *Streptococcus faecalis, Lactococcus lactis* susp. *lactis* and *Lc. lactis* subsp. *cremoris*[103] are the most common lactococcal species isolated from Domiati. The most commonly found lactobacilli are *Lb. casei, Lb. plantarum*[101,102] *Lb. brevis, Lb. fermenti*[102] and *Lb. lactis.*[103] *Leuconostoc mesenteroides* subsp. *cremoris* is also found in Domiati.[103] Several non-lactic acid bacteria have been isolated also, especially *Micrococcus luteus.*[101] Yeasts that have been isolated belong to the following genera: *Trichosporon, Sachharomyces, Pichia, Debaryomyces, Hansula, Torulopsis, Endomycopsis* and *Cryptococcus.*[104,105] The following moulds have been isolated also; *Aspergillus, Penicillium, Cladosporium* and *Geotricum* spp.[106]

4.8.2 Effect of Manufacturing and Storage Conditions

Raw milk cheese generally has higher bacterial counts than the pasteurized milk cheese during the first month of ripening,[44] but both cheeses have similar counts thereafter. The total microbial count increases rapidly, reaching a maximum after a week of storage and then declines.[44] Lactococci behave similarly but disappear after 2–3 months of pickling.[44,102] Lactobacilli reach a maximum after 2–4 weeks and then decrease gradually[44] but the count of the non-lactic acid bacteria decreases steadily during pickling.

The high salt content added to the cheese milk reduces the totals of microbial count and different microbial groups.[101] Micrococci and lactobacilli share dominance in Domiati with high salt content.[102]

4.8.3 Starter

Traditionally, starters are not used in the manufacture of Domiati cheese. Several non-conventional starters have been recommended for Domiati manufacture as they tolerate the high salt content and improve the quality of the cheese. These include *Streptococcus*[107,108] *faecalis, Pediococcus* sp, *L. mesenteroides* and *Lc. casei.*[109]

4.8.4 Survival of Harmful Organisms

The presence of coliforms in Domiati is related to the level of salt added to the milk for cheesemaking.[110,107] Not less than 9·5% NaCl should be added to the milk to suppress the growth of coliforms in cheese made from raw milk.[110,107] Clostridia have been found in Domiati made from pasteurized milk without the addition of starters; the species isolated were predominantly *Cl. tyrobutyricum* and *Cl. perfringens.*[111] *Bacillus cereus* has been isolated from Domiati cheese samples also.[112,113] *Staphylococcus aureus* can tolerate 2·5–15% NaCl in Domiati but staphylococcal enterotoxin was not detected in fresh or one week old cheese.[114] *Salmonella typhi* can survive for up to 16 days in Domiati made from milk containing 10% NaCl.[115]

4.8.5 Defects

Early blowing may be the principal defect in Domiati cheese, particularly that made from raw milk. It is characterized by the formation of gas holes in the cheese, a spongy texture and blowing of the cans. This defect arises from two factors; gas-forming microorganisms[107,116] and electrolytic corrosion[117] of tins by NaCl and developed acidity.

5 FETA CHEESE

5.1 Introduction

A soft white cheese made from sheep's milk appeared in the Mediterranean basin and the Balkans in ancient times. A specific variety of this cheese, named Feta, was developed in Greece.[118,119] Traditionally, the cheese was packed in wooden kegs and the shape of the cheese blocks in the kegs looked like watermelon slices. The name Feta, which means 'slice' in Greek, has probably come from the original shape of the cheese or, according to another suggestion, from the property of the cheese which enables it to be sliced without falling apart.

Feta is the principal cheese consumed in Greece where someone saying 'cheese' always has Feta in mind, just like an English person saying 'cheese' would mean Cheddar. Cheese consumption in Greece is the highest in the world,[120] reaching 22 kg per capita per year.[121] Over 60% of the consumption is Feta cheese, excluding other brined cheese varieties. During the last two or three decades white brined cheeses for export under the name Feta have been manufactured from cows' milk in many European countries. This has created problems as it has been impossible to duplicate the flavour of Feta produced from sheep's milk when using cows' milk, particularly when milk is ultrafiltrated or when the cheese is produced by techniques other than those used for Feta. White cheese from cows' milk meets with consumer complaints because the body is dry and crumbly, the flavour is too acid, unnatural and over-fermented and the colour becomes yellowish with ageing. Techniques of modern technology do not alleviate these problems substantially.

5.2 Key Points in the Traditional Manufacture of Feta Cheese[118,122]

The most suitable milk for Feta is from sheep but goats' milk may also be used or a mixture of both. The milk is standardized to a casein to fat ratio of 0·7–0·8, pasteurized and coagulated by rennet at 32°C in 45–50 min. Starters used are combinations of lactic acid bacteria, e.g. *Lc. lactis* and *Lb. bulgaricus* in a ratio of 1:3. Traditionally, the milk was not pasteurized and the rennet used was home-made from the abomasa of unweaned lambs and kids. This technique is still used to some extent. The coagulum is cut using 2–3 cm wire knife and is ladled in thin layers into perforated moulds. The moulds are cylindrical when the cheese is to be packed in barrels and rectangular when the cheese is to be packed in tin cans of dimensions 23 × 23 × 35 cm. The curd is drained at 14 to 16°C, without pressing, until it is firm enough to remove the moulds. The block of curd is cut into two or four sub-blocks of 23 × 11 × 8 cm and 11 × 11 × 8 cm, respectively, which are placed close together on a salting table and coarse salt sprinkled on the surface to allow slow penetration of the salt into the cheese. After 12 h, the blocks are turned and dry salted again on the surface. This is repeated until the cheese contains 3–4% salt. Following salting, the cheese blocks remain on the table for a few more days or until a slime starts to develop on the surface. Dry salting and slime formation are essential for the development of the characteristic Feta flavour during ripening.

The cheeses are then packed either in wooden barrels or tin cans. Although flavour develops better in wooden barrels, the filled barrels weigh about 50 kg and are difficult to handle; the filled tin cans weigh 17–18 kg, are easier to handle and transport and hence the overall cost of storage is reduced. Following packaging, 6–8% salt solution (brine) is added to cover the cheese in the containers. The cheeses are then kept at 14–16°C for 10–15 days until a pH of 4·6 and a moisture content less than 56% are reached. At this point, the cheeses are transferred to new containers, brine is added and the containers are sealed and stored at 4–5°C. Care should be taken so that the cheeses are always kept completely submerged in brine to avoid mould growth on the surface. The quantity of brine added is about 1·5–2 kg per can. The cheese may be marketed two months after manufacture, but may be stored for well over a year at 2°C.

5.3 Organoleptic Characteristics and Grades

The flavour of typical Feta is mildly rancid, slightly acid and salty and is frequently described as flavourful and appetizing. The body and texture are firm, smooth and creamy, which makes the cheese sliceable. No gas holes should be present, but irregular small mechanical openings are desirable. The colour is snow white both inside and on the surface of the cheese.

The grades of Feta cheese in Greece are 'excellent' and 'first quality'. The former contains maximum moisture and a minimum fat in dry matter of 54% and 46%, respectively, and the latter 56% and 43%, respectively.[123]

5.4 Gross Composition

The general composition of Feta according to a market survey of the Greek Ministry of Commerce,[124] is: moisture: $54.34 \pm 2.86\%$; fat-in-dry matter: $49.76 \pm 3.59\%$; total protein: $17.59 \pm 1.15\%$; lactose: $0.53 \pm 0.28\%$; ash: $4.12 \pm 0.54\%$; salt-in-moisture [salt content % $\times$ 100/(salt % + moisture %)] $4.39 \pm 0.78\%$ and acidity, as lactic acid: $1.86 \pm 0.66\%$. It is worth noting the large standard deviations and that 20% of the cheeses included in the survey exceeded the legal standard of 56% moisture. Also, the salt-in-moisture is generally below the limit of 5.0% which is considered necessary for proper ripening fermentations and organoleptic characteristics for this kind of cheese.[125] Other data from experimentally produced Feta by several workers[126–129] fall within these limits, but their standard deviations are not large.

The moisture content of 1 day old Feta seems to be about 60% which decreases to about 55% in 10 days[126,128,130] or even less and sooner[129] due to salting and acid development. During storage and ripening, further moisture losses normally occur, but some moisture increase is often observed for Feta[129] and other white soft cheeses kept in brine.[131,132] The latter can be attributed to peptide bond cleavage and the formation of new ionic groups which take up water[133] from the surrounding brine.

During storage some salt diffuses from the surrounding brine into the cheese as the salt concentration in the brine is 1–2% higher than brine concentration within the cheese. If the concentration of the salt solution added to the can is lower than salt-in-moisture within the cheese, absorption of the brine by the cheese may occur and will lead to defects.[122]

5.5 Acid Development

Feta cheesemaking is similar to that of Teleme cheese as regards the starters used. Acid development is slow when yoghurt is used as starter,[134] particularly at levels below 0.75%.[135] Several combinations of lactic acid cultures have been suggested for starters at rates from 0.5–2.0%.[136,137] Satisfactory results have been obtained with lactococci added at a level of 0.5% 1 hour before renneting or with one part of *Lc. lactis* and three parts of *Lb. bulgaricus* added at a level of 0.5% 20 min before renneting. For proper drainage and ripening it is necessary that acid develops to pH 5.0–5.2 within 6–8 hours and to pH 4.8–5.0 by the next morning. Further acid development to pH 4.6 is necessary during the next few weeks before the cheese is sealed in containers and placed in the cold store. Slight changes in pH may occur during storage, but pH values over 5.0 might be due to defects.[122]

5.6 Proteolysis

Protein degradation in Feta can be attributed to the coagulant used in cheese manufacture, the proteinases produced by the microflora within the cheese, the slime formed on the surface of the cheese before it is placed in brine and the indigenous milk proteinases, especially plasmin.

Traditionally, Feta was manufactured using home-made lamb and kid rennets which contained several enzymes besides chymosin. Modern dairies use calf rennet, but some use it mixed with lamb and kid rennet preparations to impart a flavour closer to that of the traditional cheese. The use of commercial rennets of different origin, with or without the simultaneous use of other enzymes, was not successful in substituting the traditional coagulants with respect to flavour. The microbial proteinases in the cheese depend on the microflora within the cheese and on its surface. The microflora in the cheese mass is derived from the starter, the milk microflora that survive pasteurization and the contaminants of milk and curd during cheesemaking. A characteristic feature of Feta is slime formation on its surface which consists of a variety of microorganisms. Although the surface of the cheese is washed before storage in brine, the enzymes penetrate into the cheese and contribute to proteolysis, as well as to other changes, during ripening. Slime microflora and contaminants are random, coming from the environment of the dairy and therefore hydrolysis may vary from one dairy to another, giving rise to variants of Feta cheese.

The contribution of plasmin to proteolysis in Feta cheese is not expected to be extensive, extrapolating from other types of cheese.

The total nitrogen (TN) of the cheese decreases during the first stages of ripening,[128] probably due to salting and drainage of the cheese until sealed in containers. Later losses may be attributed to protein breakdown and diffusion of water-soluble nitrogen into the brine. The rate and extent of hydrolysis can be estimated by the level of nitrogenous substances formed during ripening. Table X shows that in

Table X

Changes of Non-Casein Nitrogen (NCN), Trichloroacetic Acid-Soluble Nitrogen (TCA-N) and Phosphotungstic Acid-Soluble Nitrogen (PTA-N), Expressed as Percentage of Total Nitrogen (TN) During Ripening of Feta Cheese

Age, days	NCN			TCA		PTA-N
	(A)[126]	(B)[128]	(C)[129]	(B)[128]	(C)[129]	(C)[129]
1	6·20	—	—	—	—	—
5	—	—	14·14	—	4·80	0·93
10	16·38	17·47	—	9·64	—	—
20	—	—	18·77	—	7·18	0·97
30	—	16·89	—	—	—	—
35	20·27	—	—	9·48	—	—
40	—	19·59	—	—	8·24	1·44
60	—	16·60	—	8·73	—	—
65	23·09	—	—	—	—	—
80	—	—	21·19	—	12·06	2·07
100	24·32	—	—	—	—	—
120	—	21·50	26·94	14·52	15·33	5·00
125	25·58	—	—	—	—	—
240	—	—	32·37	—	17·44	6·46

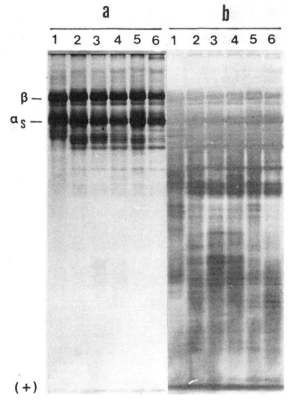

Fig. 3. Polyacrylamide gel electrophoretogram of Feta cheese. Tracks 1, 2, 3, 4, 5, 6; cheese aged 5, 20, 40, 80, 120, 240 days, respectively. (a) Gels stained with Coomassie Brilliant Blue G-250. (b) Gels stained with silver stain.

4 months-old cheese, about 25% of the TN is in the form of non-casein nitrogen (NCN) (pH 4·6–soluble N), 14–15% is soluble in 12% trichloroacetic acid (TCA-N) (i.e. peptide and amino acid nitrogen) and 5% is soluble in 5% phosphotungstic acid (PTA-N) (i.e. amino acid nitrogen). In market samples, water-soluble nitrogen was 21·95% of TN, called the coefficient of ripening.[124] The free amino acid (mg/100 g dry matter) content of Feta cheese increased during ripening to 71·2, 695·3 and 1196·0 at 1, 60 and 120 days, respectively.[128] The amino acids showing the greatest increase during storage were leucine, lysine, phenylalanine, valine and alanine. The electrophoretogram of Fig. 3 shows numerous protein degradation products, derived mainly from α_s-casein. As curing continues after 20 days, β-casein degradation products are also significant.[129]

5.7 Lipolysis

The flavour of Feta, being mildly rancid and slightly acid, is critically affected by high levels of short chain fatty acids in the cheese. Although all reports agree

Table XI
Free Fatty Acids (FFA) in Feta Cheese (mg/100g Dry Matter)

FFA	References		
	128	129	138
C_2	104	84	102
C_3	14	41	18
C_4	76	37	172
C_6	18	9	9
C_8	8	5	28
C_{10}	15	18	128
C_{12}	22	43	67
C_{14}	33	28	64
C_{16}	91	98	95
C_{18}	164	37	39
$C_{18:1}$	90	37	157
$C_{18:2}$	—	—	—
$C_{18:3}$	121	1	ND[a]
Total FFA	756	438	879
C_2 as % of total FFA	13·8	19·2	11·6
C_2 as % of total VFA (C_2 to C_8)	47·3	47·7	31·0

[a]Not detected.

that the dominant acid is acetic, the levels and the percentage of acetic acid in the total FFA or volatile fatty acids (VFA) differ considerably. According to the data in Table XI, acetic acid in Feta represents from 11·6 to 19·1% of the total FFA content, which varies widely from 439 to 879 mg/100 g dry matter, and from ca. 30–50% of the VFA (C_2 to C_8). Values of acetic acid as high as 80–92% of the VFA have been reported for Feta of non-Greek origin.[5,138,139]

Butyric acid is also present at high levels (37–172 mg/100 g dry matter) in Feta and along with the other VFA, may contribute to the characteristic flavour of Feta. Higher concentrations of volatile acids develop in ewes' milk cheese than cows' milk cheese,[140] which is also an indication of the better suitability of ewes' milk for Feta cheese. Although the addition of certain lipases during cheesemaking favour the development of Feta flavour, it is certain that the quality and intensity of flavour also depend on other factors, such as acidity and protein degradation.[138]

It is worth noting that although lipolysis is associated with the presence of lipases, the acetic acid is produced from several fermentation pathways. In Feta cheese, acetic acid is produced during the early stages of ripening and lactose is present even in mature cheese, so it is assumed that acetic acid is produced mainly through the fermentation of lactate. Also, the high percentage of free butyric acid (13–19% of the total FFA), which is more than four times the reported average for the esterified acid, might indicate selective lipolytic acitivity[138] and probably other fermentation pathways. Although short chain fatty acids have a major effect on flavour development, marked flavour differences between Feta samples are not always explained by variations in the concentrations of acetic, propionic and butyric acids and rancid flavour seems to be related to high concentrations of FFA in the cheese.[138] From the available data on lipolysis, it may be concluded that acetic acid is a major flavour and aroma component of Feta cheese, but the typical Feta flavour is the result of numerous other components, in quantities and proportions yet to be determined.

5.8 Other Flavour Components

Feta flavour has not been studied intensively and few specific compounds have been suggested to be involved in a balanced flavour. Vafopoulou et al.[129] reported that diacetyl is not present in Feta while the concentration of acetaldehyde increases to a maximum at 4 months and decreases to minimum levels at 8 months (1·53 mg and 0·34 mg/100 g dry matter, respectively). Therefore, acetaldehyde is thought not to be crucial for balanced flavour. Headspace analysis of Feta by Horwood et al.[139] revealed many volatile components, including ethanol, propan-1-ol, butan-2-ol and butan-2-one, in relatively large quantities. The authors consider that these compounds very probably contribute to Feta flavour. For proper flavour development it is necessary that fat hydrolysis goes along with proteolysis to ensure a balanced production of flavour compounds.[129] This means that accelerated ripening using proteolytic enzymes will require the simultaneous use of appropriate lipases and/or appropriate starter cultures.

5.9 Microorganisms[129,141]

The numbers of aerobic bacteria, lactic acid bacteria and proteolytic bacteria reach maxima of (mean $\log_{10}$ counts): 9·0, 7·7 and 8·0 at 5, 20 and 45 days, respectively. They decrease 10-fold at 3 or 4 months and remain almost constant thereafter. The lipolytic count ranges from log 5·5 to log 6·5, showing a slight increase during ripening and storage. Psychrotrophs increase to maximum numbers (log 4·6–5·4) in 5–20 days and exhibit fluctuations during storage. The maximum number of microorganisms correlates with sharp increases of compounds formed by proteolysis and lipolysis during the early stages of ripening. Breakdown products continue to be formed even when the numbers of microorganisms decrease. Apparently, microbial enzymes accumulate during the early stages of ripening, when microbial counts are high, and cause proteolysis and lipolysis later, after cell lysis. On the other hand, some microbial growth continues and slowly increases the breakdown products during storage. Yeasts are present in Feta in low numbers (< log 3·0) throughout ripening and coliforms may be found in negligible numbers in fresh cheese.

6 OTHER PICKLED SOFT CHEESE VARIETIES

6.1 General Remarks

The common feature of this group of cheeses is that they are salted, ripened and stored in brine until consumed.

They are manufactured from ewes', cows', goats' or buffalo's milk or from a mixture of two types of milk. Usually, a milk with high fat and total solids content (ewes' or buffalo's) is blended with cows' or goats' milk. In this way, not only is the high fat content of rich milks lowered but also syneresis is helped.

Because most brined cheeses are traditional products, they originally were made from raw milk. But for a long time, they have been made from pasteurized milk, although pasteurization of cheese milk for some of them is not obligatory. The cheese milk may be pasteurized at various temperature and time combinations, but the most common are: HTST, 68°C for 10 min or 63–65°C for 30 min.

For acid and flavour development, mesophilic starters are used usually, although for acid development some thermophilic cultures are also commonly used. Starters usually include a *Lactococcus* together with a *Lactobacillus* strain. The proportion of the starter culture added to cheese milk is within the range of 0·1 to 2%.

For milk coagulation, calf rennet in powdered or liquid form is added, but small artisanal cheesemakers use a locally-made rennet paste (usually from kid abomasa). The temperature for milk coagulation varies but mostly is in the range 28–32°C. Curd setting time varies widely, from less than 30 min to more than 1 or even 2 h.

The strength of the salting brine varies from 12 to 24% NaCl. For pickling, a brine of lower strength (8–12% NaCl) is usually used, but there are exceptions. Pickling temperature varies but it ranges from 14 to 18°C. Some types of these cheeses are pickled in sour salted whey (deproteinated or not).

Brined cheeses have no rind. Their shape is mostly rectangular or rectangular with rounded corners, although a wedge shape is not uncommon when cheeses are packaged in barrels (wooden or plastic). Cheese dimensions varied widely: $11 \times 11 \times 8$–10 cm, $7 \times 7 \times 7$ cm or even 21–25×13–16×17–19 cm.

The colour of the cheeses is clean pure white when they are made from ewes', goats' or buffalo's milk. When cows' milk is used, either alone or in mixtures with the above milks, the cheeses have an off-white or yellowish colour. Some of those cheeses were originally made from ewes' milk and consumers became accustomed to the white colour. Nowadays, the same cheeses are made from cows' milk and, in order to correct the colour, manufacturers subject the cheese milk to different treatments, the most common being the addition of a chlorophyll preparation. However, it should be noted that the resultant colour is not pure white, as the original colour of these cheeses.

Manufacturing conditions for the principal types of this group of cheeses are summarized in Table XII.

Table XII
Technological Differences Between Some Pickled Cheese

	Teleme (a)[134,142]	Bjalo salamureno[143]	Beyaz peynir[144,145]	Beli sir u kriskama[146]	Brinza (Israel)[147,148]	Chanakh[149]
1. Type of milk	ewes', cows', goats' buffalo's, mixed	ewes', cows'	ewes', cows' buffalo's, ewes' cows' or goats'	ewes', goats', mixed	ewes', ewes' + 25% cows' or goats'	ewes', cows'
2. Heat treatment	Optional or HTST, 63–65°C for 30 min	68°C for 10 min or HTST	HTST or 65°C for 30 min	HTST	Optional or HTST	Optional or HTST
3. Starter	Yoghurt, Lc. lactis + Lb. casei, Lc. lactis + Lb. bulgaricus (0·1–1%)	Lc. lactis + Lb. casei (0·1–0·4%)	Yoghurt, Lc. lactis + Lc. cremoris (0·15–1%)	Lc. lactis + Lb. casei Lc. lactis + Lc. cremoris (0·1–0·3%)	Lc. lactis, butter culture (1–3%)	butter culture (0·4–1%)
4. Ripening time of milk, min	30–120	30	—	—	—	—
5. Rennet	Calf	Calf	Calf	Calf	Calf	Calf
6. Renneting temperature,°C and time, min	30–32 60	28–32 60–90	28–32 70–135	28–32 60–90	28–32 40–90	30–33 25–30
7. Cutting	cubes, 2–3 cm	cubes, 2–3 cm	cubes, 3×3 cm	Crushing the curd or cubes, 1·5–2 cm,	cubes, 1–1·5 cm	cubes, 1–1·5 cm
8. Lading	into hoops	into frames + cheese cloth	into frames + cheese cloth	into frames	into frames + cheese cloth	into frames + cheese cloth
9. Pressing	1–1·5 kg/kg curd for 1–2 h	1–1·5 kg/kg curd for 4–6 h	10 kg/kg curd for 30 min	2 kg/kg curd for 1–2 h then 4 kg/kg curd for 1 h	light for 2–3 h	1–1·5 kg/kg curd
10. Salting						
brine concentration, %	14–18	22–24	14–16	20–24	12–24	16–22
time, h	12–16	12–24	6–8	8–20	48–72	12–15 days
temperature,°C	14–18	14–16	12–26	12–15	—	12–15
11. Pickling						
brine concentration, %	8–12	10–12	14–16	12–15	10–14	16
temperature,°C	14–18	14–15	—	—	—	—
time, days	7–15	12–15	—	—	—	—
12. Storage temperature, °C	4–8	3–4	4–10	<10	4–10	8–10

Table XIII
Average Composition of some Pickled Cheeses

Type of cheese	Moisture (%)	Fat in dry matter (%)	Total protein %	Salt in moisture (%)	pH	Acidity % or °T
Beli sir u kriskama[150]	54·44	47·78	19·19	4·08	—	252(°T)
Beyaz peynir[151]	58·18	—	—	6·12	4·68	—
Beyaz peynir[144]	62·52	43·78	13·78	7·03	—	
Bjalo salamureno[143]	48–52	48–50	15·5–20	6–7	—	180–300 (°T)
Bjalo salamureno[152]	51·5–55·1	54·3–57·2	—	7·4–7·7	4·4	300(°T)
Brinza[153]	56·1	—	—	7	4·36	1·32
Chanakh[149]	49–50	40–50	—	8–12	—	150–160(°T)
Teleme[a]	54·41	49·7	16·42	5·99	4·77	1·20

[a] Average from analysis of 120 samples from Greek markets during 1986–1989.

It is difficult to give an accurate composition of the cheeses under discussion. Most of them are made from more than one kind of milk and in many cases the casein to fat ratio is not standardized, a fact which, among others, affects the fat content of the cheese. The situation is more complicated in cases of cheeses which have the same name but are manufactured in different countries with different national standards and different consumer habits. A good example is Teleme which is consumed widely in Greece and Romania. Moreover, from literature data, it is clear that a high percentage of market samples analysed in various countries do not meet their national standards.

Table XIII lists the composition of some types of brined cheeses, but it is far from complete and may even contain some inaccuracies as it is difficult to obtain reliable data for all of them.

6.2 Changes in Cheeses During Manufacture, Ripening and Storage

Numerous changes occur during maturation of the cheese. Undoubtedly some changes start during curd formation, but most of them become evident during ripening and storage. Composition, structure and organoleptic properties are greatly altered due to biochemical, chemical, microbiological and physical changes, which involve mainly the conversion of lactose, protein and fat. Unlike other cheeses, pickled cheeses, as a rule, are salted, ripened and stored in brine until consumed and therefore they exchange soluble substances with the surrounding brine. This exchange is most pronounced during ripening and slows down during storage of the cheese at lower temperature, but it may never terminate.

6.2.1 Fermentation of Lactose

Formation of lactic acid by the starter bacteria is probably more vital for the preservation of pickled cheeses than for other groups of cheeses. Production of lactic acid and subsequent reduction of the pH to about 5·2 in about 24 h after curd manufacture is a necessity for the pickled cheeses. Because of the high level of moisture (whey) retained in the curd due to light pressing and to early brining (usually 2–3 h after cutting and hooping), the curd and the young cheeses contain enough lactose for fermentation by the starter microorganisms and thus the pH is decreased to 4·8 and later to 4·2–4·5. At the same time, however, coliforms and other undesirable lactose-fermenting microorganisms may multiply to high numbers if the pH of young cheeses is not in the range mentioned above, causing some of the major defects of pickled cheeses (see defects).

Starter bacteria used for pickled cheese manufacture may differ greatly as to growth rate, the maximum number to which they multiply in curd and cheese and the rate at which they slow down or lose viability in the high acid and high salt environment of the cheese. At high levels of salt-in-moisture (> 5%), the metabolism of most starter bacteria slows down or completely stops and residual lactose is fermented by the abundant secondary flora. The fermentation of residual lactose by secondary flora, which includes lactic acid bacteria (lactobacilli, pediococci, etc), is in part beneficial. These microorganisms metabolize lactose heterofermentatively and produce, among others, acetic acid which is the major volatile acid of this group of cheeses.

To help keep the pH value lower than 5, some pickled cheese varieties are packed with acidified (pH < 5·0) brine or acidified whey.[154]

6.2.2 *Proteolysis and Amino Acid Catabolism*

Data on proteolysis of the various types of pickled cheeses are difficult to compare, not only because of differences in the milks used and in technologies of cheesemaking followed, but also due to the various methods used to determine the nitrogenous compounds and estimate proteolysis. Generally, the reported data on nitrogen substances of pickled cheeses during ripening and storage may be summarized as follows:

1. Pickled cheeses undergo continuous proteolysis during ripening and storage in pickle. The total nitrogen content of the cheeses gradually decreases owing to diffusion of degradation products into the pickle. However, despite diffusion, soluble nitrogenous substances increase in the cheese.
2. The contribution of the milk coagulants to the proteolysis is evident from the modification of protein (qualitative as well as quantitative). The contribution of the cheese microflora (starter and secondary flora) is evident from studies on different factors affecting the formation of water-soluble N, e.g. ripening temperature, pH and NaCl concentration.

When discussing proteolysis and amino acid catabolism in pickled cheeses, one must keep in mind the pH of these cheeses (which is close to the isoelectric point of the caseins), the high salt content and the fact that they are salted, ripened and stored in brine with which they exchange soluble substances.

The milk clotting enzymes contribute much to the level of proteolysis, due to their retention of high moisture curd and to the ripening and storage of some of them in salted whey which contains residual clotting enzymes.

From the available literature, it is clear that α_S-casein of ewes' and cows' milk is hydrolysed extensively and is the main source of peptides in these cheeses. β-Casein resists hydrolysis by chymosin in pickled cheeses. which may be due to the high salt content which reduces very significantly the hydrolysis of β-casein and favours that of α_s-casein.[84] Alichanidis et al.[132] found that in 4 months-old Teleme made from ewes' milk, ~74% of α_s-casein was hydrolysed while β-casein hydrolysis was limited to ~13%. Mansour & Alais[155] found that para-κ-casein in Syrian white pickled cheese made from cows' milk, resisted hydrolysis throughout storage.

Polyacrylamide gel electrophoresis has been used extensively to study the proteolysis in pickled cheeses. In such electrophoretograms, a number of fast and slow moving bands are apparent. The bulk of the breakdown products of α_{s1}-casein have higher mobilities than α_{s1}-casein and are comparable to those formed by the action of chymosin on this casein. Mulvihill & Fox[156] found that the formation of the α_{s1}-casein peptides, α_{s1}-VII and α_{s1}-VIII, in solution is stimulated by and may depend on the presence of NaCl (5%). These peptides have lower mobilities than the parent casein and are formed in Teleme from cows' milk.[157] Slow-moving bands are comparable in electrophoretic mobility to γ-caseins produced from β-casein by the action of the indigenous alkaline milk proteinase (plasmin). However, plasmin activity in pickled cheeses may be questionable because of their low pH and high salt content. This point needs to be investigated.

As with other cheeses, lactic acid bacteria contribute to proteolysis mainly by producing small peptides and free amino acids. However, the high salt content and low pH of pickled cheeses do not favour growth of starter bacteria and their activity is diminished quickly. Secondary lactic flora (mainly lactobacilli) are much better adapted to the cheese environment and contribute to proteolysis during ripening and storage although the peptidases of both groups of microorganisms usually have pH optima near neutrality and far removed from the pH of pickled cheeses.[158] When raw milk is used, some microorganisms of the natural flora of milk (enterococci, micrococci, etc.) are certainly involved in proteolysis.

Proteolysis is not very intense in pickled cheeses. Although it is difficult to compare results obtained for different types of pickled cheeses by various authors, soluble nitrogen (SN) in ripened cheeses does not generally exceed 25% of total nitrogen (TN) and is mainly in the range of 12–20%. This soluble fraction contains a large number of small peptides and the non-protein nitrogen (NPN, nitrogen soluble in 12% TCA) is about 10 to 15% of TN. The nitrogen of very small peptides and free amino acids (PTA-N, phosphotungstic soluble nitrogen) is in the range of about 3 to 5% of TN.

There is a wide variation in the total free amino acid content of pickled cheeses: highest values (948 mg/100 g) were reported for Brinza cheese[159] and lowest (227 mg/100 g) for Chanakh cheese.[160] Values for ewes' milk Teleme[161] were 403 mg/100 g. Beyaz peynir[162] made from ewes' milk contained 853 mg/100 g while the same cheese from cows' milk contained 698 mg/100 g cheese. Despite these quantitative differences, most of the results agree that the major free amino acids in pickled cheese are leucine, valine and phenylalanine.

Citrulline and ornithine found in pickled cheeses result from the breakdown of arginine.[161] In some of these cheeses, a substantial reduction of the concentration of glutamic acid was observed, accompanied by the formation of γ-aminobutyric acid through a deamination reaction. Amino acid breakdown in pickled cheeses also leads to the production of volatile compounds such as ammonia, aldehydes, alcohols, acids, amines and other products. More than 35 volatile amines were isolated from Chanakh cheese, of which di-n-butylamine was one of the main components of the total amines.[163] Also, the non-volatile amine, tyramine, was found in small quantities in Beyaz peynir.[164] Tyramine may be produced by tyrosine decarboxylase of faecal streptococci, e.g. *Str. faecalis*, commonly found in pickled cheeses.[151] Ammonia is a normal constituent of these cheeses but quantities are much less (< 1% of total N in Teleme cheese)[165] than those found in mould-ripened cheeses[166] (7–9% of total N). In pickled cheeses, NH_3 is easily washed into the brine so that it increased 18-fold in the brine during the 12-week storage of Bulgarian white soft cheese.[167] Amino acids may also give rise to acetic and propionic acids, the former being a very significant contributor to the flavour of pickled cheeses.

6.2.3 Lipolysis and Production of Free Fatty Acids

Most of the pickled cheeses are traditional products and were made generally from raw ewes' or goats' milk using rennet pastes from kids or lambs for milk coagulation. These pastes, in addition to coagulating enzymes, also contained pregastric esterases. Thus, it is to be expected that their flavour is connected at least to some extent with lipolysis, although the main flavour characteristics are acidity and saltiness. A few of them have a sharp flavour after normal ripening (e.g. Chanakh), but most of them become sharp after advanced ripening, especially when they are stored at room temperature.

The major volatile fatty acid of pickled cheeses is undoubtedly acetic acid. Volatile fatty acids (VFA) in Teleme and Bjalo salamureno cheese are presented in Table XIV. Dilanyan & Magakyan[169] found 1181 mg/kg of total VFA, of which 85% was acetic acid, in Chanakh cheese. Buruiana[170] reported that acetic acid accounted for 67·8% of VFA in 60-day old Teleme made from cows' milk. In market samples of Israeli pickled cheese, VFA analysis showed only the presence of

Table XIV
Volatile Free Fatty Acids in Teleme and Bjalo Salamureno Cheese (mg/kg)

Acid	Teleme[168]	Teleme[138]	Bjalo salamureno[139]	Bjalo salamureno[138]
C_2	491	669	604	692
C_3	78	43	12	30
C_4	101	150	50	76
C_6	58	3	26	2
C_8	165	8	14	5
C_2 (% C_2 to C_8)	55	77	86	86

acetic acid and traces of propionic acid.[153] Because the short-chain fatty acids are water-soluble, part of them can be transferred to the pickling brine. Obretenow *et al.*[171] showed that the concentration of C_2–C_5 fatty acids in the pickle were higher than in the cheese, while the reverse was found in the case of C_6 and C_8 fatty acids due to differences in their solubilities in aqueous solution and in cheese fat.

Volatile fatty acids are part of the total free fatty acids (FFA) of pickled cheeses. Studies on Teleme,[168] Bjalo salamureno[138,140] and other[172] pickled cheeses showed that the average content of total FFA is between 1000 and 2500 mg/kg cheese. Higher values found may be associated with advanced storage at high temperatures or with the rennet pastes[138] used as milk coagulant or with the direct addition of pregastric esterases.

The origin of the lipases responsible for fat hydrolysis and the factors affecting the accumulation of FFA may be summarized as follows:

1. Milk and milk treatment. Milk lipase can play a role in the production of mainly VFA when cheeses are manufactured from raw milk. However, the activity of milk lipase is reduced because of the high salt content and the low pH of the cheese.[173] The use of ewes' milk may enhance the production of higher levels of short chain fatty acids.[140]

2. Starters. Those commonly used in pickled cheese manufacture may add little to lipolysis, mainly by producing FFA from mono- and diglycerides formed by milk lipase and/or other microbial lipases.[174]

3. Lipases in rennets. These lipases are responsible for the wide variation in lipolysis observed even in the same type of cheese due to differences in concentration and specificities of contaminating lipases. The use of rennet paste from kids and lambs produces high contents of C_4–C_8 fatty acids, since the lipases present preferentially release short chain fatty acids.

4. Conditions of ripening. Ripening and storage temperature and brine concentration can affect the accumulation of FFA. Mladenov[175] reported that lipolysis increases with an increase in storage temperature or a decrease in brine strength; however, changes in cheese stored at 3–5°C were slow and independent of brine concentration. Lipolysis in Syrian white pickled cheese was marked in cheese stored at 10–20°C but it was limited at 5°C. Increasing the salt content in the brine had a limited effect on lipolysis in this type of cheese.[155] The age of the brine also had a beneficial effect on Chanakh cheese quality by increasing its VFA.[176]

Although there are no available data, it seems that lipases from secondary flora, mainly bacteria, and to a limited degree, yeasts or moulds, which contaminate the cheeses, may contribute to the accumulation of FFA in pickled cheeses.

6.3 Microorganisms

In mature cheese (two to eight months old), the log_{10} counts of total aerobic bacteria, lactic acid bacteria, proteolytic, lipolytic, psychrotrophic and yeasts range between 7·0–9·5, 7·8–8·0, 6·0–8·5, 6·0–7·0. 2·0–6·0 and 2·5–4·5, respectively.[129,134,153,177]

The predominant lactic acid bacteria are streptococci (*S. faecalis, S. faecium, S. durans*), *Leuconostoc* (*L. dextranicum*) and lactobacilli (*Lb. plantarum, Lb. casei, Lb. brevis*).[178–181] The principal psychrotrophs found are from the genera *Pseudomonas, Aeromonas* and *Acinetobacter*.[182] Coliform bacteria are often reported in high numbers, over 10^5/g cheese, due mainly to the use of unpasteurized milk and poor sanitary conditions during cheesemaking,[183,151] but they decrease and eventually disappear upon ageing.[134] Enterotoxin producing strains of staphylococci are also found in brine cheese,[179] which are also good substrates for the survival of *Listeria monocytogenes*,[184] *Yersinia enterocolitica*[185] and aflatoxin production when they remain out of brine.[186] Sometimes, moulds can be detected in the cheese.[151]

6.4 Defects

A major defect of brined cheese is the 'early blowing' characterized by small or large gas holes, spongy texture and blowing of the cheese. This normally appears a few days after cheesemaking and is due to coliforms and/or yeasts growing in excessive numbers. 'Late blowing' may also occur, causing blowing of the containers and is attributed to either coliforms or heterofermentative lactic acid bacteria. Butyric acid fermentation is less likely to cause this defect. Late blowing is unlikely to occur when the cheese has a normal pH and is sealed in the containers after complete ripening.[187]

Ropy brine is often observed, making the appearance of the cheese unsightly. Slimy substances are produced mainly by *Lactobacillus* spp. and ropiness of brine has been attributed to the fermentation of lactose by *Lb. plantarum* var. *viscosum*.[188]

Excessive softening of the cheese during storage has been attributed to insufficient acidity, low storage temperature and low salt content in the brine.[148] In our experience, this may also occur when the brine added to the cheese before sealing the containers is of lower salt concentration than the salt in moisture content of the cheese.

REFERENCES

1. Zaky, A. & Iskander, Z., 1943. *Annales de Services des Antiquites de l'Egypte*, **41**, 295.
2. Hansen, R., 1977. *North European Dairy J.*, **43**, 304.
3. Jensen, G.K., Olsen, P.M. & Hyldig, G., 1988. *Scand. Dairy Industry*, **3**, 39.
4. Abd El-Salam, M.H., El-Shibiny, S. & Fahmi, A.H., 1976. *N.Z. J. Dairy Sci. Technol.*, **11**, 57.
5. Lloyd, D.G.T. & Ramashaw, E.H., 1979. *Aust. J. Dairy Technol.*, **34**, 180.
6. Abou-Donia, S.A., 1986. *N.Z. J. Dairy Sci. Technol.*, **21**, 167.
7. Abou-Donia, S.A., 1981. *Indian J. Dairy Sci.*, **34**, 136.
8. Mashaly, R.I., Abou-Donia, S.A. & El-Soda, M., 1983. *Indian J. Dairy Sci.*, **36**, 93.
9. El-Gendy, S.M., 1983. *J. Food Prot.*, **46**, 358.
10. Sirry, I. & Rakshy, S.D., 1954. *Indian J. Dairy Sci.*, **8**, 9.

11. Anifantakis, E.M. & Kaminarides, S.E., 1983. *Aust. J. Dairy Tech.*, **38**, 29.
12. Ahmed, T.E.K., 1987. M.Sc. Thesis, University of Khartoum, Sudan.
13. Abou-Donia, S.A. & Abdel Kader, Y.I., 1979. *Egyptian J. Dairy Sci.*, **7**, 221.
14. FAO, 1982. *Food Composition Tables for the Near East.* Food & Agric. Organization, United Nations, Rome.
15. Sharara, H.A., 1958. *Indian J. Dairy Sci.*, **11**, 175.
16. Puri, B.R. & Parkash, S.J., 1965. *J. Dairy Sci.*, **48**, 611.
17. Abd El-Salam, M.H., Osman, Y.M. & Nagmoush, M.R., 1978. *Egyptian J. Dairy Sci.*, **6**, 9.
18. Saito, Z. & Hoirose, M., 1972. *Bull. Faculty Agric.*, Hirosaki University, Japan, **18**, 35.
19. Saito, Z., Igarashi, Y. & Nakasato, H., 1972. *Bull. Faculty Agric.*, Hirosaki University, Japan, **18**, 22.
20. Schmidt, D.G., 1982. In *Advances in Dairy Chemistry*, Vol. I, Proteins, ed P.F. Fox, Elsevier Applied Science, London, pp. 61–86.
21. Pestskas, D. & Ramanauskas, R., 1974. *Proc. XIX Intern. Dairy Congr., Delhi*, **B5**, 169.
22. Sirry, I. & Shipe, W.F., 1958. *J. Dairy Sci.*, **41**, 204.
23. Abd El-Hamid, L.B., Amer, S.N. & Zedan, A.N., 1981. *Egyptian J. Dairy Sci.*, **9**, 137.
24. Ibrahim, M.K.E., Amer, S.N. & El-Abd, M.M., 1973. *Egyptian J. Dairy Sci.*, **1**, 127.
25. Ling, E.R., 1963. *A Textbook of Dairy Chemistry*, Vol. 1, Theoretical, 3rd edn, Chapman & Hall, London, pp. 140.
26. Gal, S. & Bankay, D., 1971, *J. Food Sci.* **36**, 800.
27. Gal, S. & Hunziker, M., 1977. *Makromol. Chem.*, **178**, 1535.
28. Hardy, J.J. & Steinberg, M.P., 1984. *J. Food Sci.*, **49**, 127.
29. Ramananskas, R., 1978. *Proc. XX Intern. Dairy Congr.*, Paris, **B** 265.
30. Geurts, T.J., Walstra, P. & Mulder, H., 1974. *Neth. Milk Dairy J.*, **28**, 46.
31. Moneib, A.F., 1962. PhD Thesis, Meded Landbouw, Wageningen, The Netherlands.
32. Ling, E.R., 1966. *J. Dairy Res.*, **33**, 151.
33. Fahmi, A.H. & Sharara, H.A., 1950. *J. Dairy Res.*, **17**, 312.
34. El-Sokkary, A.M., El-Sadek, G.M. & Hamed, M.G., 1957. *Ann. Agric. Sci.*, University of Ain-Shams, **2**, 123.
35. El-Sadek, G.M. & Abd El-Motteleb, L., 1958. *J. Dairy Res.*, **25**, 85.
36. El-Shibiny, S., Abd El-Salam, M.H. & Ahmed, N.S., 1972. *Milchwissenschaft*, **27**, 217.
37. Hamed, M.G., 1955. M.Sc. thesis, Ain-Shams University, Cairo, Egypt.
38. Ahmed, N.S., Abd El-Salam, M.H. & El-Shibiny, S., 1972. *Indian J. Dairy Sci.*, **25**, 246.
39. Tawab, G.A. & El-Koussy, L.A., 1975. *Egyptian J. Dairy Sci.*, **3**, 359.
40. El-Abd, M.M., Ibrahim, M.K.E., Amer, S.N. & Mostafa, S., 1975. *Egyptian J. Dairy Sci.*, **3**, 195.
41. Abd El-Salam, M.H. (unpublished).
42. Goodenough, E.R. & Kleyn, D.H., 1976. *J. Dairy Sci.*, **59**, 45.
43. Shehata, A.E., El-Sadek, G.M., Khalafalla, S.M. & El-Magdoub, M.N.I., 1975. *Egyptian J. Dairy Sci.*, **3**, 139.
44. Naguib, M.M., El-Sadek, G.M. & Naguib, Kh., 1974. *Egyptian J. Dairy Sci.*, **2**, 55.
45. Omar, M:M. & Buchheim, W., 1983. *Food Microstructure.*, **2**, 43.
46. El-Safty, M.S., 1969. M.Sc. Thesis, Ain-Shams University, Cairo, Egypt.
47. Hagrass, A.B., 1971. M.Sc. Thesis, Ain-Shams University, Cairo, Egypt.
48. Abd El Salam, M.H., El-Abd, M.M., Nagmoush, M.R. & Saleem, R.M., 1978. *Egyptian J. Dairy Sci.*, **6**, 221.
49. Abd El-Salam, M.H., Saleem, R.M. & Nagmoush, M.R., 1978. *Egyptian J. Dairy Sci.*, **6**, 187.
50. Moneib, A.F., Abo El-Heiba, A., Al-Khamy, A.F., El-Shibiny, S. & Abd El-Salam, M.H., 1981. *Egyptian J. Dairy Sci.*, **9**, 237.

51. Dawood, A.E., 1964. M.Sc. Thesis, Cairo University, Cairo, Egypt.
52. Teama, Z.Y., 1967. Ph.D. Thesis, Ain-Shams University, Cairo, Egypt.
53. El-Kousy, L.A., 1966. Ph.D. Thesis, Ain-Shams University, Cairo, Egypt.
54. Abd El-Salam, M.H., El-Shibiny, S., Ahmed, N.S. & Ismail, A.A., 1981. *Egyptian J. Dairy Sci.*, **9**, 151.
55. Goma, E.A., 1990., Ph.D. Thesis, Michigan State University, East Lansing, U.S.A.
56. Al-Khamy, A.F., 1988. Ph.D. Thesis, Al-Azhar University, Cairo, Egypt.
57. Gewaily, E.M., 1968. M.Sc. Thesis, Ain-Shams University, Cairo, Egypt.
58. Shahara, H.A., 1962. *Alexandria J. Agric. Res.*, **10**, 127.
59. And El-Salam, M.H. & El-Shibiny, S., 1982. *Asian J. Dairy Res.*, **1**, 187.
60. Abd El-Salam, M.H., El-Shibiny, S., El-Koussy, L. & Haggag, H.F., 1982. *Egyptian J. Dairy Sci.*, **10**, 237.
61. Khalafalla, S., El-Sadek, G.M., Shehata, A. & El-Magdoub, M., 1973. *Egyptian J. Dairy Sci.*, **1**, 163.
62. Sirry, I. & Kosikowski, F.V., 1959. *Proc. XV Intern. Dairy Congr.*, **2**, 812.
63. Ganguli, N.C., Prabhakaran, R.J.V. & Iya, K.K., 1964. *J. Dairy Sci.*, **47**, 13.
64. Abd El-Salam, M.H. & El-Shibiny, S., 1977. *J. Dairy Sci.*, **60**, 1519.
65. Abd El-Salam, H.H. & El-Shibiny, S., 1976. *J. Dairy Res.*, **43**, 448.
66. El-Shibiny, S., Abd El-Salam, M.H. & Ahmed, N.S., 1973. *Egyptian J. Dairy Sci.*, **1**, 56.
67. Abd El-Salam, M.H. & El-Shibiny, S., 1973. *Egyptian J. Food Sci.*, **1**, 225.
68. Ahmed, N.S., Abd El-Salam, M.H. & El-Shibiny, S., 1973. *Sudan J. Food Sci. Technol.*, **5**, 18.
69. Ismail, A.A., El-Hifnawi, M. & Sirry, I., 1972. *J. Dairy Sci.*, **55**, 1220.
70. Kamaly, K.M., 1978. M.Sc. Thesis, El-Menoufia University, Shebeen El-Koam, Egypt.
71. Shehata, A.E., Magdoub, M.N., Fayed, E.O. & Hofi, A.A., 1983. *Ann. Agric. Sci.*, Ain-Shams Univ., **28**, 737.
72. El-Shibiny, S., El-Koussy, L.A., Girgis, E.A. & Mehanna, N., 1983. *Egyptian J. Dairy Sci.*, **11**, 215.
73. Askar, A.A., Gaafar, R.H., Magdoub, M.N. & Shehata, A.E., 1982. *Egyptian J. Dairy Sci*, **10**, 73.
74. Abdou, S., Ghita, I. & El-Shibiny, S., 1976. *Egyptian J. Dairy Sci.*, **4**, 147.
75. Edelsten, O., Hamdy, A. & El-Koussy, L.A., 1969. *Yearbook Royal Vet. & Agric. Univ.*, Copenhagen, 201.
76. El-Safty, M.S. & El-Shibiny, S., 1980. *Egyptian J. Dairy Sci.*, **8**, 41.
77. Hamdy, A., 1970. *Proc. XVIII Intern. Dairy Congr., Sydney*, **IE**, 350.
78. Hamdy, A., 1972. *Indian J. Dairy Sci.*, **25**, 73.
79. Abd El-Salam, M.H., El-Shibiny, S. & Mehanna, N., 1983. *Egyptian J. Dairy Sci.*, **11**, 291.
80. Mehanna, N.M., El-Shibiny, S. & Abd El-Salam, M.H., 1983. *Egyptian J. Dairy Sci.*, **11**, 167.
81. Abd El-Salam, M.H. & El-Shibiny, S., 1972. *J. Dairy Res.*, **39**, 219.
82. El-Shibiny, S. & Abd El-Salam, M.H., 1976. *Milchwissenschaft*, **31**, 80.
83. El-Shibiny, S. & Abd El-Salam, M.H., 1974. *Egyptian J. Dairy Sci.*, **2**, 168.
84. Fox, P.F. & Walley, B.F., 1971. *J. Dairy Res.*, **38**, 165.
85. Ramos, M., Foutecha, J., Juarez, M., Amigo, L., Mahfouz, M.B. & El-Shibiny, S., 1988. *Egyptian J. Dairy Sci.*, **16**, 165.
86. Eigel, E.N., 1977. *Intern. J. Biochem.*, **8**, 187.
87. Creamer, L.K. & Olson, N.F., 1982. *J. Food Sci.*, **47**, 631.
88. Pfeil, R. & Abd El-Salam, M.H. (Unpublished).
89. El-Erian, A.F.H., Farag, A.H. & El-Gendy, S.M., 1974. *Agric. Res. Rev.*, **52**, 190.
90. Mehanna, N.M., Antila, P. & Pahkala, E., 1989. *Egyptian J. Dairy Sci.*, **17**, 19.
91. El-Shibiny, S., Abd El-Baky, A.A., Farahat, S.M., Mahran, G.A. & Hofi, A.A., 1974. *Milchwissenschaft.*, **29**, 666.

92. Hamed, A.I., Farag, S.I. & Abou Zied, N.A., 1987. *Egyptian J. Dairy Sci.*, **15**, 209.
93. Precht, D. & Abd El-Salam, M.H., 1985. *Milchwissenschaft*, **40**, 213.
94. Magdoub, M.N., Shehata, A.E., Fayed, E.O. & Hofi, A.A., 1983. *Ann. Agric. Sci.*, Ain-Shams Univ., **28**, 761.
95. Sabry, Z.I. & Guerrant, N.B., 1958. *J. Dairy Sci.*, **41**, 925.
96. Khattab, A.A. & Zaki, N., 1986. *Egyptian J. Dairy Sci.*, **14**, 165.
97. El-Shibiny, S., Ahmed, N.S. & Abd El-Salam, M.H., 1973. *Egyptian J. Food Sci.*, **1**, 107.
98. Abd El-Salam, M.H. & El-Shibiny, S., 1973. *J. Dairy Res.*, **40**, 113.
99. Knoop, A.M., Omar, M. & Peters, K.H., 1976. *Milchwissenschaft*, **30**, 745.
100. Kerr, T.J., Washam, C.J., Evans, A.L. & Todd, R.L., 1981. *J. Food Protection*, **44**, 496.
101. Shehata, A.F., Magdoub, M.N.I., Fayed, E.O. & Hofi, A.A., 1984. *Egyptian J. Dairy Sci.*, **12**, 47.
102. Helmy, Z., 1956. MSc Thesis, Cairo University, Cairo, Egypt.
103. Naguib, M.M., 1965. MSc Thesis, Ain-Shams University, Cairo University, Cairo, Egypt.
104. Ghoneim, N.A., 1968. *Milchwissenschaft*, **23**, 482.
105. Seham, M., Sheleih, M.A. & Saudi, A.M., 1982. *J. Egyptian Vet. Med. Assoc.*, **42**, 5.
106. Mahmoud, S.A.Z., Moussa, A.M., Zein, G.N. & Kamaly, K.M., 1979. *Res. Bull. Fac. Agric.*, Ain-Shams University, No. 1033.
107. Hegazi, F.Z.M., 1972. MSc Thesis, University of Assiut, Assiut, Egypt.
108. Abo El-Naga, I.G., 1971. *Tejiper*, **20**, 27.
109. El-Gendy, S.M., Abdel-Galil, H., Shahien, Y. & Hegazi, F.Z., 1983. *J. Food Protection*, **46**, 335.
110. El-Sadek, G.M. & Eissa, A., 1957. *Indian J. Dairy Sci.*, **10**, 184.
111. Naguib, Kh. & Shauman, T., 1973. *Zentbl. Bakt. Parazit. Kde.*, **11**, 84.
112. El-Naway, M.A., El-Mansy, M.A. & Khalafalla, S.M., 1982. *Zentralblatt fur Bakt. Mikrobiol. & Hygiene*, **1**(3), 561.
113. Helmy, Z.A., Abd El-Malek, Y. & Mahmoud, A.A., 1975. *Zentralblatt fur Bakt. Paras. Infek. & Hygiene*, **II**, 130, 334.
114. Ahmed, A.A., Mostafa, M.K. & Marth, E.H., 1983. *J. Food Prot.*, **46**, 412.
115. Naguib, M.M., Sabour, M.M. & Nour, M.M., 1979. *Archiv. fur Lebensm.*, **30**, 150.
116. El-Shibiny, S., Tawfik, N.F., Sharaf, O.M. & Al-Khamy, A.F., 1988. *Egyptian J. Dairy Sci.*, **16**, 331.
117. Abo Elnaga, I.G., 1968. *Milchwissenschaft*, **23**, 198.
118. Zygouris, N., 1952. *The Dairy Industry*, 2nd ed., Ministry of Agriculture, Athens, pp. 391–427.
119. Walter, H.E. & Hargrove, R.C., 1953. *Cheese varieties*, USDA, Bulletin No. 54, Washington D.C.
120. Eekhof-Stork, N., 1976. *The World Atlas of Cheese*. Paddington Press Ltd, UK.
121. *Greek National Bulletin of Statistics*, 1982.
122. Zerifidis, G.K., 1989. *Technology of Milk Products, Cheesemaking*. Thessaloniki, Greece, pp. 123–155.
123. *Greek Standards for Foods and Drinks*, 1988. Ministries of Economics and Agriculture, Athens.
124. *Bulletin No. 41*, 1964. Ministry of Commerce, Greece, Athens.
125. Manolkidis, K.S., Zerfiridis, G.K. & Karazanos, G.P., 1974. *Geoponica*, **21**, 295.
126. Veinoglou, B.K., Kalatzopoulos, G.K., Stamelos, N.K. & Anifantakis, E.M., 1969. *Bulletin of Agricultural Bank of Greece*, **168**, 1.
127. Anifantakis, E.M. & Kandarakis, J.G., 1983. *Le Lait*, **63**, 416.
128. Alichanidis, E., Anifantakis, E.M., Polychroniadou, A. & Nanou, M., 1984. *J. Dairy Res.*, **51**, 141.
129. Vafopoulou, A., Alichanidis, E. & Zerfiridis, G., 1989. *J. Dairy Res.*, **56**, 285.
130. Anifantakis, E.M., 1980. *Le Lait*, **60**, 525.

131. Manolkidis, C., Polychroniadou, A. & Alichanidis, E., 1970. *Le Lait*, **50**, 38.
132. Alichanidis, E., Polychroniadou, A., Tzanetakis, N. & Vafopoulou, A., 1981. *J. Dairy Sci.*, **164**, 732.
133. Creamer, L. K. & Olson, N.F., 1982. *J. Food Sci.*, **47**, 631.
134. Zirfiridis, G.K., Alichanidis, E. & Tzanetakis, N.M., 1989. *Lebensm. Wiss. u. Technol.*, **22**, 169.
135. Pappas, C.P. & Zerfiridis, G.K., 1989. In *Seminars in Dairy Technology*, Greek National Dairy Committee, Athens, pp. 77–125.
136. Veinoglou, B.C., Boyatzoglou, E.S. & Kotouza, E.D., 1979. *Dairy Ind. Intern.*, **44**, 29.
137. Efthymiou, C.C. & Mattick, J.F., 1964. *J. Dairy Sci.*, **57**, 593.
138. Efthymiou, C., 1967. *J. Dairy Sci.*, **50**, 20.
139. Horwood, J.F., Lloyd, G.T. & Stark, W., 1981. *Austr. J. Dairy Technol.*, **36**, 34.
140. Prodanski, P.G. & Dzhordzorowa, O.D., 1969. *Milchwissenschaft*, **24**, 734.
141. Vafopoulou-Mastrojiannaki, A., Litopoulou-Tzanetaki, E. & Tzanetakis, N., 1990. *Microbiologie-Aliments-Nutrition*, **8**, 53.
142. Buruiana, L.M. & Farag, S.I., 1983. *Egyptian J. Dairy Sci.*, **11**, 53.
143. Dimov, N., Salichev, Y. & Mineva, P., 1975. *Mlekarstvo. Drtchavno Izdatelstvo za Selskostopanska Literatura*. Sofia, 224.
144. Yildiz, F., Kotcak, C., Karasabej, A. & Gürsel, A., 1989. *Turk. Vet. Ve Hay.*, **13**, 384.
145. Tekinsen, O.C., 1983. *Ankara Univ. Vet. Fak. Derg.*, **30**, 449.
146. Zivkovic, Z., 1971. *Mljekarstvo*, **21**, 8.
147. Yanai, Y., Rosen, B., Pinsky, A. & Sklan, D., 1977. *J. Dairy Res.*, **44**, 149.
148. Markin, L. & Pollack, F.G., 1959. *Proc. XV Intern. Dairy Congr.*, **2**, 887.
149. Dilanyan, Z. Ch., 1957. *Technology of Milk and Milk Products*, Grosudars. Izdatelstvo Selskokhzaizvennoi Literature, Moskow, 424.
150. Carić, M., Milanovic, S., Gavaric, D. & Španovic, A., 1979. *Zbornik, Radova Techn. Fac.* Novi Sad., **10**, 29.
151. Turantaş, F., Ünlütürk, A. & Göktan, D., 1989. *Intern. J. Food Microbiol.*, **8**, 19.
152. Antonova, T., Daov, T., Nachev, L., Bodurska, I. & Nikoevska, Ts., 1985. *Acta Microbiol. Bulgarica*, **16**, 73.
153. Yanai, Y., Rosen, B., Pinsky, A. & Sklan, D., 1976. *J. Milk Food Technol.*, **39**, 4.
154. Kerimov, G.G., 1962. *Proc. XVI Intern. Dairy Congr.*, **B**(IV), 101.
155. Mansour, A. & Alais, C., 1972. *Le Lait*, **52**, 515.
156. Mulvihill, D.M. & Fox, P.F., 1980. *Irish J. Food Sci. Technol.*, **4**, 13.
157. Kalogridou-Vassiliadou, D. & Alichanidis, E., 1984. *J. Dairy Res.*, **51**, 629.
158. Cliffe, A.J. & Law, B.A., 1979. *J. Appl. Bacteriol.*, **47**, 65.
159. Omar, M.M. & El-Zayat, A.I., 1987. *Nahrung*, **31**, 783.
160. Agagabyan, A., Saakyan, R. & Amirkhanyan, R., 1970. *Proc. XVII Intern. Dairy Congr.*, **1E**, 389.
161. Polychorniadou, A. & Valchos, I., 1979. *Le Lait*, **59**, 234.
162. Ücüncü, M., 1981. *Molkerei-Zeitg Welt der Milch*, **35**, 634.
163. Magakyan, D.T., Zhuravleva, I.L., Dilanyan, Z.Ch. & Golovnya, R.V., 1976. *Prickladnaya Biokhimiya i Mikrobiologiya*, **12**, 253. Cited from *D.S.A.*, 1976. **38**, 5913.
164. Kayaalp, S.O., Renda, N., Kaymakralan, S. & Ozer, A., 1970. *Toxic. Appl. Pharmac.*, **16**, 459.
165. Manolkidis, C., Polychroniadou, A. & Alichanidis, E., 1970. *Le Lait*, **50**, 128.
166. Lenoir, J., 1963. *Ann. Technol. Agric.*, **12**, 51.
167. Zikovic, Z., 1963. *Arch. Poljoper Nauke*, **16**, 92.
168. Alichanidis, E., 1981. Privat Dozent Thesis, University of Thessaloniki, Greece.
169. Dilanyan, Z.Ch. & Magakyan, D.T., 1978. *Proc. XX Intern. Dairy Congr., Paris*, 1978, **1E**, 295.
170. Buruiana, L.M. & El-Senaity, M.H., 1986. *Egyptian J. Diary Sci.*, **14**, 201.
171. Obretenow, T., Dimitroff, D. & Obretenowa, M., 1985. *Milchwissenschaft*, **33**, 545.

172. Koçak, C., Gürsel, A., Ergül, E. & Gürsoy, A., 1987. *GIDA*, **12**, 179.
173. Driessen, F.M., 1983. Doctoral Thesis, Agricultural University, Wageningen, The Netherlands.
174. Stadhouders, J. & Veringa, H.A., 1973. *Neth. Milk Dairy J.*, **27**, 77.
175. Mladenov, M., 1970. *Nauchi Trudove Vissh. Veterinaromeditsinskii Institute* 'Prof. Dr. G. Pavlov', **22**, 531. Cited from *D.S.A.*, 1973. **35**, 2665.
176. Magakyan, D.T., 1982. *Proc. XXI Intern. Dairy Congr., Moscow*, **1**(1), 430.
177. Tzanetakis, N., Litopoulou-Tzanetaki, E. & Vafopoulou-Mastrojiannaki, A. (in press). *Lebensm. Wiss. u. Technol.*
178. Ozer, I., 1964. *Ankara Univ. Vet. Fak. Yayinl.* Cited from *D.S.A.* 1966. **28**, 928.
179. Petrica, L., Pambucol, Z., Purcel, M. & Rosu, A., 1968. *Lucr. Stiint Inst. Agron. Timisoara*, **11**, 241. Cited from *D.S.A.*, 1971. **33**, 4226.
180. Chomakov, Kh., Kirov, N., 1973. *Izvestiya Nauchnoizsledovatelski Inst. po Mlechna Promishlenost*, **7**, 171. Cited from D.S.A., 1975. **37**, 1300.
181. Aleksieva, V., 1980. *Veterinarnomendusinksi Nauki.*, **17**, 85. Cited from D.S.A., 1981. **43**, 3737.
182. Kalogridou-Vassiliadou, D., Manolkidis, K. & Kouraki-Dimopoulou, A., 1981. *Food Technol. Hygiene Review*, **3**, 13.
183. Manolkidis, C., Litopoulou-Tzanetaki, E., Kalogridou-Vassiliadou, D. & Tzanetakis, N., 1975. *Arch. Inst. Pasteur. Hellenique*, **XXI**, 53.
184. Sipka, M., Zakula, S., Kovininc, I. & Stajner, B., 1974. *Proc. XIX Intern. Dairy Congr., Dehli*, **1E**, 157.
185. Karaioannoglou, P., Koidis, P., Papageorgiou, D. & Mantis, A., 1985. *Milchwissenschaft*, **40**, 204.
186. Zerfiridis, G.K., 1985. *J. Food Prot.*, **48**, 961.
187. Chomakov, C., 1966. *Milchwissenschaft*, **21**, 215.
188. Chomakov, C., 1967. *Milchwissenschaft*, **22**, 569.

12

Mozzarella and Pizza Cheese

PAUL S. KINDSTEDT

Department of Animal Sciences, University of Vermont, Burlington, USA

1 INTRODUCTION

Mozzarella is a prominent member of the pasta filata, or stretched curd, cheeses that originated in Italy. Pasta filata cheeses are distinguished by a unique plasticizing and kneading treatment of the fresh curd in hot water, which imparts to the finished cheese its characteristic fibrous structure and melting and stretching properties.

Although Mozzarella originated in Italy, the United States has become its principal producer. Mozzarella cheesemaking technology was brought to the US by Italian immigrants. The cheese remained an ethnic product with a limited market until around World War II, when Italian cuisine in general, and pizza pie in particular, began its meteoric rise in popularity that continues to the present.[1] In 1989, 723 600 tonnes of Mozzarella were produced in the US, representing 28% of the nation's total cheese production.[2] The popularity of pizza is growing rapidly in the United Kingdom and northern Europe, and with it Mozzarella cheese manufacturing, notably in the UK and Denmark.[3]

In the US, Mozzarella cheese is divided into four separate categories defined by standards of identity on the basis of moisture content and fat-in-dry-matter (FDM), as indicated in Table I.[4] Mozzarella and part-skim Mozzarella are high in moisture (> 52%), soft bodied, and are often consumed fresh as table cheeses.

TABLE I
Compositional Standards for Mozzarella Cheese in the United States[4]

Type	Moisture (%)	FDM[a] (%)
Mozzarella	> 52 but ≤ 60	≥ 45
Low-moisture	> 45 but ≤ 52	≥ 45
Low-moisture Part-skim	> 45 but ≤ 52	≥ 30 but < 45
Part-skim	> 52 but ≤ 60	≥ 30 but < 45

[a] Fat-in-dry-matter.

337

They are rarely used in food service as an ingredient for pizza due to their poor shredding and clumping properties and limited shelf life. In contrast, low-moisture and low-moisture part-skim Mozzarella have much lower water content (typically 47–48%), longer shelf life, firmer body, good shredding properties, and are used primarily as ingredients for pizza and related foods. In the US, the term 'pizza cheese' was used to designate low-moisture Mozzarella until 1964 when the current standards of identity were adopted.[1] Over 75% of all Mozzarella produced in the US in 1985 was used for pizza.[5] This chapter will focus mostly on the low-moisture Mozzarella types (pizza cheese) in view of their dominant market status.

2 MANUFACTURE

High and low moisture Mozzarella differ somewhat in their manufacture. In the US, both must be made from pasteurized milk.

2.1 Low-Moisture Mozzarella

Traditional manufacture of low-moisture Mozzarella, as outlined in Fig. 1, is quite similar to that of Cheddar, with some notable exceptions. Usually, a mixed culture consisting of *Streptococcus salavarius* ssp. *thermophilus* (formerly *Streptococcus thermophilus*) and *Lactobacillus delbrueckii* ssp. *bulgaricus* (formerly *Lactobacillus bulgaricus*) or *Lactobacillus helveticus* is used. These thermophilic bacteria thrive at the higher scalding and cheddaring temperatures (e.g. 42°C) used compared to Cheddar cheese manufacture. The curd is cheddared to pH 5·2, and then subjected to a kneading and plasticizing process in hot water at approximately 70°C. The hot plastic curd is moulded into the desired shape, cooled briefly in chilled water and salted in chilled brine. Detailed descriptions of the cheesemaking process are given in Christensen,[6] Kosikowski[7] and Reinbold.[8]

Several noteworthy modifications of the traditional manufacturing process have been used widely in the industry. In one patented process, the cheddaring step is eliminated and, instead, the granular curd is soaked in warm water until the target pH (e.g. pH 5·2) is attained.[9] In a later patent by the same company,[10] the partially acidified granular curd is washed in cool water and then transferred to cold storage overnight, during which acidification to pH 5·3 is completed. In both cases, the acidified curd is stretched in hot water by the normal process. Alternatively, Mozzarella can be made by a stirred curd process in which cheddaring is replaced by continuous stirring of the drained curd until pH 5·3 is reached.[11] The curd may then be processed in hot water in the usual way or, in the case of unstretched Mozzarella, the curd is steamed to a semi-plastic state and then pressed into block form.

2.2 High-Moisture Mozzarella

Manufacturing procedures for high-moisture Mozzarella are described in detail by Christensen,[6] Kosikowski,[7] and Scott.[12] A mesophilic lactic starter culture such as *Lactococcus lactis* ssp. *lactis* or *cremoris* (formerly *Streptococcus lactis* or

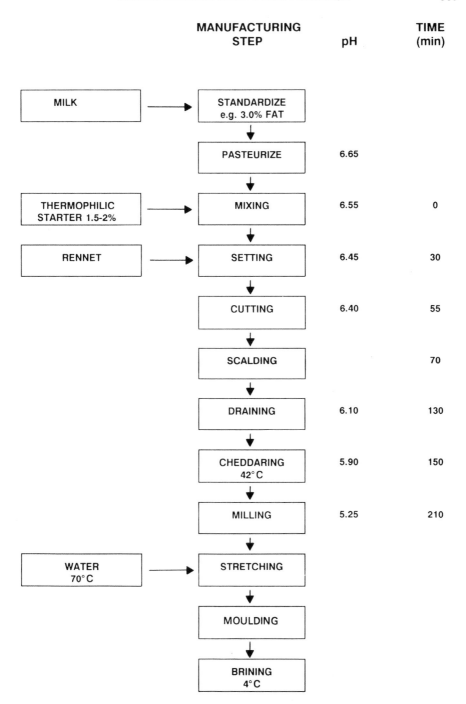

| | MANUFACTURING STEP | pH | TIME (min) |

Fig. 1. Example of flow diagram for the manufacture of low-moisture Mozzarella cheese by a traditional process.

cremoris) may be used, although some procedures specify a thermophilic culture consisting of *Streptococcus salavarius* spp. *thermophilus* and *Lactobacillus delbrueckii* ssp. *bulgaricus*. In either case, starter inoculum size is small (e.g. 0·1%) and vat temperature is kept low (e.g. 32–35°C) to facilitate slow curd acidification and high moisture retention. In traditional practice, the whey is drained off shortly after cutting and the curd is allowed to mat together. The partially acidified curd is then washed in cold water, placed in refrigerated storage and allowed to drain overnight, or longer, in cloth bags. After draining, the curd is warmed to room temperature and acidification to a target pH of 5·2 is completed. The curd is then processed in hot (e.g. 70°C) water into a plastic mass, moulded into a desired shape, cooled in chilled water and brined.

2.3 Directly Acidified Mozzarella

Both high and low-moisture Mozzarella cheeses can be manufactured without the use of a starter culture through direct addition of an organic acid. According to this method, cold (4°C) pasteurized, standardized milk is firstly acidified directly to pH 5·6 with a food-grade organic acid, such as citric or acetic. The acidified milk is warmed to about 37°C and coagulated in the usual way with rennet. After cutting, the curd may be scalded, depending on desired moisture content and then the drained curd is subjected to the normal hot water kneading and plasticizing process. Procedures for directly-acidified Mozzarella are described by Kosikowski,[7] Breene *et al.*,[13] Larson *et al.*[14] Patel *et al.*[15]

2.4 Mozzarella from Recombined and Ultrafiltrated Milk

Several alternatives to using fresh milk for Mozzarella cheesemaking have been explored extensively. Flanagan *et al.*[16] and Thompson *et al.*[17] described the manufacture of Mozzarella using reconstituted low-heat skim milk powder and fresh cream as the fat source. Cheeses had acceptable melting properties but poor shreddability due to gumminess. Replacement of as little as 12% of the low-heat powder with high-heat powder had an adverse effect on the melting properties of the cheese. Demott[18] manufactured a Mozzarella-like product from recombined skim milk powder and cream by direct acidification. Lelievre *et al.*[19] used recombined low-heat skim milk powder and anhydrous butteroil, homogenized under conditions of low (400 kPa) or high (6700 kPa) pressure, to manufacture Mozzarella cheese. Low pressure homogenization produced cheeses with acceptable melt and stretch properties but homogenization at high pressure had a detrimental effect on melting properties.

The manufacture of Mozzarella cheese from ultrafiltration (UF) retentates has received considerable attention. Reports in industry trade journals indicate that ultrafiltration is being used commercially to manufacture Mozzarella cheese in the US,[20,21] although one commercial venture proved unsuccessful due to technical problems.[20] In early studies, Covacevich & Kosikowski[22] prepared a pre-cheese mix from skim milk retentate supplemented with freeze dried retentate and plastic cream. The resultant cheese had poor melting and stretching properties. However, when ultrafiltration was carried out using diafiltration and simultaneous fermentation, the resultant cheese had acceptable melt and stretch. Covacevich[23] modified the process by acidifying the milk to pH 6·0 with HCl prior to diafiltration. The resultant UF retentate was then further concentrated to a pre-cheese mix using a swept surface vacuum evaporator. Jensen et al.[24] used glucono-delta-lactone to manufacture directly-acidified Mozzarella from non-acidified retentates and retentates that had been pre-acidified to pH 6·3 or 5·8 before ultrafiltration. Pre-acidification improved meltability, but UF cheeses were inferior to the control Mozzarellas in all cases. Friis[25] described a process for the manufacture of Mozzarella using UF retentates.

Fernandez & Kosikowski[26-28] concentrated cheesemilks by ultrafiltration to relatively low levels (up to 2 : 1) and then used conventional cheesemaking practices to manufacture cultured[26] and directly acidified[27] Mozzarella cheeses. In an alternative approach, the same authors supplemented whole milk with highly concentrated UF retentates to concentration levels up to 2:1.[28]

The authors reported that the quality of UF cheeses was acceptable in each case. The use of ultrafiltration technology in cheesemaking, including Mozzarella, was reviewed recently.[29]

3 CHEMICAL COMPOSITION

A number of surveys of the chemical composition of commercial Mozzarella cheeses have been conducted.[30-36] Data on gross composition, taken from several

TABLE II
Chemical Composition of Commercial Mozzarella Cheeses Surveyed in the United States

Cheese Type[a]	No. of Samples		Moisture	FDM	Salt	Protein	Ref. No.
				%			
LM	16	X̄	50·8	48·6	1·6	22·8	33
		RANGE	45–56	47–56	0·8–2·7	18–28	
LMPS	36	X̄	49·6	41·2	2·0	26·6	33
		RANGE	45–56	32–53	0·8–3·3	20–30	
M	2	X̄	54·0	50·0	2·4	21·5	33
		RANGE	52–56	47–53	1·2–3·5	21–22	
PS	2	X̄	54·9	47·5	1·0	22·0	33
		RANGE	54–56	45–50	0·7–1·2	22–22	
LM	21	X̄	46·4	46·1	—	—	32
		RANGE	44–51	35–51	—	—	
LMPS	39	X̄	47·7	39·4	—	—	32
		RANGE	44–56	32–36	—	—	
LMPS	34	X̄	48·5	37·9	1·65	26·1	35
		RANGE	44–55	31–48	0·8–2·6	21–30	
LM	22	X̄	47·7	47·0	—	—	36
		RANGE	44–53	45–50	—	—	
LMPS	22	X̄	48·1	36·0	—	—	36
		RANGE	45–51	33–46	—	—	
M	10	X̄	52·6	45·9	—	—	36
		RANGE	48·60	44·49	—	—	
PS	10	X̄	54·7	39·1	—	—	36
		RANGE	51–56	30–44	—	—	

[a]LM = low moisture; LMPS = low moisture, part-skim; M = Mozzarella; PS = part-skim.

of the more recent surveys, are summarized in Table II. In general, large variations in moisture, salt and FDM contents were found among commercial cheeses within each of the four Mozzarella categories. Perhaps this should come as no surprise, in view of the wide range of composition permitted under US standards of identity (Table I).

A very detailed tabulation of the chemical composition of Mozzarella cheeses is given in USDA Handbook No. 8.[37] Unfortunately, the data are from a very small number of cheese samples. Other specific aspects of composition that have been surveyed are amino acid,[38] fatty acid[39,40] and mineral[41] contents.

4 MICROFLORA

The microflora in Mozzarella cheese have been studied extensively. Nilson and LaClair[42] enumerated total bacterial counts and total counts for *Lactobacillus, Streptococcus*, and *Staphylococcus* species, coliforms, yeasts and moulds in commercial low-moisture Mozzarella cheeses during ageing for up to 5 months. Total bacterial counts on the day of manufacture ($\bar{x} = 3.8 \times 10^5$ cfu/g) were low relative to Cheddar cheese, presumably due to the stretching process, during which the curd temperature may range from 49 to 60°C for up to several minutes. The same authors[32] conducted a national survey of Mozzarella cheese quality in the United States. Sixty cheeses were analysed for counts of total (aerobic) bacteria, coliforms, yeasts and moulds. In 1978, Irvine[43] presented data from several large surveys of the bacteriological quality of Canadian Italian cheeses, including Mozzarella. Bacteriological enumeration included total coliforms, faecal coliforms and *Staphylococcus aereus*. The Canadian government established microbiological standards for cheese in 1978, based in part on the results of these studies. Nine years later, Irvine[44] reported on a follow-up survey of Italian cheeses in Canada to evaluate industry compliance with the standards.

The microbiological safety of Mozzarella cheese has been a concern. In 1989, an outbreak of food poisoning in Minnesota and Wisconsin was linked to Mozzarella cheese contaminated with *Salmonella javiana*. This prompted Eckner *et al.*[45] to investigate the survival of *Salmonella javiana* during the manufacture, brining and storage of Mozzarella-type cheese. The organism, which was pre-inoculated into the cheesemilk at levels ranging from ca. 1×10^4 to 1×10^6 cfu/ml, grew through the acid-ripening phase, but the high temperature attained in the curd mass during stretching and moulding (60°C) killed all *Salmonella* present.

The production of staphylococcal enterotoxin A in Mozzarella cheese was investigated by Tatini *et al.*[46] *Staphylococcus aureus*, inoculated into cheesemilk at levels ranging from 1×10^5 to 7.6×10^5 cfu/ml, attained levels of $\sim 2 \times 10^8$ cfu/g of curd at milling. However, counts were reduced to levels ranging from 2×10^4 to 5.8×10^5 cfu/g during stretching. No subsequent growth and no enterotoxin was detectable in the cheeses.

Hashisaka *et al.*[47] studied the survival of *Listeria monocytogenes* pre-inoculated into Mozzarella cheese prior to gamma irradiation treatment. *Listeria monocytogenes* had a lower resistance to radiation than *Bacillus cereus* var. *metieus* spores tested under similar conditions.

5 Microstructure

The microstructure of Mozzarella varies considerably depending on whether it is made by the stirred curd (unstretched) process or by conventional hot water stretching. Using scanning electron microscopy (SEM), Kalab,[48] and later Paquet & Kalab,[49] observed distinct curd granule junctions in stirred curd Mozzarella (see Fig. 2) which were not present in conventional stretched Mozzarella. The junctions appeared to be depleted of fat and contained a higher concentration of lactic acid bacteria than the interior of granules. Stretched Mozzarella displayed an oriented structure of fat and protein (see Fig. 3), while no orientation was evident in the stirred curd cheese. Fat globules in both stretched and stirred curd cheeses occurred in clusters. Yiu[50] confirmed the presence of curd granule junctions in stirred curd Mozzarella using fluorescence microscopy.

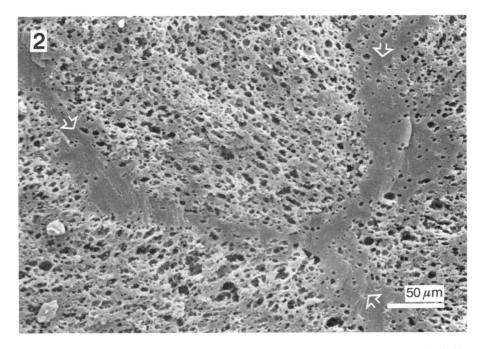

Fig. 2. Curd granule junctions (arrows) in stirred-curd Mozzarella cheese. (From Ref. 49, reproduced by permission of O. Johari, *Food Microstructure.*)

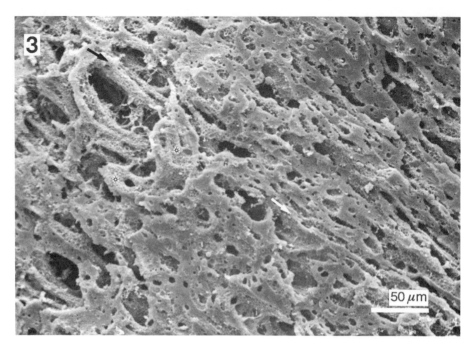

Fig. 3. Parallel orientation of protein fibres (arrows) in stretched Mozzarella cheese is evident from a longitudinal freeze-fracture. Fat globules have been removed during the preparation of the samples for SEM from places marked with asterisks. (From Ref. 49, reproduced by permission of O. Johari, *Food Microstructure*.)

Round or kidney-shaped crystalline structures, tentatively identified as calcium phosphate, were also observed in the granule junctions. In contrast to most reports, Taranto *et al.*[51] failed, using SEM, to find an oriented structure in stretched Mozzarella and reported little aggregation of fat globules. The fat globules were larger in Mozzarella than in Cheddar cheese and the protein matrix appeared more compact.

Electron micrographs of Mozzarella curd before and after hot water stretching clearly demonstrate the transformation that occurs from a non-oriented matrix of protein and fat globules to a highly oriented fibrous structure.[52] Curd microstructure undergoes further transformation during baking, as in the making of pizza.[49] Fat globules agglomerate into large pools of lipid, leading to a collapse of the protein structure and loss of original orientation as the cheese melts and flows (see Figs 4 and 5). The effects of microwave baking are similar but not as severe.

The microstructure of imitation Mozzarella containing 1 or 2% added calcium caseinate was clearly distinguishable from that of natural Mozzarella.[53] Fat globules in the former were randomly dispersed and often agglomerated into large bodies, such that the curd matrix contained large areas of protein without lipid and other areas where fat was plentiful.

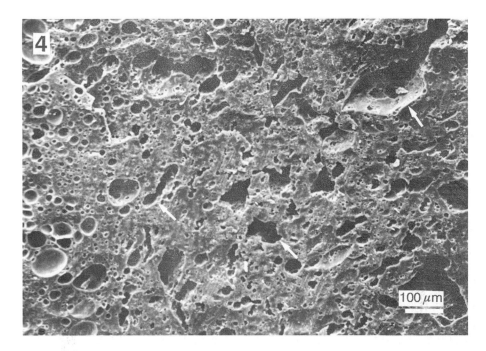

Fig. 4. Microstructure of stirred-curd Mozzarella cheese that had been baked in a conventional oven. Fat globules are aggregated in large fat particles which acquired irregular shapes (arrows). (From Ref. 49, reproduced by permission of O. Johari, *Food Microstructure.*)

6 PHYSICAL PROPERTIES — UNMELTED CHEESE

Although most Mozzarella is used for melting purposes, the physical properties of the unmelted cheese are also very important. When used for pizza and most other applications, the cheese must first be cut, diced or shredded into discrete particles of uniform size to facilitate even distribution and melting. Therefore, shreddability and resistance to clumping after shredding are major determinants of overall cheese quality. Research that specifically addresses these functional properties is scarce.

6.1 Rheological Characterization

Several researchers have characterized the rheological behaviour of Mozzarella and other cheeses using the Instron Universal Testing Machine.[54][56] Relative to most other cheeses tested, Mozzarella consistently showed a high degree of elasticity and a low level of hardness when subjected to compression under various test conditions. In addition, Chen et al.[54] reported that Mozzarella (type not specified) was among the top three of 11 cheese varieties for chewiness, cohesiveness, gumminess and adhesiveness. Overall, the rheological measurements for

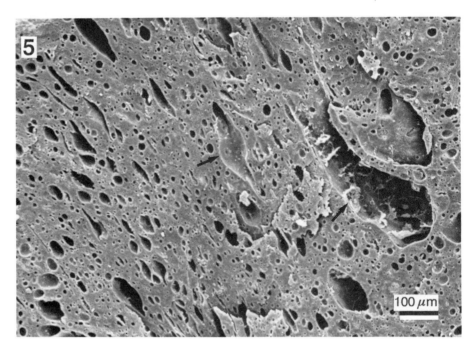

Fig. 5. Microstructure of stretched Mozzarella cheese that had been baked in a conventional oven. The orientation of the protein matrix is apparently due to the flow in the oven rather than due to the original stretching. Fat globules are aggregated in large fat particles which acquired irregular shapes (arrows). (From Ref. 49, reproduced by permission of O. Johari, *Food Microstructure*.)

Mozzarella in this study were closer to those for Swiss than any other variety. It should be noted, however, that the above studies[54-56] encompassed a very limited number of samples of each cheese variety and that sample age and composition, strong determinants of cheese textural properties (see chapter 8, volume 1), were not documented. Both Chen *et al.*[54] and Lee *et al.*[56] reported strong correlations between rheological measurements by the Instron test and the sensory assessments of a panel.

Test conditions are critical for valid rheological measurements and their interpretation. Casiraghi *et al.*[57] compared the behaviour of several cheeses, including Mozzarella, in lubricated, bonded, and non-lubricated non-bonded uniaxial compression. They concluded that it is essential to perform cheese compression tests under either bonded or lubricated conditions to eliminate frictional effects.

The rheological properties of natural and imitation Mozzarella have been characterized using a mechanical spectrometer.[58] The viscosity of low-moisture part-skim Mozzarella measured at room temperature was sensitive to additions of 1–2% calcium caseinate during processing, thus providing an objective basis for distinguishing between natural and imitation cheeses.

A sliding pin consistometer was used to differentiate between Mozzarella, Cheddar, processed and cream cheeses at temperatures ranging from 0·8 to 25°C.[59] Average force values decreased as temperatures increased.

6.2 Effect of Composition

The effects of various aspects of composition on the texture of Mozzarella cheese have been studied. Taranto et al.[51] related the rheological properties of Mozzarella and Cheddar cheeses to differences in microstructure and composition. Mozzarella had a lower value for hardness and higher values for work ratio (cohesiveness) and springiness (elasticity) than Cheddar, in agreement with others.[54-56] They postulated that these rheological differences could be due to the higher water content and the greater compactness of the protein matrix in Mozzarella.

Masi & Addeo[52] reported an inverse relationship between the ratio of fat/solids-not-fat and the modulus of elasticity in experimentally manufactured Mozzarella cheeses. Their data showed a linear increase in cheese firmness as fat content decreased (Fig. 6). However, an interacting effect that the authors did

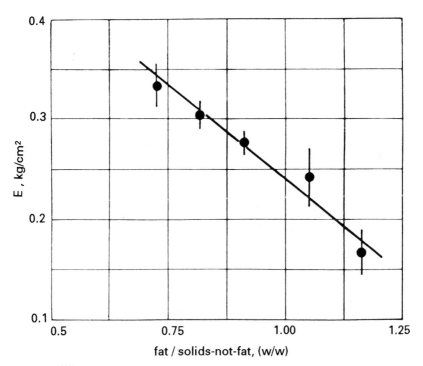

Fig. 6. Effect of composition on the compressive modulus of elasticity, E, of Mozzarella cheese. Points represent average experimental values of five runs. Bars indicate range of data points. (From Ref. 52, reproduced by permission of P. Masi and F. Addeo, *J. Food Engineering.*)

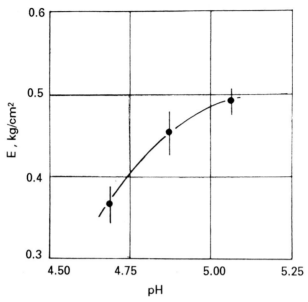

Fig. 7. Effect of curd ripening (pH) on the compressive modulus of elasticity, E, of Mozzarella cheese. Points represent average experimental values of five runs. Bars indicate range of data points. (From Ref. 52, reproduced by permission of P. Masi and F. Addeo, *J. Food Engineering.*)

not take into account was cheese moisture content, which also increased as the fat content of the cheese increased. Thus, the effects of moisture and fat on cheese firmness were undoubtedly confounded. The same authors showed that the modulus of elasticity (firmness) decreased with decreasing cheese pH (Fig. 7) and calcium content. Keller *et al.*[60] also reported direct relationships between the calcium and phosphate contents and the modulus of elasticity (firmness) in directly-acidified Mozzarella cheeses.

Moisture and FDM contents in low-fat Mozzarella cheeses were inversely related to hardness, gumminess and chewiness, while springiness showed an inverse relationship with moisture only and cohesivenesss showed an inverse relationship with FDM only.[61] In directly-acidified Mozzarella, moisture content was inversely related to cheese firmness.[60,62] The effect of salt content on the texture of Mozzarella was investigated by Cervantes *et al.*[63] Cheese firmness was not significantly affected by salt immediately after manufacture; however, after 39 days in refrigerated storage, high-salt Mozzarella (2·4%) was significantly firmer than low-salt Mozzarella (0·27%), presumably due in part to the inhibition of proteolytic enzymes at the higher salt concentration. The same authors reported that rapid (i.e. within 6 h) freezing and thawing did not affect cheese firmness. However, freezing and thawing at slower rates produced a pronounced lack of cohesiveness which persisted for at least seven days after thawing.[64]

Active residual coagulant may play a role in textural changes during storage and ageing. Directly-acidified Mozzarella made with fungal rennet showed greater formation of pH 4·4-soluble N and formal N and a more rapid loss of curdiness during 2 months of storage at 4°C than cheeses made with veal rennet or pepsin.[65] However, Micketts & Olson[66] found no significant differences in the formation of soluble N or compressability and elastic recovery of directly-acidified Mozzarellas made with different levels of veal rennet or pepsin.

6.3 Defects

It is possible that Mozzarella cheese may develop several specific rheological defects that lead to poor shredding properties. Soft body defect is characterized by a soft pasty body and poor shredding properties that develop during ageing, especially in the cheese centre. Hull *et al.*[67] reported that the defect was associated with high levels of *Lactobacillis casei* ssp. *casei*, a common contaminant of raw milk. Strains isolated from commercial Mozzarella cheeses were resistant to pasteurization, salt-tolerant and able to grow at temperatures as low as 6°C. Presumably, soft body was associated with the proteolytic activity of *Lb. casei*; however, nitrogen levels soluble in 2, 5 or 10% TCA in the soft centre and firm perimeter of defective cheeses did not differ significantly. Ryan[68] also found high numbers of *Lb. casei* as well as *Lb. fermentum*, *Lb. lactis* and *Lb. helveticus* in commercial Mozzarella cheeses with soft body. Experimental cheeses made from milks inoculated with *Lb. fermentum* showed pronounced softening and the formation of gas pockets during refrigerated storage. Less pronounced softening was observed in cheeses with added *Lb. casei*. *Propionibacterium freudenreichii* has also been implicated in the development of abnormal gas production and textural defects in Mozzarella cheese.[69]

Soft rind defect is common to brine-salted cheeses, including Mozzarella. The defect is characterized by a soft, moist, sometimes slimy, surface layer that is evident immediately after the cheese has been salted in freshly-prepared brine. Geurts *et al.*[70] discovered that in Gouda cheese this defect is caused by the migration of calcium from the cheese into the brine, which results in partial solubilization of casein. Such calcium leaching of can be prevented by calcifying the brine before its first use, which prevents soft rind defect in Mozzarella.[71]

Mozzarella sometimes develops a soft, pasty, high moisture surface during ageing. This defect has been linked to the practice of salting warm curd (e.g. 54°C) in cold (e.g. 4°C) brine, which results in the retention of much higher moisture levels at the cheese surface than occurs when brining is performed at higher temperatures.[72] During ageing, diffusion of water from the low-salt centre of the cheese to the high-salt surface can lead to a high moisture, soft surface.[71,73]

7 PHYSICAL PROPERTIES—MELTED CHEESE

The physical properties of melted Mozzarella cheese are highly complex and give rise to at least five important functional attributes: meltability, stretchability, elasticity, free oil formation and browning.[74,75] Precise definitions for these attributes and standard objective methods for their measurement are lacking. In general terms, meltability refers to the capacity of cheese particles to form a uniform continuous melt; stretchability is the ability of the melted cheese to form fibrous strands that elongate without breaking under tension, and elasticity, or 'strength of the stretch', is the capacity of the fibrous strands to resist permanent deformation. Although these attributes are often treated as independent properties, in reality they overlap with no clear demarcation. Free oil formation, also called 'oiling off' or 'fat leakage', is the separation of liquid fat from the melted cheese body into oil pockets, particularly at the cheese surface. Browning, which occurs at the cheese surface during high temperature baking, is characterized by the formation of a skin-like layer containing coloured patches that may range from light or golden brown to black in extreme cases.

7.1 Analytical Tests for Melting Properties

In some respects, the most meaningful way to assess the melting properties of pizza cheese is to subjectively evaluate its performance on pizza baked under commercial conditions. This approach is the basis for the USDA specifications relating to the melting, stretching and oiling off properties of Mozzarella,[76] is commonly used by industry for quality control purposes and has been used for research.[13,77–79] However, any subjective evaluation scheme has obvious limitations for research unless used in combination with meaningful objective measurements.

Early researchers relied on so-called meltability tests to objectively evaluate the melting behaviour of Mozzarella cheese. Several widely used versions of the meltability test involve heating a standardized cylindrical cheese sample in an oven[7,31,80] or in a boiling water bath[13,30] under defined conditions. Meltability is expressed as a function of either decreased sample height or increased sample area. Park et al.[81,82] found a marked lack of correlation between the Schreiber test, which measures increased sample area, and the Arnott test, which uses decreased sample height. They postulated that 'meltability' includes more than one physical property and that the two tests respond differently to these properties. Time and temperature conditions during analysis strongly affect meltability measurements, and different cheese varieties (e.g. Mozzarella vs. Processed American) respond differently when test conditions are changed.[81,82] This makes it very difficult to compare the research of workers who use different test conditions. Meltability testing is particularly cumbersome with brine-salted Mozzarella owing to the small test sample used and the large variation in melting properties within cheeses.[83] An elaborate sampling plan is needed to ensure a representative test result. Modified versions of the meltability test have been used to measure free oil formation in melted Mozzarella.[13,32] As with the meltability tests, representative sampling for free oil in Mozzarella using these tests is difficult.[36] A quantitative test that uses centrifugal force to separate free oil from melted Mozzarella cheese has been reported.[36] The test utilizes a grated sample to facilitate representative sampling. Free oil measurement can be expressed either as percent in cheese or percent in cheese fat (free oil–fat basis).

An alternative meltability test that was developed originally for processed cheese[84] and later applied to Mozzarella[85] involves melting the sample in a test tube under defined conditions and measuring the distance over which the sample flows. An advantage is that a grated sample can be used, enabling more representative sampling. As with other meltability tests, time and temperature conditions are critical because different cheeses respond differently when conditions are changed.[71]

Olson & Nelson[86] developed a test for melting behaviour based on the tendency of a viscoelastic material to climb up a rotating rod, known as the Weissenberg effect. A modified version of this method, in which the rod was replaced by a helical screw, was reported.[87] Smith et al.[88] used capillary rheometry to study the flow properties of Mozzarella, Cheddar and processed cheese at various temperatures. The standard protocol and corrections, which were developed originally for plastic polymers, were applicable to melted Mozzarella cheese but not to the other types, in which slippage due to fat separation and other artifacts played a dominant role. Lee et al.[56] used rotational viscometry to investigate the effect of temperature on viscosity of various cheeses, including Mozzarella. Melting curves for each cheese were derived from the viscosity data. The melting behaviour of Mozzarella has also been characterized with the Brabender Farinograph, an instrument used to measure the rheological properties of bread dough.[87] The Instron has been used to directly measure the stretchability of Halloumi cheese[19,89] but apparently not of Mozzarella cheese. A procedure for measuring the apparent viscosity of melted Mozzarella cheese by helical viscometry was reported[90] and later modified.[91]

Visual descriptive analysis of Mozzarella cheese on pizza has been used by several researchers to study browning properties.[78,92] Johnson & Olson[92] developed a predictive test for browning which utilizes the Hunterlab colorimeter. In this test, a cheese sample is heated in a boiling water bath for one hour and then cooled and analysed for the three colour indices, L*, a*, and b*. The sample is then incubated at 70°C for 48 h, after which the colour indices are measured as before. For each time period, the three colour indices, expressed relative to a colour standard, are used to calculate a colour difference value, E, which corresponds to the darkness of the sample colour.

7.2 Effect of Storage/Ageing on Melting Properties

Although Mozzarella is considered an unripened or fresh type cheese, it in fact undergoes a rather dramatic and characteristic change in melted functionality

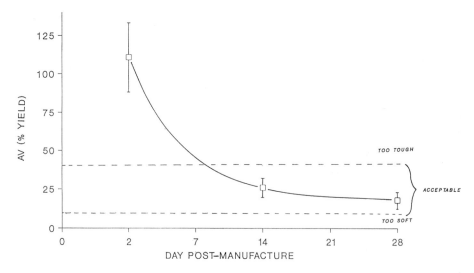

Fig. 8. Change in apparent viscosity (AV) of commercial low-moisture part-skim Mozzarella cheese during storage at 4°C. Points represent the average value of eight cheeses from different vats. Bars indicate the standard deviation.[71]

when it is stored under conditions of refrigeration (e.g. 4°C).[71,74] Initially after manufacture, Mozzarella melts to a tough elastic consistency that is unacceptable for pizza. During the first week or so of ageing, however, melted consistency 'mellows' substantially and the cheese soon melts to a desirable, moderately elastic state. Eventually, the cheese becomes excessively soft and fluid when melted and is no longer acceptable for pizza. In short, under conditions of normal refrigerated storage there is a relatively short period of acceptability, spanning perhaps 3–4 weeks, during which Mozzarella can be used for pizza.

7.2.1 Apparent Viscosity

The average change in apparent viscosity of eight commercial low-moisture part-skim Mozzarella cheeses during one month of ageing at 4°C is shown in Fig. 8.[71] Apparent viscosity was initially very high, indicating a tough, fibrous, melted consistency, but decreased markedly during the first 2 weeks of ageing and thereafter continued to decrease at a slower rate. The period of acceptability for use on pizza, indicated by dashed lines in Fig. 8, roughly corresponds to the range of 10 to 40% yield, as determined from a survey of cheese samples collected from five pizza restaurants over an 8-week period.[93]

7.2.2 Free Oil Formation

Free oil formation also shows a characteristic pattern of change during ageing, with a sharp increase during the first 1–2 weeks, followed by a more gradual increase thereafter. A typical ageing pattern of a commercial low-moisture part-skim Mozzarella is shown in Fig. 9.[94] For this cheese, approximately 25%

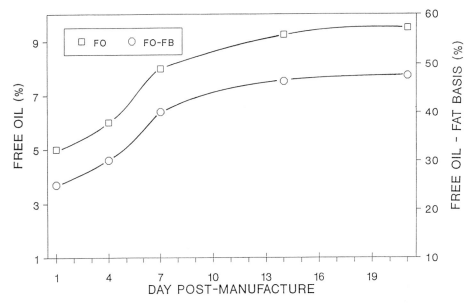

Fig. 9. Change in free oil formation of a typical commercial low-moisture part-skim Mozzarella cheese during storage at 4°C. Free oil is expressed as percent in cheese (free oil (%)) and percent in cheese fat (free oil – fat basis (%)).[94]

of the total fat formed free oil upon melting on one day post-manufacture; by 21 days' post-manufacture, free oil had doubled to almost 50% of total cheese fat.

7.2.3 Proteolysis
The striking changes in apparent viscosity and free oil formation during short term (1 month) ageing suggest that proteolytic activity may be an important factor in the development of the melting properties of Mozzarella, as postulated by Lawrence et al.[95] Creamer[96] found low levels of proteolysis in 12 weeks' old Mozzarella compared to Gouda and Cheddar cheeses of equal age. Percentages of total N soluble at pH 4·5 were 7·3, 20 and 15·5, respectively, and in 12% TCA were 2·4, 13·9 and 10·3, respectively. Gel electrophoresis revealed more intact α_{S1}-casein in Mozzarella than in Gouda or Cheddar cheese, although levels of α_{S1}-I peptide, indicative of residual rennet activity, and γ-casein levels, resulting from plasmin activity, in Mozzarella were intermediate between Gouda and Cheddar. Unfortunately, cheese composition and temperature of storage, which are strong determinants of proteolytic activity, were not reported. Presumably, the low level of proteolysis in Mozzarella is due to thermal inactivation of proteases during hot water stretching. Creamer[96] demonstrated that residual rennet activity is strongly dependent on the temperature profile of the curd during stretching. Matheson[97] confirmed the presence of low levels of active chymosin in one month-old Mozzarella cheese using an immunochemical assay.

Di Matteo *et al.*,[98] who studied proteolysis in high-moisture Mozzarella cheeses by soluble N formation and electrophoresis, found considerable breakdown of α_{s1}-casein to α_{s1}-I and α_{s1}-II in cheese stored at 11 or 20°C, suggesting residual coagulant activity. γ-Caseins were also present. These cheeses exemplified traditional high-moisture Mozzarella, which is stored in water or dilute brine and is not comparable in composition or functional properties to low-moisture Mozzarella/pizza cheese.

Kindstedt *et al.*[99] conducted electrophoretic investigations on four commercial low-moisture and low-moisture part-skim Mozzarella cheeses which were stored vacuum packaged for 4 weeks at 4°C. Intact α_{s1}- and β-caseins decreased by approximately 20–40% during storage. Farkye *et al.*[100] measured proteolysis by electrophoresis and WSN formation in four concentric layers (I–IV) within rectangular blocks of brine-salted low-moisture Mozzarella cheeses, where I = 0–1 cm from surface, II = 1–2 cm, III = 2–3 cm and IV = core. Intact α_{s1}- and β-caseins decreased by an average of 26·4 and 40·2%, respectively, in cheeses that were vacuum packaged and stored at 4°C for 14 days. Electrophoretograms revealed significant formation of α_{s1}-I peptide and γ caseins during storage, as seen in Fig. 10. Average WSN levels increased from 4·07% in 1 day-old cheese

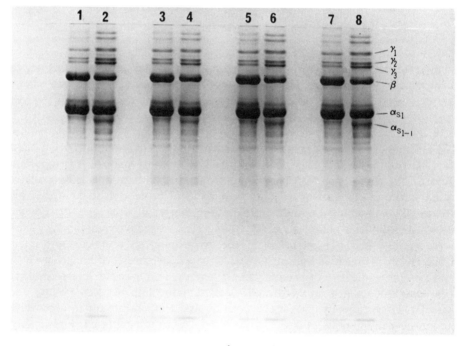

Fig. 10. Electrophoretogram of Mozzarella cheese during refrigerated storage. Lanes 1 and 2, 3 and 4, 5 and 6, and 7 and 8 represent sections I, II, III and IV, respectively, of a 2·72 kg block of cheese at one and 14 days after manufacture. Section I = 0–1 cm from the cheese surface, II = 1–2 cm, III = 3–4 cm and IV = core. (Reproduced by permission of R.L. Richter, *Journal of Dairy Science.*)

to 9·66% in 14 days old cheese. No significant differences in the rate of proteolysis determined electrophoretically or by WSN formation were found among the four locations within cheeses, despite large differences in composition.

Oberg et al.[85] reported that Mozzarella cheeses made with a starter culture containing proteinase-negative Lb. delbrueckii ssp. bulgaricus had better meltability and was less stretchable than cheeses made with the proteinase-positive parent strain. The data support the view that starter proteases affect melting properties; however, firm conclusions cannot be drawn, because cheese composition and proteolysis were not measured in this study.

7.2.4 Browning

Browning of cheese on pizza is believed to result, at least in part, from heat-induced reactions between sugars and proteins during baking, i.e. Maillard browning.[101] Alvarez[5] reported that browning and burning on the pizza is age-dependent. A high degree of burning occurs when the cheese is fresh but burning decreases dramatically during the first 2 weeks of ageing, followed thereafter by a sharp increase.

Johnson & Olson[92] demonstrated that accumulation of galactose, resulting from the incomplete fermentation of lactose by the starter culture, is a major determinant of browning. Most strains of Str. salavarius ssp. thermophilus and Lb. delbrueckii ssp. bulgaricus are unable to metabolize the galactose moiety of lactose.[102-105] In contrast, Lb. helveticus ferments lactose completely and may also ferment free galactose,[105] resulting in little accumulation of galactose when it is used in mixed-strain starter cultures.[92] Mozzarella cheeses have been made with (Gal+) strains of Str. salivarius ssp. thermophilus; however, these cheeses developed high levels of residual galactose unless the starter culture also contained (Gal+) Lb. helveticus, indicating that the latter is essential to efficient fermentation of galactose.[106] As a consequence, Lb. helveticus is gaining widespread used in the industry.

Several alternative manufacturing approaches have been used to limit browning in Mozzarella cheese. In one patented procedure,[9] the granular curd is soaked in warm water rather than cheddared to wash out excess lactose. In a different process, the curd is washed with cold water and then allowed to acidify overnight in a cold room, resulting in more complete fermentation of residual lactose.[10] A more recent approach has been to use non-traditional lactose fermenting starter species such as Pediococcus cerevisiae and Lb. plantarum as adjuncts to the traditional rod:coccus starter culture.[107]

7.3 Effect of Composition on Melting Properties

7.3.1 Mineral Content

According to Lawrence et al.,[95] the characteristic stretching properties of Mozzarella cheese are related to curd pH and the proportion of colloidal calcium phosphate retained in the curd. Kosikowsky[77] was the first to postulate that demineralization of curd caused by acidification during cheesemaking is responsible for the initial onset of stretching properties. Kindstedt[108] reported that the rate of acid production and the whey pH at draining were strong determinants of calcium and phosphorus levels in the fresh curd and its suitability for hot water stretching. For directly-acidified Mozzarella, an additional factor is the type of acidulant. Acids that are strong calcium chelators, e.g. citric, cause greater curd demineralization than non-chelating acids, e.g. acetic.[60,62] Curd pH is secondary in importance to mineral content in determining suitability for hot water stretching, as proved by the higher stretching pH used for directly-acidified Mozzarella (e.g. pH 5·6) compared to traditional cultured Mozzarella (e.g. pH 5·2).

The pH of the milk at coagulation is a strong determinant of the melting properties of directly-acidified Mozzarella due to its effect on cheese mineral content. Keller et al.[60] demonstrated that cheese calcium and phosphorus levels decreased with decreasing coagulation pH, resulting in a concomitant increase in meltability. This inverse relationship between coagulation pH and meltability was confirmed by Kim & Yu[109] and by Anis & Ladkani.[110] In the manufacture of UF Mozzarella the cheesemilk is usually acidified before ultrafiltration or during diafiltration to reduce the calcium content of the cheese and improve its meltability.[29]

7.3.2 Moisture Content

An inverse relationship between moisture content (non-fat substance basis) and apparent viscosity at 12 days post-manufacture was observed during a survey of 50 cheeses collected from two manufacturing plants over a 10 week period, as shown in Fig. 11.[111] The large variations in moisture content and melting properties among these cheeses were linked to erratic starter culture performance at one of the cheese plants.[111,112]

7.3.3 Fat Content

Fat content is a strong determinant of oiling off, as indicated from surveys of commercial Mozzarella cheeses. Cheeses with high FDM tended to yield high free oil levels, as measured by the disk fat leakage test[13,32] or by the centrifugal free oil test.[36]

Lelievre et al.[19] investigated the effect of homogenization pressure on the melting properties of Mozzarella and Halloumi cheeses made from recombined milk. Cheeses with normal flow and stretching properties were obtained when reconstituted skim milk powder and anhydrous milk fat were recombined with a low homogenization treatment (400 kPa). However, higher homogenization pressure (6700 kPa) had an adverse effect on the stretch and flow properties of the melted cheese. In Halloumi cheese, these adverse effects were overcome by adding lecithin to the recombined milk before high pressure homogenization. The data demonstrate that fat-casein interactions are important determinants of melting properties. When milk is homogenized under high pressure, the resulting fat globules predominantly contain casein at the fat-water interface, which enables them to participate in the formation of the casein matrix during cheesemaking. The authors postulated that such fat particles function as permanent crosslinks which prevent the molten cheese from flowing and stretching.

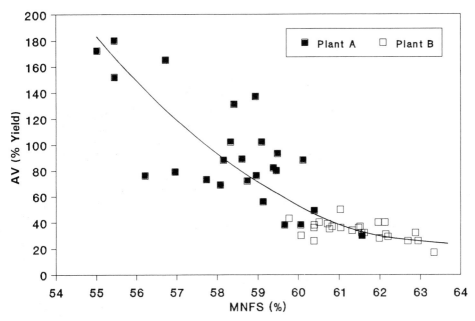

Fig. 11. Relationship between moisture content (nonfat substance basis – MNFS) and apparent viscosity of Mozzarella cheeses obtained from two commercial cheese plants over a 10 week period. Cheeses were analysed 12 days after manufacture, following storage at 4°C.[111]

7.3.4 Salt Content

Salt (NaCl) content strongly influences melting properties. Olson[64] reported that Mozzarella containing 1·78% salt was less meltable and less stringy initially after manufacture than cheese of equal age containing 1·06% salt. During ageing, elasticity of the melted cheese decreased more rapidly in the lower salt Mozzarella.

Non-uniform salt concentration persists for an extended period within brine-salted Mozzarella cheeses due to the slow inward diffusion of salt.[113] Such nonuniformity has been used to study the effects of composition on melting properties. Rippe & Kindstedt[114] reported that salt concentration decreased and free oil formation increased from the surface of brine-salted Mozzarella to the centre, both immediately after manufacture and after 16 days of refrigerated storage. In another study, Kindstedt *et al.*[71,94,115] observed less free oil and higher apparent viscosity for samples taken from the surface of brine-salted Mozzarella than at the interior, both initially and after 16 days of refrigerated storage. Salt content was higher and the calcium content was lower at the cheese surface than at the interior, but the moisture content was variable. The authors postulated that sodium from the brine exchanged with casein-bound calcium at the cheese surface, causing calcium to migrate into the brine. The emulsifying properties of

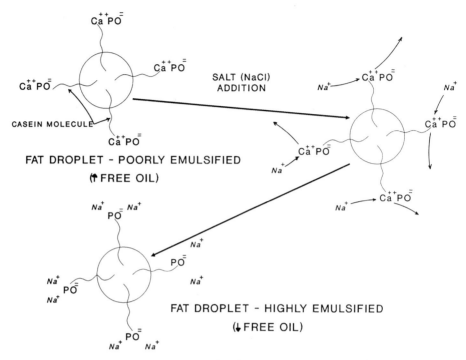

Fig. 12. Proposed model for the effect of NaCl concentration on free oil formation in Mozzarella cheese through exchange of sodium for casein-bound calcium.[115]

casein were enhanced by the sodium–calcium exchange, leading to decrease free oil formation and contributing to increased apparent viscosity. In essence, the authors proposed that sodium exchange at the cheese surface affected melting characteristics in a manner analogous to that of emulsifying salts in processed cheese, as illustrated in Fig. 12.

8   CONCLUSION

Undoubtedly, a myriad of physico-chemical, microbiological, enzymatic and structural factors influence the textural and melting properties of Mozzarella and pizza cheeses. A number of important factors have been identified, and an understanding of their individual roles in the development of cheese functionality and their interactions with other factors is beginning to evolve. Nevertheless, the systematic study of Mozzarella has lagged behind that of other major cheese varieties such as Cheddar, and large gaps in our understanding are evident. Notably, the physico-chemical changes that render cheese curd suitable for stretching in hot water and which give rise to the characteristic melting properties of Mozzarella are poorly understood. Equally incomplete is our understanding of the dramatic transformations in functionality that occur during ageing. Presumably, proteolytic activity has a major role in this process but systematic studies to differentiate the individual contributions of coagulant, starter proteases, non-starter proteases and plasmin have yet to be undertaken. Indeed, the degree to which proteases from various sources are thermally inactivated (or possibly activated in the case of plasmin) during stretching is largely unknown. Even well-established industrial practices that have been used for decades to alter the textural and melting properties of Mozzarella cheese, such as varying the rod:coccus ratio in the starter, have largely escaped serious scientific scrutiny. Future research should aim to establish a comprehensive understanding of the fundamental processes that govern this cheese and their relationship to manufacturing practices.

ACKNOWLEDGEMENTS

The author wishes to thank L.J. Kiely for his assistance during the preparation of this manuscript.

REFERENCES

1. Ferris, S. & Palmiter, H.A., 1987. *Italian Type Cheeses in the USA*, American Producers of Italian Type Cheese Association.
2. United States Department of Agriculture, *Dairy Products 1989 Summary*, Agric. Stat. Board, USDA, Washington, DC.
3. Mann, E.J., 1989. *Dairy Ind. Int.*, **54**(10), 9.
4. United States Food and Drug Administration, 1989. *Code of Federal Regulations*, Title 21, Parts 100–169, U.S. Dept. Health Human Services.
5. Alvarez, R. J., 1986. *Proc. 23rd Ann. Marschall Invit. Ital. Cheese Sem.*, p. 130.
6. Christensen, V.W., 1966. *Proc. 3rd Ann. Marschall Invit. Ital. Cheese Sem.*
7. Kosikowski, F.V., 1982. *Cheese and Fermented Milk Foods*, 2nd edn, Edwards Bros Inc., Ann Arbor, Michigan.
8. Reinbold, G.W., 1963. *Italian Cheese Varieties*, Pfizer Cheese Monographs, Vol. 1, New York.
9. Kielsmeier, L.O. & Leprino, J.G., 1970. U.S. Patent 3 531 297.
10. Kielsmeier, L.O., 1976. U.S. Patent 3 961 077.
11. Nilson, K.M. & LaClair, F.A., 1976. *Proc. 13th Ann. Marschall Invit. Ital. Cheese Sem.*
12. Scott, R., 1986. *Cheesemaking Practice*, 2nd edn, Elsevier Applied Science, London.
13. Breene, W.M., Price, W.V. & Ernstrom, C.A., 1964. *J. Dairy Sci.*, **47**, 1173.
14. Larson, W.A., Olson, N.F., Ernstrom, C.A. & Breene, W.M., 1967. *J. Dairy Sci.*, **50**, 1711.
15. Patel, G.C., Vyas, S.H. & Upadhyay, K.G., 1986. *Indian J. Dairy Sci.*, **39**(4), 394.
16. Flanagan, J.F., Thompson, M.P., Brower, D.P. & Gyuricsek, D.M., 1978. *Cult. Dairy Prod. J.*, **13**(4), 24.
17. Thompson, M.P., Flanagan, J.F., Brower, D.P. & Gyuricsek, D.M., 1978. *Proc. 15th Ann. Marschall Invit. Ital. Cheese Sem.*
18. Demott, B.J., 1983. *J. Dairy Sci.*, **66**, 2501.
19. Lelievre, J.L., Shaker, R.R. & Taylor, M.W., 1990. *J. Soc. Dairy Technol.*, **43**, 21.

20. Honer, C., 1990. *Dairy Field*, **173**(4), 30.
21. Dryer, J., 1988. *Cheese Market News*, **8**(44), 3.
22. Covacevich, H.R. & Kosikowski, F.V., 1978. *J. Dairy Sci.*, **61**, 701.
23. Covacevich, H.R., 1981. *Proc. 2nd Bien. Marschall Intern. Cheese Conf.*, p. 237.
24. Jensen, L.A., Riesterer, B.A. & Olson, N.F., 1987. *J. Dairy Sci.*, **70**(Suppl. 1), 66.
25. Friis, T., 1981. *Nordeuropaeisk Mejeri Tiddsskrift*, **47**(7), 220.
26. Fernandez, A. & Kosikowski, F.V., 1986. *J. Dairy Sci.*, **69**, 2011.
27. Fernandez, A. & Kosikowski, F.V., 1986. *J. Dairy Sci.*, **69**, 643.
28. Fernandez, A. & Kosikowski, F.V., 1986. *J. Dairy Sci.*, **69**, 2551.
29. Lawrence, R.C., 1989. *Intern. Dairy Fed. Bull.* No. 240.
30. Olson, N.F. & Neu, J., 1964. *Proc. 1st Ann. Marschall Invit. Ital. Cheese Sem.*
31. Olson, N.F., Breene, W. & Ernstrom, C.A., 1964. *Proc. 1st Ann. Marschall Invit. Ital. Cheese Sem.*
32. Nilson, K.M. & LaClair, F.A., 1976. *Proc. 13th Ann. Marschall Invit. Ital. Cheese Sem.*
33. Nieradka, T., Nilson, K.M., Duthie, A.H. & Atherton, H.V., 1979. *Cult. Dairy Prod. J.*, **14**(4), 11.
34. Kleyn, D.H. & Pintauro, N.D., 1980. *J. Dairy Sci.*, **63**(Suppl. 1), 37.
35. Barbano, D.M., 1984. *Proc. 21st. Ann. Marschall Invit. Ital. Cheese Sem.*, p. 1.
36. Kindstedt, P.S. & Rippe, J.K., 1990. *J. Dairy Sci.*, **73**, 867.
37. Posati, L.P. & Orr, M.L., *U.S. Dept. Agric., Agric. Handbook* No. 8–1, U.S. Dept. Agric., Washington, DC.
38. Hoskins, M.N., 1985. *J. Amer. Dietetic Assoc.*, **85**, 1612.
39. Feeley, R.M., Criner, P.E. & Slover, H.T., 1975. *J. Amer. Dietetic Assoc.*, **66**, 140.
40. Woo, A.H. & Lindsay, R.C., 1984. *J. Dairy Sci.*, **67**, 960.
41. Wong, N.P., LaCroix, D.E. & Alford, J.A., 1978. *J. Amer. Dietetic Assoc.*, **72**, 608.
42. Nilson, K.M. & LaClair, F.A., 1975. *Proc. 12th Ann. Marschall Invit. Ital. Cheese Sem.*
43. Irvine, D.M., 1978. *Proc. 15th Ann. Marschall Invit. Ital. Cheese Sem.*
44. Irvine, D.M., 1987. *Proc. 24th Ann. Marschall Invit. Ital. Cheese Sem.*, p. 30.
45. Eckner, K.F., Roberts, R.F., Strantz, A.A. & Zottola, E.A., 1990. *J. Food Prot.*, **53**, 461.
46. Tatini, S.R., Wesala, W.D., Jezeski, J.J. & Morris, H.A., *J. Dairy Sci.*, **56**, 429.
47. Hashisaka, A.E., Weagant, S.D. & Dong, F.M., 1989. *J. Food Prot.*, **52**, 490.
48. Kalab, M., 1977. *Milchwissenschaft*, **32**, 449.
49. Paquet, A. & Kalab, M., 1988. *Food Microstructure*, **7**, 93.
50. Yiu, S.H., 1985. *Food Microstructure*, **4**, 99.
51. Taranto, M.V., Wan, P.J., Chen, S.L. & Rhee, K.C., 1979. *Scanning Electron Microscopy*, **III**, 273.
52. Masi, P. & Addeo, F., 1986. *J. Food Eng.*, **5**, 217.
53. Tunick, M.H., Basch, J.J., Maleeff, B.E., Flanagan, J.F. & Holsinger, V.H., 1989. *J. Dairy Sci.*, **72**, 1976.
54. Chen, A.H., Larkin, J.W., Clark, C.J. & Irwin, W.E., 1979. *J. Dairy Sci.*, **62**, 901.
55. Imoto, E.M., Lee, C. & Rha, C., 1979. *J. Food Sci.*, **44**, 343.
56. Lee, C., Imoto, E.M. & Rha, C., 1978. *J. Food Sci.*, **43**, 1600.
57. Casiraghi, E.M., Bagley, E.B. & Christianson, D.D., 1985. *J. Texture Stud.*, **16**, 281.
58. Nolan, E.J., Holsinger, V.H. & Shieh, J.J., 1989. *J. Texture Stud.*, **20**, 179.
59. Davey, K.R, 1986. *J. Texture Stud.*, **17**(3), 267.
60. Keller, B., Olson, N.F. & Richardson, T., 1974. *J. Dairy Sci.*, **57**, 174.
61. Tunick, M.H., Mackey, K.L., Smith, P.W., Shieh, J.J. & Malin, E.L., 1990. *Proc. 23rd Intern. Dairy Congr.*, Montreal, **2**, 488.
62. Shehata, A.E., Iyer, M., Olson, N.F. & Richardson, T., 1967. *J. Dairy Sci.*, **50**, 824.
63. Cervantes, M.A., Lund, D.B. & Olson, N.F., 1983. *J. Dairy Sci.*, **66**, 204.
64. Olson, N.F., 1982. *Proc. 19th Ann. Marschall Invit. Ital. Cheese Sem.*, p. 1.

65. Quarne, E.L., Larson, W.A. & Olson, N.F., 1968. *J. Dairy Sci.*, **51**, 848.
66. Micketts, R. & Olson, N.F., 1974. *J. Dairy Sci.*, **57**, 273.
67. Hull, R.R., Roberts, A.V. & Mayes, J.J., 1983. *Aust. J. Dairy Technol.*, **38**, 78.
68. Ryan, J.J., 1984. *Proc. 21st Ann. Marschall Invit. Ital. Cheese Sem.*, p. 97.
69. Champagne, C.P. & Lange, M., 1990. *Sci. des Aliments*, **10**, 43.
70. Geurts, Th.J., Walstra, P. & Mulder, H., 1972. *Neth. Milk Dairy J.*, **26**, 168.
71. Kindstedt, P.S. & Kiely, L.J., 1990. *Proc. 9th Biennial Cheese Ind. Conf.*, Logan, UT.
72. Nilson, K.M., 1968. *Proc. 5th Ann. Marschall Invit. Ital. Cheese Sem.*
73. Kindstedt, P.S., Duthie, C.M. & Farkye, N.Y., 1990. *J. Dairy Sci.*, **73**(Suppl. 1), 81.
74. Kindstedt, P.S., 1991. *Cult. Dairy Prod. J.*, **26**(3), 27.
75. Pilcher, S.W. & Kindstedt, P.S, 1990. *J. Dairy Sci.*, **73**, 1644.
76. United States Department of Agriculture, 1980. *Specifications of Mozzarella cheeses*, Agric. Marketing Service, USDA, Washington, DC.
77. Kosikowsky, F.V., 1951. *J. Dairy Sci.*, **34**, 641.
78. Nilson, K.M., 1974. *Proc. 11th Ann. Marschall Invit. Ital. Cheese Sem.*
79. Nilson, K.M. & Fife, C.L., 1972. *Proc. 9th Ann. Marschall Invit. Ital. Cheese Sem.*
80. Arnott, D.R., Morris, H.A. & Combs, W.B., 1957. *J. Dairy Sci.*, **40**, 957.
81. Park, J. & Rosenau, J.R., 1984. *Korean J. Food Sci. Technol.*, **16**, 153.
82. Park, J., Rosenau, J.R. & Peleg, M., 1984. *J. Food Sci.*, **49**, 1158.
83. Kindstedt, P.S. & Rippe, J.K., 1988. *Proc. 25th Ann. Marschall Invit. Ital. Cheese Sem.*, p. 1.
84. Olson, N.F. & Price, W.V., 1958. *J. Dairy Sci.*, **41**, 999.
85. Oberg, C.J., Wang, A., Moyes, L.V. & Richardson, G.H., 1990. *J. Dairy Sci.*, **73**(Suppl. 1), 146.
86. Olson, N.F. & Nelson, D.L., 1980. *Proc. 17th Ann. Marschall Invit. Ital. Cheese Sem.*, p. 109.
87. Rippe, J.K., & Kindstedt, P.S., 1988. *J. Dairy Sci.*, **71**(Suppl. 1), 69.
88. Smith, C.E., Rosenau, J.R. & Peleg, M., 1980. *J. Food Sci.*, **45**, 1142.
89. Shaker, R.R., Lelievre, J., Taylor, M.W., Anderson, J.A. & Gilles, J., 1987. *N.Z. J. Dairy Sci. Technol.*, **22**, 181.
90. Kindstedt, P.S., Rippe, J.K. & Duthie, C.M., 1989. *J. Dairy Sci.*, **72**, 3117.
91. Kindstedt, P.S. & Rippe, J.K., 1990. *J. Dairy Sci.*, **73**(Suppl. 1), 119.
92. Johnson, M.E. & Olson, N.F., 1985. *J. Dairy Sci.*, **68**, 3143.
93. Kindstedt, P.S. & Harris, J., 1991. *J. Dairy Sci.*, **74**(Suppl. 1), 140.
94. Kindstedt, P.S. & Kiely, L.J., 1990. *Proc. 27th Ann. Marschall Invit. Ital. Cheese Sem.*, p. 49.
95. Lawrence, R.C., Creamer, L.K. & Gilles, J., 1987. *J. Dairy Sci.*, **70**, 1748.
96. Creamer, L.K., 1976. *N.Z.J. Dairy Sci. Technol.*, **11**, 130.
97. Matheson, A.R., 1981. *N.Z.J. Dairy Sci. Technol.*, **15**, 33.
98. Di Matteo, M., Chiovitti, G. & Addeo, F., 1982. *Scienza e Technica Lattiero — Casearia*, **33**, 197.
99. Kindstedt, P.S., Duthie, C.M. & Rippe, J.K., 1988. *J. Dairy Sci.*, **71**(Suppl. 1), 70.
100. Farkye, N.Y., Kiely, L.J., Allshouse, R.D. & Kindstedt, P.S., 1991. *J. Dairy Sci.*, **74**, 1433.
101. Olson, N.F., Bley, M.E. & Johnson, M.E., 1983. *Proc. 20th Ann. Marschall Invit. Ital. Cheese Sem.*, p. 1.
102. Somkuti, G.A. & Steinberg, D.H., 1979. *J. Food Prot.*, **42**, 881.
103. Tinson, W., Hillier, A.J. & Jago, G.R., 1982. *Aust. J. Dairy Technol.*, **37**, 8.
104. Tinson, W., Ratcliff, M.F., Hillier, A.J. & Jago, G.R., 1982. *Aust. J. Dairy Technol.*, **37**, 17.
105. Turner, K.W. & Martley, F.G., 1983. *Appl. Environ. Microbiol.*, **45**, 1932.
106. Hutkins, R., Halambeck, S.M. & Morris, H.A., 1986. *J. Dairy Sci.*, **69**, 1.
107. Reinbold, G.W. & Reddy, M.S., 1978. U.S. Patent 4 085 228.

108. Kindstedt, P.S., 1985. *Proc. 22nd Ann. Marschall Invit. Ital. Cheese Sem.*, p. 14.
109. Kim, Y.H. & Yu, J.H., 1988. *Korean J. Dairy Sci.*, **10**, 21.
110. Anis, S.M.K. & Ladkani, B.G., 1988. *Egyptian J. Dairy Sci.*, **16**, 267.
111. Kindstedt, P.S., Rippe, J.K. & Duthie, C.M., 1988. *Proc. 25th Ann. Marschall Invit. Ital. Cheese Sem.*, p. 59.
112. Kindstedt, P.S., Rippe, J.K. & Duthie, C.M., 1989. *J. Dairy Sci.*, **72**, 3123.
113. Lee, H.J., Olson, N.F. & Lund, D.B., 1980. *J. Dairy Sci.*, **63**, 513.
114. Rippe, J.K. & Kindstedt, P.S., 1989. *J. Dairy Sci.*, **72**(Suppl. 1), 133.
115. Kindstedt, P.S. & Duthie, C.M., 1990. *Proc. 23rd Intern. Dairy Congr.* Montreal, **2**, p. 481.

13

Fresh Acid-Curd Cheese Varieties

T.P. GUINEE

The National Dairy Products Research Centre, Moorepark, Fermoy, Co Cork, Republic of Ireland

P.D. PUDJA

Institute of Food Technology and Biochemistry, Faculty of Agriculture, University of Belgrade, Yugoslavia

&

N.Y. FARKYE

Dairy Products Technology Centre, California Polytechnic State University, San Luis, Obispo, CA 93407 USA

1 INTRODUCTION

Fresh acid-curd cheeses refer to those varieties produced by the coagulation of milk, cream or whey via acidification or a combination of acid and heat, and which are ready for consumption once the manufacturing operations are complete (Fig. 1). They differ from rennet-curd cheese varieties (e.g. Camembert, Cheddar, Emmental), where coagulation is induced by the action of rennet at pH values of 6·4–6·6, in that coagulation occurs close to the isoelectric point of casein, i.e. pH 4·6, or at higher values when elevated temperatures are used, e.g. in Ricotta (pH 6·0 to 80°C). While a very small amount of rennet may be used in the production of Quarg, Cottage cheese and Fromage frais to give firmer coagula and minimize casein losses (on subsequent whey separation), its addition is not essential.

The production of fresh acid-curd cheeses generally involves milk pre-treatments, slow acidification (via in situ conversion of lactose to lactic acid by added starter) and gelation, whey separation and/or curd treatment (Fig. 2) Many processing factors (e.g. milk pre-heat treatment, rate and temperature of acidification, level of gel-forming protein, pH) influence coagulum structure and hence the rheological and physico-chemical stability of the product in cold-pack (i.e.

Fresh acid-curd cheeses

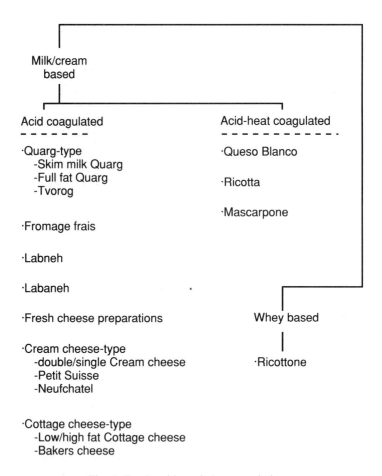

Fig. 1. Fresh acid-curd cheese varieties.

packaged without pasteurization of separated curd) products such as Quarg and Fromage frais-type cheeses. In hot-pack products, such as Cream cheese and Petit Suisse, curd treatments (i.e. pasteurization, homogenization, hydrocolloid addition) further influence structure and rheology.

Annual world production of fresh acid-curd cheeses amounts to about 3·2 million tonnes, which is equivalent to ~ 25% of total cheese (1–6, Table I). Quarg, Cottage, Cream, Fromage frais and Ricotta are commercially the most important types. Fresh cheeses as a percentage of total cheese production is highest in the former USSR (60%), Eastern Europe (40–60%) and West Germany (45%). Per capita consumption as a percentage of total cheese ranges from ~ 70% in Israel (11·3 kg) and Poland (8·2 kg) to 45% in Germany (8 kg) to 5–7% in Denmark (1·1 kg), UK (0·7 kg) and Ireland (< 0·5 kg).

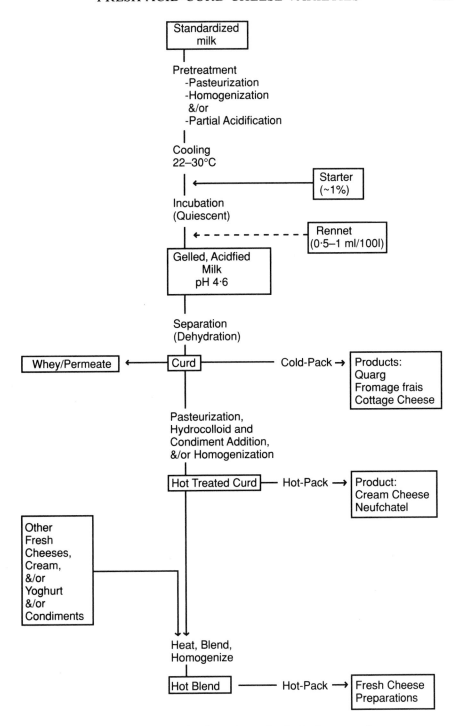

Fig. 2. Generalized production scheme for fresh cheese products.

TABLE I
Annual Fresh Cheese Production and Consumption (Approximate Values Based on Data Available for Period 1984–1989)

Country	Production ('000 tonnes)	Major types	Per capita consumption (kg/head)
Former USSR	1000	Tvorog	4·5
Germany	517	Quarg, Cream cheese	8
USA	550	Cottage, Cream cheese	2
France	381	Fromage frais, Petit Suisse	7
Italy	207	Ricotta, Quarg	4·5
Poland	120	Tvorog	8·2
Czechoslovakia	50	Tvorog	4·5
Canada	40	Cottage, Cream cheese	1·2
Belgium	26	Fromage frais	3·9
Denmark	24	Cream cheese, Quarg	1·1
Norway	20	Mysost	0·5
Irish Republic	0·2	Cottage, Cream cheese	< 0·5
Israel	—	Cottage, Quarg, Ricotta	11·3

Compiled from Refs 1–6.

The consumption of fresh cheese products has grown by ~ 4% per annum during the period 1983–87. Many factors have contributed to this increase, including:

(i) the large variety fresh cheese products offer in terms of consistency and flavour as affected by variations in processing parameters, blending of one or more cheese types to create new products and the additions of sugar, fruit purées, spices and condiments. In this respect, such cheeses are ideally suited to exploit a market where individualism of meals is increasing and the true-to-brand/-product loyalty of the consumer is disappearing.

(ii) the ease of handling the cheese which, in combination with packaging developments, helps to create attractive and conveniently sized packs suited to family use and individual 'irregular' consumers;

(iii) their soft, ingestable consistency which makes them safe and attractive to very young children. This consistency also makes them very suitable for snacking, a form of eating which is becoming more widespread;

(iv) the healthy perception of these products by the diet-conscious consumer. Such perceptions are easily understood in the case of Quarg and low fat Fromage frais products which possess clean flavours and are wholesome and nutritious without giving the feeling of satiety. On average, the fat content of these cheeses is less than that of rennet-curd cheeses (Table II); e.g. double Cream cheese which, as an exception in the group, might be considered 'fattening', has only the same fat level (%, w/w) as Cheddar.

TABLE II
Approximate Composition of Various Fresh Cheeses

Variety	Dry matter	Fat	Protein	%, w/w Lactose (lactate)	Salt	Ca (mg/100 g)	pH
Cream cheese							
Double	40	30	8–10	2–3	0·75	80	4·6
Single	30	14	12	3·5	0·75	100	4·6
Neufchatel	35	20	10–12	2–3	0·75	75	4·6
Labneh	25	11·6	8·4	4·3	—	—	4·2
Quarg							
Skim milk	18	0·5	13	3–4	—	120	4·5
Full fat	27	12	10	2–3	—	100	4·6
Cottage cheese							
Low fat	21	2	14	—	—	90	4·8
Creamed	21	5	13	—	—	60	4·8
Fromage frais							
Skim milk	14	1	8	3·5	—	0·15	4·4
Queso Blanco	49	15	23	1·8	3·9	—	5·4
Ricotta							
Whole milk	28	13	11·5	3·0	—	200	5·8
Part skim	25	8	12	3·6	—	280	5·8
Ricottone	18	0·5	11	5·2	—	400	5·3
Mysost							
Gubrandsalost	82	30	11	38	—	400	—
Floteost H$_2$O	80	19	11	46	—	—	—

Compiled from Refs 6–15

However, these cheeses on the whole are relatively low in calcium when compared to rennet-curd cheeses, such as Cheddar (750 mg/100 g) and Swiss (950 mg/100 g);

(v) ease of production of consistent, high quality products on a day to day basis.

The acid-curd cheeses generally have relatively low levels of dry matter, fat, protein and Ca and, because of the relatively high moisture levels, high levels of lactose/lactate (Refs 6–15; Table II). Mysost which may be considered as a (fat/protein enriched) concentrated, ungelled whey, has very high dry matter (82%) and lactose (37%) levels.

2 PRINCIPLES OF ACID GEL FORMATION

Slow quiescent acidification of milk, as effected by the in situ conversion of lactose to lactic acid by added starter culture, or by the addition of acidogens such as glucono-δ-lactone, is fundamental in the manufacture of fresh acid-curd products such as yoghurt and fresh cheeses. Acidification promotes two major physico-chemical changes, i.e. solubilization of micellar calcium phosphate and reduction of the negative charge on casein. These changes (which influence other related physico-chemical changes) confer metastability on the casein system, which, through structural rearrangements, reaches a new stable state in the form of a gel network. Physico-chemical changes induced by acidification will be discussed in detail below.

2.1 Casein Micelle Structure

Examination of milk, by electron microscopy and light scattering, shows that the casein in milk exists in the form of roughly spherical aggregates of colloidal dimensions, which vary in size from ~ 40 to 300 nm, known as casein micelles.[16-21] The micelles on a dry weight basis consist of $\sim 7\%$ ash (predominantly Ca and P), $\sim 92\%$ casein and $\sim 1\%$ minor compounds. The casein essentially comprises four different molecules, namely α_{s1}, α_{s2}, $\beta+\gamma$ and κ which are present at an approximate ratio of $4:1:4:1\cdot3$, respectively.[22]

The casein micelle has been the subject of extensive research, but its structure is still a topic of controversy. There are essentially two proposed models of casein micelle structure, namely the submicellar model[22-24] and the α_{s1}-casein skeleton model.[19,25] In the submicellar model it is proposed[22,23] that:

(i) the micelle is composed of submicelles which are linked together principally by colloidal calcium phosphate (CCP) which binds to the ester phosphate groups of α_{s1}-, α_{s2}- and β-caseins of the submicelles;

(ii) the hydrophobic residues of the individual caseins, i.e. α_{s1}-, α_{s2}-, β and κ are buried in the interior of the submicelle to form a submicellar apolar core. The charged groups, notably the phosphoresine residues of α_{s1}, α_{s2} and β-caseins and the almost phosphate-free but highly polar glycomacropeptide region of κ-casein, surround this core;

(iii) κ-casein rich submicelles are located principally at the outside of the micelle where the protruding glycomacropeptide chains confer stability to the micelle via hydration, charge and steric repulsion and Brownian motion.

Japanese workers[24,26,27] proposed that there are two major types of submicelles, namely F2- and F3-types. F2-submicelles are ~ 20 nm in diameter and consist of α_{s1}- and κ-caseins; F3 submicelles are ~ 10 nm in diameter and are composed of α_{s1} and β-caseins. The more hydrophobic F3 submicelles are located in the core of the micelle whereas the more hydrophilic F2-type are at the surface. Large micelles, with a relatively low surface area to volume ratio, are rich in the F2-submicelles.

In the 'skeleton' model, the micelle, rather than being composed of individual subunits (submicelles), is considered to consist of a continuous framework of α_{s1}-casein molecules to which κ-casein is bound by micellar calcium phosphate; loosely bound β-casein, attached to the framework by hydrophobic bonding, may enter and leave the micelle quite readily.

TABLE III
Various Forms of Calcium and Phosphate in Bovine Casein Micelles

Component	Form	Composition	References
CCP[a]	Complex of Ca, Mg, Pi and citrate linked to casein via phosphoseryl residues	$3Ca_3(PO_4)_2CaHCitr^-$	28,39
	[Ca]/[Pi]: ~1·44 (oxalate titration); ~1·67 (EDTA titration)	$Ca_9(PO_4)_6$	17,23 39 39
Caseinate Ca	Attached directly to casein (Cn) by the carboxyl groups of aspartate and glutamate residues	—	—
Pi[b]	Part of CCP	—	—
Organic phosphate (Po)	Phosphoseryl residues of Cn	—	—
MCP[c]	Complex of CCP and organic phosphate [CA]/[P]~1·1	$Cn-Po-CaHPO_4.2H_2O$	31

[a] Colloidal calcium phosphate
[b] Inorganic phosphate
[c] Micellar calcium phosphate

Both models of micelle structure help to explain some of the physico-chemical and structural changes in milk on slow quiescent acidification.

2.2 Physico-Chemical Changes on Slow Quiescent Acidification of Milk

2.2.1 Dissociation of Colloidal Calcium Phosphate

Bovine micelles consist of ~ 92% casein, 7% calcium and inorganic phosphorous and small amounts of Mg, citrates and other trace elements. Calcium and phosphate are partitioned in a number of forms within the micelle (Table III). Calcium exists in two main colloidal forms, namely colloidal calcium phosphate and caseinate calcium. While the composition and structure of calcium phosphate has been a subject of controversy,[28-31] it is generally accepted that colloidal calcium phosphate (CCP) is an amorphous complex of calcium (excluding caseinate Ca) and inorganic phosphate. CCP, with a Ca/P ratio of ~ 1·5, is considered to be attached to the phosphoseryl residues of α_s-, β- and κ-caseins via calcium bridges. Holt et al.[30,31,32] do not consider CCP as a discrete entity; instead they proposed that micellar calcium phosphate (MCP) with an approximate Ca:P ratio of 1·1:1 consists of CCP and the organic phosphate moiety of phosphoseryl residues.

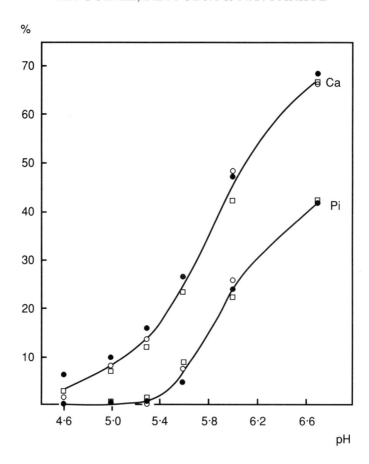

Fig. 3. Micellar calcium and inorganic phosphate in skim milk as a function of pH at 30°C. Micellar Ca and Pi (i.e. remaining with sedimented casein following ultracentrifugation at 88 000 g for 1·5 h) are expressed as % total concentration in milk. (From Ref. 37, reproduced by permission of J.T.M. Wouters, *Netherlands Milk Dairy Journal.*)

CCP progressively solubilizes on acidification of milk; plots of Ca and Pi as functions of pH are typically sigmoidal[33–36] (Fig. 3; Ref. 37). Considerable variation exists as to the pH at which CCP is fully solubilized, pH 5·2–5·3 at 20–30°C[19,37] or 5·0–4·9 at 0–20°C.[38,39] From the data of Dalgleish and Law,[40] it can be inferred, from the constancy of Pi-values of milk ultracentrifuge in the pH region 5·3–5·8, that CCP is fully soluble at pH 5·8 at 20°C. Discrepancies may be at least partially due to differences in methodology and acidification conditions: i.e. temperature, rate of acidification, time allowed for mineral and pH equilibria prior to measurement, acidulants used and their calcium chelating effects, the concentration of Ca^{2+}, and ionic strength of the serum phase. On complete solubilization of CCP, approximately 14–16% of total calcium (3·3–4·0 mole Ca/mole casein) remains associated with the micelles on ultracentrifugation;[35,37]

this calcium fraction may be considered to be part or all of the caseinate Ca. The constancy of the Ca/Pi ratio over the pH range $6.7–5.5^{35,37,39,40}$ together with the pK values of aspartic and glutamic acids (4·1 and 4·6, respectively; Ref. 41) suggest that caseinate Ca is released only after almost complete solubilization of CCP. Further pH reduction to the isoelectric point results in further release of calcium from the micelle but not of (organic) phosphate.[19,36–38,40,41]

2.2.2 Casein Dissociation and Voluminosity Changes

Solubilization of CCP on acidification of milk is paralleled by a temperature-dependent dissociation of individual caseins from the micelle to the serum phase.[37,40,42–45] Casein dissociation (detected by measuring non-sedimentable N) increases with decreasing pH from 6·7 to a maximum at pH 5·1–5·5; further reduction of pH causes a decrease in serum casein which is approximately zero at the casein isoelectric point. The pH of maximum dissociation is temperature-dependent: i.e. ~ 5·6 at 30–35°C,[37,42,43] ~ 5·4 at 20°C[43] and 5·4–5·1 at 4°C.[43–45] In the pH range 6·8–4·8, the extent of casein dissociation increases with decreasing temperature in the range 4–35°C (Fig. 4; Refs 37,42–46). The approximate

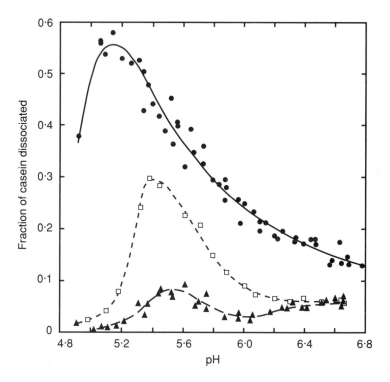

Fig. 4. Total serum casein in skim milk as a function of pH at 4°C (●), 20°C (□) and 30°C (▲). Serum casein (non-sedimentable at 70 000 g for 4 h, 2 h and 1·75 h at 4, 20 and 30°C, respectively) is expressed as a % of total casein. (From Ref. 43, reproduced by permission of F. Hancock, *Journal of Dairy Research.*)

levels of serum casein at the pH of maximum dissociation at 4, 20 and 30°C are 60, 30 and < 10% of total casein, respectively.

While all caseins dissociate on acidification, it is generally agreed that β-casein is the principal component of serum (non-micellar) casein.[37,42–46] As for whole casein, dissociation of individual caseins is strongly influenced by pH and temperature. In the pH range 6·8–5·4, κ-casein, as a percentage of total serum casein, is essentially independent of pH but strongly increases with increasing temperature with values of $\sim$ 8, 15 and 20% of total serum casein at 4, 20 and 30°C, respectively.[43] Further pH reduction towards the isoelectric point reduces the proportion of κ-casein in the serum casein at temperatures > 20°C.[43] α_{s1} and β-caseins exhibit opposite dissociation behaviour in the pH ranges 6·8–6·0 and 5·2–4·8 at 4 and 30°C (Fig. 5, Ref. 43). As for κ-casein, dissociation of α_{s1}- and β-caseins is essentially unaffected by pH in the range 6·0–6·8, but is strongly dependent on temperature with their proportions in serum casein being 8 and 75, 15 and 55, and 25 and 40% at 4, 20 and 30°C, respectively.[43]

At the pH of maximum casein dissociation (i.e. $\sim$ pH 5·4) at 20°C, the incubation temperature normally used for fresh cheese production, serum casein amounts to $\sim$ 30% total casein with α_{s1}, β and κ representing $\sim$ 30, 45 and 15%, respectively.

The increase in casein dissociation with decreasing pH and temperature (to a maximum) and the fact that β-casein is the principal component of serum casein over the pH range 6·8–4·8 may be attributed to:

(i) solubilization of the casein cementing agent, CCP, the effect of which may be considered as the predominant factor promoting casein dissociation at relatively high temperatures (i.e. > 20°C);

(ii) the decrease in the hydrophobic interactions between the caseins as the temperature is lowered[47,48] which augments the effect of CCP solubilization at low temperatures (e.g. 4°C).

The decrease in casein dissociation from a maximum at pH $\sim$ 5·4 to a minimum at the isoelectric point may be attributed to the overriding influence of charge reduction,[19,49,50] which is at least partly due to Ca^{2+}, the concentration of which increases on solubilization of CCP.

2.2.3 Reduction of Casein Charge and ζ-Potential

Adsorption of ions and ionization of amino acid residues on the surface of the micelle at normal pH, i.e. $\sim$ 6·7 at 20°C, give the micelle a net negative charge or –8 to –20 mV.[49–54] This charge, which confers stability upon the micelles by electrostatic repulsion, is pH and temperature dependent. ζ-Potential decreases gradually from pH 6·7 to 6·0, and sharply from pH 6·0 to a minimum at pH $\sim$ 5·4, but then increases to a maximum at pH $\sim$ 5·1, below which it decreases, reaching $\sim$0 mV at the casein isoelectric point (Fig. 6; Refs 41,49,50). Considering the protein component of micelles, a progressive decrease of ζ-potential would be expected with decreasing pH[50] at least to pH of $\sim$ 5·6, where glutamic acid (anionic group present at a relatively high concentration of 50 mM) begins

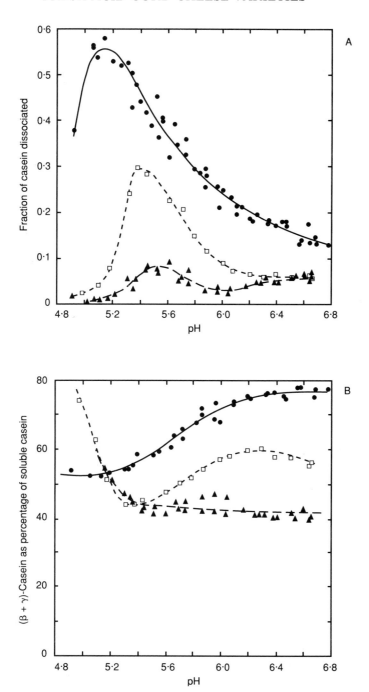

Fig. 5. Concentrations in the serum of α_{s1}-casein (A) and $\beta+\gamma$-casein (B) as a function of pH at 4°C (●), 20°C (□) and 30°C (▲). (From Ref. 43, reproduced by permission of F. Hancock, *Journal of Dairy Research.*)

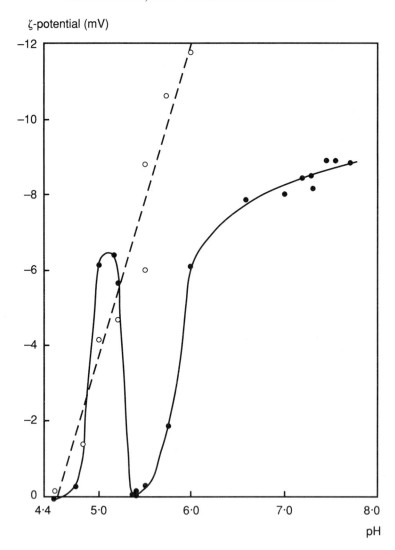

Fig. 6. Dependence on pH of the ζ-potential of washed unheated casein micelles (—) and calcium phosphate precipitates (– –). (From Ref. 50, reproduced by permission of J.T.M. Wouters, *Netherlands Milk Dairy Journal.*)

to become protonated (pK = 4·6, Ref. 41). Based on calcium adsorption observations by Ottenhof,[55] Schmidt & Poll[50] explained the minimum in the ζ-potential-pH curve on the basis of superimposition of charge neutralization by Ca^{2+} adsorption on charge neutralization by protonation. At pH 5·3–5·4 (minimum in the ζ-potential-pH curve), Ca^{2+} adsorb specifically to micelles whereas at pH > 5·4 they bind preferentially to inorganic phosphate; at pH < 5·3, no adsorption was found.

2.2.4 Solvation, Voluminosity and Size Changes

Solubilization of CCP and casein on acidification of milk are paralleled by changes in related parameters, i.e. solvation (hydration), voluminosity (v) and casein micelle size. Based on pellets obtained on ultracentrifugation (88 000–198 000 g), solvation is defined as: g H_2O/g protein[37,46,56] or g H_2O/g dry pellet[19,36] and voluminosity as the specific volume of the micelles (i.e. ml/g pellet, Ref. 57) or ml/g protein.[37,45] Reported voluminosity values for bovine casein micelles at the pH of milk vary widely, i.e. v = 1·5 to 7·1, depending on the method employed, but values obtained using any particular method agree closely.[58]

Solvation at 20°C decreases in the pH range 6·6 to 6·0–6·1, then increases gradually to a maximum at pH 5·3–5·4, below which it decreases to a minimum at the isoelectric point (Fig. 7; Refs 36,37,46,56). At the natural pH of milk, decreasing temperature and calcium concentration (by replacement of Ca with Na using ion exchange) result in an increase in solvation.[36,46] Rennet action reduces solvation in the pH range 6·6 to 4·6.[36,37]

The relationship between casein solvation and pH may be explained by the dominance of one of two opposing forces over the pH region, i.e. those promoting charge neutralization and micelle contraction on one hand and those promoting an increase in micelle porosity on the other. In this context, increased porosity is considered to be the 'loosening' of micelles (decreasing density of the micellar network) following solubilization of CCP and casein on acidification.[37,45,46,57] It is envisaged that newly created micellar interstitial space, resulting from CCP and casein dissociation, becomes partially occupied by serum; the uptake of serum, while increasing the solvation of the micelles, would not

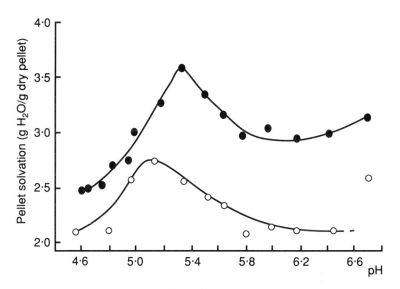

Fig. 7. Solvation of casein micelles as a function of pH at 20°C for skim milk (●) and rennet-treated skim milk (○). (From Ref. 36, reproduced by permission of H.W. Kay, *Milchwissenschaft*.)

significantly alter their size.[57] Measurements based on electron microscopy[25,45,57] show little change in micelle size on acidification to pH 5·5 while those based on ultracentrifugation (i.e. voluminosity, ml/g^{-1}) show a progressive reduction with pH decrease in the same range.[57] This apparent decrease is considered to be an artefact and effected by compaction of the micelles on ultracentrifugation.[57] Structural changes that occur within the casein micelles on further pH reduction are unclear; at pH 5·2–5·3 at 20–30°C the micelles appear, from electronmicroscopial observations, to disintegrate at first into smaller units (probably submicelles and/or new casein particles formed from aggregation of dissociated caseins) which then aggregate into a continuous gel structure.[19,25,45]

On acidification from pH 6·6 to 6·0, at 20–30°C, charge neutralization may be considered to be dominant and solvation decreases. The decrease in solvation with charge neutralization, is supported by: (i) the reduction of micelle solvation by rennet action,[36,37] which reduces the ζ-potential of micelles;[51,53,54] (ii) the inverse relationship between micelle solvation and Ca^{2+}.[46,59,60] Acidification of milk from pH 6·6 to 6·0 at 20–30°C results in ~30–40% solubilization of CCP, which causes no significant (<5%) dissociation of casein[43] and hence no envisaged large increase in micelle porosity. Further acidification to pH 5·5 (where casein dissociation and CCP solubilization are close (85%) to maxima) not only results in further charge reduction[19,25,49,50] but in a large increase in casein dissociation[40,43,59] and hence in micelle porosity.[44] In this region, forces promoting increased micelle porosity may be considered to override the solvation-decreasing effects of charge neutralization and hence solvation increases sharply. Casein solvation increases to a maximum at pH ~5·3, i.e. the pH at which CCP is fully solubilized at 20–30°C.[19,37] The fact that the pH maximum for casein solvation (~5·3) is slightly lower than that for casein dissociation (~5·4–5·5) suggests that CCP, even at a low level, has a strong influence on micelle porosity: at pH 5·5, ~25% of the CCP is insoluble and remains in the micelle.[37]

In the pH range 5·3–4·6, charge neutralization and increased hydrophobicity probably lead to a contraction of the casein particles which reduces interstitial space and expels serum. Indeed, because of the increased hydrophobicity in this region, dissociated caseins reversibly associate with the original altered casein 'micelles'[25] and/or aggregate to form new casein particles.[44] The influence of hydrophobicity on solvation is further supported by the inverse relationship between heating (which effects an increase in hydrophobicity[47,48]) and casein micelle solvation (Fig. 8; Ref. 46).

2.3 Structural Changes

Electronmicroscopic studies show that slow quiescent acidification of milk from pH 6·6 to 4·6 is accompanied by a number of concerted structural changes which are pH-related.[19,25,44,61–66] For fresh or pasteurized milk (72°C at 15 s) acidified at 20–30°C, these may be summarized as follows:

   (i) at pH values > 5·6, no major changes are observed; the individual micelles retain their shape, dimensions and integrity;[19] a wide spectrum of

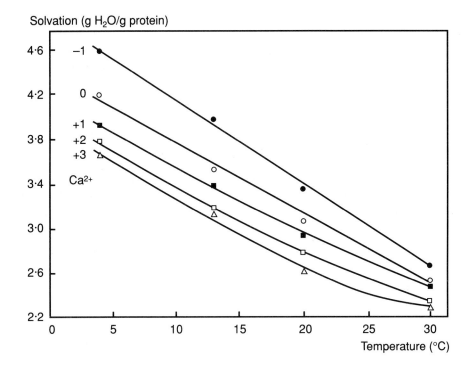

Fig. 8. Solvation of casein as a function of temperature for reconstituted skim milks with reduced −1 mmol/kg (●) and increased +1, (■); +2, (□) and + 3, (Δ) mmol/kg calcium levels. The calcium levels were reduced by ion exchange and increased by the addition of $CaCl_2$. The pH value of all milks was readjusted, after calcium alteration, to that of the control sample (○), i.e. pH 6·7. (From Ref. 46, reproduced by permission of H.W. Kay, *Milchwissenschaft.*)

particles with an average diameter of 100 nm, are present. Vreeman *et al.*[57] concluded that acidification from pH 6·6 to 5·6 results in a reduction in the number of the smallest micelles (due to dissociation to extremely small particles undetectable by electron microscopy and/or aggregation to, or condensation onto, larger micelles) whereas the average micelle size remained practically constant;

(ii) at the pH of maximum casein dissociation (~ pH 5·5), the micelles become more porous as a consequence of casein dissociation;

(iii) on further reduction of pH to 5·2, i.e. where practically all CCP is solubilized, smaller particles, in addition to the 'original' micelles, become increasingly more visible. These particles probably originate from the aggregation of already-dissociated caseins.[19,44,67] However, it is also conceivable that the smaller particles may correspond to sub-micelles lost from the micelles (especially from the outside) on complete solubilization of the cementing agent, CCP;[44]

(iv) at pH ~ 5·2, gelation commences with the formation of casein aggregates which appear as relatively large corpuscular structures. Aggregation is evident from the simultaneous appearance, on electronmicrographs, of large fields containing casein particles of widely different size and fields devoid of casein particles. Based on previously described physico-chemical changes, the unhomogeneously distributed casein aggregates may be considered to consist of a mixture of original altered micelles, altered micelles to which dissociated caseins have reattached, and new casein particles;

(v) on further pH reduction, the casein-rich areas contract and the aggregates touch to form short dangling pieces of network which, as the isoelectric point is approached, touch and crosslink to give a three-dimensional particulate gel network which extends, more or less continuously, throughout the serum phase.

Many factors (milk pasteurization and homogenization, rennet addition, incubation temperature and rate of acidification) influence not only the final gel structure but also the gelation rate, the degree of structure at different pH values, and the pH at which gelation is initiated.

2.4 Gelation

Slow quiescent acidification of milk is accompanied by two opposing sets of physico-chemical changes:

(i) a tendency towards disaggregation of the micelles into a more disordered system as a result of:

(a) solubilization of the internal micellar cementing agent, CCP, which at 20–30°C is fully soluble at pH ~ 5·2–5·3 (c.f. section 3.2.1);

(b) a pH and temperature dependent dissociation of individual caseins, especially β, from the micelles with a parallel increase in serum casein concentration.

(c) an increase in micelle solvation and porosity,[57] as a consequence of (a) and (b), over the pH range 6·7 to 5·3–5·4. (c.f. section 2.3).

(ii) a tendency for the casein micelles to aggregate due to:

(a) the reduction of negative charge on the casein molecules and increased hydrophobicity over the pH range 6·7–4·6, and the reduction in solvation from pH 5·4 to the isoelectric point;

(b) the increase in the ionic strength of the milk serum (due to solubilization of CCP and the concomitant increase in the concentrations of Ca^{2+} and phosphate ions) which has a shrinking effect on the matrix of casein micelles.[57] This effect is enhanced by protonation of the (second) ester phosphate groups of phosphoserine residues (pK 6·5) on acidification. Ionic strength increases over the whole pH region (i.e. 6·7–4·6) of calcium dissociation (both CCP and caseinate Ca) from the micelles.

Both the above factors, i.e. (ii)(a) and (ii)(b), reduce the repulsive forces and increase the cohesive attractions between micelles.

Above the pH value at which gelation begins, i.e. $\sim 5\cdot1$–$5\cdot3$ at 20–30°C,[25] dis-aggregating forces predominate and hence a gel is not formed, while below this pH, forces that promote aggregation override the disaggregating forces and gel formation proceeds. (At some pH slightly greater than at which the onset of gelation occurs, due to structural rearrangement of the physico-chemical altered caseins, the casein system may be considered as being metastable, Fig. 9.)

2.5 Prerequisites for Gel Formation

A gel is formed when dispersed particles aggregate to such an extent that they form a continuous network throughout the liquid in which they are dispersed. Acid casein gels may be considered as particulate network gels, i.e. they consist of overlapping, cross-linked strands which are composed of particles (i.e. casein aggregates) linked together by various types of bonding.[19,25,44,61,62,64–66,68–73]

Aggregation and structural rearrangement of casein, as a consequence of the acidification-induced physico-chemical changes, may result in the formation of a gel, as described, or in a precipitate, depending on the extent of aggregation. Gelation occurs when aggregation forces slowly overcome repulsive forces, resulting in the formation of relatively loose porous hydrated aggregates with a small density gradient between themselves and the serum phase in which they are dispersed. Because of the relatively small density gradient, the aggregates have sufficient time to knit together, via strand formation, to a continuous network before, otherwise, separating out of dispersion. When conditions that promote aggregation are more extreme (i.e. rapid acidification under non-quiescent conditions at high temperature), casein particles aggregate more rapidly to form smaller, less porous and less hydrated aggregates, giving a higher density gradient between themselves and the surrounding serum phase. Owing to the high density, the aggregates fall out of the dispersion as a precipitate which lacks the matrix continuity and waterholding characteristics of a gel.[71]

To obtain a gel rather than a precipitate, the number of attractive forces, and hence the surface area of contact, between the dispersed particles must be limited.[74,75] Such limited inter-particle attractions are promoted by the correct ratio of attractive to repulsive forces between the conformationally-rearranged casein particles[71] as effected by the desired rate of concerted physico-chemical changes.

The physico-chemical changes are responsible for structural rearrangement of the casein particles (i.e. as effected by solubilization of casein and CCP and the partial exchange of hydrophilic and hydrophobic regions at the micelle surface, cf. Ref. 76) and promoting the aggregating forces (such as hydrophobicity and charge reduction) which lead to their interaction and aggregation.

As the number of inter-particle attraction points is increased by increasing the rate of acidification, for example, the resulting gel becomes less structurally organized, coarser, less voluminous and closer to a precipitate in organization and structure. Alternatively, if the number of inter-particle attraction sites is

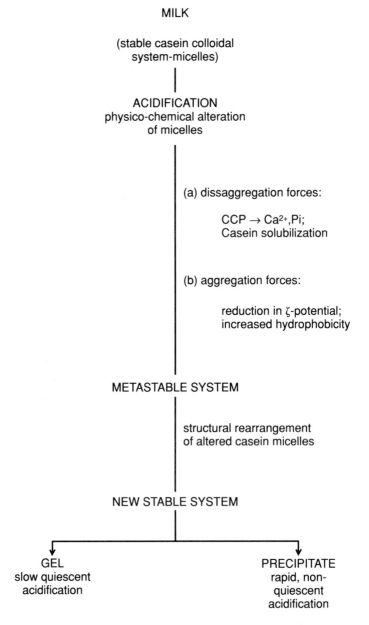

Fig. 9. Influence of acidification on milk.

lower than optimum, slowly-forming aggregates may have sufficient time to precipitate before they fuse to form into strands of a network. An example of the latter is the defect in Cottage cheese production known as 'major sludge formation' whereby phage infection of the starter, after acid development has pro-

(A)   (B)

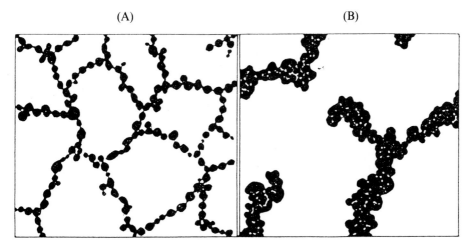

Fig. 10. Diagram of a fine structured yoghurt gel (A) made from heated (90°C for 5 min) milk and a coarse structured yoghurt gel (B) made from unheated milk. (From Ref. 61, reproduced by permission of J.J. O'Neil, *Journal of Texture Studies.*)

gressed to an advanced stage (~ pH 5·2–5·3), leads to precipitation, rather than gelation, of the entire casein system.[77]

From the foregoing it may be concluded that, within the context of slow quiescent acidification (e.g. bacterial fermentation of milk), as aggregation proceeds more slowly, a more highly ordered, narrow-stranded, highly branched network with small interstitial spaces between the network strands is formed. Alternatively, if protein–protein interactions occur rapidly the resulting gel has thicker strands, larger interstitial spaces and less continuity, i.e. it is a coarser gel network.

Consideration of Fig. 10 (where the concentrations of gel-forming protein in A and B are equal[61]) shows that the micelles in A have formed into thin strands (chains) giving a highly branched, continuous gel network. On the other hand, the micelles in B have fused to a much greater degree giving thicker strands and a gel which is more discontinuous than A and which has larger interstitial spaces. Gel A may be described as being finer and more voluminous than B. The finer more continuous structure of gel A results in a more even distribution of the gel-forming protein, smaller interstitial spaces, a superior water holding capacity and less tendency to spontaneous syneresis[61,78] (i.e. syneresis in the absence of externally applied pressure; cf. section 5). In gel B, whey can seep more easily through the large open channels between the network strands as a consequence of matrix contraction on ageing.

3   FACTORS THAT INFLUENCE GEL STRUCTURE

Gel structure, i.e. degree of fineness/coarseness, is central to the quality of fresh cheese products, especially cold-pack products such as Quarg and Fromage frais where the gel after whey separation receives no further treatments. It is a major determinant of quality aspects such as mouthfeel (i.e. smoothness), appearance (coarseness/smoothness) and physico-chemical stability (i.e. absence of wheying-off and graininess during storage). Many factors influence the structure of acid milk gels, as discussed below.

3.1 Level of Gel-Forming Protein

For a given protein type and degree of gel fineness, high levels of gel-forming protein result in a denser [greater number of strands (of equal thickness) per unit volume], more highly branched network which has a greater degree of overlapping of strands and a narrower pore size.[61,79,80] However, for a gel with a given level of protein, all other conditions being equal, the structure is strongly influenced by the proportions of different proteins (i.e. caseins: whey proteins) present.[72,79.80] Reducing the casein to whey protein ratio (from 4·6 to 3·2) results in set yoghurt with a finer, more highly branched, narrower-pore sized network which has a smoother consistency and which is less prone to spontaneous wheying-off due to its superior water-holding capacity as effected by the smaller pore size.[61,79,80]

Factors which lead to an increase in the effective protein concentration include:

(i) fortification of milk with protein stabilizers, as practised in the production of yoghurt and Fromage frais;

(ii) homogenization of fat-containing milk, as practised in yoghurt and Cream cheese production,[81] which results in conversion of fat globules to pseudo-protein particles;[41,82,83]

(iii) high heat treatment which effects denaturation and binding of whey proteins to casein;[84-88] undenatured whey proteins are soluble and do not participate in gel formation.

3.2 Heat Treatment of Milk

It is well known in practice that high heat treatment of milk, prior to culturing, in the manufacture of stirred-curd fermented milk products (yoghurt, Fromage frais, Quarg, low-fat fresh cheeses) results in a smoother and thicker consistency, although the level of the effect varies considerably with the type of preheating, i.e. whether in-vat (85°C for 10–40 min), high temperature-short time (98°C for 0·5–2·0 min) or UHT (140°C for 2–8 s) treatments.[89-92] At a molecular level, high heat treatment (90–95°C for 2–5 min) effects a high degree (> 80%) of whey protein denaturation,[89-94] a greater effective concentration of gel-forming protein and a finer-structured gel network with reduced propensity to spontaneous wheying off (Fig. 11; Refs. 61,62,64,65,89,90–93). The above effects are attributed to the formation, via disulphide interaction, of a complex between κ-casein (at the micelle surfaces) and denatured β-lactoglobulin and/or α-lactalbumin[84,95,95] and the aggregation of denatured whey proteins.[97] The κ-casein-whey protein complex

(A)

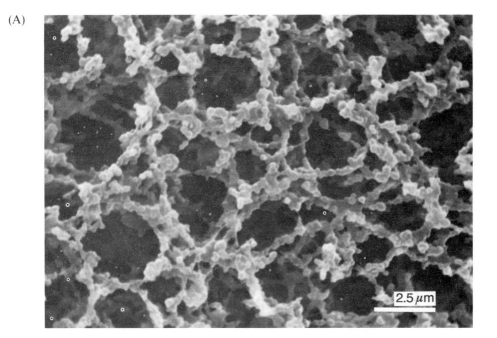

(B)

Fig. 11. Scanning electron microscopy of yoghurt made from unheated skim milk (B) and skim milk batch-heated to 90°C (A). (Reproduced by permission of M. Kalab.)

results in the formation of filamentous appendages which protrude from the micelle surfaces.[19,25,61,62,66,79] On subsequent acidification and gel formation, these appendages lead to the formation of a 'core and lining' structure [i.e. a lining of compressed β-lg-κ-casein appendages surrounding the micelle (core)] which prevents the close approach, and hence large-scale fusion, of micelles due to steric hindrance.[44,66,98] The higher effective protein concentration, in combination with the above physico-chemical/structural changes, results in a more controlled, limited aggregation of the network building units and hence a finer, more branched and narrower-pore gel which has a smoother consistency and is less prone to spontaneous syneresis than gels made from unheated milk.

An additional explanation for the finer gel network formed from high-heat treated milk is that the whey protein–casein micelle interaction results in a smaller decrease in ζ-potential[50] in the pH range 6·0–5·4 which leads to an increase in the ratio of disaggregating/aggregating forces during the earlier stages of acidification. However, high heat treatment results in the initiation (onset) of gelation at a higher pH value irrespective of whether acidification is induced by lactic culture fermentation or by glucono-δ-lactone.[25,65,99,100] In yoghurt produced at 43°C using a mixed culture of *Lactobacillus delbrueckii* subsp. *bulgaricus* and *Streptococcus salivarius* subsp. *thermophilus,* the onset of gelation for high-heated (90°C for 15 min) and unheated milks occurred at pH values of 5·35 and 5·1, respectively.[65]

The increased rate of gelation (for equal rates of acidification), despite the higher ratio of disaggregating to aggregating forces in the pH region 6·0–5·4, probably occurs due to an increased hydrodynamic radius of the casein micelles in the heated milk as effected by the protein lining. A higher dynamic radius would result in a higher volume fraction of the micelles which would in turn, all other conditions being equal, lead to a higher probability of the micelles touching.

3.3 Incubation Temperature

Increasing the incubation temperature during acidification results in: (i) the onset of gelation at higher pH values,[25] e.g. pH 5·5 at 43°C compared to 5·1 at 30°C, (ii) coarser gel networks which are more susceptible to spontaneous wheying-off on storage.[63,78] These effects may be attributed to a higher ratio of aggregating to disaggregating forces during the earlier stages of acidification owing to decreased casein dissociation from the micelles,[42–44] a reduction in repulsive forces due to increased hydrophobicity[101] and a faster rate of acidification (subject to the type of bacterial culture). The reduced contribution of decreasing ζ-potential (which increases in magnitude with temperature in the range 6–45°C over the pH region 6·7–5·4, Refs 49,50,53,54) to aggregation is probably more than offset by the decrease in micellar casein dissociation and the increase in casein hydrophobicity.

In model experiments, where raw skim milk was acidified at 0°C (by various acids to give pH values of 4·6 or 5·5 at 22°C) and subsequently heated under quiescent conditions to 40°C or 90°C, Harwalkar & Kalab[62] found that setting at the higher temperature (90°C) gave finer gels which were less susceptible to syneresis (on centrifugation) than those obtained on incubation at 40°C. These observations may at first sight appear contrary to those mentioned above. However, in the experiments of Harwalkar & Kalab,[62] whey protein denaturation (and denatured whey protein–κ-casein interaction) at the higher temperature setting of 90°C, which did not occur at the lower temperature (40°C), is undoubtedly responsible for the finer gel network at the higher temperature (90°C; Refs 61,66). In conclusion, within the normal incubation temperature range used in fresh acid cheese production (i.e. 20–45°C) an increase in incubation temperature promotes coarser gel structures; however, increasing the setting temperature to very high values, e.g. to 90°C, may or may not, depending on the level of milk heat treatment prior to incubation, rate of acidification, type of acidulant used, and level of whey protein denaturation effected by the incubation temperature *per se*, promote finer gel structures.

3.4 Rennet Addition

It is common practice during the manufacture of some fresh cheese products, such as Quarg, Fromage frais, Cottage cheese and low-fat fresh cheeses, to add a small quantity of rennet (0·5–1·0 ml single strength per 100 litres) to the milk shortly after the start of fermentation (i.e. ~ 1–2 h after culture addition when pH is ~ 6·1–6·3). Rennet action promotes the following physico-chemical changes: (i) hydrolysis of κ-casein with a concomitant decrease in ζ-potential[51,53,54,102] and (ii) a decrease in casein dissociation and in micelle solvation[36,37] over the pH region 6·6–4·6. Both these changes contribute to an enhanced aggregation of micelles and hence an increase in the ratio of aggregating to disaggregating forces during the early stages of acidification. Hence, gelation begins at a higher pH and a gel firm enough for cutting and separation is obtained at pH 4·8–4·9. In the absence of rennet, the gel is cut at pH 4·7 in order to obtain a gel of satisfactory firmness and to prevent excessive loss of fines (casein) on whey separation.[10,13]

3.5 Rate of Gelation

For a given type and level of gel-forming protein, increasing the rate of gelation results in the onset of gelation at a higher pH value, and a coarser network with a greater propensity to spontaneous syneresis.[61–63,102] Gelation rate is increased by an increased rate of acidification[61,103] as effected by increased levels of starter and/or acidulant or increased incubation temperatures in the range 20–45°C (provided starter growth is not inhibited; Refs 25,103). Indeed, rapid acidification of milk to pH 4·6 promotes rapid aggregation of casein with the formation of large dense aggregates which precipitate rather than form a gel (Fig. 9, Ref. 103). However, gelation by rapid acidification is possible when the tendency of micelles to coagulate is reduced, e.g. by acidifying ~ pH 4·6 at low temperatures (0–4°C) and subsequently heating slowly (~ 0·5°C/min) under quiescent conditions to ~ 30°C.[44,62]

4   RHEOLOGICAL ASPECTS

As discussed in section 2, the basic building units of acid casein gels are aggregates of various sizes which consist of fused casein micelles/particles. During the course of gel formation, these aggregates fuse together to greater or lesser degrees to form initially relatively short strands (i.e. onset of gelation). These strands, which are particulate because of the limited area of contact between the aggregates, grow in length and overlap to form a network.[25,44,72,74,79] Viewed by electronmicroscopy, acid milk gels appear to contain dense, protein-rich areas inter-connected by strands which are only a few particle diameters thick (i.e. 100–400 nm); the large cavities between the strands and conglomerates represent the matrix pores which are filled with serum.[44] The dense casein areas, which represent the areas of strand overlap, may be considered as loose porous conglomerates (of casein aggregates), being many times thicker than the interconnecting strands. Hence, regarding the spatial distribution of the matrix building material, acid gels, as indeed rennet gels, may be considered as being quite heterogeneous.

Application of a small stress (i.e. $\ll$ yield stress) to such a gel results in recoverable stretching of the matrix strands; if the stress is increased to a value above the yield value, the strands are broken, at least temporarily. The response of the gel matrix to stress is dependent on many factors:

(i) the number of strands per unit area. Considering a gel to which a relatively small stress (i.e. much less than yield value) is applied in the direction x, the elastic modulus (i.e. ratio of stress to strain, δ/γ), G, can be related to the number of strands per unit area according to the equation of van Vliet & Walstra:[104]

$$G = CN \, d_2A/dx^2$$

where: N = number of strands per unit of stress-bearing area in a cross-section perpendicular to x;

C = characteristic length determined by the geometry of the network;

dA = change in Helmholtz energy when the particles in the strands are moved apart by a distance dx;

(ii) gel homogeneity. The homogeneity of the gel determines the number of stress-bearing strands. For a coarse gel network, for example, the number of stress-bearing strands is less than in a fine gel; the thickness, and hence the strength, of the stress-bearing strands will be, however, on average, greater in the coarser gel;

(iii) the number and type of bonds between the basic building units, i.e. aggregates, within a strand. Various bonding types between the aggregates are possible, i.e. hydrophobic bonds, van der Waals' attractive forces, electrostatic interactions, hydrogen bonds and steric interaction.[44,73,105] The role of these different types of bonds will vary with gelation conditions; the exact contribution of the different types to gel structure is unknown. The smaller the number and weaker the bonds between the aggregates of a matrix stand, the more susceptible the strand is to deformation. For a given type of inter-aggregate attractive force, the strands of a fine gel may be considered as having less stress-bearing potential than those of a coarse gel.

The above structural parameters of a gel are influenced by many factors such as milk composition (levels of fat and protein), processing parameters (heat and homogenization treatments of the milk) and conditions of gel formation (incubation temperature, rate of acidification, addition of rennet, final pH value, etc.). These factors determine the magnitude of the physico-chemical changes which in turn regulate structure formation (section 3).

Various methods have been employed to assess the rheological characteristics of fresh acid-curd products. Viscometric studies have been used to measure viscosity and stress (τ) as a function of shear rate ($\dot{\gamma}$) for fermented milk[63,89–92,106] and soft fresh cheese (12–20% DM, 7–10% protein, Refs 107,108). Measurement of stress during penetration by penetrometers of various types has been applied to acid milk gels (fermented or acidulated, Refs 62,92,109), yoghurt[80,81,110] and low (19% DM, 6·6% protein), medium (26% DM, 10·4% protein) and high solids (36% DM, 10·3% protein) fresh cheeses and Cottage cheese gels (before cutting, Refs 90,100,111). Using compression tests, the force required to compress the cheese to various levels, yield value, adhesiveness and/or other parameters have been measured for Cream cheese,[112,113] Cottage cheese[114] and acidified skim milk gels.[103]

To obtain more precise information on the rheological characteristics of viscoelastic systems, such as milk gels, it is necessary to perform the measurement with as little damage as possible to the sample. Most techniques, e.g. those mentioned above, for rheological analysis of gels cause sample damage, e.g. the transfer of a gel to the measuring unit results in breakage and syneresis. Moreover, during measurement the samples are subjected to relatively large deformations which disrupt the gel network and hence the above techniques do not provide an exact representation of the rheological behaviour of the natural, unaltered gel.[44,115–118] Hence, instruments (e.g. Den Otter and Bohlin Rheometers; low-shear 30 Son Oscillatory Viscometer) have been developed recently which provide more precise rheological data on gels; the gel is formed within the measuring instrument and subjected to low strain (< 30 millistrains), via application of low amplitude oscillations, which minimize structural damage to the gel. These instruments have been used to measure the elastic or storage modulus, G', and viscous or loss modulus, G" of acid milk gels.[44,105]

It is difficult to compare the rheological data reported by different authors for acid milk gels, because of differences in instrumentation, measuring conditions (temperature, rate of shear, compression ratio, degree of sample destruction) and the gel sample (composition, pH, degree of gel fineness/coarseness). However, certain trends on the rheological behaviour emerge from these investigations. Acid gels, like rennet gels, show viscoelastic behaviour, i.e. they possess the characteristics of both solids and liquids in that they flow on application of stress and partially relax on removal of the stress.[41,44,119] Acid milk gels, including fermented milks (natural stirred yoghurts) and soft fresh cheeses (12–20% DM, 7–10% protein, 0–40% FDM), exhibit pseudoplastic behaviour, i.e. they undergo shear thinning with increasing shear rates.[63,90–92,107,108] Moreover, the viscosity decreases with shearing time at fixed shear rates. The decrease in viscosity with shearing rate and time indicates that the relatively weak gel network is progressively broken down on prolonged shearing to strands, to strand pieces and possibly, eventually, to aggregates. Indeed, Korolczuk and Mahaut[107] found that the viscosity of soft fresh cheese (12·5% DM, 8·5% protein) decreased progressively over 48 h at 4°C; unfortunately, no electromicroscopic study was undertaken to observe structural changes induced by shearing.

Various factors influence the rheological properties of acid milk gels and fresh cheeses. Increasing protein concentration, for a given rate of gelation, results in firmer and more continuous acid gels.[41,44,80,100] However, for a given level of gel-forming protein, increasing the casein to whey protein ratio results in coarser and firmer gel networks.[72,80] Such an effect may be significant when choosing protein supplements for the manufacture of yoghurt, Fromage frais, etc. For a given level of solids, increasing the fat content gives a weaker gel because of the reduced level of gel-forming protein and the physical interference of the fat globules with the protein network. However, if the milk (or cream) is homogenized, the fat globules become partially covered with casein and whey proteins and are converted to pseudoprotein particles which then participate in network formation. Increasing the effective protein concentration in this way results in firmer gels.[41,82,93] Homogenization is practised in the manufacture of fresh cheese products where the milk protein level is marginal e.g. yoghurt and single Cream cheese.

The viscosity of soft fresh cheeses,[107] the firmness of natural, fortified yoghurt[72,80] and the storage modulus, G', of acidified reconstituted caseinate and skim milk powder gels[44] increase with increasing protein concentration. For a gel of a given degree of fineness, increasing the level of gel-forming protein may be considered as an increase in the number of stress-bearing strands per unit area. High heat treatments have a strong influence on gel structure and structure-related parameters, e.g. rheology and syneresis. It is generally accepted that high heat treatment of milk results in an increase in the shear stress (over a range of shear rates), viscosity and firmness of low-solids acid-curd products, e.g. natural yoghurts.[62,89,93,97,109] [However, Kim & Kinsella[103] found that preheating (90°C for 15 min) of milk caused a decrease in firmness and an increase in the shear stress of GDL-acidified milk gels.] This effect is attributed to the high level of whey protein denaturation and binding of denatured whey protein to the micelles which effects an increase in the level of gel-forming protein, and a finer, more highly branched continuous network. The number of stress-bearing strands (though possibly weaker) per unit volume of such a gel would be greater than in the coarser, lower-density matrix, gel of unheated milk. The extent and method of milk heat treatment, for similar levels of whey protein denaturation, appears to have a significant influence on the textural parameters of fermented milks. For similar levels of whey protein denaturation, ultra-high temperature treatments (130–150°C for 2–15 s) give natural yoghurts of lower shear stress and firmness than high temperature-short time (~ 80–90°C for 0·5–5 min) treatments which in turn give lower values than those obtained for yoghurt made using batch (63–80°C for 10–40 min) heat treatments.[89-92] While there is generally a strong positive correlation between the viscosity and firmness (i.e. resistance to penetration) of yoghurt as functions of whey protein denaturation, variations[90,93] for similar levels of denaturated whey protein may be due to the different rates of whey protein denaturation which alter their binding to casein micelles and hence the structure of the gel.[97]

Higher incubation temperatures, all other factors being equal, promote coarser gel networks (in fermented milks) which have higher viscosities and firmness values.[41,62,63,103] It is possible that in the coarser gel, the higher stress-bearing capacity of the relatively thick strands overrides the effects of a greater gel discontinuity which would, otherwise, be expected to give a weaker, more easily deformed network.

Increasing the rate of acidification results in the onset of gelation at higher pH values and firmer gels.[62,103]

The firmness of acid milk gels increases with decreasing pH towards the casein isoelectric point;[109] this is attributed to a greater number of bonds with greater strength between the strand-forming aggregates;[45] the maximum storage modulus, G', is at pH ~ 5·4.[41,44]

5 SYNERESIS

Syneresis, i.e. the outflow of the serum phase from a gel, requires rearrangement of the gel matrix into a more compact structure as effected by the breaking of bonds within strands and the consequent formation of new bonds.[75] Rearrangement necessitates stress (for strand breakage) which can be caused: (i) externally by the application of pressure, e.g. centrifugation, gravity, stirring and/or cutting; (ii) internally by the spontaneous breaking of gel strands due, possibly, to micelle flow and thermal motion of gel strands.[75] Shrinkage of the casein particles of the network, as induced by a reduction in pH and/or increase in temperature following gel formation, may enhance both types of syneresis. Shrinkage and compaction of the matrix exerts a pressure on the entrapped moisture and forces it out through the matrix pores.[120] Simultaneously, the outward flow of moisture is impeded by the sieve effect of the pores which becomes increasingly greater as progressive structural rearrangement leads to further matrix contraction. Because of the reduced syneresis for a given ΔP, due to the sieve effect, and the decrease in the net pressure acting on the entrapped liquid (due to the counteraction of syneretic pressure by elastic reaction forces building up in the matrix), the rate of outward migration of moisture decreases with time.

For unidimensional flow through a porous medium, such as an acid milk gel, the rate of syneresis may be expressed by Darcy's law:[44,75]

$$v = B\Delta P/\eta l.$$

where: v = liquid flux (i.e., volume flow rate in the direction, l, divided by the cross-sectional area perpendicular to l through which the liquid flows) (ms^{-1});

B = permeability coefficient of the matrix (m^2), which corresponds to the average cross-sectional area of the pores;

η = viscosity of liquid flowing through the matrix (Pa.s);

ΔP = pressure gradient, arising from syneretic pressure exerted on the entrapped moisture by the matrix (Pa);

l = distance over which the moisture flows (m).

The permeability coefficient, B, depends on the volume fraction, Ø, of the protein matrix and the spatial distribution of the matrix strands (i.e., gel fineness/coarseness). The greater the permeability coefficient, the less is the resistance to the flow of moisture through the gel for a given syneretic pressure.

In the manufacture of fresh acid cheeses such as Quarg, Labneh, Cream cheese and Cottage cheese, the gel, following incubation, is concentrated (i.e., whey removal) by cutting, stirring, cooking, whey drainage, and/or mechanical centrifugation. However, for other fresh cheese-type products, e.g. Laban, Fromage frais, set yoghurt and fresh cheeses made by recombination technology (i.e. standardizing to the desired product DM prior to culturing) the gel, which may be packaged, is the final product and hence whey removal is not practised. In both classes of product, spontaneous syneresis, following packaging, is undesirable but occurs frequently because of the relatively high moisture to protein ratio compared to rennet-curd cheeses (e.g. $\sim 17 \cdot 6$ g H_2O/g protein in yoghurt versus $\sim 1 \cdot 44$ g H_2O/g protein in Cheddar).

Various methods have been used to measure syneresis in acid milk gels and products, including: (i) the volume of whey expressed from the gel on centrifugation at various G-forces has been used extensively for natural, unsweetened yoghurts;[62,63,89,110] (ii) the volume of whey expressed by cut curd held under quiescent conditions for a fixed time has been used for Cottage cheese,[100] Laban[106] and natural yoghurt;[80] (iii) the decrease in height of a thin slab of gel under its own weight over time has been used for syneresis measurement in acidified skim milk gels.[121] Permeability measurements on acid casein gels (formed by acidification of skim milk or sodium caseinate dispersions in the cold and subsequent quiescent heating to, and ageing at, 20–40°C) were made by Roefs.[44]

Acid milk gels, formed *in situ* in the package, e.g. set natural yoghurt or Fromage frais, show little tendency to syneresis if left undisturbed. However, even in this situation spontaneous syneresis may occur with time to greater or lesser degrees depending on the level of fortification and processing conditions (such as preheating of milk which effects differences in porosity and gel structure). Such spontaneous syneresis may be due to slow proteolysis of the casein (by the enzymes of starter bacteria), pH decrease and temperature fluctuations. Hydrolysis of the casein, which would increase its reactivity (making it more like paracasein) and which depends on several conditions, may be responsible for the wide day-to-day variation in the syneretic properties in set-fermented milk products experienced within and between factories.[75] Disturbance of these products, by movement during cartoning and transport, which provides stress for bond breakage and matrix rearrangement, may initiate or accentuate syneresis. Indeed, it is not uncommon for the onset of syneresis (in what was a relatively 'wheyless' product) to occur in a product such as set natural yoghurt by the dipping of a spoon into the product (a pool of whey forms in the space created).

In fresh products where the gel is concentrated to obtain the correct level of DM (e.g. Quarg, Labneh, Fromage frais, Cream cheese), the application of relatively high pressure has already led to large-scale syneresis which is necessary for whey separation. However, once the gels have been concentrated to the correct dry matter (varies from a few seconds for mechanical centrifugation and ultrafiltration to hours for traditional draining methods), syneresis may continue (due to delayed network rearrangements) and in this situation is undesirable. The level of syneresis will depend on several conditions (i.e. composition and further treatments of concentrated gel) which affect structure and porosity and on the absence/presence of hydrocolloids which bind the moisture phase.

The factors that affect syneresis of unconcentrated gels (such as set acid products) will be discussed below; the effect of concentration and further curd treatments will be discussed in section 6.

For a given level of syneretic pressure, syneresis increases with increasing surface area to volume ratio of the gel. The shape of the package containing the gel may also influence syneresis; for example, in a package with sloping walls the gel may have a tendency to come loose from the walls — this leads to a stress in the gel which induces syneresis.[75] At the initial stage of syneresis, i.e., after disturbing the gel by cutting, stirring or centrifugation, higher pressures (e.g. higher g-forces) give a higher rate of syneresis (i.e. ms^{-1} cm^{-2}; Refs 62,63,110).

Increasing the level of total solids and gel-forming protein results in gels which are less susceptible to syneresis.[80,93,100,110] Harwalkar & Kalab[110] found an inverse relationship between the level of total solids and the susceptibility to syneresis for natural yoghurt (10–15% DM); however, the relationship differed for yoghurts from heated (90°C for 10 min) and unheated milks. An inverse relationship between gel firmness and susceptibility to syneresis has been found for natural yoghurt produced from reconstituted skim milk,[80,110] Cottage cheese gels ranging in solids from 8 to 15%[100] and chemically acidified skim milk gels.[62] The decrease in syneresis with the level of gel-forming protein may be attributed to a denser (greater number of strands of a given thickness per unit volume) matrix with a lower porosity. Indeed, in chemically acidified skim milk gels, aged at 30°C for 16 h, the permeability coefficient, B, decreased with increasing casein concentration; i.e., B is proportional to $[casein]^{-3.3}$.[44] However, for a given level of gel-forming protein, the susceptibility to syneresis, and indeed firmness, strongly depends on the proportions of casein and whey proteins. Modler et al.[80] investigated the relationship between gel structure and the susceptibility to syneresis (volume of whey expressed from the cut gel on holding) in yoghurts from skim milk (3·5% protein) fortified to 5·0% protein using non-fat dry milk (NFDM), caseinate (Cn), milk protein concentrate produced by ultrafiltration (MPC) or whey protein concentrates (WPC) produced by ultrafiltration (53·2% protein), ion exchange (33% protein) or electrodialysis (35% protein). Because of the differences in the composition of the stabilizers, the ratio of casein-to-non-casein protein varied from 4·56:1 (Cn), 2·85:1 (NFDM or MPC) to 1·08 in WPC-fortified yoghurts. Gel structure became progressively coarser with increasing casein to non-casein protein ratio; casein-fortified gels, similar to gels from unheated unfortified milk, showed a high degree of casein fusion. This trend suggests an insufficient coating of the micelles with denatured and bound whey protein at the higher casein to non-casein protein ratios.[64] Syneresis was lowest in casein-fortified yoghurts and highest in the WPC-fortified yoghurts (apart from the UF-produced WPC which gave intermediate syneresis values comparable to that obtained using NFDM and MPC, Ref. 80). (Gel firmness, being highest in the caseinate-fortified yoghurts, showed an opposite trend.)

It may appear contradictory that a coarser gel network with a relatively high porosity, is less susceptible to syneresis than a finer gel network on the application of an external syneretic pressure. Indeed, the effect of high heat treatment of the milk in promoting finer gel structures which are less prone to spontaneous syneresis, has been extensively reported on;[80,93,110,122] however, here the level of gel-forming protein increases with inclusion of heat-denatured whey protein. The discrepancy may be explained in the context of the type of syneresis, i.e. whether spontaneous (indigenous) syneresis or external syneresis (as effected by application of external pressure). In fine gel structures, spontaneous syneresis is lower (because of the lower porosity) than in a coarser gel. However, the finer gel will probably be more susceptible to external syneresis because the matrix strands, even though more numerous, are weaker (singly and as a whole) and broken more easily than the thicker strands of a coarser network. For a given external pressure, the finer gel will probably be broken to a greater extent and its porosity may then become greater than that of the coarser gel—hence moisture will flow out more easily. Indeed, Roefs[44] found that the permeability coefficient for acid skim milk gels with networks of varying degrees of coarseness (as affected by differences in the gelation temperature) increased with degree of coarseness (permeability measurements were performed on the undisturbed gels). For a given level of gel destruction, the coarser gel would be more prone to syneresis as effected by external pressure. Indeed, in practice when fine gels (i.e. those with low casein/whey protein ratios, i.e. 1:2) are broken it is visually difficult to detect syneresis as the exuded fluid is milky (but of a slightly more liquid consistency than the bulk phase material), indicating large-scale breakage of the gel matrix, probably to short strands. Over time the exuded fluid becomes clearer, indicating that slow network reformation occurs.

All process factors which influence gel firmness and structure (milk heat treatment, rate of gelation, etc.) influence syneresis in a similar way (cf. Sections 3 and 4). Hence, spontaneous syneresis of acid skim milk-based gels decreases with increasing milk heat treatment, milk homogenization and lower gelation temperatures.[62,63,89,90,110] Decreasing the pH at cutting of Cottage cheese gels (9–11% DM) from 4·92 to 4·59 resulted in a reduced level of syneresis[100] on holding the cut gel for 1 h at 32°C. A decrease in pH during syneresis results in greater syneresis than if the gel is brought to the same pH value before cutting (i.e. before initiating external syneresis, Ref. 75). A wide variety of hydrocolloids (including gelatin, pregelatinized starch, cellulose derivatives, alginates and carageenans) are used in practice to immobilize water and reduce syneresis in fresh acid-curd products, especially yoghurts; their effects on yoghurt quality have been extensively studied.[80,123,124] The use of slime-producing cultures in yoghurt has also been found to reduce syneresis considerably.[63]

6 FURTHER TREATMENTS OF THE GEL

In the production of many fresh cheese products, the gel produced following acidification (fermentation) is subjected to one or more further processing steps, e.g. stirring, whey separation/concentration, heating, homogenization, agitation and/or cooling (Fig. 2). Various materials such as cream, sugar, salt, fruit purées, hydrocolloids and others may be added to the curd. Such treatments will influence the structural, rheological and syneretic properties of the final product. Surprisingly little information, apart from that of a proprietary nature, is available on the influence of such parameters. The effects of various processing steps are discussed below.

Cutting the gel into cubes, as for Cottage cheese, initiates syneresis which is enhanced by cooking and stirring, as in rennet-curd cheese production. This results in contraction of the matrix, and hence an increase in dry matter and firmness of the curd particles.

Stirring the gel (as in Quarg, Cream cheese, Fromage frais) leads to breakage of the matrix strands, with the extent of breakage depending on the severity of the agitation. By this process the gel is converted into a thick viscous liquid. Increasing the temperature (25–50°C) lowers the activation energy for aggregate interaction within the broken strands and facilitates the process of subsequent whey separation. A high pH (> 4·6) at whey separation results in large N losses in the whey (more casein fines) owing to greater physical damage to the softer gel. Rennet action during acidification increases gel strength[100,125] and therefore makes it possible to separate the whey at slightly higher values, i.e. 4·7–4·8, without incurring large N-losses in the whey.[13,125] Any factor which increases the firmness of the gel at separation, e.g. proximity to the isoelectric point, rennet addition, milk homogenization, higher levels of gel-forming protein, increased temperature of gelation, will make it less susceptible to breakage for a given degree of shear. Cooling of the gel to temperatures of <20°C, to retard a further decrease in pH (before heating and whey separation), may result in more destruction of the gel for a given degree of agitation; the strength of hydrophobic bonds, which play an important role in the structure of acid-curd cheeses,[44,105] decreases with decreasing temperature.[47,48,71] Whey separation (which may be performed by pouring the hot fluid onto cheese cloths, ultrafiltration or centrifugation) causes concentration and aggregation of the broken pieces of gel to greater or lesser degrees. Collision during concentration may be expected to result in the formation of large irregular-shaped conglomerates (of varying thickness and length) which are forced into close proximity. The moisture content of the curd is closely related to the degree of aggregation; all factors which enhance aggregation (such as factors that promote a coarser gel structure, higher separating temperature in the range 25–85°C, and increased syneretic force during separation) reduce the water content and increase the coarseness and firmness of the resulting curd. Homogenization of milk before culturing leads to an increase in the effective level of gel-forming protein,[41,82] due to the formation of pseudo-protein (casein-covered fat) particles, resulting in a firmer gel and curd.[106] The degree of protein–protein and protein–pseudo-protein aggregation at this stage will influence the size of conglomerates and hence the coarseness/fineness of the curd network. Increasing coarseness is expected to result in a curd which has a coarser/rougher appearance and a granier mouthfeel.[126] Hence the matrix structure is crucial to the quality of products which are packaged at this stage, such as cold-pack Cream cheese, Fromage frais or Quarg.

Usually, gels concentrated by centrifugation at high temperature (e.g. hot-pack Cream cheese) are homogenized and/or subjected to shear; hydrocolloids, e.g. carrageenans, alginates and/or others, may also be added. Homogenization and/or shear results in destruction, to a greater or lesser degree depending on the severity, of conglomerates, resulting in a more homogeneous distribution in the size and spatial distribution of the matrix-forming material (protein/pseudo-protein particles, Ref. 127). The degree of matrix formation during cooling of the homogenized hot-packed product is uncertain; it is likely that a discontinuous matrix is formed[127] with the degree of continuity being governed by the size and spatial distribution of the matrix-forming material and the rate of cooling (which affects the rate of reactions controlling matrix rearrangement). Indeed the fact that Cream cheese has a yield value[113] indicates that the product has some degree of matrix structure. A finer matrix probably manifests itself in a product which has a smoother appearance and mouthfeel and which is less susceptible to spontaneous wheying-off on storage. The addition of hydrocolloids to the curd will also minimize syneresis; stabilizers which interact with casein, particularly κ-carrageenan,[41] may interrupt matrix formation and yield a smoother, softer product. Indeed, in many yoghurts produced by ropy or non-ropy cultures, the product with the ropy culture (which produces polysaccharides to attach itself to the casein) was consistently softer and had a lower shear stress at low shear rates.[63]

7 MAJOR FRESH ACID-CURD CHEESE VARIETIES

The processing steps for different cheeses and their effects on product quality are discussed in more detail below.

7.1 Cottage Cheese

Cottage cheese is a soft unripened, mildly acid cheese. Although its specific origin is unknown, the name 'Cottage' implies that the cheese was originally produced on family farms. The first industrial production of Cottage cheese was in the USA.

It is prepared by blending 'Dry Curd Cottage Cheese' with a cream dressing. The blended product, Cottage cheese, has at least 4% milkfat and < 80% moisture.[15] Low-fat Cottage cheese must contain 0·5–2·0% fat and > 82·5% moisture (15; c.f. Table II).

7.1.1 Milk

Cottage cheese curd is acid-coagulated from pasteurized skim milk or from reconstituted low-heat skim milk powder (RSM) of good microbial quality. In addition to the microbial quality of the milk, its dry matter (DM) content is important in relation to product quality and yield.

Milks from different breeds have different manufacturing properties resulting in differences in the concentration of the dry matter, of which citrate and β-casein appear to be most important.[128] Compared to curds made from Jersey milk (9·8% DM), curds from Friesian milk (8·7% DM) are more fragile during Cottage cheesemaking; consequently, loss of caseins, as fines, in whey is higher when manufacturing cheese from the latter type of milk.[128] Increasing the DM level of Friesian milk to at least 9% by the addition of Na citrate (0·1%) and Na caseinate (0·25–0·55%) improves its curd formation (as affected by the higher level of gel-forming protein) and reduces fines losses in the whey.[128]

White & Ryan[129] investigated the production of Cottage cheese from reconstituted skim milk powder (RSM), at solids levels ranging from 10·5–15%, and skim milk fortified with skim milk powder to 10·5% solids. With increasing level of solids in the RSM, the weight of moisture-adjusted cheese per unit volume of RSM increased due to the higher level of total solids in RSM *per se*. However, the net yield (i.e. weight of moisture-adjusted cheese solids per unit weight of RSM solids) decreased due to a lower recovery of casein. Such a decrease in net yield may be due to an inadequate curd handling system, which was incapable of cutting efficiently at the high curd firmness which led to a large amount of casein fines.

When the DM content of skim milk is >10·5%, the manufacturing time also increases as a result of the high buffering capacity of such milks. Consequently, conventional starters (5%) take longer than normal to reduce the pH in the cheese vat.[130]

There are conflicting reports on the use of lactose-hydrolysed skim milk for Cottage cheese production. Gyuricsek & Thompson[131] reported that hydrolysis of > 90% of the lactose in skim milk prior to Cottage cheesemaking reduced manufacturing time because glucose is more rapidly fermented than lactose by the starter bacteria. The reduced manufacturing time resulted in reduced curd shattering and, consequently, increased yields. Contrarily, Fedrick and

Houliman[132] found that the use of hydrolysed-lactose milk did not affect setting time, yield or quality of Cottage cheese.

7.1.2 Acidification and Coagulation (Setting)

Cottage cheese may be manufactured by short, long or intermediate set methods with starter bacteria (cultured Cottage cheese, CC) or by direct acidification using food-grade acid, acidogen or acid whey (direct-set Cottage cheese, DA). The conventional methods of making Cottage cheese have been described by Emmons & Tuckey,[133] Kosikowski[7] and Scott.[134] However, the cooling/washing, drainage, creaming and packaging operations of commercial Cottage cheese manufacture have undergone large-scale automation and semi-continuous multi-batch production is now normal practice.[135-137] Equipment has also been developed for continuous Cottage cheese manufacture. In continuous systems, coagulation of milk occurs in tubes surrounded by a heating medium (as in direct set cottage cheese) or in semi-circular trough-shaped rotating belts with mixing/dosing, coagulation and whey expulsion zones which are continuously moving forward and separated by semi-circular spacing plates; ancillary equipment includes washing/drainage belts with spray nozzles and pressing facilities to ensure adequate drainage.

Regardless of the method used, the main principle behind Cottage cheese manufacture is the coagulation of caseins at or near their isolectric pH (4·6) and cooking of the curds therefrom at similar pH values.

Starters. Strains of *Lactococcus lactis* subsp. *lactis* or *cremoris*, which are least susceptible to agglutination, are used for acid production during CC manufacture. Agglutination of starter bacteria during Cottage cheese production has been reviewed (Salih & Sandine[138]). On agglutination, starter bacteria clump and settle to the bottom of the vat,[139] resulting in localization of the lactic acid produced to give a pH difference of about 0·5 unit between milk at the bottom and top of the vat after approximately 4 h incubation.[140,141] Consequently, precipitation of caseins occurs and forms a sludge at the bottom of the vat.

Brooker[142] demonstrated that starter agglutination causes minor sludge (a thin sediment which forms at the bottom of the vat) while major sludge, which results in total vat loss, is caused by slow acid production due to the destruction of starter bacteria by bacteriophage. This view was supported by electron micrographs of samples taken from the bottom of failed vats which showed compacted casein micelles and starter bacteria at various stages of lysis and bacteriophage.[142] Minor sludge formation results in yield losses of 4–8%.[143]

Agglutination of starter bacteria can be minimized by homogenization of skim milk (154·8 bar)[144] or bulk culture (176 bar)[141] or by the addition of defatted lecithin (0·5%) to the bulk culture.[141] Homogenization of skim milk destroys milk agglutinins while homogenization or addition of lecithin to starter culture causes fragmentation of starter chains without affecting cell numbers or acid production.

Low CO_2-producing strains of *Lactococcus lactis* subsp. *diacetylactis* or *Leuconostoc* spp. are added for the production of diacetyl which is important for flavour. However, most of the diacetyl produced (> 3·2 ppm) is lost in the whey at cutting.[145] Therefore, some processors prefer to add diacetyl to the creaming mixture or to culture the creaming mixture with diacetyl-producing organisms.[146] The use of strains which produce excessive amounts of CO_2 (through citrate metabolism) results in curd flotation during cooking, and hence reduction in yield due to large fines losses.

Direct acidification. The use of food-grade acids or acidogens instead of starter bacteria, which is the subject of many patents,[147-150] is an alternative method of manufacture described in the standards of identity for Cottage cheese.[150] Essentially, it involves the addition of a combination of lactic and phosphoric acids to cold (2–12°C) milk to achieve a pH of approximately 5·2. This is followed by the addition of glucono-delta-lactone (GDL) which is hydrolysed slowly to gluconic acid, resulting in a gradual reduction of pH to 4·6–4·8 within one h. The slow acidification of milk by GDL causes casein micelles to aggregate into a network with fewer links compared to clustered casein micelles observed (by electron microscopy) during rapid acidification with HCl or lactic acid.[62]

The obvious advantage of direct acidification is efficiency of cheese manufacture and the elimination of starter-associated problems such as agglutination, curd flotation and starter failure due to bacteriophage or antibiotics.[151] The total processing time from the acidification of skim milk to the end of curd washing during continuous Cottage cheese manufacture by direct acidification, with HCl, is about 35 min.[135] However, direct set Cottage cheese produced by the continuous process has a poorer texture than that of DA made by conventional methods.[135,152] Pre-culturing milk to pH 5·5 prior to direct acidification significantly improves the texture and body of the finished cheese.[135] A method for making Cottage cheese using a combination of starter bacteria and direct acidification has been patented.[153] Other patented methods for Cottage cheese manufacture by direct acidification describe the use of HCl,[154] an aliphatic dione (C_2–C_8 glyoxal) and H_2O_2.[155]

Rennet and calcium chloride. Very low levels of rennet are added to the cheese milk within one hour after starter addition for CC and DA Cottage cheeses. The pH of coagulation increases with the amount of rennet [0·2–1·0 ml (single strength)/454 litres milk] used. If the gel is to be cut at pH 4·8, rennet at a level of 0·2 ml/454 litres milk is most satisfactory[156] for CC. However, for DA, more rennet is necessary and attempts to manufacture DA without rennet have been unsuccessful due to a weak coagulum, which shatters easily on cutting and agitation (Gillis & Bourgouin, pers. comm.). Bishop *et al.*[157] observed that the average size of casein micelles in CC curd made with microbial rennet was twice that in curd made without rennet, indicating that rennet promotes aggregation of casein micelles and shortens coagulation time.

U.S. standards permit the addition of < 0·02% $CaCl_2$ to skim milk to improve curd firmness at cutting during Cottage cheese production. However, Emmons et al.[156] reported that added $CaCl_2$ had no effect on curd strength or on the quality of Cottage cheese, indicating that $CaCl_2$ does not play a role in the coagulation of casein micelles during Cottage cheese manufacture; at pH values < 5 most, if not all of the colloidal calcium phosphate in milk is dissolved from the casein micelles at 20–30°C (cf. Section 2.2.1).

7.1.3 Cutting and Cooking the Curd

Cottage cheese is often characterized as 'large-curd' or 'small curd', depending on the cut size; cut sizes of 0·95 cm and 0·64 cm correspond to large and small curds, respectively. The optimum cutting time depends on pH and is probably the most critical step in Cottage cheese manufacture. The desired cutting and cooking pHs during Cottage cheese manufacture are 4·75–4·8 and 4·55–4·6, respectively.[130,158] At constant cooking temperature in the range 46–60°C, the firmness and cohesiveness of dry-curd Cottage cheese and its dry matter content increases with the pH at cutting in the range 4·6–4·9.[130,159] However, when the coagulum is cut at pH 4·9, the curds have a tendency to matt during cooking.[130] A soft coagulum,[156] which holds a high level of moisture after cutting and cooking,[160] is obtained when heat treatment of milk prior to cheesemaking is more severe than normal pasteurization (72°C for 15 s). To obtain a firm, cuttable coagulum from milk which has been severely heated (e.g. 80°C for 30 min), more rennet (15–20 ml/454 litre) is used.[156] Consequently, the pH at cutting is high (5·1–5·2) and the quality of the cheese is reduced.[156,161]

After cutting, the curds are left undisturbed in the whey for about 15 min to 'heal' (i.e. contract and expel some whey) before commencing agitation and heating (cooking) to 52–60°C, over 1·5–2 h. Collins[162] recommended cooking to a minimum temperature of 55°C and holding at this temperature for at least 18 min to kill coliforms and psychrotrophic bacteria since D-values for *Escherichia coli, Pseudomonas fragi, Lc. lactis* subsp. *cremoris* and *Lc. lactis* subsp. *lactis* in whey (pH 4·6) at 55°C are 4·3, 1·88, 0·64 and 4·57 min, respectively.

High cooking temperatures also increase firmness and the dry matter of the curds.[130,163] However, at a constant cooking temperature in the range 46–60°C, DM levels in the curd increase with cooking pH (4·5–4·8).[130] When rennet is used to aid coagulation and the pH at cooking is < 4·5, proteolysis of the caseins results in a soft curd, even when cooked to a high temperature (60°C).[130,158]

Provided that agitation is not too rapid to shatter the curd, the optimum heating rate is the fastest at which no matting occurs.[133] An increase in heating rate from 0·11°C/min at the start to 0·3°C/min at the end of cooking, so that a temperature of 51·6–54·4°C is reached within 2 h, has been suggested for good quality Cottage cheese.[133,158] The firmness and dry matter content of Cottage cheese curd increase with heating rate in the range of 0·18–0·50°C/min.[163]

Tuckey[158] proposed that if the heating rate is too rapid, a surface layer of protein forms a film, which acts as a semi-permeable membrane, around each curd particle and causes whey to be entrapped in the curd. Scanning electron microscopy of Cottage cheese does not clearly demonstrate the presence of a continuous protein film at the surface of curd particles[16,164] but rather a

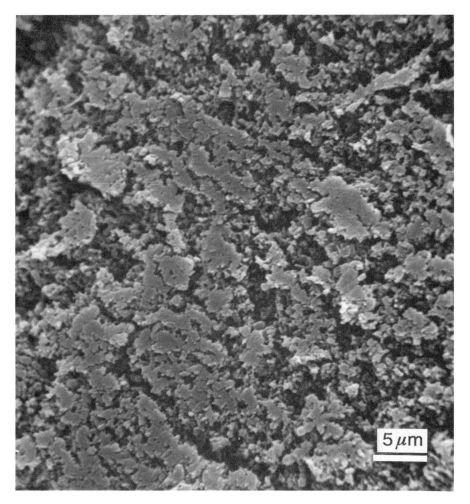

Fig. 12. Scanning electron micrograph of surface of Cottage cheese curd (× 3000). (Reproduced by permission of M. Kalab.)

non-homogeneous distribution of micelles into areas of relatively low and high density; the micelles, especially in regions of high density, appear to be compacted to give a more or less discontinuous film (Fig. 12). While the interior of a dehydrated curd granule also shows areas of varying casein density, there are no traces of a semi-continuous film (cf. Fig. 13). The formation of a semi-continuous film at the surface of curd particles, as accentuated by cooking, impedes increasingly the outflow of whey during cooking. Using transmission electron microscopy (TEM), Glaser et al.[164] showed a distinct difference between the surface and interior of Cottage cheese curd particles and observed that a surface skin appeared only after severe heat treatment (70°C for 1 h), which is unusual in Cottage cheese manufacture.

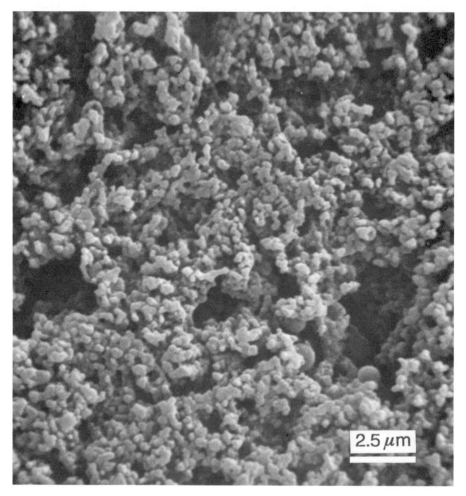

2.5 μm

Fig. 13. Scanning electron micrograph of the interior of dehydrated Cottage cheese curd granule (× 7000). (Reproduced by permission of M. Kalab.)

7.1.4 Washing and Creaming of Curd

After cooking, the whey is drained and the curds are washed two to three times with water to remove excess lactose and lactic acid, thereby stabilizing curd pH. Washing also cools the curds and retards bacterial growth. To achieve a final curd temperature of < 4°C, wash water at ~ 30, 16, 10 or < 4°C for two or three successive washings, respectively, is used. The volume of wash water varies (4·7–15·2 l/kg curd, i.e. 30–100% milk volume, depending on the equipment used) among manufacturers;[165] a typical value is 80% of the original volume of milk.

The wash water obtained after the first 20 min of washing contains about 87 and 93%, respectively, of the DM and lactose present after 60 min of washing.[166] Therefore, washing for not longer than 20 min may be economically and nutritionally advantageous.

The quality of water used to wash curd affects the keeping quality of Cottage cheese. Washing curds with high pH water dissolves the casein on the surface of the curds, making them sticky and slimy in appearance. Therefore, wash water is chlorinated (5–8 ppm) and then acidified (pH 5·5–6·0) before use;[167] it may also be pasteurized.

Cream dressing is added to, and mixed with, the dry curds after washing. Methods for preparing cream dressing and the ratio of dressing to curds needed to control the fat content of Cottage cheese have been described by various authors.[168–170] Flavouring materials may be added at creaming.

The final pH of Cottage cheese should be 5·2; pH values < 5·0 may promote separation of whey from the dressing and the appearance of free whey in the package while pH values > 5·2 lead to reduced keeping quality.

7.1.5 Physical Structure of Cottage Cheese

The microstructure of DA and CC Cottage cheeses are similar except for slight differences observed during the early stages of coagulation.[171] During Cottage cheesemaking, the number-average diameter of the casein micelles increases from ~ 88 nm in milk to 182 nm during early gelation, 185 nm at the end of healing to 207 nm at the end of cooking.[171] However, the average diameter of the micelles is larger in cheese made with rennet than those in rennet-free cheese (Bishop et al.[157]), suggesting that rennet aids in the fusion of micelles. Typical electron micrographs (Fig. 14) of Cottage cheese curds show a porous mass of casein particles in the form of clusters and short chains; starter bacteria trapped in the curd are anchored to the casein particles by filaments (Fig. 15).

7.1.6 Yield

Typical yields (kg cheese at 80% moisture per 100 kg skim milk) of Cottage cheese made from skim milk subjected to the standard HTST treatment (72°C for 15 s) are 15–17%, depending on the protein (casein) content of the milk. Except for the results of White & Ray,[160] yields reported for DA are, generally, slightly higher[151,152,172] than those for CC. This may be attributed, in part, to loss of soluble peptides produced through casein proteolysis by starter bacteria. Yield loss due to the proteolytic activity of starter bacteria can be minimized by using proteinase-negative starters.[173]

Other methods reported to increase the yield of Cottage cheese include:

(i) heating of skim milk to temperatures higher than minimum HTST;[156,160,161]

(ii) thermization (74°C for 10 s), followed by storage at 3°C for 7 days, then HTST treatment;[174]

(iii) addition of sodium hexametaphosphate (SHMP);[175]

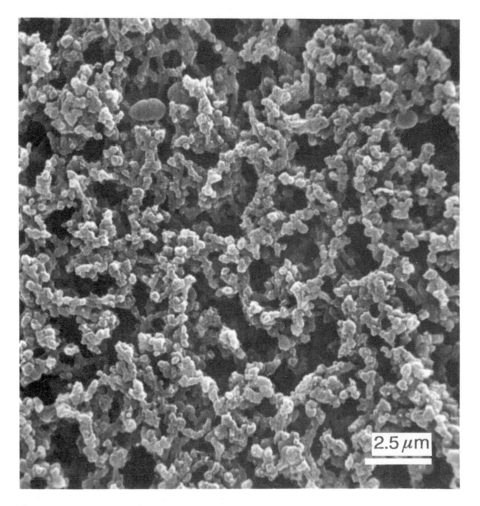

Fig. 14. Scanning electron micrograph of Cottage cheese curd (× 7000). (Reproduced by permission of M. Kalab.)

(iv) addition of iota-carrageenan plus SHMP;[176,177]

(v) ultrafiltration; while ultrafiltration (UF) is used widely on a commercial scale for the production of high moisture, non-granulated fresh cheeses such as Quarg, Fromage frais and Cream cheese, it is not, to our knowledge, used in the Cottage cheese industry. However, several investigations have been undertaken[178–180] on the manufacture of Cottage cheese from skim milk concentrated to various protein levels (6·4–15%) using UF. These studies have concentrated on obtaining the correct Ca/casein ratio in the curd by various means (pre-acidification of skim milk, diafiltration, addition of calcium-chelators) in order to eliminate defects (e.g. excessive

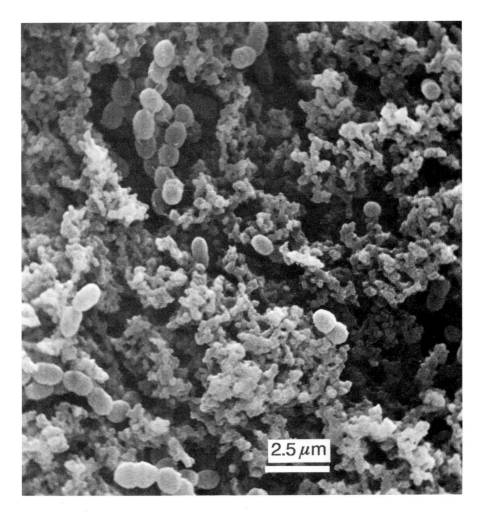

Fig. 15. Scanning electron micrograph of Cottage cheese curd ($\times$ 7000) showing chains of *Lactococci*. (Reproduced by permission of M. Kalab.)

curd elasticity and poor absorption of cream dressing) and improve product yield and quality.

It is not known whether the methods listed above have been adapted for the commercial production of Cottage cheese.

7.1.7 Nutritional Quality

Nutritionally, Cottage cheese is a wholesome low-calorie food, however, it contains lower levels of Ca (about 30 mg/100 g for dry curd, 60 mg/100 g for creamed curd and 68 mg/100 g for low-fat creamed curd) than most rennet-coagulated cheeses which have Ca levels in the range 700–950 mg/100 g. These values suggest that approximately 50% of the Ca in Cottage cheese comes from the cream dressing. The calcium content of large curd Cottage cheese is about 60% higher than that of small curd cheese because of the higher surface area to volume ratio of the latter, which effects a greater loss under similar conditions of washing.[181] Addition of $CaCl_2$ to milk does not affect the Ca content of Cottage cheese; however, the calcium level may be increased by the addition of calcium salts (chloride, lactate or phosphate) to cream dressing without adverse effects on sensory and microbiological quality.[182]

The concentration of Na in Cottage cheese is about 4 mg/g,[183,184] most of which is added with the cream dressing since only about 3% of the Na in milk (~ 0.5 mg/g) is retained in Cottage cheese curd (~ 0.11 mg/g) after three washings.[181] Low-Na Cottage cheese with sensory qualities similar to regular Cottage cheese can be made by reducing the Na content of the dressing by 25%[185] or by replacing up to 50% of the NaCl by KCl.[184]

Generally, dairy products are poor sources of dietary iron. Wong et al.[186] reported the concentrations (μg/100 g) of Fe, Zn, Cu and Mn in Cottage cheese as 174·1, 482·6, 2·6 and 3·7, respectively. Addition of Fe as ferric ammonium citrate to skim milk (to a final concentration of 20 μg Fe/ml milk) resulted in 58% recovery in washed curds without adverse effects on cheese quality.[187] Fortification of Cottage cheese with vitamins A and C has been reported, with satisfactory results.[188]

7.1.8 Microbiological Quality

Due to the high moisture levels ($\sim 80\%$) and relatively high pH (~ 5.2), the shelf-life of Cottage cheese is limited to 2–3 weeks of storage at 5–7°C[189] (Table IV).

The most frequently found spoilage microorganisms in Cottage cheese are psychrotrophic bacteria (*Pseudomonas, Achromobacter, Flavobacterium, Alcaligenes, Escherichia, Enterobacter*), yeasts and moulds.[193,194] However, *Pseudomonas fluorescens, Ps. putida* and *Enterobacter agglomerans* are the major causes of spoilage.[195]

Psychrotrophic bacteria and their proteases cause defects such as surface discolouration, off-odours and off-flavours, e.g. bitterness, in Cottage cheese.[196–198]

The shelf-life can be greatly extended (> $\sim 75\%$) by adding sorbic acid or potassium sorbate (0·075%, w/w); these materials inhibit psychrotrophic bacteria and

TABLE IV
Shelf-life of Cottage Cheese

Country	Shelf-life	Reference
Canada	> 16 days	Roth et al. (190)
US	17·8 days	Hankin et al. (191)
UK	9–15 days	Brocklehurst & Lund (192)

moulds without producing objectional flavours.[195,199-201] Sorbic acid is most effective as an antimicrobial agent in its undissociated form, which represents about 31–59% of the concentration used in Cottage cheese with pH values of 4·6–5·1.[195]

Other 'natural' ways of extending shelf-life are the addition of bifidobacteria to inhibit *Staphylococci*[202] or a pre-cultured skim milk product, Micrograd™.[203] The inhibitory effect of Micrograd™ is due to a heat-stable, low-molecular weight (about 700 daltons) peptide (Sandine, W.E., pers. comm.). It is claimed in a recent patent[204] that addition of *Propionibacterium shermanii* NRRL–B-18074 (10^6–10^7 cfu/g) plus *Lc. lactis* subsp. *diacetylactis* NRRL-B-15005, NRRL-B-15006, NRRL-B-15018 or ATCC 15346 (10^6–10^7 cfu/g) to Cottage cheese dressing inhibits psychrotrophic bacteria and moulds by producing propionic and acetic acids which act as antimicrobial agents. Earlier work[145] showed that addition of acetic or propionic acid to Cottage cheese dressing retarded slime formation. Lactococci have been shown to produce bacteriocins also;[205,206] the patented strains[207] of *Lc. lactis* subsp. *diacetylactis* exert their preservative effect without metabolizing lactose or citrate because they are *Lac⁻*, *Citr⁻*.

The survival of *Listeria monocytogenes* (strains Scott A or V7) during the manufacture and storage of CC was studied by Ryser *et al.*,[208] who found that small numbers (< 100 cfu/g) of these organisms survived the cheesemaking process. However, listeria-free DA was made when acidification was with GDL rather than with HCl because undissociated organic acids are more readily soluble in bacterial cell membranes and are more bacteriostatic than dissociated acids.[209]

7.1.9 Cottage Cheese Flavour

The acidic taste of Cottage cheese is due largely to lactic acid, which is present at concentrations ranging from 124 to 452 mg/kg; other acids present include formic (23–306 mg/kg) and acetic (11–292 mg/kg) and low concentrations (< 1 mg/g) of propionic and butyric acids.[195] Formic, acetic, propionic and butyric acids are volatile, and therefore also contribute to the aroma of Cottage cheese.

The most distinctive flavour compound in Cottage cheese is diacetyl, which is produced by oxidative decarboxylation of α-acetolactic acid. Diacetyl production is pH dependent, occurring at pH values > 5·5.[210] Acceptable levels of diacetyl in Cottage cheese are estimated at about 2 ppm[211] but a diacetyl/acetaldehyde ratio of 3–5 it is desirable for good flavour.[212] Ratios of diacetyl/acetaldehyde > 5 or < 3 result, respectively, in harsh or green flavour defects. Lack of diacetyl results, in part, from its oxidation to acetoin by some starter or contaminating bacteria (e.g. coliforms, *Pseudomonas* and *Alcalagenes*) which possess diacetyl reductase.[213] Since oxidation increases with temperature,[214] storage of Cottage cheese at refrigeration temperature is important for the retention of diacetyl flavour.

7.2 Quarg and Related Varieties

Quarg is a soft, homogeneous, mildly-supple white cheese with a smooth mouth-feel and a clean, refreshing, mildly acidic flavour. The product is shelf-stable for

2–4 weeks at < 8°C; stability refers to the absence of: (i) bacteriological deterioration; (ii) wheying-off (syneresis); (iii) development of grainy texture and (iv) over-acid or bitter flavours during storage. Quarg is sometimes loosely referred to as the German equivalent of Cottage cheese; however, while being related in the sense that they are both fresh acid-curd cheeses of similar composition, they are quite different from a production viewpoint and in sensory aspects; Cottage is a (dressed) granular cheese with the granules ideally having a chewy, meat-like texture.

Also referred to as Tvorog in some European countries, Quarg is a cheese of major commercial significance in Germany where the annual per caput consumption is ~ 8 kg.[11] Quarg is normally made from pasteurized (72–85°C for 15 s) skim milk which is cooled (20–23°C) and inoculated at a rate of 1–2% with an 0-type culture (*Lactococcus lactis* subsp. *lactis* and/or *Lc. lactis* subsp. *cremoris*).[10,11,125,215,216] The milk (20–23°C) is held for 14–18 h until the desired pH of 4·6–4·8 is reached; shortly (1–2 h) after culture addition a small quantity of rennet [0·5–1·0 ml (35–70 chymosin units) per 100 litre] is added when the pH is ~ 6·3–6·1. Rennet action helps, via partial hydrolysis of κ-casein (which promotes higher aggregation forces during the earlier stages of acidification and the onset of gelation at a higher pH, Ref. 217) to give a firmer curd at a higher pH;[100] its addition thus minimizes casein losses on subsequent whey separation[13,125] and reduces the susceptibility to over-acid products (in the absence of rennet, a lower pH is required to obtain the same degree of curd firmness).

The fermented gelled milk, gently stirred (100–200 rpm) into a smooth flowable consistency, is pumped to a nozzle centrifuge where it is separated into curd (i.e. Quarg) and whey containing ~ 0·65% whey protein and 0·2% non-protein N. The Quarg, as it exits the separator, is cooled (< 10°C) on its way to the buffer tank feeding the packaging machine.

Various methods have been employed to reduce the loss of whey proteins and to increase yield:[10,11,215]

(i) Westfalia Thermoprocess, where, (a) the milk is pasteurized at 95–98°C for 2–3 min, (b) the gelled milk (pH 4·6) is heated to 60°C for 3 min, and then cooled (25°C) to the separation temperature. This process gives recovery of 50–60% of the whey proteins in the cheese;

(ii) Centriwhey process, where whey is heated (95°C) to precipitate the whey proteins. The denatured whey proteins are recovered by centrifugation as a concentrate (12–14% solids) which is added to the milk for the next batch of Quarg;

(iii) Lactal process, where the whey is heated (95°C) to precipitate the whey proteins which are allowed to settle; by partial decantation of the serum, a concentrated whey of ~ 7–8% solids is obtained; this is further concentrated, using a nozzle centrifuge, into a whey Quarg (17–18% solids) which is blended at a rate of ~ 10% with regular Quarg;

(iv) Ultrafiltration (of the gelled milk) is now being used on a large scale for the commercial production of Quarg and other fresh cheese varieties.[8,10,11,218-223] This method gives complete recovery of whey proteins in the cheese; however, the non-protein N, which amounts to 0·2-0·3% (w/w) of the milk,[224] is lost in the permeate.

In skim milk Quarg (produced from milk heated to 72°C for 15 s) containing ~14% protein and 18% DM, the casein and whey protein levels of products produced by the separator (centrifuge) and ultrafiltration techniques are ~13·4 and 0·6% and 11·2 and 2·8%, respectively.[218] Whey proteins in the native, un-denatured state do not gel under the heating and acidification conditions used in standard separator Quarg production. Suppliers of UF equipment,[219] therefore, recommend a high milk heat treatment (i.e. 95°C for 3–5 min) in order to achieve extensive complexing of denatured whey proteins with the casein and thus to bring the level of gel-forming protein to the same level as in separator-produced Quarg. Otherwise, the product, though containing the correct level of total protein, has a relatively thin consistency due to the lower level of gel-forming protein. Quarg produced by the recommended UF-procedure (i.e. high milk heat treatment prior to culturing) has sensory characteristics as good as those produced using the standard separator process.[8,125]

Due to its relatively high moisture (82%) and low protein (14%) levels, the shelf-life of Quarg is limited to two to four weeks at < 8°C because of microbial growth, syneresis and off-flavour (especially bitterness) defects. Microbial quality can be improved by various methods, including the addition of sorbates,[225] modified atmosphere packaging, thermization (58–60°C) of the broken gel prior to separation[226] and product thermization (with added hydrocolloids, Refs 227,228). Korolczuk et al.[226] described a method for extending the shelf-life of Quarg by reducing counts of Escherichia coli and moulds. Milk was acidified to pH 3·1–3·7 by Lactobacillus delbrueckii subsp. acidophilus starter (43°C for 16–18 h) and then pasteurized (65°C for 30 min), cooled to 25–30°C and then mixed with unacidified milk (at a ratio of one part of acidified to two to three parts unacidified milk) until the pH reached 4·5–4·6. Butter starter (up to 10%) may be added to the mixture or to the unacidified milk to impart flavour to the product. Excess rennet [> 10 ml (> 775 CU) per 1000 litres] while increasing yield, caused excessive bitterness in Quarg after 4 weeks' storage at 5 or 10°C;[13] a level of 5 ml/1000 litres (i.e. 338 CU) gave the optimum compromise between yield and absence of bitter flavour.[13] As rennet addition is not necessary for the production of Quarg by ultrafiltration (as yield is not effected in any case), it is easier to prevent bitterness in UF-produced Quarg. Quarg produced from lactose-hydrolysed milk was found to be sweeter and had a yellower colour than that produced from normal milk.[229]

Further processing (hydrocolloid addition, heating, homogenization and/or aeration) and addition of various materials (spices, herbs, fruit purées, cream, sugar, other fresh, fermented milk products of different fat levels) to Quarg give rise to a range of Quarg-based products such as half fat (20% FDM) and full fat (40% FDM) Quarg, fruit and savoury Quargs, Shrikhand, dairy desserts and fresh cheese preparations (Frischekäsezubereitungen).[216,230,231-234]

Labneh, Labeneh, Ymer and Fromage frais are quite similar to Quarg;[12,106] these products are versions of concentrated, natural stirred-curd yoghurt which represent the interface between the classical fresh cheeses (i.e. standard separator Quarg, cream cheese) and yoghurt. Similar to yoghurt, the milk is subjected to a high heat treatment (~95°C for 5 min) in order to effect a high degree of β-lactoglobulin-κ-casein interaction which in turn leads to a finer gel network, which manifests itself in the form of a product with a smoother mouthfeel and one with the ability to occlude more water. In contrast to yoghurt, the milk is normally unfortified (though it may be in the case of Fromage frais). The gelled milk is concentrated by various means (pouring onto cloth bags as in traditional Labeneh manufacture, Quarg-type separator or ultrafiltration), the product is not treated with a hydrocolloid stabilizer (normally, stirred-curd yoghurt is stabilized using starches, gums and/or pectin present in the added fruit preparation).

Production generally involves: (i) standardization and heat treatment of the milk; (ii) acidification using an 0-type starter culture to pH 4·6; (iii) concentration of the gelled milk, and (iv) homogenization of curd using an ALM-type homogenizer.

These products may be flavoured by the addition of sugars, fruit purées and other condiments, which are blended in prior to homogenization. While acceptable products in their own right, they are (like Quarg and Cream cheese) often blended with yoghurts and other fresh cheeses for the production of 'new' fresh cheese products with different compositional, textural and flavour attributes.

7.3 Cream Cheese and Related Varieties

Cream cheese (hot pack) is a cream–white, slightly acid-tasting product with a mild diacetyl flavour; its consistency ranges from brittle, especially for double Cream cheese (DCC), to spreadable, e.g. single Cream cheese (SCC). The product, which is most popular in North America, has a shelf-life of ~3 months at <8°C.

Cream cheese is produced from standardized (DCC, 9–11% fat; SCC, 4·5–5·0% fat), homogenized, pasteurized (72–75°C for 30–90 s) milk.[7,14] Homogenization is important in the following respects: (i) it reduces fat losses on subsequent whey separation; (ii) it brings about, via the coating of fat with casein and whey protein, the conversion of naturally emulsified fat globules into pseudoprotein particles which participate in gel formation on subsequent acidification. The incorporation of fat by this means into the gel structure gives a smoother and firmer curd and therefore is especially important to the quality of cold-pack Cream cheese where the curd is not further treated.

Following pasteurization, the milk is cooled (20–30°C), inoculated at a rate of 0·8–1·2% with a D-type starter culture (*Lc. lactis* subsp. *lactis, Lc. lactis* subsp. *cremoris, Lc. lactis* subsp. *lactis* biovar *diacetylactis*) and held at this temperature until the desired pH value, ~4·5–4·8, is reached. The resulting gel is gently agitated and heated (batchwise in the ripening tank or continuously in a heat exchanger) and concentrated by various methods:

(i) in the traditional batch method, the hot (75–90°C) product is placed in muslin bags and allowed to drain at room temperature over 5–12 h;

(ii) continuous concentration in a centrifugal cheese separator at 70–85°C or by ultrafiltration at 50–55°C.

In the batch method, the curd, which is cooled to ~ 10°C to prevent excessive acid development, is treated with salt (0·5–1·0%) and hydrocolloid stabilizer (e.g. sodium alginate, carrageenan, etc.) The treated curd may be then:

(i) packaged directly as cold-pack Cream cheese; this product has a some-what spongy, aerated consistency and coarse appearance;

(ii) heated (70–85°C) and 'homogenized' batchwise in a processed cheese-type cooker with relatively high shear rates for 4–15 min; the extent of 'cream-ing' (emulsification and thickening)[235,236] as influenced by the degree of heat and shear and duration of cooking, has a major in-fluence on the consistency of the final product. Increasing the latter two parameters, while keeping temperature constant, generally results in an increasingly brittle texture. The hot, molten product, known as hot-pack Cream cheese, has a shelf life of ~3 months at 4–8°C.

In the continuous production method, curd from the separator is treated con-tinuously with stabilizer via an on-line metering/mixing device, homogenized and fed to the buffer tank feeding the packaging machine.

Owing to the thick, viscous consistency of Cream cheese, concentration by UF necessitates a two stage process (stage one: standard modules with centrifugal/ pos-itive displacement pumps; stage two: high-flow modules with positive pumps) in order to maintain satisfactory flux rates and to obtain the correct dry matter level.

The flavour diversity of Cream cheese may be increased by adding various flavours, spices, herbs and fish.[237,238] Cream cheese-type products (45% DM, 30% fat) were prepared by blending one of two protein bases (Ricotta or Queso Blanco) with cultured buttermilk, pasteurization, homogenization and hot pack-aging.[239] These cheeses compare well with commercial double Cream cheese in all quality aspects. Partial replacement (60%) of butter fat with vegetable fat does not affect the sensory attributes of Cream cheese.[240]

The manufacture and sensory attributes of other cream cheese-type products, such as Neufchatel and Petit Suisse, are similar to double Cream cheese; they differ mainly with respect to composition. However, Mascarpone differs from other cream cheese-type products in that acidification and coagulation are brought about by a combination of chemical acidification (using food-grade organic acids) to ~ pH 5·0 and heat (90–95°C) rather than by starter fermentation at 20–45°C. The hot, acidified cream (40–50% fat) which is Mascarpone cheese, is packed in cartons or tubs and stored at ~ 5°C. The product, which has a shelf life of 1–3 weeks, has a soft homogeneous texture and a mild buttery, slightly tangy flavour.[241]

7.4 Queso Blanco

Queso blanco (white cheese) is the generic name for white, semi-soft cheeses, produced in Central and South America and which can be consumed fresh;

however, some cheeses may be held for periods of 2 weeks to 2 months before consumption.[242] Elsewhere in the world, similar cheeses include Chhana and Paneer in India, Armavir in Western Caucasus, and Zsirpi in the Himalayas and low salt (< 1%, w/w), high moisture (> 60%) unripened cheeses (e.g. Beli sir types) in the Balkans. (The latter cheeses, i.e. Beli sir types, may also be salted and ripened in brine for up to 2 months to give white pickled cheeses, usually known under local names, e.g. Travnicki sir and Sjenicki sir.)

In Latin America, Queso blanco covers many white cheese varieties which differ from each other by the method of production (i.e. acid/heat or rennet co-agulated), composition, size, shape and region of production.[242,243] Examples include Queso de Cincho, Queso del Pais and Queso Llanero which are acid/heat coagulated, and Queso de Matera and Queso Pasteurizado which are rennet co-agulated. The use of high heat (80–90°C) during the production of the acid/heat-coagulated white cheeses was in traditional times a very effective way of extending the keeping quality in warm climates.

In general, Queso blanco-type cheeses are creamy, high-salted and acid in flavour; the texture and body resembles those of young high-moisture Cheddar and the cheese has good slicing properties.[7,242] The average composition of the fresh cheese is 50–56% moisture, 21–25% protein, 15–20% fat, 2·5–2·7% lactose, 341 mg% Ca, 357 mg% P and 665 mg% Na; the pH is in the range 5·2–5·5.[7,241,244]

7.4.1 Manufacturing Methods

A factory method for the manufacture of a version of Queso blanco (Queso del Pais or cheese of Puerto Rico) was first developed in the US by Weigold.[245] Since then, much work has been carried out on the optimization of the production conditions for Queso blanco as made in the US; the findings of these investigations are discussed hereunder.

The production method of acid/heat-coagulated Queso blanco varies[7,242,246–248] but generally involves heating of the standardized milk to ~ 82°C; addition (via layering onto the surface) of acidulant (acetic, citric and tartaric acids, phosphoric, lime juice, lactic culture) to the milk while gently agitating; settling of the co-agulum; drainage of the whey; dry stirring and trenching of the curd (~ pH 5·3); addition of salt to the curds and pressing overnight; the pressed cheese is then cut into consumer-size portions which are vacuum wrapped and stored at 4–8°C; the product is shelf-stable at this temperature for 2–3 months.

Variations in losses during production and quality of the cheese (texture and taste) are affected by many process parameters including: milk composition, milk homogenization, type and level of acidulant used, salting temperature and the addition of condiments (e.g. peppers, garlic, fruit, etc.) at salting.[246–248,249,250]

Milk. With whole milk, containing 3–6% fat, variations in fat recovery (60–80%) lead to yields ranging from ~ 11·5 to 22%.[246,250–252] An upper limit of 4·5% fat, or a protein/fat ratio of 1:2 is recommended for the production of acceptable Queso blanco with satisfactory yields.[251] However, in India, Paneer is traditionally manufactured from buffalo milk standardized to 6% fat.[242,244,253]

Fat recovery increases to > 93% when the milk is homogenized (141 kg/cm²); however, homogenization results in cheese which has a soft body and poor slicing properties.[246,252]

Heat treatment of milk.   The manufacture of Queso blanco from milks heated to various degrees has been investigated.[245–247,251] It appears that heat treatment of milk at 85°C for 5 min results in the production of Queso blanco with the most desirable qualities.[252] In the manufacture of Paneer, buffalo milk is heated to 90°C without holding and then cooled to 70°C for acidification, to prevent excessive curd firmness.[253]

Acidification.   The coagulation of proteins from the hot (82–85°C) milk is achieved using food-grade acids, e.g. HCl,[247] H_3PO_4,[254] lactic, tartaric, citric and glacial acetic acid,[246,247,254] fruit juices,[7] and acid whey concentrates.[255] However, citric and glacial acetic acids are used most frequently.

The amount of acid required for coagulation depends on the buffering capacity of the milk.[251,254] To achieve a final pH of 5·2–5·3 in Queso blanco, the addition of ~ 0·27% (w/w) glacial acetic acid[246] or 0·34% (w/w) citric acid monohydrate[251] to the milk is required; the acid is normally diluted to 1:10 prior to the addition to the milk.[246,247,256]

7.4.2 Sensory Properties and Uses

One of the interesting properties of acid-coagulated Queso blanco is its melt resistance (due to inclusion of whey proteins); this makes the cheese suitable for use in deep-fried snack foods such as cheese sticks in batter.

The texture, and hence the sliceability, of Queso blanco is influenced by the moisture content[247] and the age[244] of the cheese.

Queso blanco is traditionally consumed fresh because the processing conditions allow for very few biochemical changes during storage. However, Torres & Chandan[244] reported that starter bacteria (*Lactobacillus* spp.) or exogenous lipases can be added to the dry curd before salting and pressing to improve the flavour of the cheese during ripening. Major volatile compounds contributing to the flavour and aroma of Queso blanco include acetaldehyde, acetone, ethyl, iso-propyl and butyl alcohols and formic, acetic, propionic and butyric acids.[257]

Unlike most cheese varieties, the pH of Queso blanco decreases from approximately 5·2 to 4·9 during ripening; this may be due to the fermentation of residual lactose to lactic acid by heat-stable indigenous bacteria in milk, which survive cheesemaking, or post-manufacture contaminating bacteria.[244]

Microbial quality.   Information on the microbiological quality of Queso blanco as made in the US or Canada by methods described above is limited, even though it has been reported that the keeping qualities of such cheeses are poor.[258,259] In commercial Venezuelan Queso blanco made without exogenous acids or starter bacteria, microorganisms enumerated include *Salmonella*, *Escherichia coli*, *Staphylococcus aureus*, *Bacillus cereus*, *Clostridium perfringes*,

Lactobacillus plantarum, L. casei, yeasts and moulds.[259] Those cheeses were made under poor sanitary conditions and had pH values > 5·3. In warm regions where the milk had high populations of indigenous lactic acid bacteria, it was traditional to make cheese without adding starter culture; this practice still exists in some farmhouse cheesemaking operations in these regions, e.g. the Balkans.

7.5 Ricotta/Ricottone

Ricotta is a soft, cream-coloured, unripened cheese, with a sweet-cream and somewhat nutty/caramel flavour and a delicate aerated-like texture. The cheese, which was traditionally produced in Italy from cheese whey of ewes' milk, now enjoys more widespread popularity, in particular in North America, where it is produced mainly from whole or partly skimmed bovine milk, or whey/skim mixtures.[7,260-265]

In the batch production method, the milk is directly acidified to pH ~ 5·9–6·0 by the addition of food-grade acids [acetic, citric, lactic, starter culture (~ 20% inoculum) or acid whey powder (~ 25% addition)]. Heating of the milk to ~ 80°C, via direct steam injection, induces coagulation of the casein and whey proteins and thus results in the formation of curd flocs in the whey after ~ 30 min. After the appearance of the flocs, the steam is switched off and the curd particles, now under the quiescent conditions, begin to coalesce and float to the surface where they form into a layer. Indirect steam (one to three bar), applied to the vat jacket, together with manual movement of curd from the vat walls towards the centre, initiates the process of 'rolling' whereby the curds roll concentrically from the vat walls towards the centre where they form into a layer which is easily recovered by scooping (using perforated scoops). The curd is filled into perforated moulds and allowed to drain for 4–6 h at < 8°C.

The above procedure gives only partial recovery of the whey proteins. A secondary precipitation, whereby the whey from Ricotta cheese manufacture is acidified to pH 5·4 with citric acid and heated (80°C) and treated as for Ricotta, is sometimes practised in order to recover remaining whey proteins in the form of Ricottone cheese. Due to its relatively hard and tough consistency, Ricottone is usually blended with Ricotta in an attempt to moderate its undesirable features.

Ricotta, because of its relatively high pH, high moisture (Table II) and open manner of moulding and cooling (by the above method) is very susceptible to spoilage by yeasts, moulds and bacteria, and hence has a relatively short shelf-life, 1–3 weeks at 4°C.

Significant advances have been made in the automation of Ricotta cheese production with the view to improving curd separation, cheese yield and microbial quality.[260-262] Maubois & Kosikowski[263] produced excellent quality Ricotta using ultrafiltration: whole milk was acidified with acid whey powder to pH 5·9 and ultrafiltered at 55°C to 11·6% protein (29·1% DM); the retentate was heated, batchwise, at 80°C for 2 min to induce coagulation (without whey separation). The coagulum, which was hot-packed, had a shelf-life of at least 9 weeks at 9°C.

Another process based on ultrafiltration is that of Skovhauge.[266] In this method, milk and/or whey are standardized, pasteurized at pH 6·3, cooled to 50°C and ultrafiltered to 30% DM; the retentate is heated to 90°C and continuously acidified, using lactic acid, to pH 5·75–6·0, at a pressure of 1–1·5 bar; by means of a regulator, pressure is reduced to induce coagulation without whey separation and the curd is cooled stepwise to 70°C and hot packed.

In a process developed by Modler,[264] a 20:80 blend of whole milk and concentrated whey (neutralized to pH 6·9–7·1) was heated from 4 to 92°C in a three-step process, pumped to a 10 min holding tube to induce whey protein denaturation, acidified to pH 5·3–5·5, via on-line dosing, with citric acid (2·5%, w/w), to induce coagulation which occurred in the holding tube, the curd was separated from the 'deproteinated' whey on a nylon conveyor belt. This process gave excellent fat and protein recoveries (99·6 and 99·5%, respectively).

Other processes employed to automate Ricotta production include filtration of whey after curd removal,[265] the use of perforated tubes or baskets in the bottom of the curd-forming vat to collect the curd after whey drainage,[267] and exit ports on the vat walls for the continuous removal of curd.[268]

Ricotta cheese, while being a very acceptable product itself, has many applications, including a base for whipped dairy desserts, use in confectionery fillings and cheesecakes and as a base for products such as Cream cheese and processed cheeses.[7,264]

REFERENCES

1. Milk Marketing Board UK, 1990. *EEC Dairy Facts and Figures*, Thames Ditton.
2. *International Dairy Federation*, 1986. Bulletin No 203, The World Market for Cheese.
3. Ramet, J.P., 1990. *Dairy Ind. Intern.*, **55**(6), 49.
4. Jøssang, K., 1987. *Scandinavian Dairy Ind.*, **1**(2), 84.
5. Abrahamsen, R.K., 1986. *IDF Bulletin* No. 202. Production and utilization of ewes' and goats' milk — production of brown cheese.
6. Otterholm, A., 1984. *IDF Bulletin* No. 171, 21. 'Cheesemaking in Norway'.
7. Kosikowski, F.V., 1982. *Cheese and Fermented Milk Foods*, F.V. Kosikowski & Associates, New York.
8. Patel, R.S., Reuter, H. & Prokopek, D., 1986. *J. Soc. Dairy Technol.*, **39**, 27.
9. United States Department of Agriculture, Agricultural Research Service, 1976. *Agriculture Handbook* No. 8, 1.
10. Winwood, J., 1983. *J. Soc. Dairy Technol.*, **36**, 107.
11. Jelen, P. & Renz-Schauen, A., 1989. *Food Technol.*, **43**(3), 74.
12. Tamime, A.Y., Davies, G., Chehade, A.S. & Mahdi, H.A., 1989. *J. Soc. Dairy Technol.*, **42**, 35.
13. Sohal, T.S., Roehl, D. & Jelen, P., 1988. *J. Dairy Sci.*, **71**, 3188.
14. Strniska, J. & Hbrek, V., 1984. *Prumysl Potravin*, **34**, 631. Cited from *Dairy Sci. Abstr.*, 1985, **47**, 555.
15. Food and Drugs Administration, 1987. *Code of Federal Regulations*, Parts 100–169, Office of the Federal Register, National Archives & Records Administration.
16. Kalab, M., 1979. *J. Dairy Sci.*, 1979, **62**, 1352.
17. Schmidt, D.G., 1980. *Neth. Milk Dairy J.*, **34**, 42.
18. Brooker, B.E. & Holt, C., 1978. *J. Dairy Res.*, **45**, 355.

19. Visser, J., Minihan, A., Smits, P., Tjan, S.B. & Heertje, I., 1986. *Neth. Milk Dairy J.*, **40**, 351.
20. Schmidt, D.G. Walstra, P. & Buchheim, W., 1973. *Neth. Milk Dairy J.*, **27**, 128.
21. Lin, S.H.C., Dewan, R.K. & Bloomfield, V.A., 1971. *Biochem.*, **10**, 4788.
22. Walstra, P., 1990. *J. Dairy Sci.*, **73**, 1965.
23. Schmidt, D.G., 1982. In *Developments in Dairy Chemistry* — Vol. I, Proteins, ed. P.F. Fox. Elsevier Appl. Sci. Publ., London, p. 61.
24. Ono, T. & Obata, T., 1989. *J. Dairy Res.*, **56**, 453.
25. Heertje, I., Visser, J. & Smits, P., 1985. *Food Microstructure*, **4**, 267.
26. Ono, T. & Takagi, T., 1986. *J. Dairy Res.*, **53**, 547.
27. Aoki, T., 1989. *J. Dairy Res.*, **56**, 613.
28. McGann, T.C.A., Buchheim, W., Kearney, R.D. & Richardson, T., 1983. *Biochim. Biophys. Acta*, **760**, 415.
29. Holt, C., 1982. *J. Dairy Res.*, **49**, 29.
30. Holt, C., Davies, D.T. & Law, A.J.R., 1986. *J. Dairy Res.*, **53**, 557.
31. Holt, C., van Kemenade, M.J.J.M., Nelson, L.S.Jr., Sawyer, L., Harries, J.E., Bailey, R.T. & Hukins, D.W.L., 1989. *J. Dairy Res.*, **56**, 411.
32. Holt, C., 1989. *Ann. Research Report*, Hannah Research Institute, p. 51.
33. Evenhuis, N. & de Vries, Th.R., 1959. *Neth. Milk Dairy J.*, **13**, 1.
34. Brule, G. & Fauquant, J., 1981. *J. Dairy Res.*, **48**, 91.
35. Chaplin, L.C., 1984. *J. Dairy Res.*, **51**, 251.
36. Creamer, L.K., 1985. *Milchwissenschaft*, **40**, 589.
37. van Hooydonk, A.C.M., Hagedoorn, H.G. & Boerrigter, I.J., 1986. *Neth. Milk Dairy J.*, **40**, 281.
38. Davies, D.T. & White, J.C.D., 1960. *J. Dairy Res.*, **27**, 171.
39. Pyne, G.T. & McGann, T.C.A., 1960. *J. Dairy Res.*, **27**, 9.
40. Dalgleish, D.G. & Law, A.J.R., 1989. *J. Dairy Res*, **56**, 727.
41. Walstra, P. & Jenness, R., 1984. *Dairy Chemistry and Physics*, John Wiley, New York.
42. Rose, D., 1968. *J. Dairy Sci.*, **51**, 1897.
43. Dalgleish, D.G. & Law, A.J.R., 1988. *J. Dairy Res.*, **55**, 529.
44. Roefs, S.P.F.M., 1986. PhD thesis, Wageningen Agric. Univ., Wageningen, The Netherlands.
45. Roefs, S.P.F.M., Walstra, P., Dalgleish, D.G. & Horne, D.S., 1985. *Neth. Milk Dairy J.*, **39**, 119.
46. Snoeren, T.H.M., Klok, H.J., van Hooydonk, A.C.M. & Damman, A.J., 1984. *Milchwissenschaft*, **39**, 461.
47. Tanford, C., 1980. The effect of temperature, In *The Hydrophobic Effect, Formation of micelles and biological membranes*, 2nd edn. John Wiley, New York.
48. Hayakawa, S. & Nakai, S., 1985. *J. Food Sci.*, **50**, 486.
49. Darling, D.F. & Dickson, J., 1979. *J. Dairy Res.*, **46**, 441.
50. Schmidt, D.G. & Poll, J.K., 1986. *Neth. Milk Dairy J.*, **40**, 269.
51. Green, M.L. & Crutchfield, G., 1971. *J. Dairy Res.*, **38**, 151.
52. Payens, T.A.J., 1966. *J. Dairy Sci.*, **49**, 1317.
53. Pearse, K.N., 1976. *J. Dairy Res.*, **43**, 27.
54. Darling, D.F. & Dickson, J., 1979. *J. Dairy Res.*, **46**, 329.
55. Ottenhof, H.A.W.E.M., 1981. *Internal Report. NIZO*, 786. Cited by Schmidt, D.G. & Poll, J.K., 1986. *Neth. Milk Dairy J.*, **40**, 269.
56. de la Fuente, B.T. & Alais, C.J., 1975. *J. Dairy Sci.*, **58**, 293.
57. Vreeman, H.J., van Markwijk, B.W. & Both, P., 1989. *J. Dairy Res.*, **56**, 463.
58. Walstra, P., 1979. *J. Dairy Res.*, **46**, 317.
59. Holt, C., Davies, D.T. & Law, A.J.R., 1986. *J. Dairy Res.*, **53**, 557.
60. Sood, M.S., Gaind, D.K & Dewan, R.K., 1979. *N.Z. J. Dairy Sci. Technol.*, **14**, 32.
61. Harwalkar, V.R. & Kalab, M., 1980. *J. Texture Studies*, **11**, 35.

62. Harwalkar, V.R. & Kalab, M., 1981. *Scanning Electron Microsc.*, **III**, 503.
63. Schellhaass, S.M. & Morris, H.A., 1985. *Food Microstructure*, **4**, 279.
64. Kalab, M., Allan-Wotjas, P. & Phipps-Todd, B.E., 1983. *Food Microstructure*, **2**, 51.
65. Kalab, M., Emmons, D.B. & Sargant, A.G., 1976. *Milchwissenschaft*, **31**, 402.
66. Davies, F.L., Shankar, P.A., Brooker, B.E. & Hobbs, D.G., 1978. *J. Dairy Res.*, **45**, 53.
67. Rollema, H.S. & Brinkhuis, J.A., 1989. *J. Dairy Res.*, **56**, 417.
68. Ferry, J.D., 1948. *Adv. Protein Chem.*, **4**, 1.
69. Davis, E.A. & Gordon, J., 1984. *Food Technol.*, **38**(5), 99.
70. Dickinson, E., 1990. *Chem. & Ind.*, No. 19, 595.
71. Kinsella, J.E., 1984. *CRC Crit. Rev. Food Sci. Nutr.*, **21**, 197.
72. Tamime, A.J., Kalab, M. & Davies, G., 1984. *Food Microstructure*, **3**, 83.
73. Knoop, A.M., 1977. *Dte. Milchwirtschaft*, **28**, 1154.
74. Walstra, P. & van Vliet, T., 1986. *Neth. Milk Dairy J.*, **40**, 241.
75. Walstra, P., van Dijk, H.J.M. & Geurts, T.J., 1985. *Neth. Milk Dairy J.*, **39**, 209.
76. Hokes, J.C., Mangino, M.E. & Hansen, P.M.T., 1982. *J. Food Sci.*, **47**, 1235.
77. Grandison, A.S., Brooker, B.E., Young, P., Ford, G.D. & Underwood, H.M., 1986. *J. Soc. Dairy Technol.*, **39**, 119.
78. Green, M.L., 1980. *Food Chem.*, **6**, 41.
79. Modler, H.W. & Kalab, M., 1983. *J. Dairy Sci.*, **66**, 430.
80. Modler, H.W., Larmond, M.E., Lin, C.S., Froehlich, D. & Emmons, D.B., 1983. *J. Dairy Sci.*, **66**, 422.
81. Tamime, A.Y. & Deeth, H.C., 1980. *J. Food Protect.*, **43**, 939.
82. van Vliet, T. & Dentener-Kikkert, A., 1982. *Neth. Milk Dairy J.*, **36**, 261.
83. Green, M.L., Marshall, R.J. & Glover, F.A., 1983. *J. Dairy Res.*, **50**, 341.
84. Baer, A., Oroz, M. & Blanc, B., 1976. *J. Dairy Res.*, **43**, 419.
85. Doi, H., Tokuyama, K, Kuo, F.H., Ibuki, F. & Kanamori, M., 1983. *Agric. Biol. Chem.*, **47**, 2817.
86. Doi, H., Ideno, S., Kuo, F.H., Ibuki, F. & Kanamori, M., 1983. *J. Nutr. Sci. Vitaminol.*, **29**, 679.
87. Doi, H., Ideno, S., Ibuki, F. & Kanamori, M., 1983. *Agric. Biol. Chem.*, **47**, 407.
88. Jang, H.D. & Swaisgood, H.E., 1990. *J. Dairy Sci.*, **73**, 900.
89. Parnell-Clunies, E.M., Kakuda, Y., Mullen, K., Arnott, D.R. & de Mann, J.M., 1986. *J. Dairy Sci.*, **69**, 2593.
90. Parnell-Clunies, E.M., Kakuda, Y. & de Man, J.M., 1986. *J. Food Sci.*, **51**, 1459.
91. Labropoulos, A.E., Lopez, A. & Palmer, J.K., 1981. *J. Food Protect.*, **44**, 874.
92. Labropoulos, A.E., Palmer, J.K. & Lopez, A., 1981. *J. Texture Studies*, **12**, 365.
93. Beyer, H.J. & Kessler, H.G., 1988. *Dte. Milchwirtschaft*, **39**, 992.
94. Mulvihill, D.M. & Donovan, M., 1987. *Irish J. Food Sci. Technol.*, **11**, 43.
95. de Wit, J.N., 1981. *Neth. Milk Dairy J.*, **35**, 47.
95. Dalgleish, D.G., 1990. *Milchwissenschaft*, **45**, 491.
97. Morr, C.V., 1985. *J. Dairy Sci.*, **68**, 2773.
98. Motar, J., Bassier, A., Joniau, M. & Baert, J., 1989. *J. Dairy Sci.*, **72**, 2247.
99. Grigorov, H., 1966. *Proc. XVII Intern. Dairy Congr.*, Munich, **E/F**, 643.
100. Emmons, D.B., Price, W.V. & Swanson, A.M., 1959. *J. Dairy Sci.*, **42**, 866.
101. van Vliet, T., 1977. PhD thesis, Wageningen, Wageningen Agric. Univ., Wageningen, Netherlands.
102. Bringe, N.A. & Kinsella, J.E., 1986. *J. Dairy Res.*, **53**, 371.
103. Kim, B.Y. & Kinsella, J.E., 1989. *J. Food Sci.*, **54**, 894.
104. van Vliet, T. & Walstra, P., 1985. *Neth. Milk Dairy J.*, **39**, 115.
105. van Vliet, T., Roefs, S.P.F.M., Zoon, P. & Walstra, P., 1989. *J. Dairy Res.*, **56**, 529.
106. McKenna, A.B., 1987. *N.Z. J. Dairy Sci. Technol.*, **22**, 167.
107. Korolczuk, J. & Mahaut, M., 1989. *J. Texture Stud.*, **20**, 169.

108. Korolczuk, J. & Mahaut, M., 1990. *Lait*, **70**, 15.
109. Harwalkar, V.R., Kalab, M. & Emmons, D.B., 1977. *Milchwissenschaft*, **32**, 400.
110. Harwalkar, V.R. & Kalab, M., 1983. *Milchwissenschaft*, **38**, 517.
111. Korolczuk, J. & Mahaut, M., 1990. *J. Texture Stud.*, **21**, 107.
112. Hori, T., 1982. *J. Food Sci.*, **47**, 1811.
113. Imoto, E.M., Lee, C.H. & Rha, C., 1979. *J. Food Sci.*, **44**, 343.
114. Emmons, D.B. & Price, W.V., 1959. *J. Dairy Sci.*, **42**, 553.
115. Korolczuk, J. & Mahaut, M., 1988. *Lait*, **68**, 349.
116. Zoon, P., van Vliet, T. & Walstra, P., 1988. *Neth. Milk Dairy J.*, **42**, 249.
117. Bohlin, L., Hegg, P.O. & Ljusberg-Wahren, H., 1984. *J. Dairy Sci.*, **67**, 729.
118. Dejmek, P., 1989. *J. Dairy Res.*, **56**, 69.
119. Mitchell, J.R., 1980. *J. Texture Stud.*, **11**, 315.
120. Pearse, M.J. & Mackinlay, A.G., 1989. *J. Dairy Sci.*, **72**, 1401.
121. van Dijk, H.J.M., 1982. PhD thesis, Wageningen Agric. Univ., Wageningen, Netherlands.
122. Hashizume, K. & Sato, T., 1988. *J. Dairy Sci.*, **71**, 1439.
123. Radema, L. & van Dijk, R., 1973. *Verikkingsmidelen voor yoghurt, NIZO-Medelingen*, No. 7, p. 51.
124. Kalab, M., Emmons, D.B. & Sargant, A.G., 1975. *J. Dairy Res.*, **42**, 453.
125. Siggelkow, M., 1984. *Dairy Ind. Intern.*, **49**(6), 17.
126. Modler, H.W., Yiu, S.H., Bollinger, U.K. & Kalab, M., 1989. *Food Microstruc.*, **8**, 201.
127. Kalab, M. & Modler, H.W., 1985. *Milchwissenschaft*, **40**, 193.
128. Mutzelburg, I.D., Dennien, G.J., Fedrick, I.A. & Deeth, H.C., 1982. *Aust. J. Dairy Technol.*, **37**, 107-112.
129. White, C.H. & Ryan, J.M., 1983. *J. Food Protect*, **46**, 686.
130. Emmons, D.B. & Beckett, D.C., 1984. *J. Dairy Sci.*, **67**, 2192.
131. Gyuricsek, D.M. & Thompson, M.P., 1976. *Cult. Dairy Prod. J.*, **11**(3), 12.
132. Fedrick, I.A. & Houliman, D.B., 1981. *Aust. J. Dairy Technol.*, **36**, 104.
133. Emmons, D.B. & Tuckey, S.L., 1967. *Cottage Cheese and Other Cultured Milk Products*. Pfizer Cheese Monographs, Vol. 3. Chas. Pfizer and Co., Inc. New York.
134. Scott, R., 1986. *Cheesemaking Practice*, 2nd edn. Elsevier Appl. Sci. Publ., London.
135. Ernstrom, C.A. & Kale, C.G., 1975. *J. Dairy Sci.*, **58**, 1008.
136. Jensen, E.M., 1983. *North Eur. Dairy J.*, **49**(1), 1.
137. Anon., 1988. *Dairy Ind. Int.*, **53**(9), 28.
138. Salih, M.A. & Sandine, W.E., 1980. *J. Food Protect.*, **43**, 856.
139. Emmons, D.B., Elliott, J.A. & Beckett, D.C., 1966. *J. Dairy Sci.*, **49**, 1357.
140. Salih, M.A. & Sandine, W.E., 1984. *J. Dairy Sci.*, **67**, 7.
141. Milton, K., Hicks, C.L., O'Leary, J. & Langlois, B.E., 1990. *J. Dairy Sci.*, **73**, 2259.
142. Brooker, B.E., 1986. *J. Soc. Dairy Technol.*, **39**, 85.
143. Grandison, A.S., Brooker, B.E., Young, P. & Wigmore, A.S., 1986. *J. Soc. Dairy Technol.*, **39**, 123.
144. Emmons, D.B & Elliott, J.A., 1967. *J. Dairy Sci.*, **50**, 957.
145. Mather, D.W. & Babel, F.J., 1959. *J. Dairy Sci.*, **42**, 809.
146. Anon., 1973. *Federal Register*, **38**, 6886.
147. Hammond, E.G. & Deane, D.D., 1981. US Patent, 2 982 654.
148. Corbin, E.A. Jr., 1971. US Patent 3 620 768.
149. Loter, I., Dissly, H.G. & Schafer, R.E., 1975. US Patent, 3 882 250.
150. Gilliland, S.E., 1972. *J. Dairy Sci.*, **55**, 1028.
151. Sharma, H.S., Bassette R., Metha, R.S. & Dayton, D., 1980. *J. Food Protect.*, **43**, 44.
152. Geilman, W.G., 1981. MS thesis, Utah State Univ., Logan, UT.
153. Reddy, M.S., Muller, J., Washman, C.J., Brown, C.G. & Hunt, C.C., 1990. US Patent, 4 959 229.

154. Ernstrom, C.A., 1963. US Patent, 3 298 836.
155. Metz, F.L., 1980. US Patent, 4 199 609.
156. Emmons, D.B., Swanson, A.M. & Price, W.V., 1959. *J. Dairy Sci.*, **42**, 1020.
157. Bishop, J.R., Bodixle, A.B. & Jouzen, J.J., 1983. *Cult. Dairy Prod. J.*, **18**(3), 14.
158. Tuckey, S.L., 1964. *J. Dairy Sci.*, **47**, 324.
159. Perry, C.A. & Carroad, P.A., 1980. *J. Food Sci.*, **45**, 794.
160. White, C.H. & Ray, B.W., 1977. *J. Dairy Sci.*, **60**, 1236.
161. Durrant, N.W., Stone, W.K. & Large, P.M., 1961. *J. Dairy Sci.*, **44**, 1171.
162. Collins, E.B., 1961. *J. Dairy Sci.*, **44**, 1989.
163. Chua, T.E.H. & Dunkley, W.L., 1979. *J. Dairy Sci.*, **62**, 1216.
164. Glaser, J., Carroad, P.A. & Dunkley, W.L., 1979. *J. Dairy Sci.*, **62**, 1058.
165. Dunkley, W.L. & Patterson, D.R., 1977. *J. Dairy Sci.*, **60**, 1824.
166. Bressan, J.A., Carroad, P.A., Merson, R.L. & Dunkley, W.L., 1982. *J. Food Sci.*, **47**, 84.
167. Angevine, N.C., 1959. *J. Dairy Sci.*, **42**, 2015.
168. Manus, L.J., 1957. *Milk Prod. J.*, **48**, 56.
169. Kemp, A.R. & Schultz, R.J., 1979. *Cult. Dairy Prod. J.*, **14**(1) 15.
170. Lundstedt, E., 1980. *Cult. Dairy Prod. J.*, **15**(2), 8.
171. Glaser, J., Carroad, P.A. & Dunkley, W.L., 1980. *J. Dairy Sci.*, **63**, 37.
172. Satterness, D.E., Parsons, J.G., Martin, J.H. & Spurgeon, K.R., 1978. *Cult. Dairy Prod. J.*, **13**(1), 8.
173. Stoddard, G.W. & Richardson, G.H., 1986. *J. Dairy Sci.*, **69**, 9.
174. Dzurec, D.J. & Zall, R.R., 1982. *J. Dairy Sci.*, **65**, 2296.
175. Dybing, S.T., Parsons, J.G., Martin, J.H. & Spurgeon, K.R., 1982. *J. Dairy Sci.*, **65**, 544.
176. Manning, D.W. Jr., 1985. Increasing cheese yields with carrageenan, *Proc. 22nd Marschall Invitation Cheese Seminar*, pp. 49–54, Madison, WI.
177. Manning, D. Jr., Witt, H. & Ames, J., 1985. *Gums and Stabilisers for the Food Industry* 3, G.O. Phillips, D.J. Wedlock, & P.A. Williams (eds), Elsevier Appl. Sci. Publ., London, pp. 379–385.
178. Matews, M.E., So, S.E., Amundson, C.H. & Hill, C.G., Jr., 1976. *J. Food Sci.*, **41**, 619.
179. Ocampo, J.R. & Ernstrom, C.A., 1987. *J. Dairy Sci.*, **70** (suppl. 1), 67 (abstr).
180. Covacevich, H.R. & Kosikowski, F.V., 1978. *J. Dairy Sci.*, **61**, 529.
181. Wong, W.P., LaCroix, D.E., Mattingly, W.A., Vestal, J.H. & Alford, J.A., 1976. *J. Dairy Sci.*, **59**, 41.
182. Shelef, L.A. & Ryan, R.J., 1988. *J. Dairy Sci.*, **71**, 2618.
183. Bruhn, J.C. & Franke, A.A., 1988. *J. Dairy Sci.*, **71**, 2885.
184. Demott, B.J., Hitchcock, J.P. & Sanders, O.G., 1984. *J. Dairy Sci.*, **67**, 1539.
185. Wyatt, C.J., 1983. *J. Food Sci.*, **48**, 1300.
186. Wong, N.P., LaCroix, D.E. & Vestal, J.H., 1977. *J. Dairy Sci.*, **60**, 1650.
187. Sadler, A.M., LaCroix, D.E. & Alford, J.A., 1973. *J. Dairy Sci.*, **56**, 1267.
188. Sweeney, M.A. & Ashoor, S.H., 1989. *J. Dairy Sci.*, **72**, 587.
189. Emmons, D.B., 1963. *Dairy Sci. Abstr.*, **25**, 175.
190. Roth, L.A., Clegg, L.F.L. & Stiles, M.E., 1971. *Can. Inst. Food Sci. Technol. J.*, **4**, 107.
191. Hankin, L., Stephens, G.R. & Dillman, W.F., 1975. *J. Milk Food Technol.*, **38**, 738.
192. Brocklehurt, T.F. & Lund, B.M., 1988. *Int. J. Food Microbiol.*, **6**, 43.
193. Cousin, M.A., 1982. *J. Food Protect.*, **45**, 172.
194. Witter, L.D., 1961. *J. Dairy Sci.*, **44**, 983.
195. Brocklehurst, T.F. & Lund, B.M., 1985. *Food Microbiol.*, **2**, 207.
196. Marth, E.H., 1970. *Cult. Dairy Prod. J.*, **5**, 14.
197. White, C.H. & Marshall, R.T., 1973. *J. Dairy Sci.*, **56**, 849.
198. Stone, W.K. & Naff, D.M., 1967. *J. Dairy Sci.*, **50**, 1497.

199. Bradley, R.J., Harmon, L.G. & Stine, C.M., 1962. *J. Milk Food Technol.*, **25**, 318.
200. Collins, E.B. & Moustafa, H.H., 1969. *J. Dairy Sci.*, **52**, 439.
201. Bodyfelt, F.W., 1979. *J. Food Protect.*, **42**, 836.
202. Brivosa, G.V., 1987. *Izvestia Vyssvikh Uchebnykh Zavedenii Pischevaya Technologiya*, **3**, 61. Cited from *Dairy Sci. Abstr.*, 1988, **50**, 584.
203. Salih, M.A., Sandine, W.E. & Ayres, J.W., 1990. *J. Dairy Sci.*, **72**, 887.
204. Boundreaux, D.P., Lingle, M.W., Vedamuthu, E.R. & Gonzales, C.F., 1988. US Patent, 4 728 516.
205. Branen, A.L., Go, H.C. & Genske, R.P., 1975. *J. Food Sci.*, **40**, 446.
206. Babel, F.J., 1977. *J. Dairy Sci.*, **60**, 815.
207. Gonzales, C.F., 1986. US Patent, 4 599 313.
208. Ryser, E.T., Marth, E.H. & Doyle, M.P., 1985. *J. Food Protect.*, **48**, 746.
209. El-Shenawy, M.A. & Marth, E.H., 1990. *J. Dairy Sci.*, **73**, 1429.
210. Collins, E.B., 1972. *J. Dairy Sci.*, **55**, 1022.
211. Hempenius, W.L., Liska, B.J. & Harrington, R.B., 1965. *J. Dairy Sci.*, **48**, 870.
212. Lindsay, R.C., Day, E.A. & Sandine, W.E., 1965. *J. Dairy Sci.*, **48**, 863.
213. Seitz, E.W., Sandine, W.E., Elliker, P.R & Day, E.A., 1963. *J. Dairy Sci.*, **46**, 186.
214. Pack, M.Y., Vedamuthu, E.R., Sandine, W.F., Elliker, P.R. & Leessment, H., 1968. *J. Dairy Sci.*, **51**, 339.
215. Kroger, M., 1980. *Cult. Dairy Prod., J.*, **15**(3), 11.
216. Lang, F., 1980. *Milk Ind.*, **82**(11), 21.
217. Guinee, T.P., 1990. Physico-chemical aspects of fresh cheese. In *Proc. 2nd Cheese Symposium, Moorepark, Fermoy, Ireland*, 31.
218. Puhan, Z. & Gallamann, P., 1981. *Nordeuropaeisk Mejeri-tidsskrift*, **47**(1), 4.
219. Koch International GmbH, 1987. *North Eur. Dairy J.*, **53**(3–4), 3.
220. Maslov, A.M., Alekseev, N.G., Silanteva, L.A., Belov, V.V. & Ivanova, L.N., 1987. *Molochnaya Promyshlenost*, (2), 15. Cited from *Dairy Sci. Abstr.*, 1990, **52**, 617.
221. Weitbrauk, H. & Krell, E., 1988. *Dte. Molkerei-Ztg.*, **109**(32–33), 992.
222. Nielsen, P.S., 1977. *Maelkeritidende*, **100**, 448. Cited from *Dairy-Sci. Abstr.*, 1988, **50**, 236.
223. Anon., 1985. *North Eur. Dairy J.*, **51**(9), 242.
224. White, J.C.D. & Davies, D.T., 1958. *J. Dairy Res.*, **25**, 236.
225. Schulz, M.E. & Thomasow, J., 1970. *Milchwissenschaft*, **25**, 330.
226. Korolczuk, J., Grzelak, D., Zmarkicki, S. & Yanicki, Q., 1983. *N.Z. J. Dairy Sci. Technol.*, **18**, 101.
227. Gutter, H., 1968. *Dte. Molkerei-Ztg.*, **89**, 1183.
228. Holdt, P., 1978. *Dte. Milchwirtschaft*, **29**, 301.
229. Sheth, H., Jelen, P., Ozimek, L. & Sauer, W., 1988. *J. Dairy Sci.*, **71**, 2891.
230. Patel, R.S. & Abd El-Salam, M.H., 1986. *Cult. Dairy Prod. J.*, **21**(1), 6.
231. Mann, J.E., 1978. *Dairy Ind. Intern.*, **43**(4), 42.
232. Mann, J.E., 1982. *Dairy Ind. Intern.*, **47**(3), 33.
233. Mann, J.E., 1984. *Dairy Ind. Intern.*, **49**(12), 13.
234. Mann, J.E., 1987. *Dairy Ind. Intern.*, **52**(8), 12.
235. Guinee, T.P., 1990. *Co-op* Ireland, Feb., 45.
236. Fox, P.F., Guinee, T.P., McSweeney, P.L.H., O'Connor, T.P., O'Brien, N.M. & Law, J., 1993. *Adv. Food Nutr. Res.*, in press.
237. Anon., 1984. *Dairy Field*, Nov., 40.
238. Kristiansen, J.R., Christiansen, P.S. & Edelsten, D., 1989. *North Eur. Dairy J.*, **55**(8–10), 181.
239. Modler, H.W., Poste, L.M., & Butler, G., 1985. *J. Dairy Sci.*, **68**, 2835.
240. Hallal, A.M. & Al-Omar, M.E., 1987. *Iraqi J. Agric. Sci.*, Zanco. Cited from *Dairy Sci. Abstr.*, 1988, **50**(3632), 408.
241. Davies, J.G., 1976. *Cheese, Vol. 3 — Manufacturing Methods*, Churchill Livingstone, London, pp. 728.
242. Torres, N. & Chandan, R.C., 1981. *J. Dairy Sci.* **64**, 552.

243. United States Department of Agriculture, 1978. Cheese varieties and descriptions, *Agric. Handbook 54*, Washington, D.C., pp. 99–100.
244. Torres, N. & Chandan, R.C., 1981. *J. Dairy Sci.*, **64**, 2161.
245. Weigold, G.W., 1958. *Milk Prod. J.*, **49**(8), 16.
246. Siapantas, L.G., & Kosikowski, F.V., 1967. *J. Dairy Sci.*, **50**, 1589.
247. Chandan, R.C., Marin, H., Nakrani, K.R. & Zehher, M.D., 1979. *J. Dairy Sci.*, **62**, 691.
248. Parnell-Clunies, E.M., Irvine, D.M. & Bullock, D.H., 1985. *J. Dairy Sci.*, **68**, 789.
249. Parnell-Clunies, E.M., Irvine, D.M. & Bullock, D.H., 1985. *J. Dairy Sci.*, **68**, 3095.
250. Siapantas, L.A. & Kosikowski, F.V., 1965. *J. Dairy Sci.*,, **48**, 764.
251. Hill, A.R., Bullock, D.H. & Irvine, D.M., 1982. *Can. Inst. Food Sci. Technol. J.*, **15**, 47.
252. Parnell-Clunies, E.M., Irvine, D.M. & Bullock, D.H., 1985. *Can. Inst. Food Sci. Technol. J.*, **18**, 133.
253. Kalab, M., Gupta, S.K., Desai, H.K., & Patil, G.R., 1988. *Food Microstructure*, **7**, 81.
254. Siapantas, L.G. & Kosikowski, F.V., 1973. *J. Dairy Sci.*, **56**, 631.
255. Hirschl, R. & Kosikowski, F.V., 1975. *J. Dairy Sci.*, **58**, 793.
256. Sawyer, W.H., 1969. *J. Dairy Sci.*, **52**, 1347.
257. Siapantas, L.G., 1967. PhD thesis, Cornell Univ., Ithaca, New York.
258. Arispe, I. & Westhoff, D., 1984. *J. Food Protect.*, **47**, 27.
259. Arispe, I. & Westhoff, D., 1984. *J. Food Sci.*, **49**, 1005.
260. Modler, H.W. & Emmons, D.B., 1984. *Modern Dairy*, **63**(4), 10.
261. Modler, H.W. & Emmons, D.B., 1989. *Milchwissenschaft*, **44**, 673.
262. Modler, H.W. & Emmons, D.B., 1989. *Milchwissenschaft*, **44**, 753.
263. Maubois, J.L. & Kosikowski, F.V., 1978. *J. Dairy Sci.*, **61**, 881.
264. Modler, H.W., 1988. *J. Dairy Sci.*, **71**, 2003.
265. Cleary, P.J. & Nilson, K.M., 1983. *Cult. Dairy Prod., J.*, **18**(1), 5.
266. Skovhauge, E., 1988. *Membrane Filtration for Acid and Heat Coagulated Cheese: Ricotta*, Pasilac-Turnkey Dairies, Ltd., Aarhus, Denmark, p. 7. Cited from *Dairy Sci. Abstr.*, 1989, **51**, 679.
267. Pontecorvo, N.E., 1974. US Patent, 3 836 684.
268. Savarese, J., 1981. US Patent, 4 254 698.

14

Some Non-European Cheese Varieties

J.A. Phelan and J. Renaud

Meat and Dairy Service, FAO, Rome

&

P.F. Fox

Department of Food Chemistry, University College, Cork, Republic of Ireland

1 Introduction

The chemistry, biochemistry and microbiology of the principal international cheese varieties (families) have been described in the preceding 13 chapters. With the exception of Domiati, all these cheese varieties are of European origin; indeed Domiati has its close counterparts in south-eastern Europe. Although not all European cheese varieties belong to one of the above families, most are generally similar to one of the principal varieties. As indicated in Table II, Chapter 1, Volume 1, little cheese is produced in Latin America, Africa or Asia, at least by European and North American standards. However, a number of cheeses, often in small quantities, are produced in these regions and it was considered worthwhile describing some of them, which are produced on a very small scale, frequently on farmsteads, or at herd or flock level by nomads or transhumant shepherds. Very little scientific information is available on most of them; some are produced by rennet coagulation, others by acid precipitation so that the general principles of coagulation, described previously, apply. In many cases, defined starters are not used so that acid production, by indigenous microflora or by addition of acidified milk or of whey from previous batches, is probably rather variable.

2 Asia

2.1 Afghanistan

Karut is a very hard cheese made from skimmed milk.

Kimish Panier is made all over Afghanistan. It is similar to Panir, which is made in India, i.e. a semi-hard, unripened cheese, usually obtained by acid coagulation of buffalo, cow, goat or sheep milk or their mixtures. Traditionally, milk is coagulated by adding cheese whey in the proportion of 25% of the volume of milk.

2.2 Bangladesh

Chhana, one of the two main cheese varieties made in India, is also produced in Bangladesh and Nepal. The technology is the same.

Ponir is a semi-hard cured cheese. It is spherically shaped and its body is white with holes. Milk is heated at 65–70°C for 30 min, 0·5 to 1 litre of mesophilic starter is added per 100 litres and coagulation is obtained in one hour with 20 to 30 ml of rennet. After coagulation, the coagulum is ladled in layers into a bamboo frame lined with cheese cloth. The cheese is pressed for 2 h, then removed from the form and cut in cubes of 10 to 12 cm and dipped in chilled water to firm the curd. Subsequently, salt is spread on the curd pieces which are kept in pots made of bamboo and covered with polyethylene film. It is cured for 4 to 6 weeks and can be stored for 6 months.

2.3 Buthan

Chhana is also made in Buthan according to the same technology as in India.

Churtsi is a hard cheese made from yak and chauri milk. Actually, it is obtained from Dartsi, which is a type of soft Cottage cheese similar to Serkam. It is smoked and looks like a large flat stone; it is kept in a leather or calf-skin bag. It has a shelf-life of several years.

Durukhowa is made from yak and chauri milk. It is also produced in Nepal where it is called Chugga or Chhurpi. It is very hard and rubbery.

2.4 India

Cheese production in India is quite limited, which is perhaps surprising for a country with a relatively strong dairying tradition. This may be regarded as the consequence of several combined factors, e.g. milk is produced mostly in small quantities of 2–4 litres by marginal farmers and from time immemorial, processing has been oriented towards the production of fermented milk (Dahi), both for the producer's family and for marketing through traditional channels. Furthermore, a strong demand for liquid milk has developed, particularly in the major cities, thus limiting the quantities of milk available for cheesemaking. Commercial production of cheese in India dates back to the early 1960s and is currently confined to 10 major producers, with almost half of the production represented by a single brand.

There is an interesting contrast between the traditional methods used in Egypt and India for preserving milk as concentrated products; in the former, extensive use is made of coagulation (rennet or acid) and salt for preservation whereas in India, concentration by boiling, frequently with the addition of sugar, is widespread.

No production data for India are listed by the FAO[1] but the IDF[2] gave production at 700 tonnes in 1977 and 1000 tonnes in 1980. In the Third Annual Edition[3] of 'Dairy India 1987' it was stated that cheese production in India is 3000 tonnes annually. In the past, processed cheese, Gouda and Cheddar, in that order, appeared to the most important varieties and there was no import or export of these. However, the production of local varieties has developed, i.e. Verka cheese, Tamil Nado, Kodai's cheese, Gujarat cheese, Aravali cheese, Abhish cheese and Abli cheese. Mozzarella cheese spread and Spice cheese continue to be produced. However, there are two main cheese varieties: Channa, a sour milk cheese made from cows' milk, and Panir (Surati, Bandal (West Bengal); Decca is a variant produced in Pakistan), which is a fresh rennet cheese made from cows' or buffalo's milk. The manufacture of these cheeses is described in some detail by De[4] and briefly by Walter & Hargrove[5] and by Davis[6] who also refers to a Burmese cheese, Deinge, made from curd prepared by boiling buttermilk from ghee (it appears to be rather like Ricotta or Anari).

Panir or Paneer was probably introduced to India from the Middle East, as a similar type of cheese, called Paneer Khiki, is produced by the nomads in Iran. Paneer is normally made from buffalo, cow, sheep and goat milk. The milk is heated to about 78°C for a short time and subsequently cooled to 35°C. About 0·5% of lactic starter and then 10% of lukewarm brine are added. Coagulation is effected by adding calf rennet. After coagulation, the rennet gel is stirred out but not cooked. If brine were not added before renneting, NaCl at 2·5% of the weight of the milk is added to the curd and whey mixture and stirred in. Curd is drained in cheese cloth without pressing and is placed in moulds, with turning, for 2 h. The cheese is then sliced into pieces of the desired size and is ready for consumption. The shelf-life is 2 to 3 days. Traditionally, the uncut gel is ladled into wicker baskets, salt being sprinkled between successive layers. The typical composition of Paneer is: 70% moisture and 13% fat.

Chhana or Chhanna is made from cows' or buffalo's milk. It is the acid coagulum of boiled hot milk obtained by adding lactic or citric acid or acidified whey from a previous batch. It may be produced on a very small farmstead scale or on a commercial, fairly highly mechanized basis. In order to obtain the desirable body and texture of Chhana, the pH at coagulation should be adjusted to about 5·4, the temperature should be 80°C and coagulation should be completed in less than 1 min. The coagulum is collected in a muslin cloth which is hung for draining off the whey; no pressure is applied. Chhana is often used as the basic material for the preparation of Bengali sweets, sweatmeat or cooked vegetable dishes. The typical composition of Chhana cheese is: 53% moisture, 25% fat, 17% protein, 2% lactose, 2% ash.

Two fermented milk products, Dahi and Srikhand, are widely produced in India. Dahi is the most important fermented milk product consumed in India. It is also produced in the Union of Myanmar, in Buthan, in Nepal and in Pakistan. It is a buttermilk-type (sweet) or yoghurt type (sour) product, depending on the culture use. Actually, in the traditional method (household scale), milk is heated and then cooled to body temperature; Dahi or buttermilk from the previous day is added at the rate of 0·5 to 1% and the milk is left overnight to sour and coagulate. In industrial scale production, milk is homogenized and lactic starters are added, i.e. *Str. thermophilus* and *Lb. bulgaricus*. Dahi may also be used as an intermediate in the manufacture of indigenous butter (makkan) or ghee. It is estimated that >40% of total milk production in India is converted into Dahi. There is also a sweetened version of Dahi (known by various names, e.g. misti dahi, hal dahi or payodhi) for the manufacture of which 6·5% sucrose is added to the milk before heating and fermentation of the milk.

Srikhand appears to be more like a fresh cheese than a fermented milk product. Dahi is partly strained through cloth to yield a solid product called Chakka (63% H_2O, 15% fat, 0·8% lactic acid) which is mixed with the desired amount of sucrose to produce Srikhand. The latter may be further dehydrated by heating in an open pan to make Srikhand Wadi (6·5% H_2O, 7·5% fat, 7·5% protein, 16% lactose, 63% sucrose).

2.5 Indonesia

Tahu Susu Atau Dadih is a soft fresh cheese made from cows' and buffalo's milk and prepared by coagulating the milk with vegetable rennet. Milk is heated to 75°C for 15 min; no starters are used. Coagulation by means of bromelain (obtained from pineapple) takes 15 to 30 min at 30–40°C. The coagulum is then broken and put in a cheese cloth which is squeezed to expel whey. Then, the pressed curd is cooked and pressed again by hand. The cheese is then salted in brine at room temperature.

2.6 Nepal

Chhana is also made in Nepal using the same technology described earlier.

Chhurpi is similar to Durkhowa. It is a precipitate of the proteins of buttermilk alone or blended with some skimmed or whole milk. The first stages of manufacture are similar to Serkham technology. The curds are strained from the whey and put into wooden moulds where they are pressed with stones. The curd is then taken out of the form, cut or broken into small pieces, which are threaded on a string and sun-dried or dried in front of a fire. Hard Chhurpi may be ground and the powder used in soups.

Serkam and its other denominations are also made in Nepal. The technique used is similar to that used in Buthan.

Shosim is a soft cheese made from yaks' and chauris' milk. Milk is first processed as in Serkam which is subsequently put into a wooden or earthenware pot, previously used but not washed. Cheese is left in the pot for at least 2 months, and up to 8 months, for curing. Shosim is consumed with soup or mixed with pickle to get Chatani.

Langtrang cheese is a semi-hard cheese made from yaks' and chauris' milk in the high mountainous areas. It is also known as Thodung or Pike cheese which are the other two areas where it is made. This type of cheese was developed between 1957 and 1970 by FAO cheese specialists in cooperation with the Swiss Association for Technical Assistance. It is the first example of improved technology at village level. Milk is coagulated in cheese vats, like those used in Europe to make Emmental, at about 35°C. The coagulum is cut and stirred to remove whey and the temperature is increased to 60–65°C. The curd is moulded and pressed and the cheeses salted in brine. The cheese is cured in a central store at Thodung.

2.7 Pakistan

Panir is a soft cheese made in remote mountainous areas of Pakistan where the transport of fresh milk is difficult. It is made from buffalo's, cows' or sheep's and goats' milk. When buffalo's milk is used, the cheese is white but it is yellowish when made from cows' milk. Milk is heated to about 78°C for a short time and then cooled to 35°C. Lactic starters are added in the proportion of 0·05% of the quantity of milk. Lukewarm brine is added at the rate of 10% of the milk. Coagulation occurs in about 60 min at 35–37°C. The coagulum is transferred to cheese-cloths and whey is drained off without pressing. The shelf-life is 1–2 days.

Peshawari cheese is a semi-hard cheese made from whole or partly skimmed cows' milk. The milk is heated to 63°C for 30 min and then cooled to 32°C. A culture of *Lactobacillus*, i.e. Lassi (tradition yoghurt/dahi mixed with cold water) is added in the proportion of 5–6% of the quantity of milk. Rennet, 10 ml diluted in water, is added for each 500 litres of milk. Coagulation is obtained at 32°C in about one hour. The coagulum is cut into small slices and put on a cheese cloth to drain off the whey.

2.8 The Philippines

No cheese production is reported by the FAO in its Animal Production Year Book. Therefore, it should be understood that production is limited. Dr C.L. Dawide (pers. comm.) estimates that cheese production in The Philippines is about 100 tonnes p.a., made in the provinces of Lyzon and the Visayan Islands. The cheese is made on a farmstead scale from buffalo's, cows' or goats' milk or their mixtures by coagulation with rennet, vinegar or a mixture of these. Only fresh, soft cheeses are made which are generally called 'Kesong Puti' (also Queso; Keso; Kesiyo; Kesilyo; white cheese). Sometimes, the cheeses are ripened for a few days and fried. Efforts are being made at the University of The Philippines, Los Banos, to expand cheese production and to introduce new varieties.

Kesong Puti is a soft fresh cheese made from Carabao and cows' milk. It can be made according to two techniques, i.e., with or without heat treatment. When raw milk is used, a piece of abomasum and some cheese whey from previous batches are added to the milk. Coagulation takes about 2·5 h at room temperature; the coagulum is transferred to a container, salted, stirred and placed in moulds made from bamboo or tinned cans lined with banana leaves. When heat treatment is applied, milk is heated up to 95°C for about 10 min. No starter is used. Coagulation is obtained with vinegar added to hot milk in 30 to 40 min. The coagulum is moulded and the cheese is salted in brine before being wrapped in banana leaves. The weight of each cheese is about 100 g when made without heat treatment, whereas cheeses made from heated milk weigh about 25–30 g. The shelf-life is four to five days at room temperature and up to 2 weeks in the refrigerator.

2.9 China

The FAO[1] reported that 102 400 tonnes of cheese were produced in China in 1980; from the same source in 1989 this production had reached about 141 800 tonnes. Information on the type of cheeses made is scarce. Like other countries where efforts are concentrated on improving the milk supply to the urban population, it seems that cheese production was established mainly on a small-scale. In this connection Long Giang cheese, a type of Edam, was reported by Mr J.C. McCarty (pers. comm.) to be made mainly for local consumption. However, although the Chinese dairy development plan gives priority to the supply of milk to the main cities, diversification of products is included and in the future cheese production may gain further momentum.

2.9.1 Mongolian Cheese-like Products
A number of fermented milk-based foods are produced in Inner Mongolia which can be regarded as forms of acid cheese (Professor Jin Shilin, pers. comm.).

Hurood is produced during the summer from raw or boiled cows', sheep's, goats' or camels' milk and may be flavoured with sugar, herbs or berries. Whole or partially skimmed milk (by gravity separation) is fermented with *L. bulgaricus* or a mesophilic lactic starter below 37°C until the acidity reaches 0·7–0·75%. The coagulum is broken and heated to 60–70°C. The whey is drained off and the curd moulded into many shapes and sun-dried to yield 'Fine Hurood'. Alternatively, the whey may be drained off only partially and the curd-whey mixture sun-dried to yield 'Crude Hurood'. The compositions of some Hurood products are summarized below.

(i)   Whole milk Hurood, without added sugar:

Fat	18–34%	Acidity	0·8%
Protein	30–35%	Colour:	Yellowish/White
Lactose	1·8–3·6%	Shape:	Rectangular
Ash	1·2–5·5%		
Moisture	7·5–45%		

(ii) Whole milk Hurood, with added sugar:

Fat	14·5–28%	Acidity	0·8%
Protein	26·5–5·32%	Colour:	Yellowish/White
Lactose	1·5–3%	Shape:	Rectangular
Ash	1·8–2·5%		
Moisture	11·5–49%		

(iii) Skim milk Hurood, without sugar:

Fat	1·5–2·5%	Acidity	2·0%
Protein	41·5–44%	Colour:	White
Lactose	1·5–2%	Shape:	Rectangular
Ash	1·8–3%		
Moisture	50–52%		

(iv) Skim milk Hurood, with sugar:

Fat	1·4–2%	Acidity	1·0%
Protein	36·5–50%	Colour:	White
Lactose	12·5–16·5%	Shape:	Rectangular
Moisture	53–54%		

Aarul is prepared by extruding a mixture of kefir, sucrose and dextrin into a thin ribbon of curd which is cut into short rods and dried in a hot air cabinet. Its composition is:

Moisture	15–18%
Total solids	82%
Sugar	25–30%
Milk fat in T.S.	29%
Ash	3·5–4·8%
Protein	55–58%
Acidity	1·6–2·25%
Colour:	White

Biaslag is prepared from Edem (a type of liquid yoghurt containing *L. bulgaricus, Str. thermophilus, Lc. lactis* and *Lc. cremoris*) by heating and draining. The curd is moulded into blocks which are held at room temperature for 1–2 weeks. The blocks are sliced to give soft Biaslag or sun-dried to give a hard-type Biaslag. The composition is variable:

Fat	5–30%
SNF	20–60%
Moisture	8–50%
Acidity	0·6–0·8%
Colour:	White or yellowish brown

Ezgi is also produced from yoghurt by essentially the same procedure as Biaslag and may be prepared as soft or hard (sun-dried) versions. Its composition is:

Fat	16–33%
Protein	41–60%
Lactose	5–15%
Ash	4·5–7%
Moisture	8–23%
Acidity:	1–3·5%
Colour:	White or Yellowish white and yellowish brown

2.10 Japan

The dairy industry in Japan has developed since World War II. Until the 1950s there was no cheese industry; however, since then production has become significant and cheese consumption developed among the Japanese population. Cheese production, according to FAO sources, was about 68 000 tonnes in 1980. In 1989, production was estimated to be 83 000 tonnes. At the same time, imports of cheese in 1988 were about 114 000 tonnes and exports were negligible (6 tonnes). Cheddar, Edam, Gouda and Mozzarella are said to be the most popular types of cheese.

3   AFRICA

Total recorded cheese production in Africa in 1989 was 472 000 tonnes, of which Egypt accounted for 315 500 tonnes, i.e. ~67% of total recorded production in the continent. In 1989, cheese production in Sudan was in second place in Africa at 63 450 tonnes. South Africa, with a production of 34 715 tonnes in 1988, was the third largest producer and Niger was the fourth largest producer with 16 835 tonnes.

In South Africa, cheeses are exclusively European types: Cheddar, Gouda, processed cheese and all other types collectively represented 49, 42, 4 and 5%, respectively, of total production in 1982.

Although statistically they do not seem to be important, numerous types of cheese are made in Africa, in general at a village level or on a domestic scale in a number of African countries. Their technology, generally, is not well known to us, but in many instances the cheeses are derived from coagulated-acidified milks, which is almost a constant feature of cheesemaking in most regions of Africa.

3.1 Algeria

Cheesemaking in the Ahaggar region of southern Algeria was described by Gast *et al.*[7] In this mountainous desert region, two types of cheese are produced by nomadic people from goats' milk: a rennet-coagulated cheese, Takammart, made from whole goats' milk (92–93% solids, 42–44% FDM, 34–37% protein in DM, 0·5% NaCl) and a sour-milk cheese, Aoules, produced from naturally-soured milk from which most of the fat has been removed for butter production (composition: 87–92% TS; 11–20% FDM; 48–56% protein in DM; 1–1·5% NaCl). Both cheeses are very small, 15–90 g, and are air/sun-dried. Similar types of rennet cheese are produced in Niger.

Walter & Hargrove[5] mention what appears to be a basically similar cheese, Toureg, also called Tchoukou in Niger, which is made by Berber tribes from the Barbary States to Lake Chad. Skim milk is coagulated by animal rennet or by a rennet preparation from the leaves of the Korourou tree.

Aoules is a dry cheese made from goats' milk and obtained by heat precipitation of sour buttermilk. This flat cheese is very hard, due in particular to the absence of fat. Raw milk is allowed to sour naturally and is churned; subsequently, butter is removed and the buttermilk is poured into a pot and heated on an open fire until the proteins precipitate. The coagulum is strained in a straw basket and cheese whey drained off. The curds are then kneaded in small quantities and given the shape of a flat disk. Then the cheeses are sun-dried until completely dry. The cheese is always ground and minced with date paste before consumption and consumed with a beverage.

Takamart is a dry uncured cheese made from whole goats' milk. It has a square shape and a dark brown colour. Raw milk is coagulated in a wooden container, using a small piece of dry kid stomach. After a few hours, the coagulum is ladled onto a mat for whey drainage and subsequently is transferred to another mat from an aromatic plant (wild fennel) to give some flavour to the cheese. The cheese is sun-dried on these mats for 2 to 3 days and then finished off in the shade. It is kept in goat-skin bags or wrapped in antelope or cattle/calf skins.

3.2 Benin

Dairying is a traditional practice among the Fulani tribes, not only in Benin but all over the Sahelian region in Chad, Burkina Faso, Togo, Nigeria, Niger, Ivory Coast, Ghana, Mali and even in Mauritania. Traditionally, herds are kept by Fulani under semi-sedentary or transhumance systems. The animals belong to the shepherds or are entrusted to them. Each week, one day's milk production belongs to the shepherd and represents his reward for taking care of the animals. The milk is processed into fermented milk or into cheeses which can be found all over the region under the same name or, according to countries, bear a different name although they are identical or almost identical.

Woagachi is made in Benin by transhumant Fulani. It is also called Wagassirou (Bariba name) or Gassigue (Peuhl name). It is a soft fresh cheese without rind, made from whole cows' milk. The milk is boiled for about one hour and coagulated in about 30 min with the sap of *Callotropis procera* which is obtained by crushing leaves of the plant and added before boiling. No rennet of animal origin or starters are used. Subsequently, the coagulum is ladled into baskets used as strainers to drain off the whey. The curd is never pressed or cut. After the cheese is removed from the basket it is dipped for about 30 min in boiling water to which salt has been added. Wagassirou is preserved by salting, smoking or drying. It keeps for about a month and is considered as a staple food.

In Benin, another name for Wagassirou/Woagachi is Wagashi, although its technology is slightly different. Wagashi is made also in Burkina Faso, Ivory Coast, Ghana, Mali, Mauritania, Niger, Nigeria and Togo. Wagashi is a soft, brine-pickled cheese with no rind but small holes, made from whole cows' milk. The milk is boiled for about 5 min and coagulated by adding the juice of crushed stems of *Bryophylum*. No animal rennet or lactic starter is used. After coagulation, the coagulum is transferred to a cheese cloth. The whey is expelled by hand squeezing and the curd is formed into small balls of various sizes. They are salted and subsequently put in brine containing 25% salt where they keep for about 2 weeks. Wagashi is almost always fried before consumption.

3.3 Chad

Pont Belile is a fresh cheese made from goats' or sheep's milk. The milk is boiled for about three min and then cooled to 30–35°C. Rennet (as tablets) is used for coagulation but no starters are added. Coagulation requires 20 to 30 min at ambient temperature. The curd is then pressed using weights and salted in brine. Subsequently, it is dipped into boiling water to be stretched and plaited. Pont Belile is a 'pasta filata' type of cheese.

3.4 Egypt

Cheesemaking is a well-established tradition in Egypt. About eight types of cheeses are produced in Egypt, all of which originated from and are still produced in rural areas, although Domiati is also made industrially at milk plant level. Domiati (which is also called Gibbneh Beda, Damiati or Damieta), is a highly salted cheese and is considered in detail in Chapter 11. Cephalotyre 'Ras' is a hard bacterially-ripened cheese; Karish or Kareish cheese is a fresh, relatively low-salt type of cheese; Kishk, which may not be properly classified as a cheese, is produced nevertheless by a rather interesting method.[84] Brinza, a highly-salted Feta-type cheese of Russian origin, and small amounts of Kashkaval are also produced; these varieties are reviewed in Chapters 11 and 9, respectively. The production of Daani and Mish cheese should also be mentioned.

Daani is a soft cheese made from sheep's milk or from a mixture of sheep's and goats' milk. The milk is not heat treated and is allowed to acidify naturally until it reaches 20–25°C, i.e. for up to 30 min. Rennet is added, at 10 to 20 ml per 100 kg, and coagulation requires up to 2 h. The curd is cut and moulded in cheese-cloths on mats and left to drain for up to 2 days; then it is cut and dry-salted on the surface at the rate of 5%. It can be cured in brine (up to 18% salt), in which case it keeps for up to 4 months.

Kishk is produced from the fermented milk, Laban Khad, or its partially dehydrated variant, Laban Zeer (dehydrated during storage in porous earthenware jars through which water evaporates). Laban Khad is produced by fermenting milk in skin bags; after the milk has coagulated, the bags are shaken to gather the fat, which is removed, leaving behind Laban Khad in the bags. Apparently, rennet may be used to coagulate the milk in cold weather.

Kishk is manufactured by mixing two, three or more parts of Laban Khad or Laban Zeer with one part of wheat flour or par-boiled wheat and the mixture is then boiled and sun-dried. The product is non-hygroscopic and may be stored in open jars for 1–2 years without deterioration. The composition appears to be quite variable: 9–13% moisture, 2–12% fat, 9–24% protein, 31–65% carbohydrate and 6–10% ash.

Karish is a soft acid cheese made from skimmed cows' milk, buffalo milk or buttermilk from sour cream; apparently it is made only on farmsteads. Initially, it is made from Laban Khad (i.e. fermented buttermilk) or from sour defatted milk, Laban Rayed. The latter is prepared from fresh whole milk placed in earthenware jars and left undisturbed; the fat rises to the surface and the partly skimmed milk beneath sours. After 24 to 36 h, the cream layer is skimmed off and the clotted, skimmed milk (Laban Rayed) is poured on to reed mats or into small cheese moulds. After a few hours, the ends of the mat are tied and some whey squeezed out; the pressed curd is permitted to drain further and the squeezing process repeated until the desired texture is obtained; the curd is then cut into pieces and salted. Increased demand has led to the commercial production of Karish cheese which, under such conditions, is frequently made from pasteurized and/or homogenized milk or reconstituted milk using L. bulgaricus as starter and usually with rennet (3 ml/100 kg) rather than acid as coagulant. The approximate composition of Karish cheese is: 31% total solids, 17% protein, 6% fat and 4·5% NaCl.

Another technique is also used: coagulation of the blend of milk/buttermilk occurs naturally in 1–3 days. The coagulum is ladled onto mats (shanda) and the curd is left undisturbed for several hours; a small quantity of salt is sprinkled when the curd is firm enough. The mat, together with the curd, is rolled to facilitate wheying off. The curd is never pressed. This long cylindrical cheese is cut into pieces of equal size. If pickled in brine in earthenware pots, it keeps for up to a year; if it is intended for consumption as fresh cheese, its shelf-life is 1 to 2 weeks.

Cephalotyre/'Ras', of Greek origin (Kefalotyri) and known in Egypt as Ras, is the most popular hard cheese in Egypt. El-Erian et al.[10] concluded that the microflora of ripe Kashkaval (Cashkaval) and Ras cheeses are very similar. However, Kashkaval is a pasta filata-type cheese whereas Cephalotyre 'Ras' is not. In this review, Kashkaval and Ras are treated as separate varieties and the former is covered under Balkan cheeses where it is a major variety. Ras is also known as Romy (Romi) and a variety called Memphis is similar, if not identical. However,

Naghmoush *et al.*,[11] in a study on starter selection for Memphis cheese manufacture, give the impression that Memphis and Ras are distinct varieties.

Although an MSc. thesis was written on the manufacture of Ras cheese in Egypt by Tawab in 1963, the first published account on the subject appears to be that of Hofi *et al.*[12] and this appears to be the method used by most subsequent investigators although manufacturing procedures used in Egypt are not standardized. The manufacturing procedure described by Hofi *et al.*[12] is generally similar to that for Gouda except that the curd is cooked to 44°C in 15 min and held at this temperature for 30 min; a further difference is that Ras curd is salted in two stages: after cooking, the whey is drained off to the level of the curd and salt at 1–2% of the weight of the original milk is added to the curd, mixed and held for 15 min; the curd is then hooped and pressed overnight and the cheeses further salted by surface application of dry salt for up to 12 days. The cheese is normally ripened at 15–18°C for 4 to 6 months, during which extensive proteolysis and lipolysis occur. The cheese is not normally waxed or otherwise packaged, so considerable moisture loss occurs through the rind.

A considerable amount of information has been accumulated on various aspects of the ripening of Ras cheese. Data from Hofi *et al.*[12] on some of the gross changes that occur during the ripening of Ras cheese made from raw or pasteurized milk are summarized in Table I; changes, especially lipolysis, were considerably more extensive in the raw milk cheese during ripening. Formation of pH 4·6–soluble N, NPN, free amino acid profiles and volatile fatty acids in Ras cheeses made when using calf rennet, pepsins or microbial rennet substitutes have been reported.[13,14] Four main classes of carbonyl compounds were identified in mature Ras cheese: four methyl ketones, four alkanals, three 2-enals and two 2,4-dienals; pentanone, pentanal, 2-heptanal and 2,4-undecadienal were the principal carbonyls.[15] No evidence was presented on the significance of the carbonyls to Ras cheese flavour but it was assumed that they are significant contributors.

A surprising amount of activity has been directed toward accelerating the ripening of Ras cheese and various approaches have been employed, e.g. addition of hydrolysed casein or whey protein,[16] inactive dry yeast or yeast hydrolysate,[17] trace elements,[18-20] ripened cheese slurries (30°C for 7 days),[21] autolysed starter,[22] animal or fungal proteinases and lipases.[23-28] There appears to be general agreement that ripening may be accelerated and flavour intensified by all the above methods although some techniques led to off-flavours on extended ripening. The influence of coating materials on compositional changes and ripening reactions in Ras cheese have been studied.[29,30]

Because of the shortage of fresh milk in Egypt, which has a very rapidly increasing population, it is not surprising that there has been quite a lot of interest in the manufacture of Ras cheese from reconstituted skim-milk powder plus cream or butter oil. While the product appears to be acceptable, it is inferior to that made from fresh milk and ripens more slowly.[31-34] There is also an interest in using soybean milk to extend cow or buffalo milk; use of up to 20% soy milk gives satisfactory results and was reported to improve the quantity of Ras or

TABLE I
Analytical Data for Cephalotyre 'Ras' Cheese (from Ref. 10)

Age	Moisture (%)	Fat (%)	Dry matter (%)	pH	Salt (%)	Dry matter (%)	Protein				Lactose (%)	Volatile fatty acids (%)
							Total (%)	Dry matter (%)	Soluble (%)	Soluble protein coef. (%)		
Raw milk												
1 day	39·03	34·00	50·57	4·97	0·74	1·21	21·35	35·02	0·90	4·36	0·4	56·68
2 week	34·70	36·70	56·20	4·90	1·18	1·81	22·62	34·62	2·30	10·17	0·2	66·13
1 month	33·77	37·88	57·20	4·90	1·76	2·66	23·87	36·04	3·10	12·99	0·2	75·58
2 month	32·84	38·08	56·70	4·83	2·25	3·35	21·66	36·72	3·60	14·00	0·15	167·08
3 month	31·92	38·36	56·31	5·05	2·28	3·35	26·25	38·56	4·50	17·14	0·1	190·36
6 month	30·03	39·26	56·11	5·03	2·03	2·90	27·18	38·85	5·30	19·50	0·1	222·44
Pasteurized milk												
1 day	45·46	29·75	53·13	5·50	1·06	1·94	22·36	41·00	0·67	3·09	0·5	47·74
2 week	40·78	31·25	52·14	5·32	1·73	2·92	24·66	41·64	1·59	6·08	0·3	48·73
1 month	38·17	32·93	52·19	5·25	2·51	4·06	26·49	42·84	2·80	10·57	0·2	51·71
2 month	37·15	33·83	52·30	5·23	2·80	4·58	26·97	42·91	3·00	11·12	0·2	51·20
3 month	34·95	34·57	51·87	5·25	2·61	4·01	28·16	43·29	3·60	12·78	0·2	55·69
6 month	34·20	34·90	52·37	5·53	2·89	4·39	28·95	44·00	4·50	15·54	0·15	62·65

Cheddar cheeses made from buffalo's milk but quality deteriorated on addition of >30% of soy milk.[35] An alternative approach towards extending the available milk supply is the manufacture of processed cheese foods incorporating non-cheese ingredients. Shehata *et al.*[36] report the successful manufacture of such a product based on Ras cheese and skim-milk powder.

Mish (M.M. Hewedi, pers. comm. and Ref. 9) is a rather interesting, popular and apparently unique Egyptian cheese product. It is used as a savory or appetizer by better-off people but it is also used by the poor and in rural areas as a significant source of dietary protein. Mish is a pickling medium in which Karish cheese is stored for ripening; Karish cheese, after ripening in Mish, is called 'Mish cheese'.

Although details of Mish manufacture vary depending on the available ingredients, the general principles of its manufacture are roughly similar and may be summarized as follows:

1. The daily production of Laban Khad is collected and preserved by adding salt.
2. Mourta, the non-fat product from ghee manufacture, is added to and blended with the Laban Khad.
3. The mixture is concentrated by heating until it becomes viscous (pasty) and a reddish-brown colour develops (this colour is transferred to the cheese during pickling).
4. When required, this paste is diluted with whey or Laban Rayeb or Laban Khad and some Mish from a previous batch (as starter), flavourings and colouring materials are added.
5. This mixture, 'primary Mish', is used as a pickling solution for 'Karish cheese' in earthenware jars, 'Ballas'. The flavouring and colouring materials vary according to their availability but the most common flavourings are 'Murta', cinnamon, chilli and pepper, and the colouring agents are saffron and annato.

During ripening, the lactococci die and lactobacilli and spore-formers become dominant. The level of total solids increases due to the disintegration of small pieces of cheese, but the extent of the increase depends on the original cheese and Mish composition and ripening conditions as influenced by time and temperature.

Due to the many factors which influence the changes during ripening, Mish shows considerable variation in chemical composition (Table II).

The chemical and microbiological changes which occur in Mish cheese during ripening lead to the development of a typical and desirable flavour, which is a combination of Cheddar and Roquefort type flavour and a characteristic aroma of butyric and caproic acids (cf. Ref. 9 for further references on Mish cheese).

Mish is usually made without any particular or defined specification and this results in a wide variation in composition. The following are procedures by which it is possible to produce Mish with a unique or standard specification and with a typical desirable flavour.

TABLE II
Chemical Composition of Mish

	Minimum %	Maximum %
Moisture	54·76	75·68
Total solids	24·32	45·24
Fat	00·50	04·60
Protein	06·95	13·13
Ash	11·13	19·79
Calcium	00·229	00·403
Phosphorus	00·180	00·215
NaCl	12% (average)	

1. By mixing minced ripened hard cheese with milk or whey at an appropriate ratio; the flavouring and colouring agents to be added as previously described.
2. By mixing fresh cheese with whey or milk and adding 'old cheese flavour' to give the typical taste of Mish. The rest of the ingredients are as above.
3. By mixing yoghurt and adding 'old cheese flavour' (ripened cheese flavour), salt, spices and colouring materials as above.

3.5 Ethiopia

Apparently, the only indigenous cheese variety produced in Ethiopia is Ayib.

Ayib is made from sour milk or, more commonly, buttermilk and/or of their mixture. O'Mahoney (pers. comm.) described the production of Ayib as follows: surplus milk is accumulated each day and allowed to ripen naturally. When sufficient milk has accumulated, it is churned; the acidity of the milk varies between 0·85 and 1·1%. The butter is removed and the skim-milk (buttermilk) heated slowly to 35–40°C min; coagulation of protein and residual fat occurs in 20–40 min, depending on the acidity of the milk and the heating temperature. The mixture is held at 35–40°C for some time to complete coagulation; it is then allowed to cool and the whey and curd separated. The practice of smoking dairy utensils imparts a smoky flavour to the cheese, which has a short shelf-life because of its high moisture content (Table III).

3.6 Kenya

Mboreki Ya Iria seems to be the only type of cheese made in Kenya. It is a fresh soft cheese made by smallholders in the western part of Kenya from their surplus of cows' and/or goats' milk. The process may vary according to the type of milk being used. Milk is heated to boiling point and cooled subsequently to ambient temperature. Fermented milk is added as starter, together with salt. No

TABLE III
Chemical Composition of Ayib Cheese

Constituent	g/100 g
Water	71·5–76·6 (73·7)
Protein	12·5–16·9 (15·2)
Fat	4·7–11·1 (6·4)
Carbohydrate	2·6–3·6 (2·9)
Ash	0·8–2·5 (1·2)
Calcium	0·08–0·13 (0·11)
Phosphorus	0·18–0·22 (0·20)

Source: Ethiopian Nutrition Institute

rennet is used. Once a coagulum is obtained it is put into a cheesecloth and hung for draining until the next day.

3.7 Madagascar

Two types of cheeses, a semi-hard and a soft fresh type, are made in Madagascar. The generic names 'Fromage' and 'Fromage Blanc' are used, respectively.

Fromage (cheese) is made from whole cows' milk. It is a semi-hard cheese with a hard rind. The temperature of the milk is adjusted at 30°C, no starter is used and acidification develops naturally. Coagulation is brought about by adding rennet and it requires 1 day. The cheese curd is drained in perforated plastic moulds and pressed by hand. Then the curd is cut into 0·5 to 1 kg pieces and salted in brine at ambient temperature in wooden barrels. The pieces are then removed and placed on draining shelves until they are marketed.

Fromage Blanc (white cheese) has been introduced only recently. It is a fresh soft cheese made from skimmed cows' milk and is not cured. The milk is boiled for 5 min and cooled to 33°C. No starter and no additives are used. Rennet is added and after about one hour the coagulum is cut and transferred to cheese moulds, generally made of aluminium, where it is pressed for 18 to 20 h at 30–33°C. It is salted on both sides and subsequently cut into 100–200 g pieces before being wrapped in cellophane.

3.8 Mali

Wagashi cheese, which has been described under Benin, is made all over the Sahel. It is a staple food, in particular for Muslims and northern tribes. In general, it is fried before consumption.

3.9 Niger

Tchoukou is the traditional cheese made by the Foulse and Touaregs. It is a hard, sun-dried cheese made from cows', sheep's, goats' or camels' milk or their mixtures. It looks like a thin rectangular sheet, 2 mm thick, 30 cm long and 16 cm wide; when dry it has a yellowish colour. Milk is not heat-treated and is coagulated at ambient temperature. Traditionally, the rennet used is obtained by soaking lamb or kid abomasum in strained water; alternatively, commercial rennet may be used. After coagulation, the curd is ladled onto straw mats, pressed slightly and turned over on the main mat to remove as much whey as possible. The cheese is transferred to another mat where it remains for 1–1·5 days for sun-drying. It is never salted and is kept in baskets where air is allowed to circulate freely. The shelf life is about 3–6 months.

3.10 Nigeria

In Nigeria, as in the other Sahelian countries where herds are kept by Fulani tribes, whether they are nomads, settled or semi-settled, milk processing, although elementary, is a common practice. The systems of cattle keeping among these tribes were described by Waters-Bayer[32] and some specific features have already been mentioned. Milk provides about 10% of the energy requirements of settled Fulanis and somewhat more of their protein requirements. Milk is mostly processed into butter, ghee and a skimmed fermented milk, called 'Nono', which together with millet flour, provides the base for making their national dish, 'Fura'. In the same way as full cream sour milk, 'Kindirmo' is also mixed and cooked with cereal flour to get 'D'ambou'. Overall, cheese production remains limited.

A soft, white unripened cheese (Wara or Awara; similar to Wagashi and Tchjoukou made by other Nigerian tribes) is also made in central Nigeria in the wet season when milk yields are relatively high. Whole fresh milk is heated in a pot over an open fire until the milk almost reaches the boiling point; some sour skim-milk is added (1:6 ratio). When the curds have formed, the mixture is placed in a shallow basket to drain, with turning, for about 24 h. The drained cheese is cut into 10–15 g cubes and fried in palm oil. The cheese is usually sold at local markets and is dipped in a chilli pepper sauce before eating (A. Waters-Bayer, pers. comm.).

A drier form of a similar cheese, 'Chukumara', is made in northern Nigeria and Niger from cows' milk but more commonly from camels' milk. The cheese, which has a fairly tough texture, is transported and stored in one piece from which portions are broken off and eaten as it is or, more commonly, after frying (A. Waters-Bayer, pers. comm.).

In the Katsina area of northern Nigeria, a cheese, Dakashi, is made by heating colostrum to the point of coagulation. No additions are made, before or after heating, and the curds are not drained (A. Waters-Bayer, pers. comm.).

In the Wasa and Wava/Zange grazing reserves a new cheese was introduced recently by the FAO/UNDP project for rural development and is developing steadily; it is called 'country cheese' and its technology is described below.

Country cheese is a hard cheese; up to 58–60% dry matter and 45% fat. Production is being developed at village level or in small dairies. Whole raw cows' milk is coagulated by means of powdered rennet in about 30 min at 32–34°C. The curd is cut until the curd gains reach the size of beans and is then stirred for 20 to 30 min at 32–34°C. About 40% of the whey is removed and is replaced by water at 40°C. When the curd grains are dry enough, the curd is left to settle. Most of the whey is removed and the cake of curd is pressed (10 kg/kg of curd) for about 40 min. Then, it is cut into 1 kg pieces and put into moulds, which are immersed in hot whey at 80–85°C for 30 min. When the cheeses are completely cooled, they are removed from the moulds and salted. The next day the cheeses are smoked for 2 to 4 h. Once cooled, they are wrapped with plastic film under a vacuum and may be stored for several weeks. Production is confined to the peak season for milk production.

3.11 Sudan

The main cheese made in Sudan is Karish which is similar to that made in Egypt, but at least four other types of cheese are made: Braided cheese, Gibbna, Mudafara cheese and white cheese.

Braided cheese is a semi-hard variety made from cows', goats' or sheep's milk. Milk is heated to 35–40°C and rennet is added; coagulation is completed in about 5 h. The coagulum is transferred to a cheese-cloth on a wooden frame for whey draining. Salt is added (10–15%) together with black cumin. The curd is heated to develop its elasticity, cut in strips of 5 × 15 cm and braided. These pieces of cheese are salted in brine and transferred to a 10–15 kg tin which is filled with brine and sealed. This cheese, which is very popular, is consumed as a staple food and is made at family level or in small dairy plants.

Gibbna resembles Feta or Domiati cheeses. It is a soft white cheese pickled in brine. After the milk has acidified naturally without addition of starters, 10% salt is added and the milk is coagulated with commercial rennet. Coagulation requires 5 to 6 hours at ambient temperature and the curd is transferred to large moulds/frames lined with cheese-cloths. It is pressed overnight and subsequently cut into blocks of 10 × 10 × 10 cm or 10 × 10 × 20 cm, i.e. of 1 or 2 kg. These blocks are transferred to tins filled with brine made from cheese whey. Another variety called Gibbna Roumi (Greek cheese) is produced in the same way but it is made into spherical shapes and it is also stored in brine.

Mudafera cheese is a semi-hard cheese made from whole cows' milk. The milk is heated at 37–38°C, i.e is kept at body temperature, and rennet tablets are added. No starters are added and acidification occurs naturally. After coagulation, whey is removed and the curd is heated close to melting point in hot water to develop its elasticity. It is worked and stretched into long threads which are given the shape of wool hanks. The hanks are put in 10–15 kg tins which are filled with brine and welded. The cheese keeps for several months at ambient temperature.

White cheese is the traditional type of cheese found in the Middle East and the Eastern Mediterranean basin. It is a soft white cheese made from cows' milk or its mixture with goats' or sheep's milk. The milk is salted and heated close to 49°C. Rennet tablets are added when the temperature has cooled to about 37°C; coagulation requires 3 to 5 h. The coagulum is transferred to cheese-cloths placed inside a wooden frame. Whey is drained off until the volume of the curd decreases to 50%. Then the cake of curd is pressed overnight and cut into 5–8 cm pieces, which are kept in brine in welded tins.

3.12 Zaire

On the basis of information available, it appears that only one variety of cheese is made in Zaire.

Mashanza is a soft fresh cheese made from raw whole cows' milk. Acidification develops naturally. No rennet is used for coagulation, which takes about 5 days if milk is poured all at once in the large container where acidification occurs. It is reported that acidification proceeds more rapidly when milk is added after each milking. The coagulum is transferred to a jute bag, whey is removed and the curd is kept in a wooden container or wrapped in banana leaves. The shelf-life is limited to about 1 week.

4 NEAR EAST — MIDDLE EAST

Cheesemaking has been practised for centuries in the Near East and Middle East, originally as a convenient way of preserving milk produced by nomad and transhumant tribes; once settled, herders continued to make cheese which, together with soured milks, had become a traditional component of people's diet. The technology remains somewhat primitive although observation and ancestral habits in this respect have made milk processing a pragmatic ritual which is scrupulously observed by milk processors, usually women. Numerous products are identical or very similar, although they bear different names. This again is probably a consequence of nomadism.

4.1 Cyprus

Anari is a soft cheese made from a mixture of cheese whey (obtained when making hard cheese) with a small quantity of sheep's or goats' milk. Cheeses can be rectangular or hemispherical and always weigh less than 500 g. Its composition is subject to variation: dry matter content is said to be 30 to 50%, of which fat may vary between 10 and 30%. No starter and no rennet are used. The mixture of milk and cheese whey is left to acidify for about 30 min and then heated slowly until the temperature reaches 80–90°C. During this phase, gentle stirring is applied. Coagulation starts after about 30 min and is complete after about

1 h. Whey is removed from the curd by heating and gently pressing. The curd is cut and either put in cheese-cloths or in woven baskets: in the first case, it is pressed for about 30 min, whereas in the baskets no pressure is applied. Subsequently, the curd is cut into rectangular blocks and dry salted; cheeses formed in baskets are never salted.

Graviera is a hard cured cheese with small eyes, made from sheep's milk. Dry matter is in the region of 60% and fat content may vary from 40 to 50%. Milk, with a fat content adjusted at 3%, is not heat-treated and no starters or additives are used. Fresh sweet milk is coagulated with rennet (about 2 g/100 litres) in 30 min at 35°C. The coagulum is cut into small pieces, the size of a bean and stirred for 30 min at 35°C. The temperature is then raised, first to 50–52°C and then to 58–59°C with stirring for 30 and 20 min or so, respectively. As in the case of Gruyère or Emmental, the curd is collected in a large cheese-cloth and pressed for about 15 h, the pressure being increased progressively. This cheese is dry-salted on both sides for about 20 days and cured in a room at 10–18°C and a humidity of 85–95% for 3 to 4 months. Sometimes, the rind is rubbed with olive oil.

Halloumi is the principal cheese traditionally made in Cyprus. It is a semi-hard cheese preserved in brine and made from sheep's milk. The manufacture of Halloumi is described by Davis,[6] Scott[38] and Anifantakis & Kaminarides.[34] The unique features of its manufacture are: (1) after separation of the curds, the whey is heated to 80–90°C (depending on acidity) to coagulate the whey proteins which are recovered and used in the manufacture of Anari, a cheese generally similar to Ricotta; (2) the deproteinized whey is maintained at 90–95°C and pieces (10 × 10 × 3 cm) of pressed curd which are held for about 30 min are added to it (Fig. 1). The cooked curds are not stretched as in Pasta Filata type cheeses, but are drained, cooled and sprinkled with salt (5%). The cheese is then ready for consumption or it may be stored in cans filled with salted whey.

The composition of commercial Halloumi cheese is quite variable (Table IV) (cf. Ref. 39). The microbiology and biochemistry of the ripening of this cheese, which appears to be a hybrid between the high-salted Feta and the Pasta filata varieties, common in the Middle East, have received little attention to date, but studies have been initiated by E. Anifantakis and colleagues of the Agricultural College, Athens.[40-42]

Nowadays, Halloumi cheese is also made industrially from pasteurized milk; industrial production tends to replace traditional Halloumi cheese. In Cyprus, one can find a kind of Halloumi moulded in wicker baskets under the denomination of 'Cyprus'. In Lebanon, Helloum, a cheese of this type, is spherically shaped by hand; it is cured for several days before it is cooked.

Kashkaval is a semi-hard cheese of the 'Pasta Filata' type, made from sheep's milk. Raw whole milk is held for up to 30 min to ripen; calcium chloride (10–15 g) and a mesophilic lactic starter (0·5–1·5%) are added per 100 litres of milk. Coagulation is complete in 30 to 50 min with about 35 ml of rennet. The

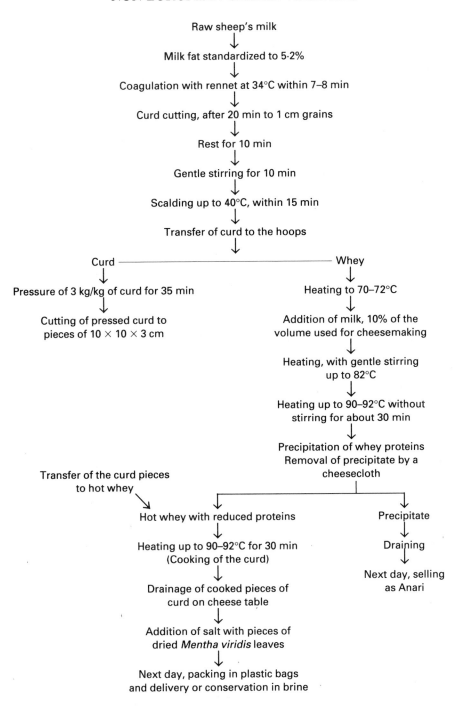

Fig. 1. Technology of making Halloumi cheese from sheeps' milk (from Ref. 39) (reproduced by permission of the *Australian Journal of Dairy Technology*).

TABLE IV

Composition of Halloumi Cheese made from Sheep's Milk and of Halloumi Cheese from the Cyprus Market (Mean of 13 and 17 Observations Respectively) (from Ref. 49)

Components	Halloumi from sheep's milk			Halloumi from the Cyprus market		
	Mean	Values range	Standard deviation	Mean	Values range	Standard deviation
Moisture (%)	42·15	39·04–43·64	1·39	42·53	35·46–48·56	3·75
Fat (%)	27·85	26·25–29·25	1·76	25·57	20·00–29·50	3·14
Fat in dry matter (%)	48·09	46·10–49·99	1·95	44·52	37·95–50·48	3·98
Protein (total N% × 6·38)	23·71	21·95–25·02	1·02	24·46	20·86–30·45	2·28
Protein in dry matter (%)	41·02	39·22–44·26	2·37	42·53	40·11–48·62	2·30
Soluble protein (%) (soluble N% × 6·38)	0·76	0·64–0·89	—	1·15	0·83–1·55	1·14
NaCl (%)	1·44	1·05–2·05	0·28	3·54	2·31–5·65	0·91
pH	5·86	5·30–6·10	0·22	—	—	—

coagulum is cut into 7–8 cm pieces and then left to settle for about 5 min: it is cut further until the size of the grains is about 0·5 cm. It is stirred vigorously for 20 min while the temperature is raised to 40–42°C. The curd grains are left to settle to the bottom of the vat, the cake is pressed for 1 to 2 h and the whey removed. The curd is then cut into pieces which are held for up to 10 h at 20°C until the pH reaches 5·2. At this stage, the curd is sliced and cooked in brine for up to 5 min; subsequently, it is kneaded and put into moulds. Every second day, the cheeses are dry salted for about three weeks and then waxed. Curing takes up to 2 months at 12–16°C. This cheese is made also in Lebanon and Tunisia.

Kefalotyri is a hard, cured cheese made from raw sheep's milk, often blended with goats' milk. Thermophilic and propionic starters may be added. Coagulation occurs in 20 to 30 min at 32–35°C. The coagulum is cut into 1 cm pieces and stirred for 30 to 40 min, while the temperature is raised to 43–45°C; stirring is continued for 15 min at that temperature and the curd is then moulded and pressed for up to 12 to 18 h. It is salted in brine for 48 h or dry-salted. The cheeses are waxed or coated with plastic film before curing for 4 to 5 months at 15°C. Cured cheese is wrapped in paper or plastic film and stored at 5°C.

Paphitico is a hard, cured cheese made from raw sheep's milk or from a mixture with goats' milk. No starters or additives are used. In the traditional process, rennet from lamb or kid abomasum is used. Coagulation takes 30 min at 33–35°C. The coagulum is ladled into moulds and pressed gently by hand; more curd is added until the moulds are full. The cheeses are turned over and put again into the moulds (often baskets). Pressing is by hand and then by piling up the cheeses on top of each other. Pressure is adjusted to three times the weight of the cheeses for a few hours. After removal from the moulds, the cheeses are dry-salted by rubbing with coarse salt and cured for up to 3 months.

4.2 Iran

FAO statistics indicate that cheese production was 112 303 tonnes in 1989. In 1981, according to the same sources, production was estimated at 107 000 tonnes. However, Kosikowski[43] reported a figure of nearly twice that for pickled white cheese alone.

Liqvan cheese is famous in Iran and known as the best white cheese made in the country. Its production is limited to the Liqvan valley near Tabriz. It is similar to Feta or Bulgarian white cheese which are described in Chapters 9 and 11, Volume 2. Actually, Iran, which previously imported Bulgarian cheese to supplement national production of white cheese, has become a major importer of Feta-type cheese from European countries. Liqvan cheese is made, in general, from sheep's milk and occasionally from surplus cows' milk. Traditionally, curing in brine was carried out in rooms dug in the ground, with access through a narrow gallery just wide enough for a man.

White cheese, known also as Iranian cheese, is made throughout Iran in the mountainous areas where sheep are kept. Its technology is similar to Liqvan and Bulgarian cheeses. Milk is heated to 65°C or above for 30 min. After it has cooled to 30–38°C, mesophilic starters may be added. It coagulates in about 1 h. It is left for 2–3 h to ripen and is removed from the vat in a large cheese-cloth to remove the whey. It is slightly pressed at the rate of 4–5 kg/kg of cheese. Then it is cut into square pieces and dry-salted; it may also be stored in saturated brine.

Similar cheeses are made in other countries in the Near East, with some variations in the technology, shape or presentation. As examples, one should mention: Akawieh, Baladi (Baida), Chelal, Hamwi, Na'aimeh in Lebanon and Syria, white cheese in Qatar and Turkey and soft cheese in Iraq and Jordan.

4.3 Iraq

Awshari, the principal cheese in northern Iraq, is produced on a farmyard scale from a mixture of sheep's and goats' milk in the proportions available. The manufacture, composition and characteristics of this cheese have been described by Dalaly et al.[44] Directly after milking, the milk is coagulated using a home-produced rennet extract (prepared by extracting a mixture of dried lamb's stomach, sugar, alum, black pepper, zingibel and cloves with a 5% NaCl brine). The coagulum is cut, NaCl is added to a level of about 1% and the mixture is heated to 45°C with stirring for 15 min. The whey is drained off and the curd transferred to long muslin bags which are pressed continuously by hand. As the curd accumulates at one end of the sack, the sack is twisted to compress it into a sphere. The drained whey is collected for making 'Jagi'.

Dry salt is rubbed on the surface of the cheeses twice daily for 4 days. The cheeses are then placed in sheep or goat skins, a measured amount of dry salt is added, the skins are closed and stored in a cool place for 3 weeks. Expressed whey is drained off, further salt is added and ripening continued for a further 6 weeks with intermittent removal of whey.

TABLE V
Composition of Ashwari Cheese (5 Samples)[a]

Moisture	(%)	37·55–47·19
Fat	(%)	12·60–33·80
Total N	(%)	3·42–4·06
Soluble N	(%)	0·51–0·71
Amino acid N	(%)	0·30–0·50
Acidity of cheese	(%)	0·49–2·12
pH		5·15–6·28
Acid value of fat		4·49–16·27
Total volatile acids		73–124
Salt	(%)	2·62–4·14
Salt in moisture	(%)	5·55–10·89

[a]From Ref. 44; the author comments that the cheeses may have been soaked in water and hence some of the soluble constituents may have been removed.

After 60 to 70 days, the cheese, which has become hard, is transferred to a dry skin in which the spaces between the cheeses are filled with Jagi. The cheeses are stored thus for at least 2 weeks and perhaps for as long as 6 months. Old cheeses may be very hard and it is common practice to soak them in water for 2 to 7 days, after which the surface of the cheeses is coated with Jagi and stored in clean dry skins.

Jagi is prepared by either of two method:

1. Fresh milk (about 5%) is added to whey collected from cheese manufactured the previous day, and the mixture heated at 70–90°C until the milk and whey proteins precipitate. The precipitate is recovered by filtration and mixed with diced garlic or kurrat in the proportion of 5:1.

2. The buttermilk obtained on churning sour milk is boiled over a direct flame; the curd which forms is skimmed off and mixed with wild herbs (garlic, onion, dry roses) and salt.

TABLE VI
Composition of Jagi (from Ref. 44)

		I	II
Moisture	(%)	48·25	49·09
Fat	(%)	8·42	13·45
Total protein	(%)	31·98	24·81
Soluble N	(%)	0·896	0·868
Amino acid N	(%)	0·868	0·840
Acidity	(%)	2·34	1·33
Total volatile acids		118	40
pH		5·65	5·85
Salt	(%)	3·71	5·09

TABLE VII
Chemical Composition of Iraq White Soft Cheese
(from Ref. 45)

Moisture	(%)	46·69–60·54
Fat	(%)	20·25–28·25
Acidity	(%)	0·52–1·85
pH		5·30–4·40
Total N	(%)	2·75–3·33
Soluble N	(%)	0·16–0·40
Lactose	(%)	0·50–0·95
Ash	(%)	1·61–4·60
NaCl	(%)	0·13–2·83
Herbs	(%)	1·48–2·68

Mature Awshari cheese has a sharp peppery, sometimes rancid, flavour and a hard, rather brittle texture with some cracks and mechanical openings. The viable bacterial counts in five commercial samples were very low, ranging from 450 to 1220 cfu/g; the microflora was mixed, consisting of lactococci, micrococci, lactobacilli and spore-formers. The chemical composition is variable; the range of composition for five commercial samples analysed by Dalaly et al.[44] are shown in Table V. The composition of Jagi is shown in Table VI.

The composition (Table VII) of farmers' soft cheese, produced on farms in Iraq, was described by Saleem et al.[45] It is not stated whether the cheese was produced by acid or rennet coagulation. The cheese is sold fresh, usually with added herbs, especially garlic, and is frequently referred to as garlic cheese.

Davis[6] does not mention Awshari but includes brief descriptions of 'Roos' and 'Meria', the method of manufacture for which appears to be somewhat similar to that of Awshari, although no mention is made of the use of Jagi; he also mentions a fresh acid cheese, Biza or Fajy.

Hallom is similar to Halloumi made in Cyprus.

Soft cheese is also produced in Iraq; it is similar to white cheese and Iranian cheese.

4.4 Israel

Consumption of cheese in Israel is well up to European standards at 14 kg/caput[46] (see Table III, Chapter 1, Volume 1); fresh cheeses, at 10·6 kg/caput, represent the majority of the cheese consumed. According to Davis,[6] most of the hard cheeses produced in Israel are European types, e.g. Edam, Emmental, Provolone, Kashkaval, Blue, Bel Paese, Brinza (Feta type).

4.5 Jordan

Djamid is a hard sun/air dried cheese made from sheep's or goats' buttermilk obtained previously through churning acidified cream. No rennet is used.

Acidified buttermilk is heated to 50–60°C for 30 to 60 min. Coagulation of protein starts after about 30 min and develops as the temperature is increased. Cohesion of the precipitate is obtained by stirring and heating. It is then collected in a cheese-cloth and after pressing, it is made into a spherical shape by hand. It keeps for up to 12 months in earthenware containers at ambient temperature. .

Similar cheeses are produced in other Arab countries and are called Djemid, Djibdjid, Djibdjib or Kaschkajal.

Shankalish is a hard cheese made from sheep's and goats' milk. Fat is removed by natural skimming and the skimmed milk is coagulated with rennet. After 24 h, the curd is transferred to a bag for draining. It is salted (up to 25–27% of salt is added). Spices are also added to the curd (anis seeds, cumin, paprika, thyme) and it is formed into small spheres. Curing takes 1 month. The cheeses are cleaned and either put under olive oil or sun-dried.

Soft cheese, similar to white cheese, is also made in Jordan.

4.6 Lebanon

Akawieh belongs to the group of fresh cheeses. It is a soft cheese made from whole goats', sheep's or cows' milk. The process used to make Akawieh is similar to that used for white cheese.

Baladi or Baida is a type of white cheese with a square shape made in Syria also.

Chelal cheese, which is also made in Syria, has the shape of strings or ropes.

Hamwi is similar to Baladi but has a cubic shape.

Helloum is similar to Akawieh, but after drainage of the whey, without pressing the curd, sesame seeds are added to give the cheese its typical flavour.

Fresh cheese is a soft cheese made from cows', goats' or sheep's milk. It is coagulated at milking temperature by rennet (tablets) or by using a piece of stomach from a young kid or lamb. The milk coagulates in 15 to 30 min; the coagulum is broken, the curd left to settle and shaped into balls. When forming the balls, a hole is made with the thumb and a little salt is put into the hole. No curing is reported and it keeps for about 1 week. It can also be kept in a bed of dry salt until it becomes hard; then it is transferred to brine.

Karichee is a soft cheese made from cheese whey from cow and goat cheese. The whey is heated at 80°C for at least 1 h until the proteins coagulate. They are removed from the surface and whey is drained off. It is similar to Ricotta, although no skimmed milk is added before heating. It must be consumed fresh.

Umbris is a soft, spreadable cheese made only from raw goats' milk. The milk is poured into an earthenware jar with a small hole in its lower part which is closed with a wooden plug. Mineral salt is added and the jars are stored for about 1 week in a dark, cool place. When coagulation is complete and the whey separates from the curd, the plug is removed and replaced by wheat straws to allow the whey to drain off. Then the jar is filled up again with milk and the same process is repeated throughout the season, or until the jar is full of curd. The curd is then removed and stored in wide-necked containers, covered with olive oil, or drained in bags and stored in brine.

4.7 Qatar

White cheese, which has been mentioned previously, is made also in Qatar.

4.8 Syria

According to Abou Donia and Abdel Kader,[47] most cheese is Syria is produced by women at home, from sheep's milk. They list four types of indigenous cheeses: Hallomm (a soft cheese) and three hard varieties, Mesanarah, Medaffarah and Shankalish, the manufacture and composition of which are described below. Walter & Hardgrove[5] list Labneh as a sour milk cheese of major importance in Syria. However, Labneh is usually regarded as a concentrated yoghurt rather than a cheese; obviously, the dividing lines are blurred. The preparation and characteristics of various forms of concentrated yoghurts are described by Tamine & Robinson.[48]

For the preparation of Mesanarah, whole sheep's milk is renneted for 4 h; the coagulum is placed in cheese-cloth bags and left to drain for 8 h. The drained curd is cut into small pieces ($3 \times 3 \times 2$ cm), sprinkled with dry salt and held for 18 to 24 h. The curd pieces are then scalded by boiling in brine (10% salt) for 5 min, during which much fat is lost. Nigella grains are pressed into the hot pieces of curd which are then immersed in saturated brine at room temperature for 1 week. The salted, spiced pieces of cheese are removed from the brine and sun-dried for 2 to 3 days prior to storage in tight containers. The cheese pieces are soaked in water for 24 h before consumption.

For Medaffarah, whole sheep's milk is coagulated with rennet and the coagulum drained as for Mesanarah. The drained coagulum is cut into large pieces and pressed until sufficient acidity has developed (as judged by stretchability after heating a sample to 75°C). When the curd is sufficiently acid, it is cut into small pieces ($3 \times 3 \times 3$ cm), warmed in water at 75°C for 3 min, kneaded and pulled to form 'cords', three of which are braided into a tress which is cut into pieces about 8 cm long. The cheese tresses are immersed in saturated brine for 1 week and then sun-dried for 2 to 3 days. Obviously, this is a Pasta filata type cheese.

Shankalish: this cheese, which is also produced in Jordan, is made from renneted partially-skimmed milk (gravity creaming and skimming). The coagulum is drained in cheese-cloth bags for 24 h and the curd then salted at the rate of 7%

TABLE VIII
Chemical Composition of Mesanarah, Medaffarah and Shankalish Cheese (from Ref. 47)

Constituent	Mesanarah	Medaffarah	Shankalish
pH	5·2	5·0	4·5
Moisture %	26·6	28·6	30·2
Fat (% in dry matter)	26·2	27·1	17·7
Protein (%)	34·3	35·6	46·6
Ash (%)	6·9	7·0	7·0
NaCl (%)	5·6	5·6	5·5

and spiced with thyme, aniseed, paprika, nigella and cumin. The salted, spiced curds are kneaded into balls, 3–4 cm in diameter, which are ripened in a dark, moist room for one month, during which a sharp flavour develops. The cheeses are then cleaned and either stored by immersion in olive oil or sun-dried for 2 to 3 days, both of which stop ripening.

The pH and composition (Table VIII) of samples of the three cheeses after ripening are reported by Abou Donia & Abdel Kader[47] but no further information on ripening appears to be available. Apparently, no starters are used for any of the cheeses; presumably, fermentation is by indigenous microorganisms and is probably variable.

The authors note that the composition of these cheeses is similar to those of other primitive varieties produced in several countries in Asia, Africa, Spain and Portugal, as reported by Davis.[6]

4.9 Turkey

FAO[1] estimated that cheese production in Turkey was 142 430 tonnes in 1989 and reported production in 1982 to be 128 000 tonnes. The principal variety is the Feta type, Teleme.

Beyaz Peyneri is a semi-hard cheese cured in brine and made from sheep's milk only or mixed with goats', cows' or buffalo's milk. Milk is pasteurized at 62–65°C for 5 to 15 min. Calcium chloride and lactic starters are added and the milk ripened for 15 to 20 min before renneting; coagulation with rennet takes 1·5 to 2·5 h. The coagulum is cut into 2 cm pieces and left under the whey for 30 min. Then it is transferred to a cheese-cloth and pressed in bulk until all the whey is removed. The curd cake is cut into pieces of about 0·5 kg, salted in brine for 4 to 6 hours and then put in tins filled with brine and stored for 4 to 6 months.

Kasar Peyneri is a hard cheese of the Pasta Filata type made from raw sheep's milk only or mixed with raw goats' milk. No starters or additives are used. Co-agulation takes 1 to 1·5 h at 28–33°C. The coagulum is cut into pieces of 0·5–1 cm and left to settle. Then it is gathered into a cheese-cloth and pressed between two boards for up to 4 h. Some pressure is maintained for 24 h. The curd is cut then into pieces which are immersed in hot water (65–75°C) in small baskets. The curd is formed into spheres by hand when still hot. After cooling, cheese is dry-salted daily for up to 60 days and cured at 12–19°C for 2 to 3 months.

Mihalic Peyneri is a hard cheese made from raw sheep's milk and cured in brine. Coagulation by rennet takes 2 h at 25–35°C, after which the coagulum is cut into pieces of 0·5 cm and stirred for 10 min at the same temperature before it is heated to 40–50°C in not more than 30 min. Curd grains are left to settle and are then collected in cheese-cloths and held for 4 to 8 h. The curd is cut into pieces of suitable size and salted on the surface and in brine for up to 10 days. The pieces of cheese are arranged in casks and kept in brine for 60 to 110 days at a temperature which may vary from 15 to 18°C.

Teleme, also called Brandza de Braila, is a pickled cheese made from goats' or ewes' milk. It is very much like Greek Feta. Fresh milk is set with rennet at about 30°C; the coagulum is broken up, and the curds collected in a cloth, covered, and pressed with a weight. The pressed curd is cut into pieces about 7–10 cm square and 3–5 cm thick, which are immersed in salt brine for one day. The cheeses are cured for 8 to 10 days, either in dilute salt brine in a cask or packed in layers in a metal container, with salt between the layers. The cured cheese is white and creamy. About 10 kg of cheese are obtained from 50 kg of ewes' milk and 75 kg of goats' milk. The analysis of the content is as follows: moisture, 28·3%; fat, 37·5%; protein, 30%; and salt, 2·4%. Teleme is made also in Bulgaria, Greece and Rumania.

Tulum Peyneri is a hard cheese made from sheep's milk or from its mixture with goats' and/or buffalo's milk. Skimmed milk or milk with adjusted fat content can also be used. Coagulation takes from 75 to 100 min at 30–40°C. After coagulation, the coagulum is cut into pieces of 0·5–3 cm, heated to 50°C for 10 min and the curd collected in a cheese-cloth which is hung for draining. The curd is then broken into 5 cm pieces and dry-salted. It is put subsequently into a goat skin (or in some cases in the rumen of a sheep or goat) and kept under anaerobic conditions for 3 to 4 months.

Zomma is a plastic curd which is very much like Katschkawalj.

4.10 Saudi Arabia

Ekt. Abou Donia[49] claims the first reported study of Ekt cheese produced in the Saudi Arabian desert and also in surrounding countries, Kuwait, Iraq, Jordan and Yemen.

Ekt is produced traditionally from sheep's milk which is boiled for about 30 min to evaporate some of the water. The heated milk is cooled to 45°C, inoculated with a small piece of cheese or curd and left for 4 to 6 h. The coagulated milk is put into a goat-skin churn and agitated to produce buttermilk and butter, which is removed. The curdled buttermilk is ladled out of the 'churn' and individual ladle-fulls placed on pieces of special matting and drained for about 10 h. The drained pieces of curd are kneaded by hand and placed on fennel mats from which they absorb some aroma. The curd mass is broken into small pieces (2–3 × 1–2 × 0·5 cm) and sun-dried for 2 to 3 days and later in the shade to the desired degree of dryness. Before consumption, the pieces of dried cheese are reduced to a powder which is dispersed at about 20% in cold water and fermented (type of fermentation not described).

The microflora of the cheese is reported[41] and the average chemical composition is: pH = 4·2; % moisture = 11·0; % fat-in-dry matter = 15; % protein = 49; % ash = 4·5; % NaCl = 1·2.

4.11 Yemen

A very brief report on Taizz cheese, made only in the town of Taizz, North Yemen, was made by El-Erian *et al.*;[50] development of a successful standard manufacturing procedure was claimed, but details of this or of the normal method were not given. The commercial and modified cheeses contained 32·2 and 31·5% moisture, respectively.

5 LATIN AMERICA

Recorded cheese production in the whole of Latin America was 628 561 tonnes in 1981 and 705 251 tonnes in 1989 (Table IX).[1] Considering the size of the population, this production is very small compared with many European countries, e.g. the Netherlands (561 275 tonnes, 14·8 million people). External trade in cheese is also very small: imports, 58 115 tonnes and exports, 17 356 tonnes. However, Argentina is a very significant cheese producer and is third only after the USA and Canada in the entire American continent.

TABLE IX
Cheese Production (Tonnes) in Latin America, 1989 (from Ref. 1)

Central America		*South America*	
Costa Rica	5,980	Argentina	281,000
Cuba	16,700	Bolivia	7,180
Dominican Republic	2,500	Brazil	60,150
El Salvador	15,397	Chile	24,700
Guatemala	16,119	Columbia	51,000
Haiti	—	Ecuador	13,181
Honduras	8,305	Peru	16,870
Jamaica	—		
Mexico	106,410	Suriname	—
Nicaragua	9,079	Uruguay	15,500
Panama	250	Venezuela	55,000
	180,740		524,581
Total	705,321		

According to Kosikowski,[43] many of the cheeses produced in Latin America are of European origin, mainly Gouda, Romano, Mozzarella, Ricotta and Manchego. Local names are given to many of the cheeses, even though their characteristics and manufacturing procedures are similar to those of their European counterparts, e.g. Oaxaca (also called Asadero) of Mexico and Chiclosa of El-Salvador are similar to Mozzarella. However, there are a number of traditional cheeses in Latin America which can be classified in two main groups, i.e. the soft and the hard/semi-hard types. At least five cheeses belong to the 'pasta filata' type and processed cheeses are produced also.

Kosikowski[51] and W. L. Dunkley (pers. comm.) comment on the problem of classifying Latin American cheeses and on the need for such a classification scheme as an aid to trade. They also commented on the range of sophisticated technology used in cheese manufacture — while most cheese is produced on farmsteads or in very small, primitive factories, there are also several modern, highly mechanized factories, especially in Argentina, Chile and Uruguay.

Within each of the aforementioned groups of cheeses, differences can be observed from one cheese to another which relate principally to the nature of each group. Differences can be observed which relate principally to the nature, composition or properties of the milk as well as to the scale and conditions of manufacture. These differences relate also to variations in technological process, similar to those one would observe in other countries when comparing the techniques used to make the same type of cheese, or very similar types, either in different regions or under variable climatic conditions, for instance. When going into detail, one will notice that these differences refer to whether the milk is heat treated or not, the coagulation temperature and time, the kind and quantity of rennet, whether the coagulum is cut or not, the size of curd grains, the conditions of stirring and heating the curd, whether pressing occurs or not, salting, curing and storage.

Have any of the cheeses produced in Latin America an indigenous origin, *sensus stricto*? The matter may be debated; however, most of them were introduced by the Spaniards and Portuguese when they colonized this part of the world and, therefore, have been produced for several centuries. Others, e.g. those produced industrially in Argentina, Chile or Uruguay, have a more recent origin and were introduced by settlers.

The oldest cheese is probably the Altiplano which was already made in Bolivia, on the Altiplano, at the time of the Incas.

Kosikowski[51] listed 24 indigenous varieties of which Queso Blanco is the most important; Walter & Hardgrove[5] listed a further 15 cheeses not mentioned by Kosikowski[51] and FAO[52] identified some additional cheeses (Table X).

Whenever possible, a description is given hereafter of the technology used for the production of some of the cheeses listed in Table X. Details referring to the process and information on the chemical and microbiological changes which occur during the maturation of those cheeses which are ripened is not readily available. However, one of the exceptions in this respect is Tafi cheese; some of the compositional and microbiological changes which occur in this cheese during

TABLE X
Names of some Native Latin American Cheeses

Anego	Quesillo	Queso de Maltera
Altiplano	Quesillo de Honduras	Queso de Perija
Asadero	Queso Andino	Queso de Prensa
Chanco	Queso Beniano	Queso de Puna
Chiahuahua	Queso Blanco	Queso de la Tierra
Colona	Queso Chaqueno	Queso del Pais
Cotija	Queso d'Austin	Queso Descremado
de Mano	Queso de Apoyo	Queso Enchilado
Goya (like Asiago)	Queso de Bagaces	Queso Estera
Guyanes	Queso de Bola	Queso Fresco
Hand	Queso de Cavalto	Queso Huloso
Minas Freschal	Queso de Cincho	Queso Llanero
Minas Gerais	Queso de Coalho	Queso Metida
Palmito	Queso de Crema	Queso Oaxaca
Panela (like Gouda)	Queso de Freir	Requejao
Paraguay	Queso Fresco	Requeson
Patagras (like Gouda)	Queso de Hoja	Tafi
Queijo de Coalho	Queso de Mano	Yamandu
Queijo de Manteiga	Queso de Muracay	Yearly

Compiled from Refs. 5, 51 and 52

ripening were described by de Giori *et al.*[53] The microflora of the cheesemilk and whey starter have been classified by de Giori *et al.*[54,55] but there is no information on the biochemical changes during ripening (de Giori, pers. comm.)

Special attention will be paid to the description of the technology of Queso Blanco (white cheese) which is known under various names in a number of countries in South and Central America.

5.1 Argentina

Goya is made both in Argentina and Uruguay, in particular in the Sorduo area. Goya is a hard, ripened cheese made from whole cows' milk; some 2000 tonnes are produced annually, mostly for export. Milk is pasteurized at 70–72°C for 15 s and cooled to renneting temperature, i.e. 32°C. About 3% of fermented whey is added as starter. Calcium chloride is added before coagulation, which is achieved in 15 to 20 min with powdered rennet. The coagulum is cut in small grains of 0·5 cm. At the same time, the temperature is raised to 49°C in about 20 min and this temperature is maintained for 30 min, the curd being stirred continuously.

The curd is pre-pressed and subsequently moulded in cylindrical moulds. It is pressed for 5 h and turned over every hour; each time the pressure is increased. The cheese is salted in brine at 12°C for 6 days and cured for 90 days at 14 to 15°C.

Tafi is a semi-hard, ripened cheese made from raw whole cows' milk at farmstead level. Its rind is covered with moulds. Cheese whey from the previous day's manufacture is added as starter. Coagulation using rennet takes 2 h at 30–31°C. The coagulum is cut and the mixture of curd and whey is heated to 39–40°C, with stirring for 3 h. Most of the whey is drained off and salt is added. The salted curds are placed in cylindrical metal moulds and pressed for 14–16 h. The cheese (1–1·25 kg), which has a smooth, close-knit body, a maximum of 50% moisture and a minimum of 35% fat is ripened for at least 60 days at 22°C and acquires a surface fungal flora (C.S. de Giori, pers. comm.).

5.2 Bolivia

Altiplano is a soft fresh cheese made from raw whole cows' and sheep's milk. This cheese was known at the time of the Incas and was made on the Altiplano. Its production has spread to lower areas but the technology has remained almost unchanged.

Raw milk is coagulated with calf rennet in about 2 h at 32°C. The curd is salted, ladled into straw moulds and pressed for 3 h. This cheese, which is not packed, is kept in salt at 6–12°C.

Quesillo is a fresh, unripened soft cheese made from whole or partly skimmed cows' milk or from a mixture of sheep's and goats' milk. Quesillo is made also in Chile and Ecuador and is known in Mexico as Banela, in Paraguay as Paraguay and in Nicaragua as Queso Blanco. Actually, this cheese is to some extent similar to Queso Blanco, which is described, with perhaps a slightly lower dry matter content.

Two types of technology are applied, depending on the scale of production, i.e. small or semi-industrial scale.

In traditional technology, raw milk is coagulated with powdered or liquid rennet in 45–60 min at 30–32°C. The coagulum is cut into 2 × 2 cm pieces and stirred gently for a while. Some (70–35%) of the cheese whey is removed and replaced by brine; the curd is transferred to moulds lined with a cheese-cloth and pressed by hand. It is sold on the same day and the shelf life is only 2 to 4 days. In general, the quality is poor.

In semi-industrial technology, the main difference is that milk is pasteurized at 72–75°C for 15 s or at 60–65°C for up to 30 min. The milk is cooled to 32–35°C, i.e. renneting temperature, calcium chloride is added and coagulation is obtained in about 30 to 40 min. The rest of the technology is similar to that used in the traditional method.

Some peculiar variations in the technology should be mentioned, e.g. the rennet used to make Paraguay cheese is obtained by soaking dried calf abomasum in lemon or bitter orange juice; Panela cheese is dry-salted by rubbing all sides of the cheese by hand with salt. The sides of Panela cheese may also be coloured using vegetable colouring substances.

Queso Benianco is known also as Quieso Chaqueño. It is a semi-hard ripened cheese made from whole cows' milk. According to traditional technology, only raw milk is used and no heat treatment is applied to the milk. Coagulation is obtained by soaking a piece of calf abomasum directly in the milk; it takes about 2 to 3 h at 25–28°C. The coagulum is not cut, but is ladled directly into moulds where it is pressed for 6 to 8 h. After removal from the moulds, the cheese is dry-salted, ripened at room temperature and sold unpacked.

In semi-industrial production, the milk is pasteurized at 65°C for about 15 min and cooled to renneting temperature (30–32°C). Coagulation is obtained in 15 min with commercial rennet and the coagulum is cut into 2 × 2 cm pieces; whey is drawn off and the curds are transferred to moulds (rectangular or circular) which have been lined with cheesecloth and pressed for 5 to 6 h. The cheeses are salted in stirred brine for 12 h at about 20°C, and ripened at room temperature for 3 weeks.

Queso Fresco is actually similar to Queso Blanco which is described in the section on cheeses of Colombia. Queso Fresco is known also in Ecuador.

5.3 Brazil

Queijo de Coalho, i.e. Rennet cheese, is a ripened, semi-hard cheese made from raw whole cows' milk, which is consumed fresh. Up to 2% of sour cheese whey is added as starter. Coagulation is achieved in 15 min at 37°C with calf rennet or with the enzyme extracted from the stomach of a type of rodent called Moco (a kind of guinea pig from the arid areas of Brazil). The coagulum is cut and stirred by hand and with a large spoon. About 75% of the volume of cheese whey is drawn off, heated and poured back on the curd to raise its temperature to 55°C and 'cook' it. All the whey is then drawn off, salt is added to the curd, which is put into wooden moulds and pressed by hand initially and then mechanically for 2 days. Curing takes up to 1 month at room temperature.

Queiso De Manteiga is a processed cheese, known also as Requeijao Baiano, Requeijao Crioulo or Requeijao Do Nordeste. Actually, Queijo de Manteiga is prepared from fresh curd of Queijo de Coalho before moulding. Salt is added to the curd which is melted in a kettle by heating and stirring at 90°C. Locally-made butter (Garrafa) is mixed with the melted curd and the mixture poured into wooden frames and allowed to cool at ambient temperature.

Queijo Minas is a semi-hard cheese made from raw whole cows' milk. Up to 2% of cheese whey is added as starter and salt may be added to the milk. Coagulation is obtained with calf rennet in 30 to 60 min at 35–37°C. The coagulum is cut into small cubes (1–3 cm) and stirred gently with a ladle for 20 to 30 min. Part of the cheese whey is removed before transferring the remainder to bottomless moulds; cheeses are inverted and pressed by hand until all the whey has been removed. Cheeses are salted overnight with coarse salt and may be cured at room temperature in which case they keep for 6 to 8 weeks or are consumed fresh. Queijo Minas is known also as Queijo de Serro.

Queijo Prato is considered as the most important cheese in Brazil. Five varieties are known, i.e. Minilanche, Lanche, Coboco, Prato and Estepe. Differences refer to shape, weight and duration of ripening: they are rectangular, square or cylindrical, weigh from 0·4–6 kg and are cured for 18 to 60 days.

Prato cheese is a semi-hard, ripened cheese made on an industrial scale from whole cows' milk; its technology seems to be an adaptation of that used for making Edam or Gouda. Standardized milk (3·5% fat) is pasteurized and cooled to 32°C. Calcium chloride and sodium nitrate are added, as well as colouring substance and lactic culture. Coagulation with commercial rennet is obtained in 45 to 60 min. The coagulum is cut into 0·5 cm cubes and stirred slowly for 20 min. Half of the whey is drawn off and stirring continued. Hot water (90–95°C) is added in order to raise the temperature of the curd and whey mixture to 40–42°C. Grains of curd must stick together when pressed and should form threads when separated from each other. Cheese whey is drawn off and the curd is pre-pressed for about 20 min. The curd is cut into pieces of a size corresponding to the moulds, placed on the moulds and pressed for 30 to 40 min, inverted in the moulds and pressed again for up to 12 h. Cheeses are removed from the moulds and salted in brine for up to 36 h at 5–15°C. Then, they are removed from the brine and left to drain at the same temperature for up to 48 h before packing in plastic film.

5.4 Chile

Farm Chanco or Chanco is a semi-hard, ripened cheese made from whole cows' milk. Until the 1950s, Farm Chanco was made at farm level or in small scale dairies; since then, its production has been industrialized. Chanco is the principal Chilean cheese and represents almost 50% of cheese consumption in Chile.

According to traditional technology, Chanco is made from raw milk heated to renneting temperature. No starters are used and acidification develops through the natural flora. Coagulation is obtained, using powdered rennet, in up to one h. In some remote areas, a piece of calf abomasum or its extract in cheese whey is used as coagulant. After coagulation the gel is cut into 1 cm cubes and is left to settle for 15–20 min. It is stirred gently and about 30% of the whey is drawn off and replaced by water at 60°C so that the temperature of the curd is raised to 36°C. The whey is then removed entirely and the curd is salted by adding salted water directly to the vat. The curds are moulded in wooden moulds lined with cheese-cloth and pressed lightly overnight. Ripening takes from 10 to 30 days at ambient temperature.

When made on an industrial scale the milk is pasteurized at 60–65°C for 20 to 30 min. Mesophilic starters (1%), as well as calcium chloride and sodium nitrate, are added. Coagulation with commercial rennet takes 40 to 60 min. The operations are similar to those described for traditional technology.

Farm Goat Cheese is a semi-hard, ripened cheese made from raw goats' milk. Its technology is very similar to that used for Quesillo. The coagulum, obtained by renneting, is cut and stirred before moulding; the curd is pressed first by hand and then mechanically. The cheeses are cured when the flocks return to the valleys. Actually this cheese, which is made by goat keepers when they move to the Andes in search of feed for their animals, is produced under very poor hygienic conditions and the resulting low quality tends to affect commercialization.

5.5 Colombia

Queso Blanco (white cheese) is a generic name and includes both rennet and acid cheeses, both of which are usually consumed fresh (Queso Fresco), but they may also be pressed and ripened, when different names are applied (e.g. Queso Prensa). In Puerto Rico, where it is also known as Queso del Pais, Queso Blanco is produced by adding food-grade acid (acetic, citric, vinegar or citrus juice) to hot milk, or by coagulation with rennet.

It is very likely that this type of cheese was introduced by the Spaniards when they colonized Latin and Central America. It can be produced in many countries under various names and with characteristics which may differ, but basically they belong to the same group of fresh pressed cheese.

Queso Blanco is an unripened fresh cheese made from whole cows' milk. It is a common cheese in Bolivia, Colombia, Cost Rica, Cuba and Honduras, as well as in the Dominican Republic, Ecuador and Venezuela. The basic technology according to a survey carried out by FAO[52] is described hereafter with some remarks as regards some features relevant in specific cases. The manufacturing protocols described by Kosikowski[51] for acid and rennet-coagulated Queso Blanco are summarized in Figs 2–4. A number of Central and South American cheeses, including some hard cheeses, were also described by Davis.[6]

The traditional technology used is as follows: Queso Blanco is made from raw milk which is not subjected to any heat treatment and no starters or cheese whey from previous batches are used. Coagulation is obtained with commercial rennet in 45 to 60 min at 32°C. The coagulum is cut with a cheese harp into 1–5 cm pieces. Part of the whey is removed and salt is added directly in the vat. The curds are filled into hoops and pressed for 10 to 24 h. Generally, the cheese is packed into plastic bags; its shelf life is about one week.

When Queso Blanco is made industrially, the main difference in the technology used refers to pasteurization of the milk. It is used in Cuba and Dominican Republic in particular where the milk is pasteurized at 60–65°C for 30 min. Starters (mesophilic/thermophilic) are added in a proportion of up to 2% and also calcium chloride and potassium nitrate. Renneting time is 30 to 35 min at 32–35°C. The other phases of the process are similar to those in the traditional method.

Many variations can be observed and these are explained by the techniques used. In Venezuela, Queso Blanco is called Llanero or Americano; it is cured for 45 days at room temperature before it is wrapped and sold. In Costa Rica, most of the production is consumed fresh, but sometimes the cheese is sun-dried for almost 1 week and coated with salt before it is smoked. In the Dominican Republic, white cheese is a staple food and is always fried before it is consumed (Queso de Freir). In Honduras, fresh cheese is always stored in high density brine and keeps up to a year at 35°C; the same practice can be observed with White Cheese in Africa and the Near East.

The technique reported by Kosikowski[51] for the production of Queso Blanco/Queso del Pais, which involves acid precipitation of casein, is illustrated

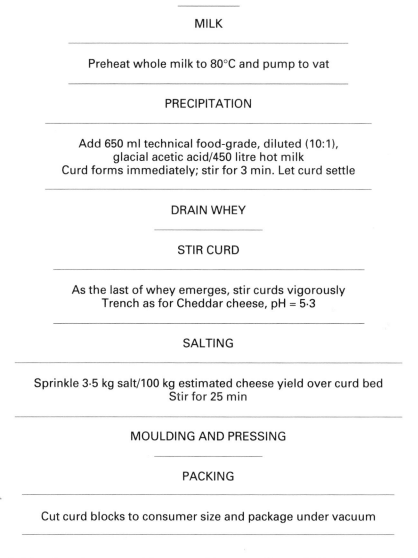

MILK

Preheat whole milk to 80°C and pump to vat

PRECIPITATION

Add 650 ml technical food-grade, diluted (10:1),
glacial acetic acid/450 litre hot milk
Curd forms immediately; stir for 3 min. Let curd settle

DRAIN WHEY

STIR CURD

As the last of whey emerges, stir curds vigorously
Trench as for Cheddar cheese, pH = 5·3

SALTING

Sprinkle 3·5 kg salt/100 kg estimated cheese yield over curd bed
Stir for 25 min

MOULDING AND PRESSING

PACKING

Cut curd blocks to consumer size and package under vacuum

Fig. 2. Manufacturing protocol for Queso Blanco, or Queso del Pais, using acetic acid (from Ref. 51).

MILK

Whole or standardized (2% fat) milk pasteurized at 71·7°C for 16 s

---

SETTING

Add 1% lactic starter to milk at 31°C
Add 80 ml of rennet extract per 450 litre of warm milk

CUT CURD

Cut soft curd with 2 cm knives
Maintain temperature at 31°C and agitate gently for 50 min

DRAINING WHEY

When whey is removed, place curds in moulds, pH = 5·7

PRESSING

Press in forms without salting for four to eight hours

BRINING

Brine to 1·5–2·0% salt by immersing in brine for four to 16 h
Remove and dry salt surfaces for one day

PACKING

Remove after two to three days on shelf at 15°C and 85% RH
and place in plastic bags for distribution

Fig. 3. Manufacturing protocol for Queso Blanco, or Queso Fresco (in Brazil known as Minas Gerais-Freschal), using rennet (from Ref. 51).

Make Queso Fresco
Use 90 ml rennet per 450 litre milk

RIPENING

After salting, seal blocks in Cryovac or coat with oil.
Place at 15–16°C and 85% RH and hold for 15–30 days.
Clean surfaces before consumer packaging in film.

Fig. 4. Manufacturing protocol for ripened Queso Blanco, or Queijo de Prensa (in Brazil known as Minas Gerais-Curado) (from Ref. 51).

in Fig. 2 whereas the production of Quesjo de Minas Gerais/Frescal in Brazil is shown in Fig. 3 and Queijo de Prensa in Fig. 4.

5.6 Costa Rica

Palmito is a pasta filata cheese made from raw whole cows' milk. Whey from the previous day is added to the milk before coagulation with commercial rennet in about one hour at ambient temperature. Most of the whey is removed from the vat without pressing the curd. Hot water is added on top of hot curd and heating is continued until the temperature of the curd reaches 70–82°C when threads begin to form; stirring and heating are stopped and the mass of curd is stretched and kneaded into long plaits/strips and rolled into spherical or oval cheese balls which are salted in brine for 6 to 8 h.

5.7 Cuba

Patagras, made from whole or standardized cows' milk, belongs to the category of semi-hard, ripened cheeses. Both its shape and the fact that its rind is coated with red wax make it look like a Gouda. Originally made at small scale level from whole milk, the production of Patagras from standardized and pasteurized milk has been industrialized. Mesophilic lactic starters (1%), calcium chloride and potassium nitrate are added and coagulation is completed with commercial rennet in about 30 min at 30–32°C. The curd is pressed in moulds for about 20 h and then salted in brine. Patagras is waxed and ripened at a controlled temperature (11–13°C).

5.8 Dominican Republic

Queso de Freir has been mentioned already as a type of Queso Blanco which is always fried before it is consumed.

5.9 Ecuador

Queso Andino was introduced in the 1970s in the mountainous areas first in Peru and subsequently in Ecuador. It is a soft, ripened cheese made from milk pasteurized at 65°C for 15 to 20 min. Lactic starters are added at 33–34°C and then coagulation is obtained by rennet powder in 30 min. The coagulum is cut and stirred for 15 min. When the curd grains are the size of beans (1·5 cm), stirring is stopped, the curds are left to settle and 30–35% of cheese whey is drawn off and replaced by the same quantity of hot water; the curds are stirred further for up to 10 min and after the whey has been removed, are pre-pressed, put in moulds and left to drain for 12 h without pressing. The cheese is then salted in high density brine and cured on wooden shelves at 12–15°C for about two weeks.

Queso Fresco is the designation for Queso Blanco in Ecuador. Its technology has been described earlier.

Quesillo is also made in Ecuador. It has been described under paragraph 5.2.

5.10 Honduras

Quesillo de Honduras, made from standardized cows' milk, is a cheese of the 'pasta filata' type which is produced mainly by small scale cheese units. Milk is standardized according to the fat content desired in the end product. Cheese whey is used as starter and coagulation with commercial rennet takes place in 10 to 15 min at 30–32°C. Whey is drained off, the curd pressed and then broken up and crumbled; salt is added and the curds heated directly until they are soft enough to withstand stretching in a uniform manner when they are made into the final shape.

5.11 Mexico

Chihuahua is a semi-hard, ripened cheese recently introduced by settlers and made from whole cows' milk, pasteurized at 60–65°C for up to 30 min. Lactic starter (2%), calcium chloride and vegetable colouring substance are added. Coagulation with rennet requires 30 to 45 min at 35°C. The coagulum is cut into 5 cm cubes and salted before it is transferred to moulds. Mechanical pressing is applied and the curd is then packed in plastic bags and ripened under controlled temperature (5°C) for about 30 days.

Cotija is a hard, cured cheese made from whole cows' milk or from goats' milk. The milk is heated close to boiling temperature (above 90°C) or between 65 and 70°C for 30 min. Lactic starter (2%), calcium chloride and a natural colouring substance of vegetable origin, called 'Archiote', are added. Coagulation is obtained with rennet in about 40 min at 32–35°C. The curd is cut until it is the size of wheat/rice grains; stirring is stopped and the curd is allowed to settle. Whey is drawn off and salt is added to the curd before moulding and mechanical pressing. The cheese is ripened for a minimum of 100 days under controlled temperature and may be kept for up to 1 year in dry cool premises.

Oaxaca is a 'pasta filata' type of cheese which is said to be derived from Italian Mozzarella cheese; however, whereas Mozzarella is spherical, Oaxaca cheese is formed into a plait. Milk is pasteurized at 60–65°C for 30 min and rennet is added to coagulate the milk in about 40 min at 35°C. The coagulum is cut into 2 × 2 cm cubes and the whey drained without pressing. The curd is worked in hot water, close to melting temperature, kneaded and stretched before being plaited. It is salted in brine or directly by applying dry salt.

Panela is a fresh, unripened cheese made from whole or partly skimmed cows' milk; a mixture of sheep's and goats' milk may also be used. Actually, Panela cheese is similar to Quesillo, which was described in Section 5.2 (Bolivian cheeses).

5.12 Paraguay

Paraguay cheese is also similar to Quesillo cheese made in Bolivia. It is consumed fresh or used for the preparation of typical dishes such as Sopa Paraguaya, a cake made from maize flour.

5.13 Peru

Queso Andino is a soft, ripened cheese made also in Ecuador and Peru. Its technology was described in section 5.9 (cheeses from Ecuador). Requeson, like Ricotta, is obtained by heat coagulation of whey proteins from skimmed or unskimmed whey. Some whey is set aside on the previous day and left to acidify for 24 h at 38°C. The next day the bulk of the whey, which was kept cool, is heated to about 70°C, acidified whey is added and mixed well with 'sweet' whey and heating is continued; as the temperature of the bulk increases, the albumin coagulates and floats on the surface of the liquid. As soon as boiling starts the whey is removed from the source of heat and allowed to rest for a while; the coagulated proteins coalesce and albumin soon forms clusters or a continuous cake on the surface of the liquid. The albumin is collected by skimming the surface or by pouring liquid and coagulum through a cheese cloth placed in a container, e.g. a bucket. The four corners of the cloth are tied and hung for 6 to 8 h to drain off the remaining deproteinated whey. Salt or sweetening agents may be added and Requeson is ready for consumption.

5.14 Uruguay

Colonia is a semi-hard, ripened cheese made from whole cows' milk, both for the domestic market and for export. Production is important and is said to be about 1800 tonnes per annum. Milk is standardized at 2·8% fat and pasteurized at 72°C for 15 s. Lactic starters and calcium chloride are added. Coagulation with calf rennet takes place at 32°C in about 25 min. The coagulum is cut into 5 mm cubes and heated slowly in about 20 min to 48°C; this temperature is maintained for 30 min. After the whey has been drained off, the curd is pre-pressed, moulded and then pressed mechanically for 3 h. During this period, the cheeses are turned over, generally four times and at each, the pressure is increased progressively. The cheeses are salted in brine for 3 days at 12–13°C and cured for 20–25 days at 25°C.

Yamandu is a semi-hard, ripened cheese made from cows' milk with a fat content standardized at 3·2%. Milk is pasteurized at 72°C for 15 s and then cooled to 33°C. Acidified whey is added as starter, together with calcium chloride, sodium nitrate and colouring substance. Powdered or liquid rennet is used for coagulation, which takes about 40 min at 33°C. The coagulum is cut into 5–7 mm cubes, pre-pressed, moulded and subjected to a progressively increasing mechanical pressure. The cheeses are salted in high density brine for 2 days at 12°C, left to dry and packed in plastic bags or coated with wax. They are cured for days at about 16°C. This cheese keeps for about 4 months.

5.15 Venezuela

De Mano cheese is a semi-hard, unripened cheese of the 'pasta filata' type made from raw whole cows' milk. No starters are used and the temperature of the milk is adjusted to 39°C and renneted. Coagulation is obtained in about 15 h at ambient temperature. The coagulum is cut and the curds are separated from the whey, heated in salted water and worked by stretching and kneading. When the elasticity of the curd is adequate, it is formed by hand into small flat disks.

Guayanes is a semi-hard, unripened cheese made from raw whole or partly skimmed cows' milk. No starters are used and the milk is coagulated with calf rennet at room temperature. The coagulum is cut when ready; whey is partly removed and the curd is cooked in water at 90°C and during this process is stirred continuously until it is moulded. The rest of the whey is drained off but without pressing. Guayanes is usually sold unpacked but also packed in plastic bags.

Llanero, known also as Americano cheese, is similar to Queso Blanco. The same technology is applied to obtain the curd, but it differs from other types of Queso Blanco as the curd is ripened for about 6 to 7 weeks, as mentioned in Section 5.5.

6   Epilogue

It is mostly through enquiries, studies and surveys that information on these types of cheese can be improved. Some of the information contained in this chapter was obtained from personal contacts and these are acknowledged. If your favourite cheese has been omitted, or if you have further information on those that are included, please write to the editor so that a revised, more complete version may be published at some time in the future.

Although it is incomplete, it is hoped that this chapter will serve to stimulate interest in non-European cheeses and your assistance in upgrading the information base would be appreciated. Since the last edition, a study on indigenous dairy products has been carried out by the FAO[52] and this revised and expanded chapter draws heavily on that publication. However, the data on the chemical and microbiological aspects are still very limited in comparison with the major European varieties.

ACKNOWLEDGEMENTS

The authors wish to acknowledge the assistance of Dr W.L. Dunkley, Dr F.V. Kosikowski, Dr E.M. Anifantakis, Dr G.S. de Giori, Dr D.K. Dalaly, Dr A. Waters-Bayer, Dr C.L. Dawide, Dr S. Singh and Dr M.M. Hewedi.

REFERENCES

1. FAO, 1989. *Production Yearbook*, Vol. 43, p. 276.
2. IDF, 1982. *The World Market for Cheese*, International Dairy Federation, Brussels, Doc. 146.
3. *Dairy India*, Third Annual Edition, 1987, New Delhi.
4. De, S., 1980. *Outline of Dairy Technology*, Oxford University Press, Delhi, p. 382.
5. Walter, H.E. & Hargrove, R.E., 1972. *Cheeses of the World*, Dover Publications Inc., New York.
6. Davis, J.G., 1976. *Cheese*, Vol. III, Manufacturing Methods, Churchill Livingstone, Edinburgh.
7. Gast, M., Maubois, J.L. & Adda, J., 1969. *Le Lait et les Products Laitiers en Ahaggar*, Mem. Centre Recherches Anthropologiques, Prehistoriques et Ethnographiques, XIV Arts et Metiers Graphiques, Paris.
8. Abou Donia, S.A., 1984. *N.Z. J. Dairy Sci. Technol.*, **19**, 7.
9. El-Gendy, S.M., 1983. *J. Food Prot.*, **46**, 358.
10. El-Erian, A.F., Nour, M.A. & Shalaby, S.O., 1976. *Egyptian J. Dairy Sci.*, **4**, 91.
11. Naghmoush, M.R., Mewtally, M.M., Abou-Dawood, A.E., Naguib, M.M. & Ali, A.A., 1979. *Egyptian J. Dairy Sci.*, **7**, 117.
12. Hofi, A.A., Youssef, E.H., Ghoneim, M.A. & Tawab, G.A., 1970. *J. Dairy Sci.*, **53**, 1207.
13. Khorshid, M.A., El-Safty, M.S., Abdel El-Hamed, L.B. & Hamdy, A.M., 1975. *Egyptian J. Dairy Sci.*, **3**, 127.
14. Mabran, G.A., El-Safty, M.S., Abdel El-Hamed, L.B. & Khorshid, M.A., 1976. *Egyptian J. Dairy Sci.*, **4**, 13.
15. Osman, Y.M., 1979. *Egyptian J. Food Sci.*, **7**, 77.
16. Hofi, A.A., Mahran, G.A., Ashour, M.Z., Khorshid, M.A. & Faharat, S., 1973. *Egyptian J. Dairy Sci.*, **1**, 79.
17. Abd El-Rahman, N.R., Sultan, N.E. & Abd El-Kader, A.E., 1981. *Ann. Agric. Sci. Moshtohar*, **16**, 45.
18. Abd El-Salam, M.H., Rifaat, I.A., Hofi, A.A. & Mahran, G.A., 1973. *Egyptian J. Dairy Sci.*, **1**, 171.

19. Hofi, A.A., Mahran, G.A., Abd El-Salam, M.H., & Rifaat, I.D., 1973. *Egyptian J. Dairy Sci.*, **1**, 33.
20. Hofi, A.A., Mahran, G.A., Abd El-Salam, M.H. & Rifaat, I.D., 1973. *Egyptian J. Dairy Sci.*, **1**, 47.
21. Abdel-Baky, A.A., El-Fak, A.M., Rabie, A.M. & El-Neshewy, A.A., 1982. *J. Food Prot.*, **45**, 894.
22. Nassib, T.A., 1974. *Assiut. J. Agric. Sci.*, **5**, 123.
23. El-Shibiny, S., Soliman, M.A., El-Bagoury, E., Gad, A. & Abd El-Salam, M.H., 1978. *J. Dairy Res.*, **45**, 497.
24. El-Shibiny, S., Mohamed, A.A., El-Dean, H.F., Ayad, E. & Abd El-Salam, M.H., 1979. *Egyptian J. Dairy Sci.*, **7**, 141.
25. Abd El-Salam, M.H., El-Shibiny, S., El-Baboury, E., Ayad, E. & Fahmy, N., 1978. *J. Dairy Res.*, **45**, 491.
26. Abd El-Salam, M.H., Mohamed, A.A., Ayad, E., Fahmy, N. & El-Shibiny, S., 1979. *Egyptian J. Dairy Sci.*, **7**, 63.
27. Soliman, M.A., El-Shibiny, S., Mohamed, A.A. & Abd El-Salam, M.H., 1980. *Egyptian J. Dairy Sci.*, **8**, 49.
28. Nasr, M., 1983. *Egyptian J. Dairy Sci.*, **11**, 309.
29. Abdel-Hamid, L.B., Dawood, A.H., Abdou, S.M., Yousef, A.M.E. & Sherif, R.M., 1977. *Egyptian J. Dairy Sci.*, **5**, 181.
30. Abdou, S.M., Abd El-Hamid, L.B., Dawood, A.H.M., Yousef, A.M. & Mahran, G.A., 1977. *Egyptian J. Dairy Sci.*, **5**, 191.
31. Omar, M.M. & Ashour, M.M., 1982. *Food Chem.*, **8**, 33.
32. Hagrass, A.E.A., El-Ghandour, M.A., Hammad, Y.A. & Hofi, A.A., 1983. *Egyptian J. Dairy Sci.*, **11**, 271.
33. Hofi, A.A., El-Ghandour, M.A., Hammad, Y.A. & Hagrass, A.E.A., 1983. *Egyptian J. Dairy Sci.*, **11**, 77.
34. El-Ghandour, M.A., Hagrass, A.E.A., Hammad, Y.A. & Hofi, A.A., 1983. *Egyptian J. Dairy Sci.*, **11**, 87
35. El-Safty, M.S. & Mehanna, N., 1979. *Egyptian J. Dairy Sci.*, **5**, 55.
36. Shehata, A.E., Magdoub, M.N.I., Gouda, A. & Hofi, A.A., 1982. *Egyptian J. Dairy Sci.*, **10**, 225.
37. Waters-Bayer, A., 1985. *Paper 2C*, Agricultural Administration Unit, Overseas Development Institute, London.
38. Scott, R., *Cheesemaking Practice*, 2nd ed. Elsevier Appl. Sci., London.
39. Anifantakis, E.M. & Kaminarides, S.E., 1983. *Aust. J. Dairy Technol.*, **38**, 29.
40. Anifantakis, E.M. & Kaminarides, S.E., 1981. *Agric. Res.*, **5**, 441.
41. Anifantakis, E.M. & Kaminarides, S.E., 1982. *Agric. Res.*, **6**, 119.
42. Kaminarides, S.E., Anifantakis, E. & Likas, D., 1984. *J. National Dairy Board of Greece*, **3**, 5.
43. Kosikowski, F.V., *Cheese and Fermented Foods*, Edward Brothers Inc., Ann Arbor, MI.
44. Dalaly, B.K., Abdel Mottaleb, L. & Farag, M.C., 1976. *Dairy Ind. Intern.*, **41**(3), 80.
45. Saleem, R.M., Mohammed, F.O. & Abdel Mottaleb, L., 1980. *Mesopotamia J. Agric.*, **15**, 101.
46. International Dairy Federation, Doc. 160, 1983.
47. Abou Donia, Y.A. & Abdel Kader, Y.I., 1979. *Egyptian J. Dairy Sci.*, **7**, 221.
48. Tamine, A.Y. & Robinson, R.K., 1985. *Yoghurt: Science and Technology*, Pergamon Press, Oxford.
49. Abou Donia, S.A., 1978. *Egyptian J. Dairy Sci.*, **6**, 49.
50. El-Erian, A.F.M., Moneib, A.F. & Dalluol, S.M., 1979. *2nd Arab Conference for Food Science and Technology*, Riyad University Press, p. 50.
51. Kosikowski, F.V., 1979. *Proc. 1st Biennial Marschall International Cheese Conference*, Madison, Wisconsin, p. 591.

52. FAO, 1990. *The Technology of Traditional Milk Products in Developing Countries*, Anim. Production and Health Paper No. 85.

53. de Giori, G.S., de Valdez, G.F., de Ruiz Holgado, A.P. & Oliver, G., 1983. *J. Food Prot.*, **46**, 518.

54. de Giori, G.S., de Valdez, G.F., de Ruiz Holgado, A.P. & Oliver, G., 1984. *Microbiol. Alim. Nutr.*, **2**, 233.

55. de Giori, G.S., de Valdez, G.F., de Ruiz Holgado, A.P. & Oliver, G., 1985. *Microbiol. Alim. Nutr.*, **3**, 167.

15

Processed Cheese Products

MARIJANA CARIĆ

Faculty of Technology, University of Novi Sad, Yugoslavia

&

MILOSLAV KALÁB

Centre for Food and Animal Research, Agriculture Canada, Ottawa, Ontario, Canada

1 INTRODUCTION—GENERAL CHARACTERISTICS

Processed cheese is produced by blending shredded natural cheeses of different types and degrees of maturity with emulsifying agents, and by heating the blend under a partial vacuum with constant agitation until a homogeneous mass is obtained. In addition to natural cheeses, other dairy and non-dairy ingredients may be included in the blend.

Processed cheese was initially manufactured without an emulsifying agent; the first attempt was as early as 1895, but only after the introduction of citrates, and especially phosphates, as emulsifying agents did the industrial production of processed cheese become feasible. Production started in Europe on the basis of a Swiss patent, using citrates, issued in 1912. The idea possibly originated from the Swiss national dish, Fondue, for which cheese is heat-treated (melted) in the presence of wine, which contains tartrate that has an emulsifying effect. Processed cheese was developed independently in the USA a few years later (1917) by Kraft, who processed Cheddar cheese with citrates and orthophosphates as emulsifying agents. The newly developed product enabled utilization of natural cheeses which otherwise would be difficult or impossible to utilize, e.g. cheeses with mechanical deformations, localized moulds, trimmings produced during cheese formation, pressing, packaging, etc. It also solved the problem of long-term storage of hard cheeses which otherwise would undergo excessive proteolysis, lipolysis and other detrimental changes, and are thus unsuitable for prolonged storage in their natural form. The assortment was further expanded due to numerous possible combinations of various types of cheese, not only hard varieties, and the inclusion of other dairy and non-dairy components which make it possible to produce processed cheese differing in consistency, flavour, size, and shape.

467

The principal advantages of processed cheese compared to natural cheeses are:

   (i) reduced refrigeration cost during storage and transport, which are espe-
        cially important in hot climates;
  (ii) better keeping quality, with less apparent changes during prolonged storage;
 (iii) great diversity of type and intensity of flavour, e.g. from mild to sharp,
        native cheese flavour or specific spices;
  (iv) adjustable packaging for various usage, economical and imaginative;
   (v) suitability for home use as well as for snack restaurants, e.g. in cheese-
        burgers, hot sandwiches, spreads and dips for fast foods.

Processed cheeses are characterized essentially by composition, water content
and consistency; according to these criteria, three main groups may be distinguished: processed cheese blocks, processed cheese foods and processed cheese
spreads (Table I). More recently established sub-types of processed cheeses are:
processed cheese slices and smoked processed cheese. The first sub-type belongs
to the category of processed cheese blocks, while the second could be either
block or spread. In addition, another group of processed cheese products should
be mentioned, i.e. processed cheese analogues which are usually based on vegetable fat-casein blends. Finally, the most recent development in cheese processing is processed cheese with a completely new look, i.e. natural cheese-like
appearance. Developed in France,[4] this product has an open texture, similar to
traditional cheeses, with eyes of about 0·5 mm in diameter.

2 PROCESSING: PRINCIPLES AND TECHNIQUES

The manufacturing procedure for processed cheese consists of operations performed in the following order: (i) selection of natural cheese; (ii) computation of
the ingredients; (iii) blending; (iv) shredding; (v) addition of emulsifying agent;
(vi) (thermal) processing; (vii) homogenization (optional); (viii) packaging; (ix)
cooling; and (x) storage.

2.1 Selection of Natural Cheese

Proper selection of natural cheese is of the utmost importance for the successful
production of processed cheese. In some countries, processed cheeses manufactured from only one variety of cheese of different degrees of maturity are very
popular, e.g. processed Cheddar cheese in the UK and Australia; Cheddar,
Gruyère and Mozzarella in the USA and Canada; Emmental in Western Europe.
More frequently, processed cheeses are produced from a mix of various natural
cheese types. This results in easier processing and a better flavour balance. The
most important criteria for cheese selection are: type, flavour, maturity, consistency, texture and acidity (pH). Since it is possible to correct certain physical
properties by skilful blending, some defective cheeses can be used in processed
cheese manufacture. Natural cheeses with microbial defects should not be selected

TABLE I
Some Characteristics of Processed Cheese Types

Type of product	Ingredients	Cooking temperature (°C)	Composition	pH	Reference
Processed cheese block	Natural cheese, emulsifiers NaCl, colouring	71–80	Moisture and fat contents correspond to the legal limit for natural cheese	5·6–5·8	Kosikowski[1]
		80–85		5·4–5·6	Meyer[3]
		74–85		5·4–5·7	Thomas[2]
Processed cheese food	Same as above plus optional ingredients such as milk skim-milk, whey, cream, albumin, skim-milk cheese; organic acids	79–85	≤44% moisture, <23% fat	5·2–5·6	Kosikowski[1]
Processed cheese spread	Same as processed cheese food plus gums for water retention	88–91	≥44% and ≤60% moisture	5·2	Kosikowski[1]
		85–98	≤55% moisture	5·7–5·9	Meyer[3]
		90–95	As for processed cheese food	5·8–6·0	Thomas[2]
Processed cheese analogue	Sodium caseinate, calcium caseinate, suitable vegetable fats (soya-bean, coconut), emulsifying agent, salt, artificial flavour	As for processed cheese food		5·8–5·9	Kosikowski[1]

for processing; spore-forming, gas-producing and pathogenic bacteria are particularly hazardous. However, proper selection of good-quality natural cheese is not, by itself, a guarantee that the processed cheese will be of the high quality desired.

2.2 Computation of Ingredients

Computation of the ingredients is conducted on the basis of established fat and dry matter contents of the natural cheese components. Formulation of the material balance of fat and dry matter, including all blend constituents, added water and condensate from live steam used during processing, must be made in such a way as to yield a finished product with the desired composition. Additional adjustments of fat and dry matter are possible, if necessary, before processing is completed.

2.3 Blending

This operation is strongly influenced by the desired characteristics of the final products. According to Thomas,[2] there is a general formulation for processed cheese (block-type): 70–75% of mild cheese and 25–30% of semi-mature or mature cheese. For the production of processed cheese in slices, where a high content of elastic, intact (unhydrolysed) protein is necessary, this ratio is changed to 30–40% young cheese, 50–60% mild cheese and only 10% mature cheese. Kosikowski[1] suggests a similar blend composition: 55% young cheese, 35% medium aged and 10% aged cheese, in order to obtain optimum firmness and slicing qualities. However, if a processed cheese spread is to be produced, the principal raw material is semi-mature cheese of shorter structure, i.e. with partially hydrolysed protein, e.g. 30% young cheese, 50% semi-mature cheese and 20% mature cheese. The relations between different aged cheeses for the manufacture of various processed cheese types and sub-types are given in detail elsewhere.[5]

The main advantages of a high content of young cheese in the blend[2] are reduced raw material costs, the possibility of using cheeses with poor curing properties immediately after manufacture and the formation of a stable emulsion with high water-binding capacity, a firm body and good slicing properties. However, there are certain disadvantages also, e.g. the production of a tasteless cheese which may have a so-called emulsifier off-flavour, excessive swelling, a tendency to harden during storage and the presence of small air bubbles which develop due to the high viscosity of the blend.

A high content of extra-mature cheese in the blend has certain advantages,[2] e.g. the development of a full flavour, good flow properties and a high melting index; however, disadvantages include a sharp flavour, low emulsion stability and a soft consistency.

Most cheeses with rind are trimmed, i.e. the rind with a thin surface layer of the cheese is removed mechanically or manually.

In addition to natural cheeses, various other dairy and non-dairy ingredients are used in the production of processed cheese spreads and processed cheese foods, as shown in Table II.[6] Since the quality of the final product is influenced considerably by all the components present in the blend, the non-cheese components must also fulfil certain qualitative and quantitative requirements. The most frequently used non-cheese ingredients are skim milk powder, casein-whey protein coprecipitates, various whey products and milk fat products.

Skim milk powder improves the spreadability and stability of processed cheese, but, if used in quantities exceeding 12% of the total mass, it may adversely affect the consistency or may remain undissolved. However, skim milk power may be reconstituted first, its casein precipitated by citric acid or proteolytic enzymes and the resulting curd added to the blend.[2] In order to avoid discolouration of processed cheese due to the Maillard reaction, caramelization or crystal formation, attention must be paid to the total lactose content, which should be less than 6% in the final product.

Milk protein coprecipitates, if added to the blend, increase the stability of the cheese emulsion, improve the physical characteristics of the final product and even act as an emulsifying agent. Their emulsifying capacity is so high that it is possible, in their presence, to reduce the amount of emulsifying salt added. This is particularly important for dietary and special food products, where limitation of the sodium content may be desirable. Milk protein coprecipitates should not exceed 5% in processed cheeses.[7]

Casein and caseinates, if added at not more than 5–7% to the blend, do not alter the flavour of processed cheese, while blend stability and structure are improved by the increased content of intact casein. Some of these findings have been reported earlier.[8,9,10] Caseinates are also commonly used in processed cheese analogues.

Although ordinary whey powder is the most common whey product used in processed cheese, whey protein products with lower mineral and lactose contents are preferable, because they yield processed cheese with better flavours. Whey protein concentrates (WPC) of various composition, obtained by ultrafiltration (UF) and other membrane separation techniques, are now available. Though whey proteins do not undergo the same reaction with emulsifying agents as casein, there are some recently published data about the advantage of their addition to the blend.[10] Being a new field of application for WPC, further investigations would add new data to this knowledge.

All milk fat ingredients (Table II) used to adjust the fat content of processed cheese to the desired level must be of high quality and be free from off-flavours. Vegetable fats (soy, coconut, cottonseed, sunflower, etc.) are used only in processed cheese analogues.

Numerous investigations have recently been carried out in order to develop new processed cheese blends with improved characteristics and/or which can be produced at a lower cost.

Production of a cheese base for the manufacture of processed cheese by ultrafiltration and diafiltration of whole milk at 50°C was developed and patented in Germany.[11] The retentate (40% dry matter, 1·17% lactose) was pasteurized (high temperature–short-time, HTST), cooled to 30°C, inoculated with lactic starter and, after 2 h, evaporated at 42°C to 62% dry matter. It was further

TABLE II

Ingredients Used in the Manufacture of Processed Cheese (From Ref. 6, Reproduced by Permission of Scanning Electron Microscopy Inc.)

CHEESE BASE:

Shredded natural cheese

EMULSIFYING AGENTS:

Melting salts
Glycerides

MUSCLE FOOD INGREDIENTS:

Ham
Salami
Fish

VEGETABLES AND SPICES:

Celery
Mushrooms
Mustard
Onions
Paprika
Pepper
Tomatoes

BINDERS:

Locust bean gum
Pectin
Starch

MILK PROTEIN INGREDIENTS:

Skim-milk powder
Whey powder
Whey protein concentrate
Coprecipitates
Previously processed cheese

FAT INGREDIENTS:

Cream
Butter
Butter oil

PRESERVATIVES

COLOURING AGENTS

FLAVOURING AGENTS

PROCESS CHEESE BLEND

SALT

WATER

incubated at 25°C until the pH reached 5·2 and was then packaged in plastic bags under a vacuum. The cheese base can be used in processed cheese production at an 80:20 ratio with ripe cheese.

At the Utah State University in the USA, a cheese for processing was manufactured from ultrafiltered whole milk adjusted to pH 5·2–6·6.[12] The melting properties of the product improved with decreasing pH of the milk before ultrafiltration, as was expected from the reduced calcium concentration in the retentate. Chymosin treatment of the UF retentate adversely affected the meltability of the cheese produced. Yugoslav authors[13] used up to 85% of a Trappist cheese produced from ultrafiltered milk in a processed cheese blend and found no negative influence on the sensory attributes of the final product, which had a higher content of essential amino acids.

A well-known curd product, 'Schmelzpack', is manufactured in Germany especially for processed cheese production. It contains 90–100% unhydrolysed casein and can be blended with natural cheese of diverse type and maturity.[14,15] Good results on the use of chicken pepsin for the 'peptonization' of curd for incorporation into processed cheese have been reported from Bulgaria.[16]

Several enzymic methods for accelerating the ripening of natural cheeses, which are now in use,[17,18] have been reviewed in detail by Law[18] (see also Chapter 14, Volume 1). Acceleration of the ripening process in the production of 'natural' cheese destined for use in the manufacture of processed cheese is particularly interesting, both technologically and economically. For example, in the manufacture of Cheddar cheese it was possible to reduce the duration of ripening from a few months to 2 to 3 weeks.[19] In more recent investigations,[20] cheese base was produced from reconstituted skim milk powder after ultrafiltration and subjected to accelerated ripening using a commercial proteolytic enzyme. The experimental processed cheese (Cheddar), containing no more than 40% cheese base, was of good quality.

Another method for obtaining the desired ripened cheese flavour in processed cheese is the incorporation of the newly developed EMC (enzymatically modified cheese) in the blend. Flavour intensity of EMC is increased about 10 to 30 times compared to the corresponding traditional cheese. Commercial preparations of Cheddar, Emmental, Parmesan and other modified cheese flavours are available, which gave excellent results in processed cheese products.[10]

Chemical hydrolysis of cheese trimmings, intended for inclusion in processed cheese blends, has been patented in the USSR.[21] The trimmings were washed with hot water, centrifuged, comminuted and hydrolysed with 1·4–1·6 M HCl at 110–116°C for 3 to 5 h. The procedure resulted in a hydrolysate containing about 80% peptides in dry matter and 20% essential amino acids (on a total amino acid basis). Hydrolysate (about 20% dry matter) was successfully incorporated in the processed cheese blend up to a maximum of 15%, depending on the type and maturity of the natural cheese used.

Another rapid procedure for the production of a cheese product for processing, based on the acidification of milk curd, was developed and patented by Kraft & Ward[22] in the USA. Milk was coagulated, the drained curd was immersed in lactic acid (0·2–1·5% acidity) and heated, with agitation, to 43°C for 6 to 40 min. After drainage, the curd was washed with water, salted, pressed and stored at 5°C; after two days, the pressed curd can be used in processed cheese manufacture. A more recent French patent describes a procedure in which concentrated milk is acidified, the resulting curd mixed with emulsifying agent and other ingredients and further processed as desired in order to obtain processed cheese.[23]

Another dairy-based product commonly used as an ingredient in processed cheese spreads is pre-cooked cheese, or 'rework', which intensifies the creaming properties of the blend.[3,5,10,24]

All non-dairy ingredients intended for blending (muscle foods, vegetables, spices, etc.) must be sterile and of the highest quality, with typical flavour. The quantities of these must be properly prescribed for blending. In addition to the various possible non-dairy components used in processed cheese blends cited in Table II, some attempts have been made to incorporate cottonseed flour,[25] dried vegetables,[26] garlic and caraway[27] or mayonnaise.[28] In certain countries, some sweet non-dairy ingredients are also used in processed cheese, e.g. fruit syrup, cocoa, vanilla, walnut, hazelnut, coffee extract, etc.[10] For example, the processed cheese industry in the USA produces different coloured, layered torten with specific flavours. A continuous procedure for the production of layered torten was recently patented in Germany.[29]

Bearing in mind that nutritional status is a very important factor influencing the quality and duration of life, a number of dairy research scientists have tried to develop new, nutritionally improved processed cheese products. During the past few years these investigations have led to processed cheese of better nutritional quality, e.g. cheeses with a reduced sodium content, reduced lipids of animal origin, fortified with calcium and enriched with vegetable proteins (soy protein isolate, etc.).[10] A recently published procedure[30] describes the replacement of 40–50% of the lipids by hydrocolloids in processed cheese spreads.

2.4 Shredding (Grinding, Milling)

This operation enables better contact between emulsifying agent and blend ingredients during processing.

2.5 Addition of Emulsifying Agents

Addition of emulsifying agents is the last step in preparing the blend for processing. Since the effects of emulsifying agents are responsible for the unique features of processed cheese production, their type and role will be described in Section 3.

2.6 Processing

Processing means heat treatment of the blend, by direct or indirect steam, under a partial vacuum and with constant agitation. There are two basic types of cooking device: (i) round (double-jacketed kettle, up to 200 litres and (ii) tube-shape (about 4 m long, fitted with one or two mixings worms) (Fig. 1a,b). Due to their greater versatility in production, kettles are most common in Europe, while in North America, where large-scale production of processed cheese slices is dominant,

(a)

(b)

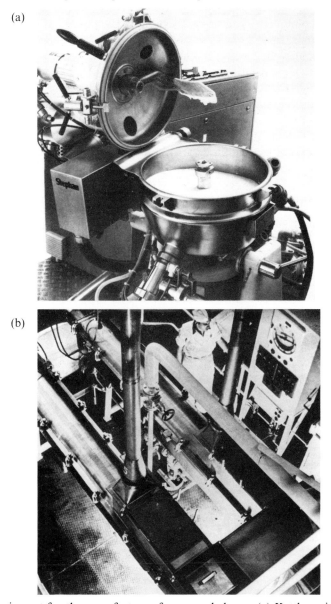

Fig. 1. Equipment for the manufacture of processed cheese: (a) Kettle, universal Stephan cutter; Courtesy A. Stephan u. Sohne GmbH & Co. (b) Programmed and electronically controlled horizontal cooking system. (Reproduced from Ref. 5, *Processed Cheese Manufacture.*)

horizontal devices are preferred. If processing is performed discontinuously, i.e. in a kettle, the temperature reached is 71–95°C for a period of 4 to 15 min (Table I), depending on various parameters;[3] this heating also provides a pasteurization effect. A newly programmed jacketed processor has been developed,[31] which is used to grind, mix and process natural cheeses with other blend components, water and emulsifiers using steam injection and a vacuum, at 75°C for 5 min. The processor is programmed via a punch-card for blend formulation and cleaning-in-place. After cooling, the blend is discharged either by tilting the processor or by aseptic pumping to a packaging machine. When continuously processed, the blend is sterilized at 130–145°C for 2 to 3 s in a battery of stainless steel tubes.[1] Zimmermann[32] patented a continuous process for simultaneous melting, homogenization and sterilization in processed cheese production without the application of pressure. A Japanese patent[33] describes a new method for the post-processing heat treatment (to 100°C) of packed processed cheese, produced in the usual way.

Chemical, mechanical and thermal parameters used in cheese processing are listed in Table III.

TABLE III

Chemical, Mechanical and Thermal Parameters as Regulating Factors in the Cheese Processing Procedures[3,5]

Process conditions	Processed cheese block	Processed cheese slice	Processed cheese spread
Raw material (a) Average of cheese	Young to medium ripe, predominantly young	Predominantly young	Combination of young, medium ripe, overipe
(b) Water insoluble N as a % of total N	75–90%	80–90%	60–75%
(c) Structure Emulsifying salt	Predominantly long, structure-building, not creaming, e.g. high molecular weight polyphosphate, citrate	Long, structure-building, not creaming, e.g. phosphate/citrate mixtures	Short to long, creaming, e.g. low and medium molecular weight polyphosphate
Water addition	10–25% (all at once)	5–15% (all at once)	20–45% (in portions)
Temperature	80–85°C	78–85°C	85–98°C (150°C)
Duration of processing, min.	4–8	4–6	8–15
pH	5·4–5·7	5·6–5·9	5·6–6·0
Agitation	Slow	Slow	Rapid
Reworked cheese	0–2·0%	0	5–20%
Milk powder or whey powder 5–12%		0	0
Homogenization	None	None	Advantageous
Filling, min.	5–15	The quickest possible	10–30
Cooling	Slowly (10–12 h) at room temperature	Very rapid	Rapidly (15–30 min) in cool air

2.7 Homogenization (Optional)

Homogenization improves the stability of the fat emulsion by reducing the average fat globule size. It also improves the consistency, structure, appearance and flavour of the processed cheese. However, since it demands additional capital to cover increased operational and maintenance costs, and prolongs the production schedule, homogenization is recommended only for blends with high fat contents.

2.8 Packaging

Hot processed cheese can be transported to filling (forming) machines by pumping through the closed pipeline system, or transferred in containers, which are not advisable, considering the risk of contamination. Processed cheese is usually packed and wrapped in lacquered foil, tubes, cups, cans, cardboard or plastic cartons and occasionally in glass jars. A newer development is the continuous formation, slicing and packing of the cheese slices, suitable for sandwiches. Slices may also be obtained by mechanically slicing rectangular processed cheese blocks.

2.9 Cooling

There is a general rule for cooling processed cheese: it should be as fast as possible for processed cheese spreads and relatively slow for processed cheese blocks (rapid cooling softens the product). However, slow cooling can intensify Maillard reactions and promote the growth of spore-forming bacteria.

2.10 Storage

The final product should be stored at temperatures below 10°C, although such low temperatures may induce crystal formation (calcium monophosphate, calcium diphosphate, calcium pyrophosphate, etc.).

3 EMULSIFYING AGENTS: TYPES AND ROLE

Emulsifying agents (melting salts) are of major importance in processed cheese production where they are used to provide a uniform structure during the melting process, and also of the products. Phosphates, polyphosphates and citrates[2,34-37] are most common but sodium potassium tartrate[2,36] or complex sodium phosphates of the general formula

$$XNa_2O.YAl_2O_3.8P_2O_5.ZH_2O$$

where $X = 6–15$, $Y = 1·5–4·5$ and $Z = 4–40$,[2] are rarely used. Sodium potassium tartrate, trihydroxyglutaric acid or diglycolic acid are sometimes used. Some characteristics of the most commonly used emulsifying agents are presented in Table IV.

TABLE IV

Emulsifying Salts Used in the Processing of Cheese[1,2,4,5,9]

Group	Emulsifying salt	Formula	Mol. mass	P_2O_5 content (%)	Solubility at 20°C (%)	pH (1% solution)
Citrates	Monosodium citrate-1-hydrate	$NaH_2C_6H_5O_7.H_2O$	232.13	—	16.8	3.75
	Trisodium citrate-2-hydrate	$Na_3C_6H_5O_7.2H_2O$	294.11	—	75.0	8.55
	Trisodium citrate-11-hydrate	$2Na_3C_6H_5O_7.11H_2O$	714.31	—	79.4	7.95
Orthophosphates	Sodium dihydrogen-phosphate	NaH_2PO_4	119.98	59.15	85.2	4.5
	—1-hydrate	$NaH_2PO_4.H_2O$	138.00	51.43	—	4.5
	—2-hydrate	$NaH_2PO_4.2H_2O$	156.01	45.49	39.9	4.5
	Disodium hydrogen-phosphate	Na_2HPO_4	141.96	50.00	9.3	9.1
	—2-hydrate	$Na_2HPO_4.2H_2O$	177.99	39.87	—	9.1
	—7-hydrate	$Na_2HPO_4.7H_2O$	268.07	26.48	—	9.1
	—12-hydrate	$Na_2HPO_4.12H_2O$	358.14	19.82	—	9.1
	Trisodium phosphate	Na_3PO_4	163.94	43.94	11.0	11.9
	—0.5-hydrate	$Na_3PO_4.0.5H_2O$	172.95	41.04	—	11.9
	—12-hydrate	$Na_3PO_4.12H_2O$	380.13	18.67	—	11.9
Pyrophosphates	Disodium pyrophosphate	$Na_2H_2P_2O_7$	221.94	63.96	13.0	4.1
	Trisodium pyrophosphate	$Na_3HP_2O_7.9H_2O$	406.06	34.95	32.0	6.7–7.5
	Tetrasodium pyrophosphate	$Na_4P_2O_7.10H_2O$	446.05	31.82	10.0	10.2
Polyphosphates	Pentasodium tripolyphosphate	$Na_5P_3O_{10}$	367.86	57.88	14.6	9.7
	Pentasodium tripolyphosphate-6-hydrate	$Na_5P_3O_{10}.6H_2O$	475.96	44.73	—	9.7
	Sodium tetrapolyphosphate	$Na_6P_4O_{13}$	469.83	60.42	170.0	8.5
	Sodium hexametaphosphate (Graham's salt)	$(NaPO_3)_n$	n.101.96	69.60	157.0	6.6

These compounds are not emulsifiers in the strict chemical sense (i.e. they are not surface-active compounds), and since emulsification is not their only purpose, melting salts are conditionally termed 'emulsifying agents'. However, true emulsifiers may be included in commercially produced emulsifying agents, which are usually mixtures of compounds, their composition being protected by the producer.

The essential role of emulsifying agents in the manufacture of processed cheese is to supplement the emulsifying capability of cheese proteins. This is accomplished by: (i) removing calcium from the protein system; (ii) peptizing, solubilizing and dispersing the proteins; (iii) hydrating and swelling the proteins; (iv) emulsifying the fat and stabilizing the emulsion; (v) controlling pH and stabilizing it; and (vi) forming an appropriate structure of the product after cooling.

The ability to sequester calcium is one of the most important functions of emulsifying agents. The principal caseins in cheese (α_{s1}-, α_{s2}-, β-) have non-polar, lipophilic C-terminal segments, while the N-terminal regions, which contain calcium phosphate, are hydrophilic. This structure allows the casein molecules to function as emulsifiers.[38-40] When calcium in the Ca-paracaseinate complex of natural cheese is removed during processing by the ion-exchange properties of melting salts, insoluble paracaseinate is solubilized, usually as Na-caseinate. The equation shown earlier[10] could be graphically presented as in Fig. 2. The affinity, i.e. sequestering ability, of common emulsifiers for calcium increases in the following order: citrate, NaH_2PO_4, Na_2HPO_4, $Na_2H_2P_2O_7$, $Na_3HP_2O_7$, $Na_4P_2O_7$, $Na_5P_3O_{10}$. During processing (prolonged heat treatment with agitation), polyvalent anions from the emulsifying agent become attached to the protein molecules, which increases their hydrophilic properties. The binding of additional quantities of water increases the viscosity of the blend, causing 'creaming'. In order to avoid emulsion destabilization, it is necessary to ensure that there is enough intact casein in the blend (a short, hydrolysed protein structure leads to phase separation).

Polyvalent anions (phosphates, citrates) have a high water-adsorption ability.

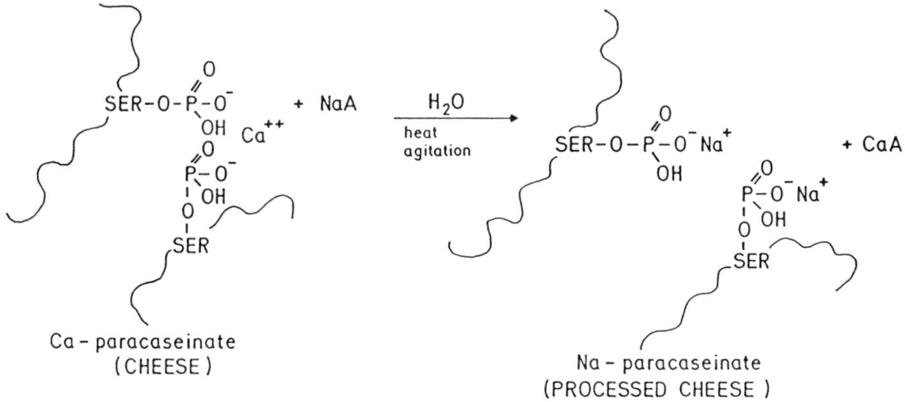

Fig. 2. Chemical reactions during cheese processing. NaA – Ca sequestering agent; A – anion: phosphate, polyphosphate, citrate, etc.

They become bound, via calcium, to protein molecules providing them with a negative charge: basic salts also increase the pH of cheese. Both changes, i.e. increased negative charge and pH, result in higher water adsorption by the proteins. The concentrations of Ca and P are about twice as high in the insoluble phase of processed cheese as in the natural cheese from which it was made. The reactivity between the emulsifier and protein is defined by the ratio of insoluble to total proteins in the natural cheese and in the processed cheese.[34] The affinity of protein for the cations and anions of melting salts is determined by the valency of the ions.[41]

Salts consisting of a monovalent cation and a polyvalent anion possess the best emulsifying characteristics. Although some salts have better emulsifying properties than others, they may have inferior calcium-sequestering abilities or may not solubilize and hydrate the protein sufficiently. It is necessary to combine two or more salts to achieve optimal emulsifying and melting characteristics simultaneously and to produce a homogeneous and stable processed cheese. An appropriate pH value is important for several reasons: it affects protein configuration, solubility and the extent to which the emulsifying salts bind calcium.[38] The pH of processed cheese varies within the range 5·0–6·5. At pH ~ 5·0, which is near the isoelectric point of the cheese proteins, the texture of the cheese may be crumbly, probably due to weakening of protein–protein interactions, but the incidence of fat emulsion breakage is reduced. At pH ~ 6·5, the cheese becomes excessively soft and microbiological problems may be encountered also. The effect of pH on the texture of processed cheese was clearly demonstrated by Karahadian,[42] using mono-, di- and trisodium phosphates, the respective pH of 1% solutions of which were 4·2, 9·5 and 13·0. Cheese made with NaH_2PO_4 (low pH) was dry and crumbly, whereas cheese made with Na_3PO_4 (high pH) was moist and elastic; the texture of cheese made with Na_2HPO_4 was intermediate. Similar pH-dependent effects apply to other emulsifiers also.[38]

Some emulsifying agents exhibit bacteriological effects. Monophosphates have a specific bacteriostatic effect which is even more pronounced with higher phosphates and polyphosphates.[1,2] Citrates lack such effects and may even be subject to bacterial spoilage. Since the usual heat treatment during processing is relatively mild, processed cheeses are not sterile; although the final product contains no viable bacteria, it may contain viable spores, including *Clostridia*, which may originate from the natural cheese or from added spices.[1,2,43] Irradiation of spices prior to usage is thus very important. Orthophosphates suppress the germination of *Cl. botulinum* spores in processed cheese whereas citrates have no effect.[44] Differences in processing conditions, e.g. type of emulsifier, pH and moisture level, also affect spore germination.

The characteristics of individually melting salts and their mixtures have been studied extensively.[1,43,44–55]

3.1 Phosphates

Two types of phosphates are used: (i) Monophosphates (orthophosphates), e.g. NaH_2PO_4, Na_2HPO_4 and Na_3PO_4 and (ii) condensed polyphosphates: (a) polyphosphates, (b) metaphosphates — rings, e.g. $Na_3P_3O_9$ and $Na_4P_4O_{12}$ and (c) condensed phosphates — rings with chains and branches.

$$NaO \diagdown \underset{NaO \diagup}{\overset{O}{\underset{\|}{P}}} \left[O \underset{ONa}{\overset{O}{\underset{\|}{P}}} O \right]_n \underset{ONa}{\overset{O}{\underset{\|}{P}}} \diagup ONa$$

The ability to sequester calcium is closely related to the ability to solubilize protein. According to von der Heide,[56] the solubilization of fat-free rennet casein was 30% with orthophosphate, 45% with pyrophosphate and 85% with polyphosphate. Similar findings were made by Daclin.[57] The concentration of soluble nitrogen increased with the concentration of polyphosphates added in the range of 1–3%.[58] Hydrolysis of polyphosphates to orthophosphates in processed cheese was evident after cooling. The calcium-sequestering ability of sodium metaphosphate is markedly lower than that of sodium tetrametaphosphate; a smooth homogeneous processed cheese is obtained with the latter salt.[59,60] Differences in depolymerization of casein and changes in the flow properties of processed cheese are related to differences in calcium complexation between mono- and tetrapolyphosphates.[61]

Melting rate, ultrafiltratable calcium concentration and textural properties (stress/relaxation, hardness, gumminess and elasticity) of processed cheese are affected more by varying the condensed phosphate than the polyphosphate concentrations.[62] Sharpf[63] suggested that the emulsifying effect of chain phosphates is associated with their interaction with paracasein in such a way that phosphate anions form bridges between protein molecules.

Newly developed phosphate/polyphosphate emulsifying agents, KSS-4 (pH = 4) and KSS-11 (pH = 11), were examined in four processed cheese plants; all the processed cheese blocks obtained, including smoked, were of excellent physicochemical, microbial and organoleptic quality.[55] Processed cheese of good quality can be produced using 1% of surface-active monoglyceride preparation in combination with 50% of the usual amount of phosphate.[64] A processed cheese with improved rheological properties and storage stability can be produced using an emulsifier consisting of tripolyphosphate and monoglycerides.[65] Addition of monoglycerides to the cheese blend increases the hydrophilic properties of the cheese immediately after processing, as well as during storage.[66]

All condensed polyphosphates hydrolyse in aqueous solutions; hydrolysis also occurs during melting and afterwards. The degradation of polyphosphates increases with the duration of processing, irrespective of the rate of stirring and the temperature used.[67] About 50% of the polyphosphates added are hydrolysed during the melting procedure and the remainder is hydrolysed after 7 to 10 weeks of storage.[68] The hydrolysis of phosphates and polyphosphates in 1% solution, as influenced by temperature, is shown graphically in a recently published book.[5]

3.2 Citrates

Of the many citrates available, only trisodium citrate, alone or in combination with other salts, is used as an emulsifying agent in processed cheese production, although citric acid may be used to correct the pH of the cheese. Potassium citrate imparts a bitter taste to the finished product. Monosodium citrate is reported[2] to cause emulsion breakdown during cheese melting because of its high acidity, while disodium citrate leads to water separation during solidification of the melt, also because of high acidity. Comparison of the effects of sodium citrate, sodium citro-phosphate and sodium potassium tartrate on chemical changes in processed cheese showed[35] that the highest and the lowest acidity were obtained with citro-phosphate and tartrate, respectively. Soluble nitrogen was higher in all three processed cheeses than in the initial natural cheese and was highest when Na citrate was used as emulsifier.

Trisodium citrate and NaH_2PO_4 have similar effects on cheese consistency and yield softer cheeses than several polyphosphates; the effect of the latter on cheese firmness increases with the degree of phosphate condensation.[69] The effect of citrate, orthophosphate, pyrophosphate, tripolyphosphate and Graham's salt $[(NaPO_3)_n.H_2O]$ on the physicochemical properties of processed cheese were examined by Kairyukshtene et al.[34] The pH values of 3% solutions of these salts were 8·16, 8·89, 6·61, 9·31 and 5·49, respectively. Soluble protein contents in processed cheese were increased and the water-binding capacity and plasticity of the cheese blends during processing were markedly improved by the use of alkaline salts. The finest fat dispersions were obtained with citrate, tripolyphosphate or orthophosphate in the processing of fresh curds or green cheese and by using citrate or Graham's salt with well-ripened cheese.

Addition of citrate, orthophosphate, pyrophosphate or sodium potassium tartrate, all at 3%, to curds obtained from concentrated milk led to products with poor sensory attributes, although citrate at 2% gave satisfactory results.[36]

3.3 Salt Combinations

As already mentioned, salt mixtures are used to combine the best effects of their individual components.[43,45–47,49–53,55] Some early results[49,51] seem to favour citrate in melting salt combinations, but more recent studies emphasize the desirable effects of phosphates.

According to Shubin,[49] a combination of sodium citrate, trihydroxyglutarate and Na_2HPO_4 gave the best results in the manufacture of processed cheese. Earlier work[51] showed that orthophosphates and pyrophosphates were generally unsatisfactory, whether used alone or in a combination, but citrate was useful to a limited extent; polyphosphates were satisfactory in every respect.

Thomas et al.[52] produced processed cheeses with a 3% addition of disodium phosphate, tetrasodium diphosphate, pentasodium triphosphate or trisodium citrate or equal quantities of sodium polyphosphate and tetrasodium phosphate. The general acceptability of all cheeses was about the same, but cheeses made with disodium phosphate, tetrasodium diphosphate or pentasodium triphosphate had elevated contents of water-soluble nitrogen compared to other cheeses. When the melting salts were used at 2 or 4%, no differences were detected in the levels of water-soluble nitrogen in any of the processed cheeses or in their stickiness, crumbliness, sliceability or general acceptability.

In general, polyphosphates yield processed cheese with superior structure and better keeping quality than other emulsifying agents,[45] apparently due to their ability to solubilize calcium paracaseinate because of their high calcium-sequestering capacity. Pyrophosphates and, in particular, orthophosphates contribute undesirable sensory attributes to processed cheese and although citrates are as efficient as emulsifiers as polyphosphates, they lack their bacteriostatic effect.

Sood & Kosikowski[53] investigated the possibility of replacing cheese solids with untreated or enzyme-modified skim milk retentates in the manufacture of processed Cheddar cheese. Casein in the retentates was largely insoluble and therefore the retentates cannot be used alone for processing. However, cheese containing up to 60% of retentate solids (treated with fungal protease and lipase preparations) had better sensory attributes than the reference cheese; a combination of sodium citrate (2·7%) and citric acid (0·3%) was the best emulsifier for retentate-containing cheese. Increasing the retentate content to 80% resulted in an unacceptable product with a hard, long-grained texture.

4 MICROSTRUCTURE OF PROCESSED CHEESE

Processing markedly changes the structure of natural cheese and results in the development of new structures in processed cheese; the changes may be studied by microscopy. Optical microscopy and electron microscopy complement each other. Relatively simple sample preparation procedures, specific staining of protein, fat, bacteria and other cheese components, and examination of the samples using affordable equipment make light microscopy the method of first choice. Preparation of samples for electron microscopy is more complex and the equipment used is considerably more expensive, but the reward is a markedly higher resolution, distinctly revealing more detail. Various aspects of cheese structure, as visualized by optical microscopy and electron microscopy, were reviewed by Brooker[70] and a review on the use of electron microscopy in food science was published by Kaláb.[71]

Boháč[72] suggested that the selection of natural cheeses for processing should be based on their interaction with melting salt solutions during heating to 85–95°C. Details of his findings were reported on earlier.[6,9] Polarized light microscopy is particularly useful in studying solubilization of melting salts during cheese processing.[72] The advantages of fluorescence microscopy, introduced to the study of cheese processing by Yiu,[73] include simplicity and specificity of the procedures. Cheese constituents in the enlarged images are characterized by their natural fluorescence and by their ability to acquire their natural fluorescence following reaction with specific fluorophores. Heertje *et al.*[74] were the first to publish micrographs of Gouda cheese obtained by confocal scanning laser microscopy. In this type of microscopy, which has only recently been introduced to food science, a laser beam is used to produce a series of micrographs at predetermined depths below the sample surface by a procedure called optical sectioning. The result is a three dimensional image of the structure of the sample.

Electron microscopy is carried out either in the scanning electron microscopy (SEM) or transmission electron microscopy (TEM) mode.[71,75] In SEM, the sample surface is scanned by the electron beam. Secondary and backscattered electrons are used to obtain an enlarged image of the surface using a cathode-ray tube. In TEM, the sample (or its replica) is placed in the electron beam and the enlarged shadow is observed on a fluorescent screen. Due to the nature of the electron beam, the samples or their replicas are examined *in vacuo*.

The structures of natural and processed cheeses differ in several respects. Characteristic features visible to the naked eye in natural cheese fixed with glutaraldehyde are curd granule junctions.[76-80] Staining with methyl violet and use of a low-magnification light microscope make the curd granule junctions even more clearly visible.[70] The patterns are simple in stirred-curd cheeses such as Brick but are more complex in Cheddar, where an additional kind of so-called 'milled curd junctions' develop as the result of milling cheddared curd.[70,81] The curd granule junctions were shown by microscopy to be areas depleted of fat[76-80] (Fig. 3). In natural cheese, fat globules retain their membranes, even when the globules are in large clusters (Fig. 4).

Heating results in separation of fat from protein in natural cheese[82] and processing destroys both the curd granule junctions and the fat globule membranes. Under the effects of melting salts, high heat and vigorous stirring, the relatively insoluble protein matrix of natural cheese is converted into a smooth and homogeneous mass of partially solubilized proteins (mostly casein) of processed cheese. The emulsifying properties of the cheese proteins, which would deteriorate on heating, are restored as the result of their reactions with the melting salts and, consequently, fat becomes emulsified into small globules[83,84] (Figs 5 and 6). New membranes on fat particle surfaces may be noticeable (Fig. 7). Extensive emulsification results in fat globules smaller than one micrometer in diameter,[20,85] particularly following homogenization of the processed cheese.

Calcium present in natural cheese reacts during processing with soluble phosphates of the emulsifying agents and forms insoluble calcium phosphate crystals[1] and their aggregates. Yiu[73] identified them by optical microscopy using alizarin red as a stain specific for calcium (Fig. 8). Crystalline inclusions already noticeable in natural cheese,[82,86-88] particularly those composed of insoluble calcium phosphate, are not affected by processing and are also found in processed cheese (Figs 5 and 6). Recrystallization may occur and anhydrous phosphates were reported to absorb water and form large aggregates[1] and additional growth of calcium phosphate aggregates was observed in processed cheese made using sodium diphosphate as the melting salt.[84]

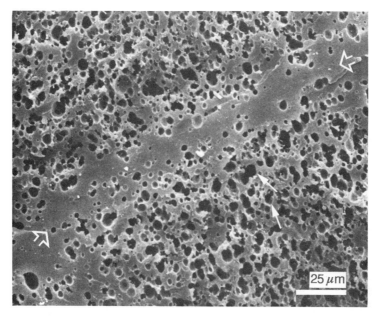

Fig. 3. Scanning electron microscopy (SEM) of Brick cheese reveals a curd granule junction (open arrows) as an area considerably depleted of fat globules.[6] Extraction of fat during preparation of sample for SEM left void spaces (small arrows) in the protein matrix.

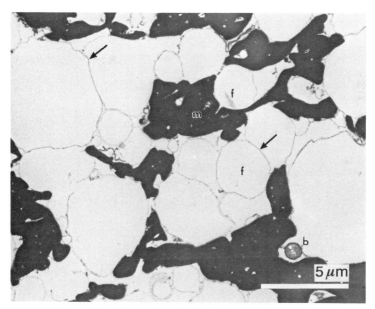

Fig. 4. Transmission electron microscopy (TEM) shows the protein matrix (m) of one-day-old Cheddar cheese interspersed with fat globules (f) encased in fat globule membranes (arrows); b = bacterium.

Fig. 5. During processing, fat (f) in the cheese becomes emulsified; crystalline sodium citrate (c) used as the melting salt gradually dissolves in the aqueous phase of the protein matrix (m). Calcium phosphate crystals (p) are insoluble. b = bacterium.[84]

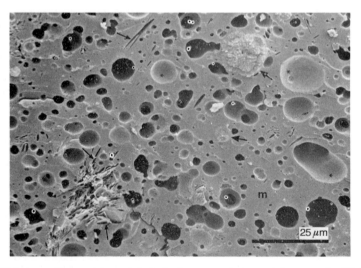

Fig. 6. SEM shows the protein matrix (m) of processed cheese interspersed with fat globules (asterisks). Insoluble calcium phosphate crystals (large arrows) differ in appearance from the imprints of soluble melting salts (small arrows).[24]

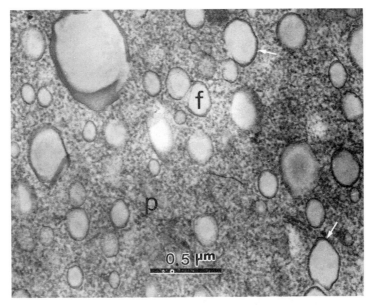

Fig. 7. Membranes (arrows) develop on the surface of emulsified fat particles (f) in the protein matrix (p) of sterilized processed cheese at low creaming time. (Reproduced with permission of I. Heertje.)

In addition to calcium phosphate aggregates, which can be as large as 30 μm in diameter, other crystalline inclusions may consist of calcium lactate in the form of randomly arranged aggregates up to 80 μm in diameter,[70,88] or crystalline amino acids such as tyrosine,[89] or calcium tyrosinate.[2] Washam *et al.*[90] presented micrographs of a variety of salt crystals found in natural and processed cheeses.

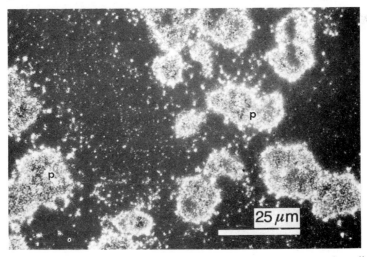

Fig. 8. Calcium phosphate crystals (p) identified by light microscopy using alizarin red staining specific for calcium.[73] (Reproduced with permission of S.H. Yiu.)

They noticed that the occurrence of crystal bloom is less troublesome in processed than in natural cheeses. Careful standardization of the moisture content and the use of phosphates as emulsifying salts are credited for a lower incidence of calcium lactate crystals in processed cheese. However, some treatments, such as smoking of the cheese, are conducive to crystal growth on the surface of processed cheese as the result of dehydration. Tinyakov & Barkan[91] established that the incidence of insoluble crystals in processed cheese made with sodium citrate as the emulsifying agent was lower and the crystal aggregates were smaller than in processed cheese which was made with sodium phosphate.

Occasionally, small white crystals, identified as calcium citrate, were found on the surface of processed cheese as early as one week after manufacture.[92] The development of such crystals was prevented by eliminating sodium citrate from the emulsifying agent used. Crystals of a tertiary sodium calcium citrate,

Fig. 9. TEM of a platinum-and-carbon replica of processed cheese shows the protein matrix (m), fat globules (asterisks) and melting salt crystals (arrows). (Reproduced with permission of A.F. Yang.)

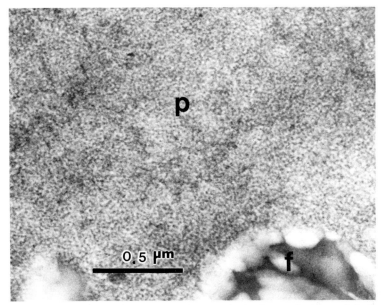

Fig. 10. The matrix of pasteurized processed cheese in the form of minute globular protein particles (p);[85] f = fat. (Reproduced with permission of I. Heertje.)

$NaCaC_6H_5O_7$, in a processed cheese were identified by Klostermeyer *et al.*[93] The composition of crystalline structures in commercial cheeses and processed cheeses was studied by Washam *et al.*[90] using SEM, energy-dispersive X-ray spectrometry (EDSA) and X-ray diffraction analysis. Carić *et al.*[6] confirmed by EDSA the declared presence of sodium–aluminium phosphate used as an ingredient in melting salts used to produce a particular variety of processed Cheddar cheese.

Melting salt crystals may be observed in processed cheese as the result of using an excessively high concentration of the melting salt or because the salt does not dissolve completely in the cheese during processing.[94] The appearance of the melting salt crystals usually differs from that of crystals which are present initially in the natural cheese (Figs 5 and 6). Being soluble in water, the crystals are washed out of the cheese during preparative steps for electron microscopy, leaving cavities (imprints) in the fixed protein matrix. Platinum-and-carbon replicas of cryofixed, freeze-fractured processed cheese, however, show the salt crystals (Fig. 9).

The dimensions and appearance of fat particles in commercial processed cheeses may differ depending on the extent of emulsification. If other parameters, such as the moisture and fat contents and pH are the same, processed cheese in which the fat occurs as large particles is softer than processed cheese containing small fat particles. The use of melting salts with a high affinity for calcium leads to the production of smaller fat particles.[6,9,38]

Differences in the structure of the protein matrix in processed cheeses may also be observed by electron microscopy, provided that a sufficiently high magnification is used.[93,95,96] Thus, the protein matrix of a particular soft processed cheese consisted predominantly of individual particles (Fig. 10) whereas

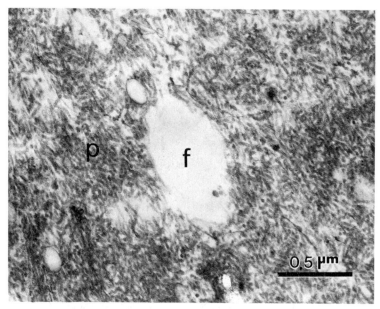

Fig. 11. The matrix of processed cheese in the form of protein strands (p);[85] f = fat.
Reproduced with permission of I. Heertje.)

a hard processed cheese contained a high proportion of long protein strands
(Fig. 11). The protein strands presumably contributed to the ability of the cheese
to retain its shape upon heating.[93,95,96] Heertje et al.[85] confirmed the existence of
protein in the form of strands in processed cheese. Abundant protein strands
were also found in a soft processed cheese made using a combined orthophosphate–polyphosphate emulsifying agent and direct steam heating.[9] Apparently,
the presence of protein strands is associated with the use of polyphosphates but
no conclusive study has been done on this subject. Tinyakov,[97] however, reported the incidence of fibrous structures in processed cheese made with sodium
citrate.

   In general, it is not possible to identify the type of cheese which has been used
to make processed cheese. White cheese and similar varieties, such as Queso
Blanco[98] and Paneer,[99] made by coagulating hot milk[100] by acidification to pH
~ 5·5, are exceptions. The casein particles of these cheeses have a characteristic
core-and-shell structure[98,99,101] which is very stable and withstands the conditions
of cheese processing. On this basis, White cheese can be detected by TEM (Fig.
12) if used as an ingredient in processed cheese at levels as low as 8%.[102,103]
White cheese in processed cheese points to a trend of making cheese processing
more economical by replacing part of the ripened cheese in the blend with less
expensive milk proteins. Other prospective ingredients are so-called cheese bases
made from retentates obtained by ultrafiltration and diafiltration of milk or
from milk fortified with milk proteins. Tamime et al.[20] reported that treatment
of the coagulum with a proteolytic enzyme gave the cheese base an open

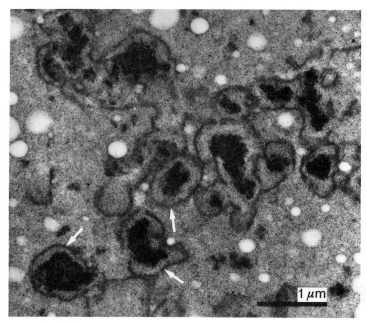

Fig. 12. Core-and-shell structures (arrows) present in processed cheese indicate that White cheese, obtained by acid coagulation of hot milk, was used as an ingredient.[102]

structure. Processed cheeses made with an excessively high proportion of a cheese base and processed cheeses made from combinations of a cheese base with very young ripened cheese had granular structures. Two cheese bases were found to be suitable as ingredients for processed cheese.[20]

The effects of rennet and acid caseins on the microstructure and meltability of processed cheeses made with trisodium citrate (CA), disodium phosphate (DSP), tetrasodium pyrophosphate (TSPP) or sodium aluminium phosphate (SALP) as melting salts were studied by Savello *et al.*[104] Processed cheese containing rennet casein and made using TSPP or DSP were highly emulsified and had poor meltability. CA or SALP caused the processed cheeses to be less well emulsified but meltability was improved. The presence of acid casein in processed cheeses made using DSP produced a highly emulsified product which melted very well.

The meltability of processed cheese is markedly reduced by its heat treatment. Excessive overheating, which may occur accidentally during continuous cheese processing, causes the processed cheese emulsion to thicken markedly. Hardened processed cheese (called 'hot melt') is removed from the pipes using compressed air and may be reused (reworked or reprocessed). It is shredded and added in small quantities to a fresh cheese blend. In general, any processed cheese that is not packaged for sale, although it meets product specifications, and is mixed with a fresh blend and reprocessed is called 'rework'. Unlike regular rework, the reworking of hot melt in processed cheese may increase the viscosity of a freshly-cooked emulsion and change the quality of the finished product in an

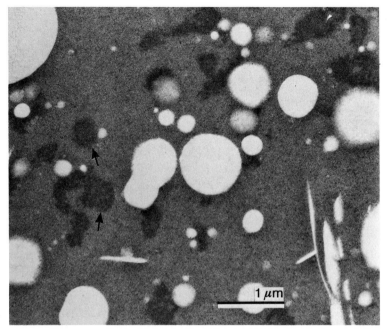

Fig. 13. Excessive heating of processed cheese leads to the development of osmiophilic areas (arrows).[24]

unpredictable way. Although relations between structure and meltability are not yet understood, structural changes in the form of dark areas (Fig. 13) have been found in thin sections of hot melt examined by TEM.[24] These areas develop depending on the extent of heating and the nature of the melting salt used. They appear darker than the rest of the sample because they contain an increased concentration of osmium absorbed during fixation. It is not known, however, whether the dark areas are caused by a higher affinity of chemically-modified protein for osmium or by local compaction of the protein. Klostermeyer & Buchheim[105] examined replicas of freeze-fractured samples and reported that the protein matrix of processed cheese heated at 140°C contained areas of compacted protein (Fig. 14). It is probable that this compacted protein is related to the osmophilic areas.

Klostermeyer & Buchheim[105] also studied relationships between the microstructure of protein matrices and creaming. Areas (1–2 μm in diameter, Fig. 15) of lower protein concentration were found in the protein matrix of processed cheese made with no creaming (melting time of four min). With mild creaming (melting time of six min), the dimensions of these areas decreased to ~ 0.5 μm in diameter and disappeared completely at optimal creaming (melting time of nine min), producing a uniform protein matrix.

It is evident that electron microscopy has contributed more than other microscopical techniques to a better understanding of cheese structure and its changes during processing. New developments in the study of processed cheese structure

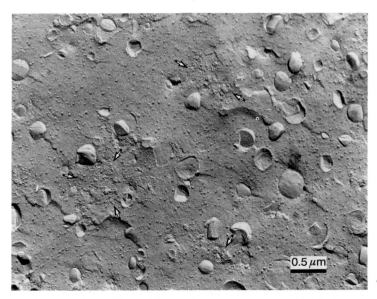

Fig. 14. TEM of a replica of freeze-fractured processed cheese which had been heated to 140°C shows areas of compacted protein (arrows).[105] (Reproduced with permission of W. Buchheim.)

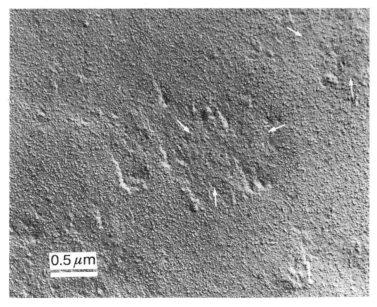

Fig. 15. Areas of lower protein concentration (arrows) found in processed cheese made with no creaming (melting time of 4 min).[105] (Reproduced with permission of W. Buchheim.)

may be anticipated with the introduction of confocal scanning laser microscopy and image analysis to this field.

5   QUALITY DEFECTS OF PHYSICO-CHEMICAL AND MICROBIAL ORIGIN

A good processed cheese should have a smooth, homogeneous structure, a uniform colour and be free from fermentation gas holes.

Various factors can cause physico-chemical or microbial defects. The most important factors are:

(i) an unsuitable blend arising from the use of poor-quality or contaminated natural cheese, a bad relationship of blend components, cheese containing poor-quality proteins, an improper protein/fat ratio, irregular quality or quantity of emulsifying agent, incorrect values for pH, moisture content, or quantity of reworked cheese;

(ii) inadequate processing, e.g. unsuitable time–temperature regimens, inadequate agitation, improper cooling or unsuitable storage.

The first step towards the solution of the problem is to identify the cause by checking all parameters which could be responsible for the defect. Although processing parameters might be changed unintentionally during processing, causing quality defects in the final product, most frequently the reason for defects is related to an improper blend.

Possible defects of physico-chemical and microbial origin in processed cheese, their causes and suggestions for their correction are presented in Table V. Many of the defects described are carry-overs from the natural cheese used in the blend and some can be avoided by proper processing. Once the cause has been established and eliminated, the defect will disappear in subsequent batches. Processed cheese with certain minor defects can be recovered by reworking small quantities of it into subsequent batches, but more severe defects cannot be corrected as they make processed cheese unsuitable for human consumption (e.g. microbial changes, presence of metal ions, excessive Maillard browning) or even dangerous to human health (e.g. *Cl. botulinum* toxin). Crystal formation and discolouration are among the most common physico-chemical defects in processed cheese.

Crystal formation, sometimes visible to the naked eye, is a serious defect in processed cheese. A common reason for crystal formation is low solubility of the emulsifying agent used. This condition is aggravated by the use of excessive amounts of emulsifying agent, a high calcium content in the natural cheese, a high pH and storage of the processed cheese at low temperatures. A poorly soluble component in the emulsifying agent either dissolves incompletely during processing or recrystallizes on cooling. Some emulsifying agents react with calcium in the cheese, producing insoluble calcium salts (e.g. calcium phosphate and calcium tartrate). Using electron microscopy, light microscopy, energy dispersive spectrometric analysis, Debye-Scherrer X-ray analysis and other instrumental methods, crystals in processed cheese have been characterized and chemically

TABLE V
Defects in Processed Cheese[1-3]

Defect	Cause	Correction (solution)
Flavour		
Mouldy	Air contamination, mouldy raw cheese	Use hermetically sealed foils; eliminate all mouldy cheeses from the blend
Empty	Too much young cheese in blend	Correct blend composition
Sharp	Too much mature cheese in blend	Correct blend composition
Bitter, rancid	Over-ripe cheese, mould-ripened cheese or off-flavour butter in blend	Correct blend composition, e.g. add young cheese
Sour, slightly bitter	Sour raw cheese in blend	Add more mature cheese (with higher pH); use emulsifying agent with higher pH
Salty	Salty raw cheese or other components; too much emulsifier	Add young, unsalted cheese or fresh curd to blend; decrease the quantity of emulsifier
Chemical flavour	Off-flavour natural cheese; certain preservatives; unclean steam; impure emulsifying agent	Remove bad-flavour components from blend; check steam for purity
Sweet	Blown raw cheese; blowing in processed cheese	Correct cheese blend composition
Sweet-salty	Too much whey concentrate or whey	Reduce quantity of whey products in blend
Soapy	High pH value (>6·2)	Add younger cheese (with lower pH value); use emulsifying agent with lower pH value
Putrid	Over-ripe and putrid raw cheese; presence of *Clostridia* in processed cheese	Properly select raw cheese blend
Metallic, oily	Traces of metal ions (Cu, Fe, etc.) which oxidize the fat; presence of other oxidants	Eliminate component(s) containing traces of metal ions
Burnt	Overheating in the presence of lactose and by indirect steam (thermal degradation of lactose)	Keep the heating temperatures <90°C and use indirect steam up to 70°C when the blend contains lactose
Burnt with browning	Maillard reaction (lactose and amino acids); usually when very young cheese or whey products are present	Use processing temperatures <90°C; cool processed cheese immediately after packaging; avoid large containers; store <30°C; avoid high pH values in final product

(continued)

TABLE V — contd.
Defects in Processed Cheese[1-3]

Defect	Cause	Correction (solution)
Texture (body, consistency)		
Too soft	High moisture; improper emulsifier; insufficient emulsifier; high pH; rapid cooling; excess ripe cheese in blend; prolonged processing. Slow agitation	Reduce water content; use suitable emulsifier; increase emulsifier content; decrease pH; slower cooling; increase proportion of young cheese in blend; reduce processing time; increase agitation speed
Too hard	Low moisture; improper emulsifier; excess emulsifier; low pH; slow cooling; improper blend; excess cream or over-creamed, reworked cheese	Increase water content; use proper emulsifier; decrease emulsifier content; increase pH; increase cooling rate; change blend composition; avoid addition of creamed or over-creamed reworked cheese
Gum-like (cheese-spread)	Excess young cheese; improper emulsifier; absence of reworked cheese in blend; all water added at once; short processing; slow agitation	Increase proportion of ripe cheese; use proper emulsifier; add reworked cheese to blend; add water in more increments; increase processing time; increase agitation speed
Hard, showing water leakage during storage	Colloidal change in cheese structure ('overcreaming'); bacteriological action leading to reduced pH	Remove all factors which affect excess creaming: choose blend components carefully; keep processing temperatures >85°C
Inhomogeneous (grainy)	Unsuitable blend; improper emulsifier; insufficient or excess emulsifier; low pH; short processing time; low processing temperature; improper amount of added water; inadequate agitation; colloidal or bacteriological changes caused by improper storage	Add younger cheese; use suitable emulsifier; correct emulsifier quantity; correct pH; prolong processing time in order to provide a homogeneous mass; increase processing temperature >85°C; increase the amount of added water; continue agitation during processing and filling; cold storage without pressure or freezing
Water separation	Colloidal change in cheese structure ('overcreaming'); bacteriological growth; unsuitable storage	Remove all factors which cause excessive creaming; choose blend components carefully; keep processing temperatures >85°C; cold storage without pressure or freezing

Fat separation	Overripe raw cheese; insufficient or excess emulsifier; hydrolysis or poor calcium-binding capacity of emulsifying agent; low moisture; processing time too long or too short	Increase proportion of young cheese; correct quantity of emulsifier; use correct emulsifying agent; increase amount of water added; use binding agents
Sticky (adhering to lid foil)	Sticky foil, insufficiently impregnated; excessively high pH; process mass left hot too long without agitation	Change aluminium foil; decrease water addition and if possible add in two portions; increase proportion of ripe cheese or cause better creaming; keep pH <6·0; continue agitation until packaging
Appearance Holes (blown)	Bacteriological changes (growth of *Clostridia*, coliform or propionic bacteria); physical changes (occluded air, CO_2 from emulsifier mixture (citrates), holes filled up with fluid from emulsifying agent having low solubility); chemical changes (hydrogen from reaction between processed cheese and Al foil)	Select cheese blend components carefully; keep processing temperatures >95°C; use a proper vacuum; preheat citrate emulsifier before processing; extend processing time; test porosity of Al foil and if necessary, change it
Crystals	Calcium diphosphate and calcium monophosphate crystals (when these anions are used in emulsifying agent); calcium crystals (when citrates are used in emulsifying agent); further crystal formation when sandy reworked cheese is used in blend; large crystals due to excess emulsifier; lactose crystal formation, caused by excess whey concentrates or low water content; light coloured, soft, grainy precipitate of tyrosine (very mature cheese in blend)	Avoid mono- and diphosphates as emulsifying agents, or combinations with higher phosphates and polyphosphates; exclude citrates from emulsifying agent; exclude sandy reworked cheese from blend; distribute emulsifying agent better; increase processing time; add emulsifying agent in solution (if necessary); use prescribed quantity of emulsifier, reduce level of whey products and increase water content; exclude raw cheese which contains tyrosine crystals
Mottles	Mechanical faults due to filling; physico-chemical changes (only when citrate is used as emulsifier)	Continuous filling; empty filler hopper before filling with a new charge; use equal mass and processing parameters in different cookers; use strong, short agitation before filling; eliminate citrate from emulsifying agent

identified as calcium phosphate, calcium citrate, sodium calcium citrate or disodium phosphate.[1,6,52,84,92,93] Crystal formation in processed cheese, as influenced by the presence of emulsifying agents, is shown in Figs 5 to 9, and is discussed in detail in Section 4.

In addition to the emulsifying agents, lactose and free tyrosine may be responsible for the development of crystals in processed cheese, particularly if these substances are present at excessively high concentrations.[1,2]

Discolouration or browning in processed cheese is a defect caused by the Maillard reaction (non-enzymic browning), when the product develops a dark brown or pink colour. It starts at elevated temperatures and continues autocatalytically. Since the main reactants in Maillard browning are amino acids and reducing sugars, the products most susceptible to these changes are blends containing high levels of young cheese (and thus a high concentration of lactose) and other lactose-containing ingredients, particularly whey powder. Thomas[2] pointed out that more intense browning occurs in processed cheese spreads than in blocks because of higher processing temperatures, longer processing times, a higher water content, a higher lactose content and a higher pH.

Bley et al.[106] found a high correlation (r = 0·929) between the galactose content and brown colour intensity in processed cheese and showed that the intensity of the browning reaction could be reduced by using a galactose-fermenting strain of Str. salivarius subsp. thermophilus, together with a mesophilic lactic starter culture, in curd production. On the other hand, a high salt concentration increased browning, possibly by suppressing the activity of the lactic acid starter culture. However, the Maillard reaction did not occur to a significant extent in processed cheese, regardless of the sugar or salt contents, if the cheese was cooled very quickly.

Borlja[107] investigated the development of non-enzymic browning in industrial processed cheese using an unusually severe laboratory heat treatment (95°C, 110°C or 120°C for 60 min) after manufacture. Increasing the heat treatment intensified the Maillard reaction, which started during processing, showing the functionality of a square polynome:

$$y = M - Nx + Ox^2$$

where M, N and O are constants, which were calculated for each sample.

The total thermal treatment, to which processed cheese containing a high level of reworked cheese is exposed, is more severe than normal and, consequently, the Maillard reaction is more extensive. The levels of melanoidins are very noticeable in sterilized processed cheese, even if it is cooled immediately after production and stored, hermetically packed, at low temperature.[108]

Microbial defects in processed cheese are caused by spore-forming bacteria, which usually originate in the cheesemilk and enter the processed cheese blend through the natural cheese used for blending; other sources of microbial contamination, e.g. water supply, equipment or additives are less common.[1-3,109,110] Contamination after processing is not common since the processing temperature is high enough to destroy vegetative bacterial cells and pasteurize the wrapping material. However, recontamination by viable bacteria or moulds can occur if, for example, the aluminium foil used for wrapping is defective or becomes damaged.

The most important spore-forming bacteria, which cause defects in processed cheese by producing gas, with or without off-odours, belong to the genera *Clostridium* (*Cl. butyricum, Cl. tyrobutyricum, Cl. histolyticum, Cl. sporogenes, Cl. perfringens*) and *Bacillus* (*B. licheniformis, B. polymixa*). Some of these species also cause discolouration or proteolysis in processed cheese, or produce toxins. Karahadian *et al.*[111] found that glucono-delta-lactone, added for acidification, delayed toxinogenesis in processed cheese. Whether produced continuously or in a batch cooker, processed cheese is not sterile. Although it is well known that heat treatment weakens spores, an increased germination ability following heat treatment, due to the destruction of inhibitory factors or due to stimulation by heat treatment has been reported. Germination of spores after processing is influenced by various factors, e.g. blend composition, NaCl concentration, type and concentration of emulsifying agent, water content, pH and the presence or absence of natural inhibitors. However, there are a number of possible ways to prevent spore outgrowth: preservatives in the cheese blend, sterilization of the processed cheese, or an increased redox potential of the blend. One of the most widely used methods of preventing spore outgrowth is the addition of preservatives into the blend. For example, a procedure patented in the USA[10] completely prevents outgrowth of *Cl. botulinum* spores and subsequent toxin formation in pasteurized processed cheese spread, inoculated with 1000 spores per g and incubated 48 weeks at 30°C, through the incorporation of nisin at the level of 250 ppm in the blend.

6 PROCESSED CHEESE ANALOGUES

Nowadays, there are two recognized groups of products which do not belong to the classical range of dairy products: (i) modified dairy products, and (ii) substitute (or imitation) dairy products.[112-114] In modified dairy products, only one dairy component, e.g. protein or fat, is substituted by a non-dairy component for economic or nutritional reasons. Imitation products are based on novel components, frequently produced by newly developed technological procedures. Some imitation products contain no dairy components. Although traditionally-oriented individuals still hold reservations on imitation products, these products have found their 'raison d'être' in both developed and developing countries. In the USA, processed cheese analogues exist on the market in parallel with 'conventional' cheese products, and have been used in the National School lunch programme. According to Kosikowski,[1] in the last decade of the 20th century we will need all the proteins and joules that can be obtained through conventional and novel sources.

Processed cheese belongs to the group of dairy products where substitution of one or more dairy ingredients does not cause technical or technological problems. From the nutritional viewpoint, the blend for modified or imitation

TABLE VI
Nutrition Information per Slice of Cheese Shown on a Label of a Commercial Pasteurized
Processed Cheese Food Substitute ('American Sandwich Slices')

Component	Pasteurized processed cheese food substitute	Pasteurized processed cheese food
Protein	3 g	3 g
Carbohydrate	1 g	1 g
Fat	4 g	4 g
Percent of calories from fat	40 %	40 %
Polyunsatured	1 g	0 g
Saturated	1 g	3 g
Cholesterol (5 mg/100 g)	0 mg (70 mg/100 g)	15 mg
Sodium	240 mg	250 mg
	Percentage of US recommended daily allowances (US RDA)	
Protein	8	8
Vitamin A	2	2
Vitamin C	0	0
Thiamine	0	0
Riboflavin	6	4
Niacin	0	0
Calcium	15	10
Iron	0	0
Calories	50	60

(simulated) processed cheese can easily be enriched with necessary or desirable microcomponents, e.g. vitamins and minerals. On the other hand, the blend can be tailored in such a way as to yield a less expensive product and, therefore, be economically attractive. Essential components are: casein and caseinates (Na-, Ca-, etc.) and also other proteins (soya, coconut, gluten, etc.), suitable vegetable fats, flavourings, vitamins and minerals, food-grade acids (e.g. lactic, citric) to correct the pH to 5·8–5·9 and emulsifying agents. Similar processing parameters (Tables I and III) and equipment are used as for traditional processed cheeses, giving a product of similar physico-chemical and microbiological characteristics. The composition of and nutritional information on a commercial processed cheese analogue launched in the USA is given in Table VI. The declared ingredients are: water, whey, partially hydrogenated soybean oil, casein, American cheese (cultured milk, salt, enzymes, artificial colour) sodium citrate, lactic acid, salt, enzyme modified cheese, modified food starch, sodium phosphate, sorbic acid (preservative), artificial colour, vitamin A palmitate, dicalcium phosphate, vitamin B_2 and vitamin B_{12}.

Numerous investigations have recently been carried out in order to identify new possibilities in this new field. Rosenau[115] developed a method for a successful pilot plant production of a processed cheese based on dairy, non-cheese components (skim milk was acidified with HCl to pH 4·6, and the coagulum was separated and processed in the usual way with emulsifying agents and other additives). Egyptian workers[116] included 20% Ras cheese (manufactured from recombined milk using a microbial enzyme) in a processed cheese blend containing the following ingredients: 20% natural cheese, 40% dried skim milk curd, 7% butter oil, water and 3% emulsifier. Modification of processed cheese by incorporating vegetable oil in the blend improved cheese flavour and resulted in a 25–50% saving of butter.[117] Japanese authors[118] used rennet casein curd (RCC) to produce imitation Mozzarella cheese. RCC improved the stringiness of processed cheese, while replacement of 25% of RCC by whey protein concentrate (WPC) affected meltability, though it was also satisfactory. Savello *et al.*[104] designed processed cheese analogues, based on acid (Group i) or rennet (Group ii) casein, with milk fat, NaCl, citrates or phosphates as emulsifying agents and various concentrations of WPC. Processed cheese analogues produced from acid casein (Group i) with disodium phosphate had best emulsification properties and melted very well. Meltability of both cheese groups (i) and (ii) decreased as WPC concentration increased. A group at Ohio State University[119] investigated the functional properties (hydration characteristics, particle swelling and viscous flow behaviour) of 12 commercial calcium caseinates for use in imitation processed cheese (IPC). From a comparison of these properties with the functional performance in IPC, it was concluded that the best cheese emulsions were associated with caseinates which hydrate and disperse quickly and which retain their pseudoplastic behaviour over an extended range of shear rates. Kirchmeier *et al.*[120] investigated the solubility, dispersibility and colloidal stability of sodium and calcium caseinate dispersions; the rheograms obtained showed that the sol state of these dispersions was the same as that of processed cheese. Various attempts have also been made to develop combined food products which contain processed cheese, but with different basic characteristics. For example, a nutritious chocolate product, composed of 50% bitter chocolate and 50% processed cheese, emulsified, solidified, frozen and coated with chocolate, was developed and patented in the UK.[121]

ACKNOWLEDGEMENTS

The authors wish to thank Dr S. H. Yiu, Mr A. F. Yang, Dr I. Heertje and Dr W. Buchheim for permission to reproduce micrographs in Figs 7–11, 14 and 15; Figs 5, 6 and 8 are reproduced with the permission of *Food Structure* and *Scanning Microscopy International* (SMI, Chicago, Illinois, USA). The authors acknowledge the assistance of Dr Dragoljub Gavarić, PhD; Spasenija Milanović, MSc; Liljana Kulić, BSc; and Zorka Kosovac, Chem. Techn., Dairy Department, Faculty of Technology, Novi Sad University, in studies of emulsifying agents and developing new commercial mixed agents for application in processed cheese production. Author M. Kaláb wishes to thank Miss Gisèle Larocque for assistance with structural studies and Dr D.J. McMahon and

Dr W.H. Modler for useful suggestions. The Electron Microscope Unit, Agriculture Canada, Ottawa, provided facilities.

This is from contribution number 876 issued by the Centre for Food and Animal Research (formerly the Food Research Centre), Ottawa.

REFERENCES

1. Kosikowski, F.V., 1982. *Cheese and Fermented Milk Foods*, 2nd edn, F.V. Kosikowski & Associates, New York, pp. 282, 470.
2. Thomas, M.A., 1977. *The Processed Cheese Industry*, Dept. of Agriculture, Sydney, Australia, pp. 1, 93.
3. Meyer, A., 1973. *Processed Cheese Manufacture*, Food Trade Press Ltd, London, pp. 58, 283.
4. Daurelles, J. & Bernard, J.Y., 1988. French Patent Application, FR 2 610 794 A1.
5. Berger, W., Klostermeyer, H., Merkenich, K., Uhlmann, G., 1989. *Processed Cheese Manufacture*, Joha Leitfaden, BK Ladenburg, p. 71.
6. Carić, M., Gantar, M. & Kaláb, M., 1985. *Food Microstruc.*, **4**, 297.
7. Thomas, M.A., 1970. *Aust. J. Dairy Technol.*, **23**, 23.
8. Gouda, A., El-Shabrawy, S.A., El-Zayat, A. & El-Bagoury, E., 1985. *Egyptian J. Dairy Sci.*, **13**, 115.
9. Carić, M. & Kaláb, M., 1987. In *Cheese: Chemistry, Physics and Microbiology*, Vol. 2, Major Cheese Groups, 1st edn., ed. P.F. Fox. Elsevier Appl. Sci. Publishers, London, p. 343.
10. Carić, M., 1991. In *Encyclopedia of Food Science and Technology*, Vol. III, ed. Y.H. Hui. John Wiley & Sons, Inc., New York, 2161.
11. Rubin, J. & Bjerre, P., 1983. German Federal Republic Patent Application, DE 32 24 364 A1.
12. Anis, S.M.K. & Ernstrom, C.A., 1984. *J. Dairy Sci.*, **67**, 79.
13. Ostojić, M. & Manić, J., 1985. *Mljekarstvo*, **35**, 355.
14. Meyer, A., 1961. *Dte Molkerei Ztg*, **82**, 531.
15. Meyer, A., 1969. *Nord Mejeritidsskr.*, **27**, 133.
16. Dimov, N. & Mineva, P. 1966. *Proc. XVII Intern. Dairy Congr.* (Munchen), Vol. D, p. 175.
17. Robinson, R.K., 1981. *Dairy Microbiology*, Vol. 2., Microbiology of Dairy Products, Elsevier Applied Sci. Publishers, London, p. 232.
18. Law, B.A., 1980. *Dairy Ind. Intern.*, **45**, 5.
19. Kimovskii, I., 1984. *Moloch. Prom.*, **15**, 22.
20. Tamime, A.Y., Kalab, M., Davies, G. & Younis, M.F., 1990. *Food Struct.*, **9**, 23.
21. Kunizhev, S.M., Kimova, E.T., Kushkhova, M.Kh. & Khunizheva, Sh.M., 1983. USSR Patent, SU1 003 795.
22. Kraft, N. & Ward, P.J., 1956. US Patent, 2 743 186.
23. Rizzotti, L.R. & Villandy, B., 1989. French Patent Application, FR 2 622 722 A1.
24. Kaláb, M. Yun, J. & Yiu, S.H., 1987. *Food Microstruc.*, **6**, 181.
25. Abou-Donia, S.A., Salam, A.E. & El-Sayed, K.M., 1983. *Indian J. Dairy Sci.*, **36**, 119.
26. Brezani, P. & Herian, K., 1984. *Zborník Prác Výskumného Ústavu Mliekarského v Žiline*, **8**, 173.
27. Lenkov, N.T., 1989. *Molochnaya i Myasnaya Promyshlennost*, **4**, 27.
28. Kuzyaeva, G.I., Yakimovich, S.I., Resh, L.I., Al'begova, T.I. & Saidov, K.M., 1990. USSR Patent, SU 1 551 322.
29. Zimmermann, F., 1989. German Federal Republic Patent Application, DE 37 27 660 A1.

30. Brummel, S.E. & Lee, K., 1990. *J. Food Sci.*, **55**, 1290.
31. Anon., 1984. *Dairy Record*, **85**, 92.
32. Zimmermann, F., 1982. German Federal Republic Patent Application, DE 31 24 725 A1.
33. Hayashi, T., Shibukawa, N., Yoneda, Y. & Musashi, K., 1982. Japanese Examined Patent, JP 57 55 380 B2.
34. Kairyukshtene, I., Ramanauskas, R., Antanavichyus, A., Butkus, K. & Lashas, V., 1973. *Trudy. Litov. Filial Vsesoyuz. Nauchno-Isled. Inst. Maslodel. Syrodel. Prom.*, **8**, 85.
35. Kapac-Paraćeva, N., 1969. *Sotsijal. Zemjodel.*, **21**, 49.
36. Lapshina, A.D. & Poplavets, P.I., 1976. *Nauch. Trudy. Omskii Sel'skokhoz. Inst. Im. S. M. Kirov*, **158**, 46.
37. Steinegger, R., 1901. *Landwirt. Jahrbuch Schweiz.*, **15**, 132.
38. Shimp, L.A., 1985. *Food Technol.*, **39**, 63.
39. Ellinger, R.H., 1972. *Phosphates as Food Ingredients*, CRC Press, The Chemical Rubber Co., Cleveland, p. 69.
40. Bonell, W., 1971. *Dte Molkerei Ztg*, **92**, 1415.
41. Lee, B.O., Paquet, D. & Alais, C., 1982. *Proc. XXI Intern. Dairy Congr.* (Moscow), **1**(1), 504.
42. Karahadian, C., 1983. MSc thesis, University of Wisconsin, Madison.
43. Carić, M., Gavarić, D., Milanović, S., Kulić, Lj. & Kosovac, Z., 1984. *Investigations of the Possibilities of Imported Additives Substitution in Processed Cheese Production*, Faculty of Technology, Novi Sad, Yugoslavia, p. 60.
44. Tanaka, N., Geoptert, J.M., Traisman, E. & Hoffbeck, W.M., 1979. *J. Food Prot.*, **42**, 787.
45. Becker, E. & Ney, K.H., 1965. *Z. Lebensmittel Unters. Forsch.*, **127**, 206.
46. Carić, M., Gavarić, D., Milanović, S., Kulić, Lj. & Radovančev, 1985. *Mljekarstvo*, **35**, 163.
47. Kapac-Parkačeva, N., 1969. *Godisen. Zb. Zemjod-Sum. Fak.*, Univ. Skopje-Zemjodel, Yugoslavia, **22**, 49.
48. Kicline, T.P., Stahlheber, N.E. & Vetter, J.L., 1967. US Patent, 3 337 347.
49. Shubin, E.M., 1961. *Izv. Vyss. Ucheb. Zaved. Pisch. Tekhnol.*, **3**, 70.
50. Shubin, E.M. & Kracheninin, P.F., 1960. *Trudy Tsentral. Nauchno-Issled, Inst. Maslodel. Syrodel. Prom.*, **5**, 75.
51. Kiermeier, F., 1962. *Z. Lebensmittel Unters. Forsch.*, **118**, 128.
52. Thomas, M.A., Newell, G., Abad, G.A. & Turner, A.D., 1980. *J. Food Sci.*, **45**, 458.
53. Sood, V.K., & Kosikowski, F.V., 1979. *J. Dairy Sci.*, **62**, 1713.
54. Bell , R.N., 1973. US Patent 3 729 546.
55. Carić, M., Kulić, Lj., Gavarić, D., Pejić, B., Stipetić, M. & Babić, I., 1989. *Mljekarstvo*, **39**, 59.
56. von der Heide, R., 1966. *Dte Molkerei Ztg*, **87**, 974.
57. Daclin, J.P., 1968. Thesis No. 96, Ecole Nat Vet, d'Alfort, France.
58. Lee, B.O. & Alais, C., 1980. *Lait.* **60**, 130.
59. Ney, K.H. & Garg, O.P., 1970. *Jahresfachheft Molkereiwesen*, **72**, 1.
60. Ney, K.H. & Garg, O.P., 1970. *Fette Seifen Anstrichmittel*, **72**, 279.
61. Kirchmeier, O., Weiss, G. & Kiermeier, P., 1978. *Z. Lebensm. Untersuch.*, **166**, 212.
62. Nakajima, T., Tatsumi, K. & Furuichi, E., 1972. *J. Agr. Chem. Soc. Japan*, **46**, 447.
63. Scharpf, I., 1971. *The Use of Phosphates in Cheese Processing*, Symposium—Phosphates in Food Processing, AVI Publishing Co., Inc., Westport, Connecticut, USA.
64. Zakharova, M.P., 1979. *Trudy. Uglich*, **27**, 105.
65. Gavrilova, N.B., 1976. *Zernoper. Pishch. Prom.*, **6**, 131.
66. Zahkarova, N.P., Gavrilova, N.B. & Dolgoshchinova, V.G., 1979. *Trudy. Uglich*, **27**, 108.
67. Glandorf, K., 1973. *Dte Molkerei Ztg*, **94**, 1020.

68. Roesler, H., 1966. *Milchwissenschaft*, **21**, 104.
69. Swiatek, A., 1964. *Milchwissenschaft*, **19**, 409.
70. Brooker, B.E., 1979. In *Food Microscopy*, ed. J.G. Vaughan. Academic Press, New York, p. 273.
71. Kaláb, M., 1983. In *Physical Properties of Foods*, eds. E.B. Bagley & M. Peleg. AVI Publishing Co., Inc., Westport, Connecticut, USA, p. 43.
72. Boháč, V., 1984. In *The Collection of Papers from the Dairy Research Institute in Prague 1978–1983*, ed. L. Forman, Technical Information Centre for the Food Industry, Prague, Czechoslovakia, p. 203.
73. Yiu, S.H., 1985. *Food Microstruc.*, **4**, 99.
74. Heertje, I., van der Vlist, P., Blonk, J.C.G., Hendrickx, H.A.C. & Brakenhoff, G.J., 1987. *Food Microstruc.*, **6**, 115.
75. Kaláb, M., 1981. *Scanning Electron Microsc.*, **III**, 453.
76. Kaláb, M., 1977. *Milchwissenschaft*, **32**, 449.
77. Kaláb, M. & Emmons, D.B., 1978. *Milchwissenschaft*, **33**, 670.
78. Kaláb, M., Lowrie, R.J. & Nichols, D., 1982. *J. Dairy Sci.*, **65**, 1117.
79. Rüegg, M., Moor, U. & Schnider, J., 1985. *Schweiz. Milchw. Forschung*, **14**, 3.
80. Taranto, M.V., Wan, P.J., Chen., S.L. & Rhee, K.C., 1979. *Scanning Electron Microsc.*, **III**, 273.
81. Lowrie, R.J., Kaláb, M. & Nichols, D., 1982. *J. Dairy Sci.*, **65**, 1122.
82. Paquet, A. & Kaláb, M., 1988. *Food Microstruc.*, **7**, 93.
83. Rayan, A.A., 1980. PhD thesis, Univ. Microfilms Intern., Ann Arbor, MI, USA, microfilm 81 04118, p. 241 & *Diss. Int. B*, 1981, **41**, 2954.
84. Rayan, A.A., Kaláb, M. & Ernstrom, C.A., 1980. *Scanning Electron Microsc.*, **III**, 635.
85. Heertje, I., Boskamp, M.J., van Kleef, F. & Gortemaker, F.H., 1981. *Neth. Milk Dairy J.*, **35**, 177.
86. Blanc, B., Rüegg, M., Baer, A., Casey, M. & Lukesch, A., 1979. *Schweiz. Milchw. Forschung*, **8**, 27.
87. Bottazzi, V., Battistotti, B. & Bianchi, F., 1982. *Milchwissenschaft*, **37**, 577.
88. Brooker, B.E., Hobbs, D.G. & Turvey, A., 1975. *J. Dairy Res.*, **42**, 341.
89. Flückiger, E. & Schilt, P., 1963. *Milchwissenschaft*, **18**, 437.
90. Washam, C.J., Kerr, T.J., Hurst, V.J. & Rigny, W.E., 1985. *Dev. Ind. Microbiol.*, **26**, 749.
91. Tinyakov, V.G. & Barkan, S.M., 1964. *Izv. Vyss. Ucheb. Zaved, Pishch. Teckhnol.*, **5**, 62.
92. Morris, H.A., Manning, P.B. & Jenness, R., 1969. *J. Dairy Sci.*, **52**, 900.
93. Klostermeyer, H., Uhlmann, G. & Merkenich, K., 1984. *Milchwissenschaft*, **39**, 195.
94. Uhlmann, G., Klostermeyer, H. & Merkenich, K., 1983. *Milchwissenschaft*, **38**, 582.
95. Taneya, S., Kimura, T., Izutsu, T. & Buchheim, W., 1980. *Milchwissenschaft*, **35**, 479.
96. Kimura, T., Taneya, S. & Furuichi, E., 1978. *Proc. XX Intern. Dairy Congr.*, (Paris), **E**, 239.
97. Tinyakov, V.G., 1970. *Izv. Vyss. Ucheb. Zaved. Pishch. Tekhnol.*, **2**, 165.
98. Kaláb, M. & Modler, H.W., 1985. *Food Microstruc.*, **4**, 89.
99. Kaláb, M., Gupta, S.K., Desai, H.K. & Patil, G.R., 1988. *Food Microstruc.*, **7**, 83.
100. Fox, P.F. In *Cheese: Chemistry, Physics and Microbiology* Vol. 2. Major Cheese Groups, ed. P.F. Fox, Elsevier Applied Sci. Publishers, London, p. 311.
101. Harwalkar, V.R. & Kaláb, M., 1981. *Scanning Electron Microsc.*, **III**, 503.
102. Kaláb, M., Modler, H.W., Carić, M. & Milanović, S., 1991. *Food Struct.*, **10**, 193.
103. Kaláb, M., 1991. In *Encyclopedia of Food Science and Technology*, Vol. III, ed. Y.H. Hui, John Wiley & Sons, Inc., New York, 1171.
104. Savello, P.A., Ernstrom, C.A. & Kaláb, M., 1989. *J. Dairy Sci.*, **72**, 1.
105. Klostermeyer, H. & Buchheim, W., 1988. *Kieler Milchwirt. Forschungsber.*, **40**, 219.

106. Bley, M.E., Johnson, M.E. & Olson, N.F., 1985. *J. Dairy Sci.*, **68**, 555.
107. Borlja, Z., 1980. MSc thesis, Faculty of Technology, Novi Sad University, Novi Sad.
108. Milić, B., Carić, M. & Vujicić, B., 1988. *Non-enzymic Browning in Food Products*, Naućna knijiga, Beograd, p. 105.
109. Kršev, Lj., 1985. *Processed Cheese Meeting*, Faculty of Technology, Novi Sad, p. 3.
110. Todorović, M., 1985. *Processed Cheese Meeting*, Faculty of Technology, Novi Sad, p. 6.
111. Karahadian, C., Lindsay, R.C., Dillman, L.L. & Deibel, R.H., 1985. *J. Food Prot.*, **48**, 63.
112. Winkelmann, F., 1974. *Imitation Milk and Imitation Milk Products*, Food and Agricultural Organisation of UN, p. 117.
113. Carić, M., 1990. *Technology of Concentrated and Dried Dairy Products*, 3rd edn, Naućna knjiga, Beograd, p. 143.
114. Petričić, A., 1984. *Fluid and Fermented Milk*, Udruženje mljekarskih radnika SRH, Zagreb, pp. 189, 200.
115. Rosenau, J.R., 1983. *Energy Management and Membrane Technology in Food and Dairy Processing*, American Society of Agricultural Engineers, Chicago, USA, p. 73.
116. Hagrass, A.E.A., El-Shendour, M.A., Hammad, Y.A. & Hofi, A.A., 1984. *Egyptian J. Food Sci.*, **12**, 129.
117. Snegireva, I.A., Volostnikova, R.V., Barkar, S.V., Darchiev, B.Kh., Morozova, R.A., Semenova, A.I., 1982. USSR Patent, SU 971 216 A.
118. Nishiya, T., Tatsumi, K., Ido, K., Tamaki, K. & Hanawa, N., 1989. *J. Agri. Chem. Soc. Jap.*, **63**, 1365.
119. Hokes, J.C., Hansen, P.M.T. & Mangino, M.E., 1989. *Food Hydrocolloids*, **3**, 19.
120. Kirchmeier, O. & Breit, F.X., 1982. *Milchwissenschaft*, **38**, 80.
121. Vajda, G., Ravasz, L., Karacsonyi, B. & Tabajdi, G., 1983. UK Patent Application, GB 2 113 969 A.

16

Cheeses From Ewes' and Goats' Milk

G.C. KALANTZOPOULOS

Department of Food Science and Technology, Agricultural University of Athens, Greece

1 INTRODUCTION

The production of cheese from ewes' and goats' milk has a very long history. In Homer's *Odyssey*, there is a very vivid description of the manufacture of cheese by the beastly Cyclops, Polyphemos, perhaps the oldest recorded cheesemaker in the world. The Cyclops, who enjoyed mythical strength from the milk and cheese of ewes and goats, is a bucolic symbol, which exists even now in some Mediterranean countries. Ipocrates (460–356 BC) also mentioned the production of cheese from goats' milk.

For countries with difficult natural conditions, seasonal and irregular rainfall and eroded soils, sheep and goats are the dominant animals for the production of milk, meat, fibre, hides and even valuable manure. The cheeses and yoghurts produced are the most important sources of protein for the local and rural population. The cheeses are, in general, made in small artisanal units, by a traditional technology, the result of a pastoral art. The products obtained have a special taste and flavour, very different from that of cheeses made from cows' milk. In Africa, Asia, South America, and in those countries where there is a goat-keeping tradition, goat flocks and the production of goats' milk are increasing steadily, whereas in Europe, except for Greece and France, the goat population is decreasing or stable.

Ewes' milk is produced mainly in the countries of the Mediterranean, Near and Middle East. One of the chief features of the utilization of ewes' milk and, to a lesser extent, goats' milk, is that its processing into cheese and cheesemaking at an industrial level is growing, especially in the Mediterranean region.

Information on the production of ewes' and goats' milk is presented in Tables I, II and III. A large part of this production is used on the producing farms or in the local region but some is sold on the national market. A high proportion of goats are raised together with sheep. In most countries, goats' milk is mixed with that of sheep and therefore, the exact quantity produced is unknown; likewise, the amount of cheese produced can only be estimated. Most goats' milk is

TABLE I
Changes in World Production of Sheep's Milk[1]

Region	Mean production tonnes × 10³ (and %)				Mean annual change, tonnes × 10³ (and %)	
	1979–81		1985–87			
Asia	3 376	(47·8)	3 632	(46·7)	42·7	(+1·3)
Europe	2 546	(36·1)	2 756	(35·4)	35·0	(+1·4)
Africa	1 032	(14·6)	1 241	(16·0)	34·8	(+3·4)
USSR	73	(1·0)	111	(1·4)	6·3	(+8·7)
America (North)	—	—	—	—	—	—
America (South)	35	(0·5)	38	(0·5)	0·5	(+1·4)
Oceania	—	—	—	—	—	—
World	7 062	—	7 778	—	119·3	(+1·7)
Mediterranean	3 585	(50·8)	3 862	(49·7)	46·2	(+1·3)
EEC-12	1 589	(22·5)	1 678	(21·6)	14·8	(+0·9)

transformed into cheese and only a limited amount is used for direct consumption and for yoghurt production, especially for consumption on the producing farms.

In recent years, due to the imposition of quotas on the production of cows' milk, the production of ewes' and goats' milk has commenced in several Western European countries, because there are no quotas for these kinds of milk. However, most of these milks continue to be produced by traditional sheep or goat husbandry systems, as follows:

(i) A 3 to 6 weeks suckling period, followed by a milking period of variable length but not more than 6 months.

(ii) Hand milking, by different methods for each production area.

(iii) The possibility of transhumance and/or nomadism.

(iv) One of rudimentary shelters and primitive flock management and milking facilities.

(v) Nutrition based on the low quality natural grazing available and some agricultural by-products.

(vi) Relatively small average flock size (15–120 individuals) and average milk yields in the range of 40 to 100 litres for ewes and a little more for goats.

TABLE II
Production of Ewes' Milk in Mediterranean Countries, Tonnes × 10³ [1]

Turkey	1 200	Spain	233
Greece	614	Yugoslavia	145
Italy	510	France	153
Rumania	455	Portugal	82
Bulgaria	313		

TABLE III
Production of Goats' Milk[2]

Region	Production, tonnes $\times 10^3$
Mediterranean:	
Africa	311
Asia	723
Europe	1 389
Total	2 423
World	7 585
Mediterranean as % of World	31·9
EEC Mediterranean as % of World	16·8
EEC Mediterranean as % of Total Mediterranean	53

Under intensive conditions of breeding and modern animal production systems, in many cases the result of genetic improvement and the introduction of well-organized selection systems is that the milk yield of ewes and goats has reached more than 400 kg and 500 kg, respectively, per lactation, e.g. in France, the mean yield of goats is 583 kg per lactation.

2 COMPOSITION OF EWES' AND GOATS' MILK

2.1 Ewes' Milk

The composition of ewes' milk differs markedly from those of other domesticated species and is very rich in cheesemaking constituents. Its composition, as reported by different authors, is presented in Table IV. The composition varies with the breed, environmental conditions, livestock husbandry conditions and stage of lactation.[3,4]

From the data in Table IV it is evident that the sheep's milk has more than twice as much fat and total protein as cows' milk. This is very important for the production of cheese.

TABLE IV
Mean Composition (%) of the Milk of Ewes of Different Breeds

Fat	Protein	Lactose	Ash	Total solids	Reference
6·88	5·74	4·59	0·95	17·80	5
8·10	5·83	4·72	0·87	19·20	5
6·04	5·35	4·51	0·96	18·89	6
7·31	5·65	4·27	0·94	17·60	7
7·19	5·69	4·66	0·90	18·40	8
7·68	6·04	4·80	0·93	19·30	9
7·20	6·00	4·90	0·93	20·20	10
7·50	6·00	4·10	1·10	18·70	11

TABLE V
Concentration of the Different Nitrogen Fractions in Ewes' Milk

Fraction	g/100 g of milk	% of total proteins	% of Total NPN
Total N	5·784	—	—
Proteins	5·515	100	—
Caseins	4·546	82·4	—
Rennet coagulable proteins	4·329	78·5	—
Soluble protein	0·970	17·6	—
NPN	0·269	—	100
Urea	0·035	—	37·9

2.1.1 Nitrogenous Compounds

The mean concentration of nitrogenous compounds in ewes' milk from 29 flocks during the lactation period is presented in Table V.

Proteins represent ~95% of total N, i.e. similar to that for cows' milk, but is not constant throughout lactation.

2.1.2 The Casein Fraction

The ratio of caseins to total proteins is constant; according to Brochet,[12] it is 82–83% but other researchers report values of 78–80%.[13,14] This factor is very important for the cheesemaking capacity of sheep's milk.

The proportions of the different caseins are presented in Table VI, with some data from different countries.[3,12,15]

Richardson & Creamer (1974) found that the micelle system of sheep's casein is similar to that of cow's casein. Both systems contain rennet sensitive fractions, κ-casein and insoluble α_s and β caseins, the insolubility of which depends on temperature. An examination of the micelles of sheep's milk by electron microscope shows that they are smaller than those of cow's milk, and most of them had a diameter smaller than 80 nm.

These data indicate that the proportions of α_{s1} and α_{s2} caseins are higher in ovine than in caprine milk but lower than in bovine milk. Also, β-caseins represent about 45% of the total casein in sheep's milk, in comparison with ~75% and 35% in caprine and bovine milks, respectively.

TABLE VI
Casein Fractions as a Percentage of Total Casein for Bovine, Caprine and Ovine Milks[3,12,15]

Casein	Bovine		Caprine		Ovine	
α_{s1}	36·0	} 45·5	—		15·5	} 30·2–47·2
α_{s2}	9·5		12·6		14·7	
$\beta1$	33·0		35·9	} 73·3	18·9	} 47·1–36·0
$\beta2$	—		39·4		28·2	
κ	9·4		8·1		7·3	10·6–12·1

Richardson & Creamer[16] resolved the caseins of sheep's milk by polyacrylamide gel electrophoresis, into six main fractions, 3 α_s, 2 β and 1 κ, compared with four in cows' milk. These fractions appear in two groups of bands (Fig. 1). The faster group contains three protein bands, with electrophoretic mobilities similar to those of bovine α_{s2}-caseins, which were designated as ovine α_{s1}-, α_{s2}- and α_{s3}-caseins on the basis of their electrophoretic mobility, amino acid composition and solubility in $CaCl_2$ solutions at 0 and 37°C. The α_{s2}- and α_{s3}-caseins are usually the principal components. They have amino acid compositions similar to bovine α_{s1}-casein, but are more sensitive to Ca^{2+}.

Two bands, of approximately equal intensity and with lower electrophoretic mobilities than the three α-caseins, were characterized as ovine β_1 and β_2-caseins; they have the same amino acid composition but β_1-casein contains one more phosphorus than β_2-casein and is more sensitive to Ca^{2+}. Both have generally similar amino acid compositions to bovine β-casein but Richardson et al.[17] showed some differences between the primary structures of ovine β_1 and bovine βA_2 caseins. The differences in structure between the caseins of these milks influence both the rennet coagulation time and the firmness of the curd. Sheep's milk coagulates 1·56 times faster and the firmness of the gel is twice that of cows' milk.[18] When cows' milk is heated at 85°C, the rennet coagulation time is increased significantly and a weak curd is produced which is not suitable for

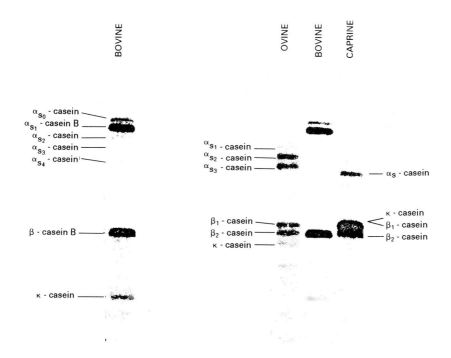

Fig. 1. Electrophoretic pattern of cows', sheep's and goats' casein on polyacrylamide gel.

cheesemaking. According to Penev,[19] prolongation of rennet coagulation is observed when sheep's milk is heated at 70–78°C, a situation that can be improved if the milk is heated at 78–85°C.

Following the action of rennet on the total caseins of cows' or ewes' milk or on their κ-caseins, a larger proportion of non-protein nitrogen is produced from the latter: whole bovine casein, 1·94%; whole ovine casein, 2·95%; bovine κ-casein, 8·5%; ovine κ-casein, 11·8%. The caseino-glycopeptide produced by the action on rennet on ovine κ-casein is similar to that from bovine κ-casein. The carbohydrate content of ovine caseins differs from that of bovine casein, e.g. it has a lower content of N-acetylneuraminic acid.[20,21]

2.1.3 Non-Casein Proteins

The composition of the non-casein proteins is given in Table VII.[12,22]

According to Brochet,[12] the non-casein proteins represent 17·6% of the total protein but other data, for local breeds, are higher. For example, for the Boutsiko breed, this percentage varied from 22·8 to 25·9 during lactation.[9] These values are supported by the data of other researchers.[5,6,23] The whey proteins are used for the production of whey cheeses in all Mediterranean countries by heating of the whey, e.g. Ricotta, Mizithra, which have a very high nutritive value.

Ewes' milk is particularly resistant to the growth of microorganisms during the first hours after milking because its immunological activity is greater than that of cows' milk.

2.1.4 Composition of Milk Lipids

It is well known that cheese made from ewes' milk has a typical taste which is more savoury, and sometimes more piquant, than that of cows' milk cheese. This characteristic is related mainly to the composition of ovine milk lipids which are particularly rich in the C_6-C_{12} fatty acids. Triglycerides and phospholipids represent 98 and 0·8%, respectively, of the lipids of ewes' milk.

The fatty acid profile of ovine milk lipids, reported by different authors from different countries,[24–27] is summarized in Table VIII. It must be noted that the fatty acid composition is influenced markedly by several factors, especially by the feed.

The most characteristic differences in fatty acids between ovine and bovine milks are:

TABLE VII
Composition of the Non-Casein Proteins of Ovine Milk

Proteins	g/100 g	% of non-casein protein
β-Lactoglobulin	0·498	51·4
α-Lactalbumin + Serum albumin	0·243	25·1
Protease-peptone	0·054	5·6
Globulins	0·173	17·6

TABLE VIII
Fatty Acids Composition of the Fat of Ovine and Bovine Milk, %

	Sheep Garcia Olmedo et al.[24]	Gattuso & Fazio[22]	Cow Routaboul[26]	Gallacier et al.[27]
C_4	3·0–6·1	3·10–5·83	2·95–5·10	3·15–4·01
C_6	2·1–5·0	2·13–3·97	2·51–4·07	1·95–2·60
C_8	1·5–5·2	1·98–3·55	2·02–3·79	1·15–1·53
C_9	—	—	0·03–0·08	—
C_{10}	3·3–13·3	5·0–9·65	5·40–10·0	2·41–3·37
$C_{10:1}$	tr-0·6	0·10–0·35	0·09–0·35	0·25–0·34
C_{11}	—	—	0·06–0·12	—
C_{12}	2·0–8·0	2·85–5·17	3·70–6·07	2·71–4·15
$C_{12:1}$	0·0–0·3	—	0·03–0·11	0·10–0·16
iC_{13}	—	—	0·04–0·14	0·06–0·13
C_{13}	—	—	0·08–0·22	0·08–0·11
iC_{14}	tr-0·5	—	0·09–0·20	0·14–0·20
C_{14}	5·3–14·4	7·99–13·44	10·11–13·03	9·50–11·90
$C_{14:1}$	0·1–1·0	0·48–1·03	0·23–0·50	1·15–1·57
iC_{15}	0·0–0·3	—	0·39–0·64	—
C_{15}	0·6–1·5	0·75–1·62	0·88–1·55	1·20–1·60
$C_{15:1}$	0·2–0·6	0·21–0·36	0·07–0·23	tr-0·15
iC_{16}	—	—	0·17–0·28	0·27–0·40
C_{16}	17·0–28·6	20·01–26·42	21·14–27·46	23·60–30·50
$C_{16:1}$	1·02–2·08	1·71–2·72	0·98–2·09	1·85–2·31
iC_{17}	—	—	0·40–0·63	0·26–0·73
C_{17}	0·2–1·0	0·67–1·09	0·73–1·27	0·69–1·35
$C_{17:1}$	0·2–0·7	0·20–0·43	0·38–0·53	0·33–0·60
C_{18}	5·6–16·4	6·83–12·13	7·74–13·65	9·55–13·10
$C_{18:1}$	13·7–36·0	16·64–27·69	17·14–27·80	21·20–28·60
$C_{18:2}$	1·03–3·2	2·80–4·25	1·54–2·54	1·52–2·05
$C_{18:3}$	0·5–4·8	2·06–4·19	—	0·62–0·96

(i) There is higher proportion, about twice as much, of the saturated fatty acids from C_6 to C_{12}, especially C_6 and C_8, in ovine milk than there is in bovine milk. During the ripening of cheese, lipolysis, with the release of fatty acids, has a major influence on the flavour, which is very characteristic. Another important feature is that the triglycerides of ewes' milk are mostly shorter than those of bovine milk

(ii) Differences in the composition of the lipids of bovine and ovine milks affect the values of some of their characteristic indices, especially the Polenski number (Table IX).

(iii) Sheep's milk contains very little β-carotene and the curd is always white.

(iv) Finally, ewes' mineral is richer in mineral constituents than bovine milk.

2.1.5 Yield of Cheese from Ewes' Milk

Owing to its composition, the yield of cheese from ewes' milk is about double that from cows' and goats' milk; the rennet-coagulable proteins are at least

TABLE IX
Some Characteristic Indices of Bovine and Ovine Milks

	Ovine	Bovine
Melting point, °C	29–31	29–34
Solidification point, °C	12–13	19–24
Reichert-Meissel number	25–31	25–33
Polenski number	4·6–6·6	1·5–3·0
Iodine number	30–35	32–42
Saponification number	230–245	220–232

78·5% of the total proteins. For Roquefort cheese, the conversion index is 75% of total nitrogen and 93% of milk lipids.[28]

3 COMPOSITION OF GOATS' MILK

It is necessary to distinguish between two types of goats' milk. The first, which is the more common, is produced from indigenous breeds, which have a low average milk yield, but the milk has a high solids content. These breeds consist of populations formed by raising animals for many years under local environmental conditions without systematic genetic selection. Significant differences are observed in milk yield and composition among the different populations in different countries. The second type of milk is produced by highly selected breeds, with high yield but with a lower solids content (e.g. Saanen, Alpine).

TABLE X
Mean Composition of Milk from Indigenous and Selected Breeds of Goat

Breed	Country	Fat	Protein, %	Total Solids, %	Ash, %	Reference
Native	Spain	5·81	3·55	14·63	0·85	23, 29
Native	Egypt	5·79	—	13·1	—	11, 29
Native	Italy	4·00	3·95	—	0·90	11
Attika natives	Greece	5·63 ± 0·31	3·77 ± 0·25	14·80 ± 0·44	0·73 ± 0·03	30
Moulawian	Moulawi	6·70 ± 0·82	3·14 ± 0·25	—	—	29
Saanen	Greece	3·00 ± 0·50	—	11·13 ± 0·69	—	30
Saanen	Israel	2·84	2·55	11·11	—	29
Saanen	Australia	—	3·16 ± 0·19	—	—	29
Alpine/Saanen	Italy	3·35	—	—	—	29
Pygmy	USA	—	5·06	—	—	21
Maltese local	Libya	—	3·76	12·94	—	29
Red Sokoto	Nigeria	—	—	15·70 ± 0·14	—	29
Malawian	Malawi	—	—	16·30	1·10 ± 0·31	29
Ionnina natives	Greece	—	—	—	0·85 ± 0·00	31
Masri	Saudi Arabia	—	—	—	0·77 ± 0·66	29

The mean composition of milk from indigenous species and selected breeds of goat are presented in Table X.

The composition, as reported by different authors, varies with the breed, environmental conditions, livestock husbandry practices and stage of lactation. Variations of the fat content among the different breeds is greater than variations in protein content.

3.1 Protein Composition

There are few quantitative data on the individual proteins in goats' milk, both caseins and whey proteins, and the data are often conflicting.[21] The proteins in the milk of Alpine and Saanen goats are summarized in Table XI.[32] These data indicate that the relative proportions of the major caseins are very different from those of bovine milk. Goats' milk has a low level of α_s-casein and often contains more α_{s2}- than α_{s1}-casein. However, the latter is present in very variable proportions, depending on the individual goat, as pointed out by Boulanger et al.[33] On the other hand, the proportions of κ- and especially β-casein, are high.

Using polyacrylamide gel electrophoresis, qualitative and quantitative differences in the occurrence of the individual caseins in individual whole caprine casein samples have been established.[34] α-Casein contains two major and four minor components but the milks of some goats were devoid of α-casein. κ-Casein consists of at least five components, differing only in the degree of glycosylation. A novel genetic variant of κ-casein has been detected in caprine milk, i.e. three genotypes of κ-casein occur in caprine milk. Qualitative and quantitative genetic polymorphism at the α_{s1}-casein locus has been confirmed; each homozygous form of α_{s1}-casein appears to contain at least three components.

TABLE XI
Proteins in the Milk of Alpine and Saanen Goats

Principal proteins	Average values		Extreme values
	Caprine	Bovine	Caprine
Total protein content, g/l	27·2	32·0	19·1–33·6
Casein, g/l	21·1	27·0	15·8–26·0
Non-protein nitrogen (% total nitrogen)	6·3	4·5	3·1–13·2
Caseins[a] α_{s1}	5·6	8·0	0–20
α_{s2}	12·2	12·0	10–30
β	54·8	36·0	43–68
κ	20·4	14·0	15–29
α-Lactalbumin + β-Lactoglobulin	0·63	0·4	0·33–1·1

[a] % of total casein

TABLE XII

Relative Content of Individual Caseins in Individual Goat Milks Selected According to the α_{s1}-Casein Genotype[35]

α_{s1}-casein variant	Sample number	α_{s1}-	α_{s2}	β-	κ-
Null	6	—	32·0	47·5	20·8
Slow	5	15·0	21·5	46·7	16·6
Fast	7	16·3	17·6	44·8	20·6
Slow/Fast	7	27·6	19·7	37·2	15·8

The α_{s2}-casein is more heterogeneous than the other caseins. Its heterogeneity has been demonstrated to arise either from genetic polymorphism or variable degrees of phosphorylation.

The resolution of caprine caseins obtained by a two-dimensional electrophoretic procedure (gel electrophoresis at acid or alkaline pH in the first dimension, followed by isoelectric focusing in the second dimension) is far superior to that obtained with other electrophoretic techniques. The mean content of the α_{s1}- α_{s2}-, β- and κ-caseins in 25 individual milks, determined by densitometry of 2-D gels, is summarized in Table XII.

3.2 Non-Casein Proteins

Data on the composition of caprine whey proteins are very variable; some Italian data[36,37] are presented in Table XIII.

The concentration of whey protein is similar to that in bovine milk but the distribution of fractions differs. Goats' milk contains four times less lactalbumin, three times less serum albumin, but more β-lactoglobulin.[21]

3.3 Composition of Goat Milk Lipids

One of the more significant differences between bovine and caprine milks is in the composition and structure of the lipids. The average diameter of the globules in goats' milk is about 2 μm compared to 2·5–3·5 μm for cows' milk fat. The

TABLE XIII

Composition of Caprine Whey Proteins[36,37]

	mg N/100 ml	% of whey protein N
Total albumin	53·9	79·03
β-Lactoglobulin	26·8	39·0
Residual albumin	27·1	39·73
Proteose-peptone	5·8	8·51
Globulin	8·5	12·46

TABLE XIV
Fatty Acids Composition of Goats' Milk Reported by Different Authors and Compared with those of Cows' Milk[23]

Acid	Goats' Milk					Cows' Milk
	Gonc et al.[38]	Garcia-Olmedo et al.[39]	Boccignone et al.[40]	Sawaya et al.[41]	Martin-Hernandez et al.[42]	Martinez-Castro et al.[43]
$C_{4:0}$	3·3–4·8	1·0–4·9	1·81	3·0	1·8–2·8	2·5–6·2
$C_{6:0}$	1·7–3·0	1·5–4·3	2·03	2·0	2·2–3·4	1·5–3·8
$C_{8:0}$	1·5–3·6	2·0–5·2	2·68	2·0	2·4–3·9	1·0–1·9
$C_{10:0}$	6·4–11·1	7·1–16·2	8·45	6·1	8·8–13·4	2·1–4·0
$C_{12:0}$	2·5–5·0	3·3–9·8	5·21	2·9	3·8–5·5	2·3–4·7
$C_{14:0}$	8·5–11·2	6·9–15·4	10·52	9·5	8·5–11·6	8·5–12·8
$C_{14:1}$	0·02–0·2	0·1–1·1	0·96	0·4	0·5–0·8	0·6–1·5
$C_{16:0}$	25·1–38·4	16·7–39·4	24·33	28·6	23·3–32·1	24·0–33·3
$C_{16:1}$	0·7–1·7	0·7–3·5	2·58	2·5	1·0–2·0	1·3–2·8
$C_{18:0}$	5·9–14·9	4·4–17·3	9·49	10·3	4·3–11·2	6·2–13·6
$C_{18:1}$	15·6–28·2	13·5–33·3	23·96	26·3	16·2–26·6	19·7–31·2
$C_{18:2}$	1·8–4·0	0·5–4·7	1·68	2·6	1·2–2·5	1·3–5·2

smaller fat globules give a better dispersion and a more homogeneous mixture of fat in the milk.

Bovine milk creams rapidly, due to clustering of fat globules promoted by an agglutinin, which is lacking in goats' milk, which therefore has a poor creaming ability, especially at low temperatures. Also, the fat globule membranes in goats' milk are more fragile, which may be related to their greater susceptibility to the development of off-flavour than cows' milk.

The fatty acid composition of caprine milk fat is summarized in Table XIV. It is apparent that the volatile water-insoluble fatty acids are approximately double those in cows' milk.[21] C_{10} and C_{12}, and also C_{14}, $C_{16:0}$, $C_{18:0}$ and $C_{18:1}$, are the principal fatty acids in goats' milk.[23]

3.3.1 Other Aspects of the Composition of Caprine Milk

The concentrations of chloride (0·18%, as Cl⁻) and potassium (0·188%) are greater in goats' than in cows' milk but the concentrations of soluble Ca and soluble citric acid are lower. Late lactation milk contains elevated levels of phosphorus and very little citrate.[4]

Goats' milk contains higher levels of many B vitamins, especially riboflavin, than bovine milk but the concentrations of vitamin B_6 and B_{12} are lower.[36]

Goats' milk also has lower activities of certain enzymes, e.g. ribonuclease, alkaline phosphatase, lipase and xanthine oxidase, than bovine milk. In contrast to bovine milk, lipase activity in goats' milk is significantly correlated with spontaneous lipolysis, probably due to the particularities of the lipolytic system, and plays a major role in off-flavour development in milk stored at 4°C.

4 PRODUCTION OF CHEESE FROM SHEEP'S MILK

4.1 Introduction

The greater part of sheep's milk produced throughout the world is transformed into cheeses, mainly homemade, traditional varieties, which constitute an important activity with good economical results in most Mediterranean countries.

These cheeses have a wide range of characteristics and quality features. They are difficult to standardize because they are homemade or made on a limited scale and are traded mainly at local or regional markets. Also, they are often made from raw milk, which can pose quality and hygiene problems in international trade on a world scale and even within the EEC.

Based on a recent survey (IDF Questionnaire 191/A) by Group A7 of the International Dairy Federation, a catalogue of 508 cheeses made from ewes' and goats' milk was compiled (see Appendix). The majority of these cheeses are made from ewes' milk or from a mixture of goats' and sheep's milks or from milk containing small quantities of cows' milk (only 50 cheeses). A classification of these cheeses is presented in Table XV. A very limited number of these cheeses are produced on an industrial scale but they have a very good reputation due to their typical and unique characteristics.

The most important of these is Roquefort, which is very well known around the world; 18 000 tonnes were sold in 1989, of which 11% was exported to more than 70 countries. There is a strong determination on the part of the producers to guarantee very high standards of quality at organoleptic, technological and hygienic levels.

The second most important is the Greek cheese, Feta, which has gained in popularity in many countries and conquered new markets throughout the world.

Italian Pecorino cheese is the third-best known cheese made from ewes' milk. This type of cheese is available in four varieties: Pecorino Siciliano, Pecorino Siciliano Pepato, Pecorino Fiore Sardo and Pecorino Romano. The last is the most important, representing 50% of total production.[11]

Kachkaval, known since the 11th century, is a cheese produced widely in Central and Eastern Europe, including the Balkan countries; it is made in many varieties under different names.

TABLE XV
Classification of Cheeses Made from Ewes' Milk

Fresh cheeses:	Type of drained yoghurt made with the addition of a small quantity of rennet.
Soft cheeses: (brined)	Different varieties of Feta cheese, Teleme, Bulgarian white brined.
Blue-veined cheese:	Roquefort.
Semi-hard cheeses:	Three categories are produced: 1. Pecorino, Kefalotyri, Ossau-Iraty Pyrenées 2. 'Pasta filata', e.g. Kaskaval and Kasseri 3. Manchego
Hard cheeses:	Graviera, Halloumi.
Whey cheeses:	Ricotta, Myzithra, Brocciu.

Manchego in Spain and Serra de Estrela in Portugal are the most popular cheeses in those countries.

All these cheeses have a long tradition; their limited production and their excellent quality lead to the production of imitation products from cows' milk. Various aspects of these varieties are reviewed in other chapters of this volume.

4.2 General Remarks Concerning the Manufacture of Cheeses from Sheep's Milk

4.2.1 The quality of raw milk for cheesemaking is not very good because the conditions of management are very different, except in the case of Roquefort.

The tradition of sheep's milk production seems to have its roots in the distant past and it is generally believed that the difficulty in finding natural pastures during spring and summer in the plains gave rise to a transhumance system and sometimes even to the settling of shepherds in the mountains in or outside the region.

The milk used for the production of traditional cheeses by shepherds is of a good quality because it is used directly after the milking. However, for industrial purposes, the collection of ewes' milk faces many problems, e.g.

(i) The large numbers of breeders with small flocks and low levels of milk production per animal.
(ii) The wide dispersal of the producers, the topography of the region and the lack of roads, water supplies, electricity, etc.
(iii) The seasonal nature of the milk production, about 50% of which occurs during April and May.

In some countries, e.g. Greece, France and Italy, there are collection systems with refrigeration stations where producers transfer the milk and from which it is transported to the factory daily. The microbiological quality of this milk is satisfactory. In Portugal, the milk is collected for industrial purposes on alternate days by using H_2O_2 as preservative. In other countries, the milk is collected without special precautions.

Concerning the major nitrogen fractions of ewes' milk, related to cheesemaking, the only data are from a study of Grappin[44] on the milk of the Lacaume breed and are taken as indicative.

Average composition, g/kg, is:

Crude protein	57·84
True protein	55·15
Casein	45·46
Coagulable protein	43·29
NPN	2·69
Urea	0·348

The most important parameters are:

NPN/casein protein	= 4·7%
Casein/total protein	= 82·4%
Coagulable protein/total protein	= 78·5%

The true protein content increases steadily from 46·5 to 61·5 g/kg during the season or lactation. While the ratios of casein to total protein and coagulable protein to total protein remain nearly constant over the production season, the ratio of NPN to casein protein decreases steadily from 5·16 to 3·55%. Depending on the month, milk composition varies between flocks by 7 to 10 g/kg for total protein, 1·9 to 2·6% for casein to total protein and from 1·5 to 2% for NPN to casein protein.[44] According to Portolano,[11] the protein to fat ratio in the milk of different breeds in Italy varies from 1 : 1·75 to 1 : 1·08.

The differences between the ovine and bovine casein systems influence both the rennet coagulation time and the firmness of the curd. Sheep's milk coagulates 1·56 times faster than the cows' milk when the same quantity of rennet is used, and the final curd is twice as firm as that from cows' milk.[18]

When cows' milk is heated at 85°C, the rennet coagulation time is increased significantly and a weak curd is produced which is not suitable for cheese manufacture.[45] According to Penev,[19] the coagulation time of sheep's milk is increased following heating at 70–78°C, a situation that can be improved if the milk is heated at 78–85°C.

After the action of rennet on the total bovine and ovine caseins or on their κ-caseins, a larger proportion of non-protein nitrogen is produced from ovine than from bovine casein: whole bovine casein, 1·94%; whole ovine casein, 2·95%; bovine κ-casein, 8·5; ovine κ-casein, 11·8%. The caseino-glycopeptide obtained after the action of rennet on ovine κ-casein is similar to that from bovine κ-casein. Milk for cheesemaking is not standardized for farmhouse production and only sometimes for industrial scale production. For example, in Greece, milk for the production of Feta is standardized to 6% fat.

The rennet used in most countries is prepared locally from the abomasa of suckling lambs and kid goats. This product, which contains mainly chymosin and only a small amount of pepsin, produces a curd which is comparable to that obtained with calf rennet.[11] The duration of coagulation, the sequence of cheesemaking operations and the transfer of proteins and other constituents of milk to cheese are similar to those for calf rennet. Cheeses made using such rennets have a stronger, spicier taste, which is preferred by many consumers, due to the increased content of short chain free fatty acids in comparison with those in cheese made using calf rennet.

In Portugal and Italy,[11] milk coagulant extracted with water from the flowers of cardoons of the genus *Cynara* is used in the manufacture of Sierra and many other cheese varieties from sheep's and/or goats' milk.[10,11]

Starters. The microflora of cheese made from sheep's and goats' milk consists of bacteria naturally present or which can enter the milk during production. In the Mediterranean region, a combination of natural lactic acid bacteria exists as a result of natural selection, which produces a special culture, entitled 'Mediterranean natural culture' used for the production of traditional cheeses. The predominant species are thermophilic bacteria, especially the *Lactobacillus* species, with the following characteristics: (i) good acidification capacity, (ii) good development in whey, (iii) resistance to phage, (iv) a small number of plasmids, (v) anti-clostridial action of some strains.

However, starters are used in the production of cheese from low-temperature pasteurized milk, e.g. 65°C for 15 min. In Greece, the survival of thermophilic microflora is sometimes enough for the preparation of all kinds of cheeses and in this case the use of homemade yoghurt as culture is the easier solution.

4.3 General Features of the Manufacture of Cheese from Ewes' Milk

Ewes' milk, during the first hours after milking, is resistant to the growth of different bacteria due to its high immunological activity. Also, due to its high mineral content, the buffering capacity is higher than that of bovine milk. The high antibacterial activity is undesirable if fresh milk is used for cultures. The quantity of rennet should be less than that used for cows' milk for the same co-agulation time, because the casein content is higher than in cows' milk. The firmness of the coagulum is higher and more resistant to the cutting knives. $CaCl_2$ is never added before renneting.

  (i) The curd is whiter than that from bovine milk.
 (ii) The syneresis time is shorter.
(iii) The stirring time is shorter.
 (iv) In the case of the production of Graviera cheese, which is similar to Gruyère cheese, the cooking temperature is 52°C instead of 56°C.
  (v) The duration of salting necessary to attain the same concentration of salt as for the same variety made from bovine milk is longer because the curd is firmer, e.g. the time required for cheese made from cows' milk is 40 days compared with 50 days for the same type of cheese from ewes' milk.

4.4 Greek Cheese Varieties Produced from Sheep's Milk

Greece has the highest per capita consumption of cheese in the world and about 85% of the cheese is produced from sheep's and goats' milk. The local production of cheese from cows' milk is very low, but large quantities are imported from other EEC countries. Approximately 65% of sheep's and goats' milk produced in Greece is transported to dairies for the manufacture of cheese, especially Feta. The rest is consumed on farms as liquid milk or used for the production of traditional cheeses and yoghurt, which are delivered to local markets within their production area.

Milk is processed in a large number of small dairies scattered throughout the country. According to data of the Ministry of Agriculture, the total number of dairies, mainly cheesemaking units, is 916, of which 172 operate throughout the year and 644 operate seasonally. Most of them, 861, are private and 55 are co-operative.

The DODONI dairy company, in northern Greece, is the largest in Europe for the collection of sheep's and goats' milk. It operates 350 collection centres,

TABLE XVI
Dairy Products Manufactured from Sheep's and Goats' Milk in Greece in 1990[46]

Product	Production, tonnes	
	from sheep's milk	from goats' milk
Cheese, soft	84 726	42 014
Cheese, hard	23 277	9 724
Whey cheese	6 067	3 452
Butter, fresh or melted	2 537	1 133
Yoghurt	68 532	32 254

where more than 10 000 producers deliver their milk daily. More than 20 000 tonnes of ewes' milk and 5000 tonnes of goats' milk are processed per annum.

The principal products manufactured from sheep's and goats' milk in Greece are listed in Table XVI.

In a book, *Greek Cheeses, a Tradition of Centuries* published in 1992 by the Greek National Committee, 17 varieties of traditional cheeses are described. Greek legislation includes standards for the following cheeses: Feta, Teleme, Graviera, Kefalotiri, Kasseri, Ladotiri, Mizithra, Anthotiros, Manouri, Xinomizithra, Galotiri, Kopanisti, Sfela, Batzos, Graviera of Crete and Formaella of Parnassos. All these cheeses are the result of a local cheesemaking tradition passed down the centuries since Homeric years.

Apart from Feta, which is described in Chapter 11, Volume 2, the best known Greek cheeses are described below.

4.4.1 Teleme
Teleme seems to have originated in Rumania, from where it has spread to the other Balkan countries, especially to Greece. It is a soft white cheese, ripened and kept in brine until it is delivered to consumers. It is manufactured from sheep's, cows' and goats' milk or from mixtures of these. The best quality is

TABLE XVII
Changes in Physicochemical Properties of Teleme Cheese, Made from Sheep's or Cows' Milk During Ripening[47,48]

Age (days)	Moisture, %		Fat, %		Protein, %		NaCl, %		pH	
	cow	sheep	cow	sheep	cow	sheep	cow	sheep	cow	sheep
1	61·65	58·79	16·52	—	13·20	17·49	3·8	—	4·95	5·0
5	—	52·26	—	—	—	17·76	—	2·81	—	4·95
6	54·18	—	20·06	—	—	—	—	4·70	4·70	—
15	—	52·50	—	—	—	17·70	—	3·17	—	4·85
30	54·12	52·65	19·94	—	15·36	17·58	3·10	3·36	4·77	4·80
60	53·61	52·49	19·93	—	15·22	17·58	3·09	3·35	4·83	4·82
90	—	52·27	—	—	—	17·66	—	3·40	—	4·76
120	53·52	—	20·11	—	15·20	—	3·05	—	4·85	—
130	—	52·16	—	—	—	17·64	—	3·45	—	4·73

made from sheep's milk; goats' milk produces a harder cheese. It is packaged in 17 kg tins. Teleme has some similarities with Feta cheese. The most significant differences in the technology are at draining of the curd; for Feta draining occurs under natural conditions but for Teleme it is accelerated by pressing. The technology of manufacture was described by Anifantakis.[46] The physico-chemical properties of Teleme during ripening are summarized in Table XVII.

4.4.2 Kefalotiri

Kefalotiri is a traditional Greek cheese made from sheep's or goats' milk or mixtures of these. The name 'Kefalotiri' stems from the Greek words 'Kefali', meaning head, and 'tiri' meaning cheese. The main characteristics are its hardness, its salty taste and strong flavour. Its quality differs more or less from one region to another. There are many varieties of Kefalotiri cheeses which are named according to the name of the region where they are manufactured (Table XVIII). The technology was described by Veinoglou et al.[49] and Anifantakis.[50] The typical

TABLE XVIII
The Technology of Kefalotiri Cheese[37,51]

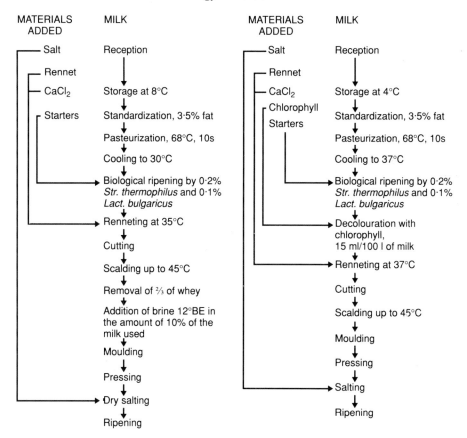

TABLE XIX
Changes in the Chemical Composition of Kefalotiri Cheese[49]

Constituents, %	Age of cheese (days)			
	4	15	30	90
Moisture	44·68	41·72	39·89	36·34
Fat	24·66	26·15	26·84	28·76
Fat in dry matter	44·58	44·87	44·65	45·26
Total proteins	24·91	24·73	24·91	26·65
Soluble proteins	1·85	3·64	4·78	6·51
Coefficient of maturation	7·34	14·84	19·38	24·50
NaCl	0·39	1·66	2·65	3·93
pH	5·28	5·14	5·10	5·04

chemical composition of Kefalotiri cheese made in Greece and Bulgaria is given in Table XIX. The fatty acids and free amino acids in mature (2 months old) cheese made from sheep's milk are presented in Tables XX and XXI.[49]

4.4.3 Graviera

The name 'Graviera' is used for various hard cheeses which are manufactured in different regions of Greece,[52] mainly from sheep's milk. The technology and characteristics are similar to Gruyère. Traditionally, Graviera cheese was made in small dairy units from raw milk. Nowadays, most the cheese is produced in small well-equipped cheesemaking units, mostly from pasteurized milk. Propionibacteria are rarely added because they occur in the milk. Information about the technology of Graviera was given by Zigouris,[52,53] Veinoglou[49,54] and Anifantakis.[46] Some analytical data for this cheese are given in Tables XXII and XXIII.

Table XXII gives the composition and some physico-chemical changes of Greek Graviera made from pasteurized sheep's[55] or cows'[56] during ripening. Water soluble nitrogen, non-protein nitrogen and amino nitrogen in 4 days-old

TABLE XX
Free Fatty Acids in Kefalotiri Cheese[49]

Free fatty acid	mg/kg of cheese	% of total
Acetic	145	3·9
Propionic	75	2·0
Iso-butyric	351	9·4
Butyric	1 020	27·4
Iso-valeric	836	22·5
Valeric	138	3·7
Caproic	799	21·5
Capylic	358	9·6
Total	3 722	100·0

TABLE XXI
Free Amino Acids in Mature (3 Month) Kefalotiri[49]

Amino acid	mg/100 g of cheese	Amino acid	mg/100 g of cheese
Lysine	262	Alanine	97
Histidine	71	Cystine	—
Arginine	23	Valine	116
Aspartic acid	37	Methionine	43
Threonine	201	Isoleucine	101
Serine	153	Leucine	232
Glutamic acid	410	Tyrosine	54
Proline	141	Phenylalanine	162
Glycine	71		

cheese was 13·9, 4·39 and 8·25% of the total nitrogen, respectively, and increased to an average 30·7, 10·88 and 16·30% at 6 months.[56]

The total and proteolytic counts range from 10^6 to 10^8/g, exhibiting maximum growth at 6 months, whereas lipolytic microorganisms increase to a maximum of 10^7/g when the cheese is 3 months-old and decrease to 10^6/g at 6 months. Lactobacilli are at 10^6/g in fresh cheese and reach maximum numbers, 10^8/g, from 15 to 45 days old. Lactococci occur in larger numbers than lactobacilli in the fresh cheese but later the counts of lactococci and lactobacilli are the same. Propionic acid bacteria reach maximum numbers, 10^6/g at 15 to 45 days and then remain constant until the cheese is 3 months old (Table XXIII).[56]

TABLE XXII
Composition and Physicochemical Changes in Greek Graviera During Ripening

Age of cheese (days)	Moisture, %		Fat, %		Salt, %		pH	CaO
	*	**	*	**	*	**	*	**
1	—	39·84	—	32·00	—	—	—	1·24
4	38·61	—	34·52	—	0·17	—	5·51	—
12	—	39·78	—	32·25	—	0·44	—	1·18
15	37·33	—	35·45	—	0·39	—	5·58	—
23	—	39·04	—	33·00	—	0·71	—	1·22
45	36·10	—	35·53	—	1·06	—	5·68	—
50	—	36·98	—	33·40	—	0·90	—	1·22
90	35·08	—	35·42	—	1·31	—	5·76	—
92	—	35·30	—	33·80	—	1·45	—	1·37
151	—	32·31	—	36·00	—	1·49	—	1·40
180	33·96	—	36·51	—	1·49	—	5·94	—

*Anifantakis[46]
**Veinoglou[49]

TABLE XXIII
Free Fatty Acids in Greek Graviera During Ripening

Free fatty acids (mg/100 g dry matter)	Age of cheese (days)			
	15	45	90	180
Acetic	70·0	106·5	188·5	287·6
Propionic	6·2	29·1	84·8	126·2
Butyric	7·2	22·0	72·9	119·6
Caproic	1·3	2·4	6·1	25·7
Caprylic	16·4	11·7	29·8	21·1
Capric	22·6	25·2	47·8	43·9
Lauric	35·9	39·9	74·3	51·1
Myristic	112·6	124·2	201·0	160·4
Palmitic	483·4	485·4	770·5	474·8
Stearic	trace	trace	trace	trace
Oleic	13·9	12·3	24·6	43·5
Linoleic + linolenic	50·2	54·9	84·1	134·1

4.4.4 Touloumotiri

This is an ancient cheese variety, perhaps that prepared by the Cyclops, Polyphemos, which is regarded as the forerunner of Feta cheese. It is ripened and preserved in the skins of goats or sheep. It is a soft white cheese made from sheep's and goats' milk with a slightly sour taste and pleasant organoleptic characteristics. Today, it is produced only in homesteads according to traditional technology.[57]

4.4.5 Kopanisti

Kopanisti is a traditional cheese made from sheep's, goats' or cows' milk or a mixture of these, especially in the Cyclades Islands.[58] Its main characteristics are an intense salty piquant taste, a soft spreadable texture and a rich flavour which approaches that of Roquefort. The main features of its manufacture are:

(i) The use of raw milk just after milking.
(ii) The curdling of milk by the combined action of rennet and acidity produced by the natural microflora of the raw milk.
(iii) During ripening, the cheese is kneaded repeatedly to incorporate the abundant flora that grow on the surface of the cheese uniformly into its mass. The formed acid curd is left to ripen, preferably in a cool, humid place. Under these conditions, a rich surface growth of a great variety of microorganisms occurs. During ripening, the soft mass of the cheese is mixed three to four times in order to incorporate the microflora into the whole mass of the cheese.

Studies on the ripening of experimental Kopanisti showed that the predominant microflora, after five days of ripening, were lactobacilli, lactococci, with yeast and moulds in high numbers at the surface. After mixing of the ripening

TABLE XXIV
Changes in Physicochemical Properties of Kopanisti Cheese During Ripening[58a,59]

Constituents	Days of ripening				
	1	8	16	32	46
Moisture, %	53·50	52·75	52·52	52·34	52·14
Fat, %	22·00	22·26	22·54	22·52	22·70
Protein (% Total N × 6·83)	17·82	18·22	18·69	19·01	19·27
Casein N	16·52	16·09	15·52	14·28	13·59
Soluble N (% of TN)	7·42	11·68	16·53	24·24	28·92
Non-protein N (% of TN)	3·89	8·62	13·30	20·37	24·62
Amino acid N (% of TN)	1·83	4·50	6·71	11·44	14·27
Ammonacial N (% of TN)	0·71	0·86	1·22	3·58	6·63
Free amino acids (mg/g of dried non-fat cheese)	48·38	136·75	209·31	355·86	421·12
Free fatty acids (g/kg of cheese)	2·26	5·69	11·92	29·93	50·16
Acid Degree Value (% of fat)	2·93	8·18	20·12	57·14	95·57
pH	4·92	4·78	4·58	4·67	4·75

mass, these organisms play important roles in maturation and the development of a rich flavour and aroma. The predominant species of yeasts are *Trichosporon cutaneum* (*T. beigli*) and *Kluyveromyces lactis* (*Kl. marxianus* var. *lactis*).[58] The characteristics of Kopanisti cheese are presented in Tables XXIV, XXV and XXVI.[58]

TABLE XXV
Changes in the Free Fatty Content (g/kg) of Kopanisti Cheese at Different Stages of Ripening[38]

Fatty acids	Days of ripening				
	1	8	16	32	46
$C_{4:0}$	0·016	0·095	0·231	0·822	1·521
$C_{6:0}$	0·009	0·084	0·222	0·696	1·253
$C_{8:0}$	0·018	0·068	0·243	0·599	0·889
$C_{10:0}$	0·078	0·252	0·521	1·104	1·614
$C_{12:0}$	0·185	0·367	0·853	1·152	1·902
$C_{14:0}$	0·281	0·708	1·505	3·587	6·230
$C_{16:0}$	0·729	1·871	3·786	9·727	16·257
$C_{18:0}$	0·242	0·473	1·172	3·068	5·465
$C_{18:1}$	0·702	1·769	3·390	9·173	15·026

Changes in the Free Amino Acids Content of Kopanisti Cheese (μmol/g of Dried Non-Fat Cheese) During Ripening[38]

Fatty acids	Days of ripening				
	1	8	16	32	46
ASP	2·18	4·23	5·25	3·84	1·68
THR	0·78	4·08	6·66	12·59	15·42
SER	3·04	2·25	0·97	2·13	6·85
ASN	0·79	6·22	10·01	15·91	29·58
GLU	1·80	12·97	12·01	4·16	5·70
GLN	1·67	6·39	11·49	17·01	23·73
PRO	5·73	5·55	14·31	24·38	28·15
GLY	0·40	2·36	4·54	11·33	15·18
ALA	3·73	8·83	10·76	31·54	45·63
AABA	—	—	—	1·84	3·88
VAL	5·74	14·91	19·96	36·12	43·75
MET	1·66	4·07	5·76	10·26	13·33
ILE	1·82	8·08	10·28	24·49	31·11
LEU	6·47	21·55	30·41	52·75	64·61
TYR	1·14	0·40	0·48	2·45	0·55
PHE	2·88	8·30	12·01	21·34	26·61
GABA	0·46	2·02	17·89	49·95	59·18
ORN	0·93	0·26	0·32	1·75	0·77
LYS	4·92	13·59	21·96	12·48	2·89
HIS	1·46	3·69	4·80	7·88	2·52
ARG	0·78	7·00	9·44	11·66	—

5 OTHER CHEESES FROM EWES' MILK

5.1 Halloumi (Cyprus)

Halloumi is produced mainly in Cyprus and exported to many European countries, especially to the UK. It is a semi-hard white cheese made either from sheep's, goats' or cows' milk or mixtures of these. Its technology is unique and is presented in Table XXVII. The available analytical data are presented in Table XXVIII.[60,61]

5.2 Ossay-Iraty Brebis Pyrénées

This is a traditional cheese made from ewes' milk in the Pyrénées Atlantique mountains. The uncooked curd is pressed and the cheeses are ripened for 90 days.[62]

TABLE **XXVII**
Protocol for the Manufacture of Halloumi Cheese from Sheep's Milk[60]

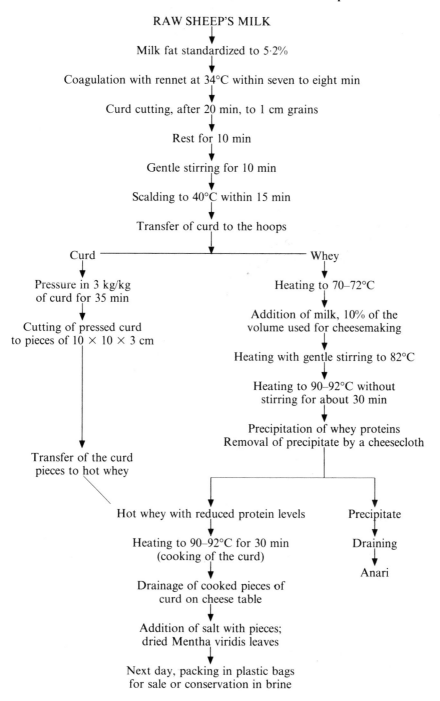

RAW SHEEP'S MILK

Milk fat standardized to 5·2%

Coagulation with rennet at 34°C within seven to eight min

Curd cutting, after 20 min, to 1 cm grains

Rest for 10 min

Gentle stirring for 10 min

Scalding to 40°C within 15 min

Transfer of curd to the hoops

Curd —————————————————— Whey

Pressure in 3 kg/kg
of curd for 35 min

Heating to 70–72°C

Cutting of pressed curd
to pieces of 10 × 10 × 3 cm

Addition of milk, 10% of the
volume used for cheesemaking

Heating with gentle stirring to 82°C

Heating to 90–92°C without
stirring for about 30 min

Precipitation of whey proteins
Removal of precipitate by a cheesecloth

Transfer of the curd
pieces to hot whey

Hot whey with reduced protein levels   Precipitate

Heating to 90–92°C for 30 min
(cooking of the curd)

Draining

Drainage of cooked pieces of
curd on cheese table

Anari

Addition of salt with pieces;
dried Mentha viridis leaves

Next day, packing in plastic bags
for sale or conservation in brine

TABLE XXVIII

Composition of Halloumi Cheese Made from Sheep's Milk and of Halloumi Cheese from the Cyprus Market (Mean of 13 and 17 Observations, Respectively)[60]

Components	Halloumi from sheep's milk			Halloumi from the Cypriot market		
	Mean	Range	Standard deviation	Mean	Range	Standard deviation
Moisture, %	42·15	39·04–43·84	1·39	42·53	35·46–48·56	3·75
Fat, %	27·85	26·25–29·25	1·76	25·57	20·00–29·50	3·14
Fat in dry matter, %	48·09	46·10–49·99	1·95	44·52	37·95–50·48	3·98
Protein (Total N% × 6·38)	23·71	21·95–25·02	1·02	24·46	20·86–30·45	2·28
Protein, % dry matter	41·02	39·22–44·26	2·37	42·53	40·11–48·62	2·30
Soluble protein, % (soluble N% × 6·38)	0·76	0·64–0·89	—	1·15	0·83–1·55	1·14
NaCl, %	1·44	1·05–2·05	0·28	3·54	2·31–5·65	0·91
pH	5·86	5·30–6·10	0·22	—	—	—

6 WHEY CHEESES FROM EWES' AND GOATS' MILK

Whey has a high content of lactose, minerals, water-soluble proteins, other nitrogenous non-protein substances of milk and fat which vary within limits, depending on the type of cheese and the manipulation of the curd, considerable quantities of vitamins and many other substances at low or trace quantities.[63,64] The chemical composition of whey from cheeses made from ewes' milk is shown in Table XXIX. It is seen from this Table that the total protein content is about twice that in whey from cows' milk cheeses.[59]

Since ancient times, the nutritional value of whey has been recognized mainly in Mediterranean countries and was used as a feed-stuff for animals and to a lesser extent as human food. In 460 BC, Hippocrates recommended drinking considerable quantities of whey for relatively long periods because he observed that it was beneficial for pathological conditions in Man. Later, during the 18th to 19th centuries AD, many curative properties were attributed to whey for many diseases, as well as skin disorders.[65]

Based on this experience in some Mediterranean countries, cheesemakers prepared cheese by heating the whey and separating the coagulated proteins, which occlude fat. These whey cheeses, which are rich in protein, have very high digestibility and very high nutritional value, because they contain all the essential amino acids, especially methionine and cystine, in quantities more than sufficient to meet the needs of Man, as shown in FAO protein requirements.[64,66]

Ricotta is an Italian whey cheese, Serac or Serai is produced in France, Broccio or Brocciu in Corsica,[67] Myzithra and Manouri in Greece. Whey cheese is also produced in other countries.[68]

TABLE XXIX
Chemical Composition of Whey from Different Types of Cheese[69]

Constituents	Total solids (%)	Moisture (%)	Fat (%)	Total protein (%)	Lactose (%)	Ash (%)	Acidity	pH
Cheese								
From cows' milk								
Feta	6·32	93·68	0·25	0·82	4·72	0·49	0·125	6·30
Graviera	6·90	93·10	0·60	0·90	4·90	0·50	0·120	6·30
Kefalotiri	6·55	93·45	0·40	0·80	4·85	0·50	0·11	6·40
Cheddar	6·30	93·70	0·50	0·80	4·85	0·50	0·15–0·19	6·3–5·7
Cottage	5·50	93·50	0·04	0·75	4·90	0·40–0·50	0·55	4·60–4·50
Emmental	7·96	92·06	0·95	0·89	5·45	0·54	0·10	—
Camembert	7·05	92·95	0·34	0·93	5·08	0·59	0·33	—
From sheep's milk								
Feta	7·87	92·13	0·39	1·61	5·33	0·78	0·18	—
Graviera	8·74	91·26	1·26	1·52	5·27	0·50	0·13	6·30
Kefalotiri	7·48	92·52	0·70	1·41	4·99	0·51	0·13	6·20

6.1 Methods for the Production of Whey Cheeses

The whey, with or without acidification, is heated to 88–90°C in 40 to 45 min in such a way as to separate the proteins and fat as curd. This incorporates some of the other constituents which, after drainage, gives a kind of cheese. If the whey is sweet, pH >6, partial acidification is required before or during heating

TABLE XXX
Chemical Composition of Different Types of Whey Cheeses[59]

Constituents	Fat (%)	Total protein (%)	Water soluble protein (%)	Lactose (%)	Ash (%)	Salt (%)	Solids (%)	pH	Total Acidity	Fat in dry matter (%)
Cheese										
Manouri	36·67	10·86	0·43	2·49	1·68	0·83	51·93	5·90	0·14	70·61
Myzithra	15·95	13·09	0·66	3·33	1·72	0·82	33·59	6·00	0·12	47·48
Dry Myzithra[60]	20·83	25·44	1·24	4·00	9·93	8·66	61·37	4·67	0·48	33·94
Ricotta	30·61	8·74	—	3·20	0·58	—	43·12	—	—	70·98
Brocciu	—	—	—	—	—	—	20	6·5–6·7	—	40·00
Broccio	30·0	11·57	—	—	3·66	3·11	48·08	4·4	—	62·40
Ricotta from cows' whey	19·6	23·08	—	4·00	2·12	—	49·10	—	0·30	39·90
Ricotta from sheep's whey	22·36	6·37	—	4·28	0·51	—	33·12	—	—	67·50
Ricotta	2·5	126·0	—	3·5	1·0	—	20·23	5·6–6·0	—	11–12·5
Serac	3–9	14–20	—	—	2·0	—	20–25	—	—	12–45·0

by adding various acids.[59] The composition of fresh and dried Mizithra and Brocciu are given in Table XXX. The yield of whey cheese varies according to the composition of the whey and added milk or cream, the technology used and the composition of the final product (moisture content).[70] The technology of different whey cheeses made in Greece are summarized in Table XXXI.

Efforts have been made to improve the manufacture of whey cheeses in order to minimize the cost of production. The whey may be concentrated by ultrafiltration to 10–20% solids or higher and mixed with cream.

A new technology has been developed at the Dairy Laboratory, Agricultural University of Athens, for the manufacture of whey cheeses from highly ultra-filtered whey. After preparation of the liquid pre-whey cheese, it is packaged in

TABLE XXXI
Protocol for the Traditional Manufacture of Whey Cheeses in Greece[59]

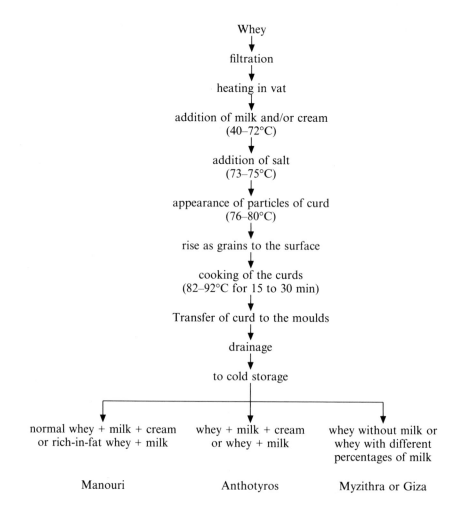

Whey
↓
filtration
↓
heating in vat
↓
addition of milk and/or cream
(40–72°C)
↓
addition of salt
(73–75°C)
↓
appearance of particles of curd
(76–80°C)
↓
rise as grains to the surface
↓
cooking of the curds
(82–92°C for 15 to 30 min)
↓
Transfer of curd to the moulds
↓
drainage
↓
to cold storage

normal whey + milk + cream or rich-in-fat whey + milk	whey + milk + cream or whey + milk	whey without milk or whey with different percentages of milk
Manouri	Anthotyros	Myzithra or Giza

plastic bags, plastic cups or cans, and heated to 90°C for 50 to 90 min, depending on the size of the containers. The proteins are denatured and the cheese which is formed attains its regular firmness after holding it at 4–6°C for 24 h. This cheese has a very good keeping quality compared with similar cheeses made by traditional technology.[31]

7 PRODUCTION OF CHEESE FROM GOATS' MILK

7.1 Introduction

The importance of goats in the life of the Mediterranean and African countries is mentioned in mythology. The milk, cheeses and meat of goats constitute a fundamental base for the nutrition of a number of peoples in the world, i.e. in areas where the husbandry of cows is very difficult.

Today, three categories of cheese are produced from goats' milk:

  (i) The first includes traditional cheeses of many varieties made on farms, which are more or less well known. It is impossible to calculate the quantity of the cheeses produced because this category includes cheeses produced for home consumption.
 (ii) The second includes traditional cheeses made partly on a farm scale under improved conditions and partly on a modern industrial scale. This is the case in France, the only country in the world with a production of more than 90 varieties of goat-milk cheese. The excellent quality of these cheeses has led to a 15% increase in their consumption in recent years and also to the export of 1860 tonnes in 1988.
(iii) The third group of cheeses are produced from mixed sheep and goat milk. In most countries, goats and sheep are farmed in mixed flocks. The production of the two types of milk is limited by the lactation period, which is 6 months or less, and also by the quantity of milk, which is low during the first and last months of lactation. For this reason, producers mix the two types of milk in order to increase the quantity for cheese-making; this is the practice in all Mediterranean countries, except France. The characteristics of cheeses produced from the mixed milk are strongly influenced by the sheep's milk, which is predominant.

7.2 General Remarks Concerning the Manufacture of Cheeses from Goats' Milk

The gross chemical composition of bovine and caprine milk are similar, but there are significant differences in physico-chemical characteristics, for example, the ratio of α_s- to β-casein is lower in goats' than in cows' milk.[16] On the other hand, there is considerable variability in the α_{s1}-casein content of milk from individual goats, related to genetic polymorphism;[25] some animals produce milk devoid of α_s-casein while for other individuals, α_s- represents 20–25% of total casein.[32]

The characteristics of the casein micelles in bovine and caprine milks are also different. The level of mineralization in goats' micelles is higher than in cows[71] and the degree of hydration is lower.[45]

Another important difference is in the rennet clotting time and in the firmness of the gel.[18] The kinetic viscosity of ewes', cows' and goats' milk is 1·47, 1·23 and 1·07 centistokes, respectively.[18]

The firmness of renneted milk gels, determined using the torsiometer of Scott-Blair, was found to be 333 degrees for sheep's milk, 164 to 176 for cows' milk and 76 to 93 for goats' milk;[18] even at the same casein content, the firmness of goat milk gels was lower than those from cows' milk. Whey drainage from goats' milk curd is more rapid than that from cows' milk. Portmann et al.[72] studied the influence of the fat and protein content of goats' milk on the yield for curd from 90 experimental batches. Statistical treatment of the data showed that cheese yield depended on the fat and protein content of the milk. The regression coefficient for protein content was highly significant; its value was four times that of the correlation coefficient for fat content.

The percentage of the milk protein which is incorporated into cheese shows that the cheese-yielding capacity of goats' milk is relatively low compared to cows' milk, i.e. 69·5 and 75%, respectively.[72]

In many instances, goat cheese is made without culture, the natural microflora being relied on to develop the acid needed for cheese production; exceptions are the cheeses made on an industrial scale in France. Many of the varietal differences between cheeses are due to the physical and chemical changes during ripening and to the manipulations during cheesemaking. Production is usually seasonal since a large volume of milk is available only from February to July or from April to September, depending on the altitude.

In France, in order to prepare fresh cheeses throughout the year, milk or drained curds are frozen and stored at –18°C in plastic bags. Comparison of the organoleptic quality of cheeses made from freshly prepared or from frozen curds showed a marked effect of frozen storage. For certain batches, oxidative defects were evident after 3 months' storage, but others did not show this, even after 6 months.[73]

Ultrafiltration has been used to produce a pre-cheese from goats' milk, which subsequently was made into cheese. It is also possible to freeze the pre-cheese.[74]

7.3 General Features of the Manufacturing Protocol for Cheeses Made from Goats' Milk

While there are wide variations in the flavour, body, texture and specific nutritional qualities of goats' cheeses, they have some characteristics in common.

It is necessary to distinguish the cheese in three groups: The first group contains cheeses made from a lactic acid curd coagulated by the indigenous lactic acid microflora. Larger factories use pure cultures of lactic acid bacteria adapted to goats' milk, in combination with a small quantity of rennet.

When fresh, these cheeses have a moisture content of 80% but they are susceptible to evaporation. In some countries with a hot climate, the fresh cheeses are exposed to the sun to dry, e.g. Djamus cheese in Syria. The curd of this group of cheeses is generally fragile and does not permit the production of large cheeses, but drainage is limited. The cheese varieties, Alicante, Cadiz and Soria of Spain belong to this group. Other members are those ripened after immersion in olive oil, e.g. Sourke (Syria), Malaga (Spain) and Ladotyri (Greece).

The second group contains cheeses made from rennet-coagulated curds which drain freshly. These cheeses contain 50% moisture, which is important for the development of the lactic acid bacteria, which are required to complete the ripening process. The cheeses of this group are divided into three types: cheese in brine, soft or semi-soft cheeses, and surface-ripened cheeses. Examples are Feta (Greece), Saint Maure and Crottin (France), Altenburg (Germany), Blanco (Libya) and Akari (Syria).

The third group contains cheese prepared from rennet-coagulated curds which are cooked to accelerate draining, e.g. Aseredo (Mexico), Salamora (Turkey) and Rumalia (Greece and Bulgaria). Similar cheeses include Gjetot (Norway) with a semi-strong flavour, Quesco Echilago (Mexico), Lighwan (Iran) and Banon (France), Caprino a pasta cruda and Caprino semicotta (Italy).

The technology and composition of most goat cheeses, especially the traditional ones, are unknown.

Martin-Hernandez et al.[51] and Ramos et al.[75] presented a table, with very limited data, on 20 goats' cheeses from around the world while Ledda[76] reported the composition of Italian goats cheese made in Sardegna.

The free fatty acids and free amino acids in two types of goats' cheese are shown in Tables XXXII, XXXIII, XXXIV, XXXV.[77] Rumelia is a pasta filata-type cheese, prepared with thermophilic cultures, while the second is a semi-hard, Edam type cheese.

7.4 Typical Traditional Cheeses

The technology used for the manufacture of most traditional cheeses is more or less unknown, except for the French varieties, which are described by Le Jaouen.[78] There has been considerable progress in this field in France and the prices of these cheeses are higher than normal. French legislation defines and protects the designation 'goats' and 'semi-goats' cheeses. Thus, the designations 'goats cheese' or 'goat blue cheese' may be used only for cheeses of various shapes and weights made solely from goats' milk. The cheeses should contain at least 45% fat after complete drying. Illustration, references, symbols, etc., suggestive of goat-raising may be used only for goat or semi-goat cheeses.

In other countries, a number of cheeses are made mainly from goats' milk, but sometimes with the addition of sheep's and/or cows' milk. Most are fresh or ripened only for a short period and rarely are smoked, e.g. Palmero and Majerero from the Canary Islands.[79] Surface mould-ripened soft cheese, with *Penicillium candidum*, are Sainte Maure, Grottin from France, Altenburg from Germany. Blue type cheeses with *Penicillium glaucum*, are Cabrales from Spain and Gemonedo from Portugal.

TABLE XXXII

Free Amino Acids of Rumelia Cheese (mg/100 g Cheese)[77]

Amino acids	1 month	% of total	2 months	% of total	3 months	% of total	4 months	% of total
Lysine	68·49 ± 4·30	11·60	161·96 ± 16·77	14·20	383·85 ± 23·92	12·13	39·21 ± 19·80	11·60
Histidine	14·79 ± 2·58	2·50	28·52 ± 3·78	2·51	82·40 ± 5·17	2·60	86·35 ± 5·20	2·60
Arginine	1·61 ± 0·49	0·27	5·95 ± 1·71	0·52	23·63 ± 3·16	0·75	25·44 ± 2·14	0·76
Aspartic acid	10·05 ± 1·23	1·70	20·51 ± 1·88	1·80	52·73 ± 3·95	1·67	53·08 ± 2·34	1·58
Threonine	63·40 ± 6·31	10·77	119·11 ± 7·84	10·47	159·83 ± 9·15	5·05	164·24 ± 8·22	4·89
Serine	49·05 ± 4·64	8·30	91·20 ± 6·12	8·00	215·02 ± 14·45	6·80	214·63 ± 9·08	6·40
Glutamic acid	91·34 ± 9·76	15·50	159·84 ± 22·57	14·05	655·01 ± 49·54	20·70	713·22 ± 33·97	21·23
Proline	50·50 ± 5·68	8·57	122·15 ± 11·44	10·74	176·03 ± 6·61	5·56	189·78 ± 6·73	5·65
Glycine	29·17 ± 3·81	4·95	35·59 ± 3·98	3·13	110·69 ± 10·48	3·50	127·78 ± 15·93	3·80
Alanine	14·02 ± 1·44	2·38	30·99 ± 1·50	2·72	105·47 ± 5·61	3·30	122·19 ± 4·27	3·64
Valine	39·90 ± 3·60	6·77	75·40 ± 5·58	6·63	263·31 ± 13·80	8·32	307·42 ± 27·18	9·15
Methionine	14·37 ± 1·45	2·44	21·51 ± 2·51	1·90	111·94 ± 8·21	3·54	116·70 ± 5·26	3·47
Isoleucine	27·15 ± 2·59	4·61	55·69 ± 3·04	4·90	203·22 ± 10·11	6·42	207·70 ± 7·37	6·18
Leucine	69·23 ± 3·27	11·75	108·99 ± 4·58	9·60	389·14 ± 15·40	12·29	401·47 ± 2·45	11·95
Tyrosine	12·90 ± 1·38	2·19	28·66 ± 1·99	2·50	49·51 ± 2·30	1·56	52·10 ± 14·99	1·55
Phenylalanine	33·73 ± 1·72	5·73	71·49 ± 6·20	6·28	183·80 ± 6·28	5·81	185·06	5·51
Total	589·68	~100	1137·55	~100	3165·56	~100	3359.37	~100

TABLE XXXIII
Volatile Free Fatty Acids in Edam-Type Cheese[77]

Fatty acid		1 month		2 months		3 months		4 months	
		g/kg	% of total	g/kg	% of total	g/kg	% of total	g/kg	% of total
Acetic	Mean	0·176	7·2	0·345	10·5	0·257	6·5	0·219	4·8
	Min–Max	0·15–0·19		0·32–0·37		0·23–0·28		0·20–0·23	
Propionic	Mean	0·122	5·0	0·189	5·7	0·154	3·9	0·105	2·3
	Min–Max	0·10–0·13		0·16–0·22		—		0·09–0·12	
Isobutyric	Mean	0·215	8·8	0·252	7·6	0·182	4·6	0·128	2·8
	Min–Max	0·20–0·23		0·22–0·28		0·17–0·20		0·11–0·31	
Butyric	Mean	0·527	21·6	0·789	23·8	1·046	26·5	1·252	27·4
	Min–Max	0·48–0·54		0·76–0·85		0·91–1·12		1·13–1·31	
Isovaleric	Mean	0·383	15·7	0·285	8·6	0·300	7·6	0·187	4·1
	Min–Max	0·35–0·41		0·26–0·30		0·28–0·31		0·17–0·19	
Valeric	Mean	0·078	3·2	0·089	2·7	0·075	1·9	0·037	0·8
	Min–Max	0·06–0·09		0·08–0·09		0·07–0·08		0·03–0·04	
Caproic	Mean	0·536	22·0	0·772	23·3	1·113	28·2	1·489	32·6
	Min–Max	0·45–0·60		0·75–0·80		1·01–1·24		1·28–1·56	
Caprylic	Mean	0·402	16·5	0·593	17·9	0·821	20·8	1·151	25·2
	Min–Max	0·38–0·43		0·54–0·63		0·74–0·89		1·06–1·29	
Total	Mean	2·439	100	3·315	~100	3·949	100	4·569	100
	Min–Max	2·26–2·67		3·12–3·58		3·75–4·12		4·36–4·79	

7.5 The Problem of 'Appelation d'Origine'

The agricultural products more or less unique to each region, like wine, cheeses, etc., are the result of a combination of the soil, climate and Man's labour.

In the occidental world in general, and in Mediterranean countries in particular, all the products were associated with the name of the region in which they are produced, e.g. Pecocino, Siciliano. With the change in socio-economic conditions and the appearance of commercial marks, the protection of 'Appelation d'Origine' becomes a necessity for agricultural products. For this reason, a certain number of regulations were introduced, e.g. The Convention of Paris (May 20, 1883), The Agreement of Madrid (1891) and more recently that of Lisbon (1958).

Under these agreements, the Appelation d'Origine (AO) consists of the name of the country, or a region or location in order to designate the origin of a product and its qualities or characteristics, which are related to the geographical origin and the natural and local human factors.

From 1960, a new notion developed in Europe, i.e. labelling of agricultural products, which means the total characterization of a product, which established a quality level and also distinguished it from mass-produced products.

For the cheeses made from sheep's or goats' milk, the problem is very important because the AO designation offers numerous advantages to the region,

TABLE XXXIV
Free Amino Acids in Edam-Type Cheese (mg/100 g Cheese)[77]

Amino acids	1 month	% of total	2 months	% of total	3 months	% of total	4 months	% of total
Lysine	51·00 ± 2·98	10·49	75·95 ± 3·15	8·06	266·98 ± 12·16	10·91	410·34 ± 15·51	14·03
Histidine	16·42 ± 1·31	3·38	34·27 ± 1·60	3·64	103·19 ± 3·52	4·22	120·25 ± 5·52	4·11
Arginine	1·26 ± 0·20	0·26	5·59 ± 0·99	0·59	11·28 ± 1·13	0·46	21·65 ± 0·65	0·74
Aspartic acid	9·94 ± 1·05	2·04	46·37 ± 1·98	4·92	89·81 ± 3·66	3·67	121·73 ± 6·23	4·16
Threonine	39·27 ± 1·99	8·08	50·15 ± 2·29	5·33	117·46 ± 6·20	4·80	140·38 ± 5·61	4·80
Serine	29·32 ± 1·44	6·03	71·25 ± 3·01	7·57	97·73 ± 8·91	3·87	106·15 ± 6·63	3·63
Glutamic acid	67·99 ± 2·30	13·98	142·63 ± 7·41	15·15	520·16 ± 25·53	21·25	591·59 ± 9·51	20·23
Proline	38·92 ± 1·18	8·00	77·76 ± 2·64	8·26	130·80 ± 5·89	5·34	171·90 ± 9·65	5·88
Glycine	15·59 ± 0·85	3·21	41·10 ± 2·78	4·36	108·60 ± 4·00	4·44	124·47 ± 5·08	4·26
Alanine	15·55 ± 0·72	3·20	51·37 ± 3·04	5·45	81·11 ± 3·10	3·31	97·01 ± 4·80	3·32
Valine	44·56 ± 1·85	9·16	81·96 ± 2·59	8·70	239·93 ± 12·88	9·80	265·43 ± 16·74	9·07
Methionine	13·89 ± 0·85	2·86	17·86 ± 2·00	1·90	86·10 ± 5·72	3·52	93·68 ± 3·69	3·20
Isoleucine	21·86 ± 1·28	4·50	63·07 ± 3·32	6·70	130·76 ± 7·14	5·34	152·42 ± 12·58	5·21
Leucine	84·86 ± 4·15	17·45	113·43 ± 6·66	12·04	295·58 ± 14·31	12·07	328·12 ± 14·10	11·22
Tyrosine	2·75 ± 0·86	0·57	10·30 ± 1·30	1·09	30·76 ± 1·90	1·26	38·55 ± 2·79	1·32
Phenylalanine	33·58 ± 1·31	6·91	58·69 ± 1·84	6·23	132·54 ± 5·76	5·41	147·57 ± 6·45	5·05
Total	486·21	~100	941·76	~100	2448·04	~100	2924·96	~100

TABLE XXXV
Volatile Free Fatty Acids in Rumelia Cheese[77]

Fatty acid		1 month		2 months		3 months		4 months	
		g/kg	% of total	g/kg	% of total	g/kg	% of total	g/kg	% of total
Acetic	Mean	0·280	9·5	0·245	6·0	0·235	5·2	0·205	4·1
	Min–Max	0·25–0·30		0·23–0·26		0·22–0·26		0·20–0·22	
Propionic	Mean	0·168	5·7	0·164	4·0	0·145	3·2	0·110	2·2
	Min–Max	0·15–0·18		0·16–0·18		0·14–0·15		0·09–0·12	
Isobutyric	Mean	0·265	9·0	0·311	7·6	0·181	4·0	·150	3·0
	Min–Max	0·25–0·28		0·30–0·33		0·17–0·19		0·13–0·16	
Butyric	Mean	0·618	21·0	0·973	23·8	1·147	25·4	1·342	26·9
	Min–Max	0·59–0·64		0·89–0·99		1·09–1·21		1·27–1·45	
Isovaleric	Mean	0·236	8·0	0·372	9·1	0·542	12·0	0·678	13·6
	Min–Max	0·22–0·26		0·36–0·39		0·48–0·58		0·58–0·70	
Valeric	Mean	0·130	4·4	0·155	3·8	0·127	2·8	0·080	1·6
	Min–Max	0·11–0·14		0·15–0·17		0·12–0·14		0·07–0·09	
Caproic	Mean	0·639	21·7	0·957	23·4	1·120	24·8	1·302	26·1
	Min–Max	0·58–0·65		0·94–0·97		1·08–1·22		1·24–1·33	
Caprylic	Mean	0·609	20·7	0·912	22·3	1·021	22·6	1·112	22·5
	Min–Max	0·58–0·61		0·85–0·93		0·94–1·13		1·09–1·24	
Total	Mean	2·945	100	4·090	100	4·372	100	4·987	100
	Min–Max	2·69–3·09		3·85–4·12		4·37–4·64		4·79–5·13	

which are up to the people of that region or country to maintain and develop. Unfortunately, too many states have refused to observe these agreements or have failed to ratify them. On the other hand, the excess of cows' milk in northern Europe induced cheese manufacturers to imitate traditional cheeses made from sheep's or goats' milk, as is the case with Feta cheese. There is an urgent need for legislators, beginning with those in the EEC, to become aware of the obligation to provide a universal guarantee for local products, which should be protected against encroachment, whether direct or indirect.

7.6 Detection of Mixtures of Milks for the Preparation of Cheeses

The excess of cows' milk and the demand for cheeses made from sheep's and goats' milk, which are more expensive than cows' milk. has resulted in the adulteration of sheep's and goats' milk with cows' milk for cheese production.

In order to ensure the authenticity of cheeses made from ewes' or goats' milk, analytical techniques for the detection of mixtures of different kinds of milk have been developed, based on the analysis of casein, fat and soluble protein in liquid milk and the cheese.[75] Since caseins are relatively stable on heating, they represent a good marker of cheese composition and electrophoretic methods based on the detection by gel electrophoresis of homologous caseins in mixtures either on the basis of their different net charge at a given pH or their isoelectric point. Para-κ-casein has been selected as a marker for bovine, caprine and ovine milks. The procedure is based on isoelectric focusing in the pH range 2·5–10 of a urea extract of the cheese.

Mixtures of ovine, caprine and bovine milks have been detected simultaneously by gel electrophoresis.[80,81] The limit of sensitivity varies according to the nature of the milk, caprine or bovine, added to ovine milk. This method can detect 20 and 10% of caprine or bovine milk, respectively, in ovine cheese. The sensitivity of the method has been improved by using a silver staining procedure of the electrophoretic bands and levels as low as 0·5% of bovine milk in ovine milk or cheese can be detected. Anifantakis et al.,[82] using an electrophoretic technique, has detected 5% added cows' milk in Feta cheese 300 days after manufacture.

Bovine milk has also been detected in ovine and caprine cheese by means of polyacrylamide gel isoelectic focusing of γ_2 caseins.[83] The method allows the detection of bovine milk at levels as low as 1–2%. A substantial increase in the sensitivity of this procedure was obtained by incubating the cheese with bovine plasmin and detection of γ_2 caseins by polyacrylamide gel isoelectic focusing.[84,85]

Immunological methods, based on antigen-antibody precipitation reactions, are particularly suitable for this type of control because of their sensitivity and specificity; they date from 1901 when Schutze[86] suggested the possibility of differentiating the proteins in the milks of different species. Since then, many publications have appeared[87–90] but the most important is that of Levieux,[2] who described a new and highly sensitive technique for quantitative analysis. The advantage of this technique was the choice of a protein not synthesized by the mammary gland, immunoglobulin G1 (IgG1), for the preparation of antiserum. The antiserum, prepared by immunizing goats or sheep, should be specific without the need for adsorption, which in practice is always difficult. Once the antiserum has been obtained, visualization can be performed using either of two methods: (i) radial immunodiffusion, or (ii) inhibition of hemagglutination. Using these techniques, it is possible to determine admixtures of cows' milk with the milk of ewes or goats or goats' milk mixed with ewes' milk.[1] Another technique for detecting mixtures of milks is based on the electrophoretic pattern of the whey proteins. Rispoli & Saugues[91] described the isoelectric focusing of whey proteins for the qualitative and quantitative determination of the cows' milk in sheep's cheeses. The same method was applied by Rispoli et al.[92] to the analysis of Roquefort cheese.

REFERENCES

1. Levieux, D., 1978. *Proc. XX Inter. Dairy Congr.* (Paris), **15**.
2. Levieux, D., 1972. *Dossiers de l'Elevage*, **2**, 37.
3. Ganguli, N.C., 1971. *International Dairy Federation Seminar on Milks other than Cows' Milk*, Madrid, p. 27.
4. Ramos, M., 1981. *International Dairy Federation Bulletin*, **140**, 5.

5. Baltadjieva, M., Veinglou, B., Kandarakis, J., Edgarian, M. & Stamernova, V., 1982. *Le Lait*, **62**, 191.
6. Dilanian, A.H., 1969. *International Dairy Federation Annual Bulletin*, Part VI, 7–13.
7. Juarez, M., Jimenez, S., Goicoechea, A. & Ramos, M., 1982. *Proc. XXI Intern. Dairy Congr.*, Moscow, **1**(2), 625.
8. Rivemale, M., 1982. *Communication Intern. Lab. Soc. des Caves Roquefort.*
9. Voutsinas, L.P., Delegannis, C., Katsari, M. & Pappas, C., 1988. *Milchwissenschaft*, **43**, 12.
10. Barbosa, M., 1990. *Production and Transformation of Sheep Milk in Portugal.* LNETI-DTIA, No. 107.
11. Portolano, N., 1986. *Tecnica Casearia.* Edagricole (Publishers), Bologna, Italia, p. 11.
12. Brochet, M., 1982. These 3^e Cycle, Universite Claude-Bernard.
13. Anifantakis, E., 1986. *International Dairy Federation Bulletin*, **202**, 42.
14. Richardson, B.C., Creamer, L.K., Pearce, K.N. & Munford, R.E., 1974. *J. Dairy Sci.*, **26**, 140.
15. Carić, M. & Djordjević, J., 1971. *Milchwissenschaft*, **26**, 495.
16. Richardson, B.C. & Creamer, L.K., 1976. *N.Z. J. Dairy Sci. Technol.*, **11**, 46.
17. Richardson, B.C., Mercier, J.C. & Ribadeau-Dumas, B., 1978. *Proc. XX Intern. Dairy Congr.* (Paris), **E**, 215.
18. Kalantzopoulos, G., 1970. PhD thesis, Agricultural University of Athens.
19. Penev, P., 1962. *Proc. XVI Intern. Dairy Congr.* (Copenhagen), **B**, 770.
20. Alais, C. & Jolles, P., 1967. *J. Dairy Sci.*, **50**, 1955.
21. Jenness, R., 1980. *J. Dairy Sci.*, **63**, 1605.
22. Gattuso, A.M. & Fazio, G., 1980. *Riv. Ital. Sost. Grasse.*, **57**, 530.
23. Juarez, M., 1985. *International Dairy Federation Bulletin*, **202**, 54.
24. Garci-Olmedo, R., 1976. *Coll. Anal. Bromatologia*, **28**, 211.
25. Grossclaude, F. & Mahe, M.F., 1986. *Llemme J. de la Rech. Ovin et Caprin*, p. 23.
26. Routaboul, M., 1981. *Communication Inter. Lab. Societe des Caves de Roquefort* (short communication).
27. Gallacier, J.P., Barbier, J.P. & Kuzdzal-Savoie, S., 1974. *Le Lait*, **54**, 117.
28. Alais, C., 1984. *Science du Lait*, Principles des Techniques Laitières. 4th edn, SEPAIC, Paris.
29. Simos, E., Voutsinas, L.P. & Pappas, C.P., 1990. *Small Ruminant Res.*, **3**, 11.
30. Anifantakis, E. & Kandarakis, J., 1980. *Milchwissenschaft*, **35**, 617.
31. Veinoglou, B., Kandarakis, J., Baltadjieva, M., Andrews, A. & Vlasseva, R., 1982. *Proc. XXI Intern. Dairy Congr.* (Moscow), **1b**, 246.
32. Remeuf, F., Lenoir, J. & Duby, C., 1989. *Le Lait*, **69**, 499.
33. Boulanger, A., Grosclaude, F. & Mahe, M.F., 1984. *Gent. Sel. Evol.*, **16**, 157.
34. Storry, J.E., Grandison, A.S., Millard, D., Owen, A.J. & Ford, G.D., 1983. *J. Dairy Res.*, **50**, 215.
35. Addeo, F., Mauriello, R., Di Loucia, A. & Chianese, L., 1988. *Proc. VI World Conference on Animal Production, Helsinki*, WAAP-88-Helsinki-Produ 8·1, GPO.
36. Castagnetti, G.B., Chiavari, C. & Losi, G., 1984. *Sci. Tech. Lat-Cas.*, **35**, 109.
37. Mariani, P., Corriani, F., Fossa, E. & Pecorari, M., 1987. *Sci. Tech. Lat-Cas.*, **38**, 7.
38. Gonc, S., Schmidt, R. & Renner, E., 1979. *Milchwissenschaft*, **34**, 684.
39. Garcia-Olmedo, R., Carballido, A. & Arnaez, M., 1979. *Anal. Bromatol.*, **31**, 209.
40. Boccignone, M., Brigiti, R. & Sarra, C., 1981. *Ann. Fac. Vet. Torino*, **28**, 3.
41. Sawaya, W.N., Khau, P. & Al Shalhat, A.F., 1984. *Food Chem.*, **14**, 227.
42. Martin-Hernandez, M.C., Juarez, M. & Ramos, M., 1986. *International Dairy Federation Bulletin*, **202**, 58.
43. Martinez-Castro, J., Juarez, M. & Martin-Alvarez, P.J., 1979. *Milchwissenschaft*, **34**, 207.
44. Grappin, R., 1985. *International Dairy Federation Bulletin*, **202**, 79.

45. Thompson, M.P., Boswell, R.T., Martin, V., Jenness, R. & Kiddy, C.A., 1969. *J. Dairy Sci.*, **52**, 796.
46. Anifantakis, E., 1986. *Cheesemaking*. Karaberopoulos (Publishers), Athens, p. 180.
47. Alichanidis, E., 1981. PhD thesis, University of Salonica, Greece.
48. Kalogridou Vassiliadou, D., 1982. Pers. Comm.
49. Veinoglou, B., Baltadjieva, M., Anifantakis, E. & Edgarian, M., 1983. *Food Technol. Hygiene Rev.*, **5**, 9.
50. Antifantakis, E., 1991. *Greek Cheeses, A Tradition of Centuries*, Greek National Dairy Committee, Athens.
51. Martin-Hernandez, M.C., Juarez, M. & Ramos, M., 1984. *Rev. Alimentacion*, **36**, 6171.
52. Zigouris, N., 1952. *The Milk Industry*. Greek Ministry of Agriculture, p. 462.
53. Zigouris, N., 1926. *Cheese of Agrafa or Cheese Graviera*. Greek Agricultural Society, Athens.
54. Veinoglou, B., Kalantzopoulos, G. & Stamelos, N., 1967. *Deltion Agrotikis Trapezas*, **154**, 3.
55. Veinoglou, B., Erland, G. & Kalantzopoulos, G., 1968. *Deltion Agrotikis Trapezas*, **159**, 7.
56. Zerfiridis, G., Vafopoulou, A. & Litopoulou, E., 1984. *J. Dairy Sci.*, **67**, 1397.
57. Polychroniadis, E., 1912. *A Guide of Cheese*. N. Chiotis (Publishers), Athens.
58. Janubarudusm, S., 1987. PhD thesis, Agricultural University of Athens.
58a. Kaminaridis, S. & Anifantakis, E., 1990. *J. Dairy Res.*, **57**, 271.
59. Kandarakis, J., 1981. PhD thesis, Agricultural University of Athens.
60. Anifantakis, E. & Kaminarides, S., 1983. *Aust. J. Dairy Technol.*, **38**, 29.
61. Kaminaridis, S. & Anifantakis, E., 1989. *Le Lait*, **69**, 537.
62. Assenat, L., 1985. *Lait et Produits Laitières*, **2**, 282.
63. Boulanger, A., 1976. These 3ᵉ Cycle, Universite Paris VII.
64. Kosikowski, F., 1977. *Cheese and Fermented Milk Foods*. 2nd edn, Edwards Brothers, Ann Arbor, MI.
65. Kosikowski, F.V., 1982. *Cheese and Fermented Milk Foods*. Edward Brothers, Ann. Arbor, MI, pp. 366–368.
66. FAO/WHO, 1985. *Codex Alimentarius*, Vol. XVI.
67. Casalta, E., Vassal, L. & Le Bars, D., 1991. *Process. Magazine*, No. 1065, Oct. p. 60.
68. Abrahamsen, R., 1986. *International Dairy Federation Bulletin*, **202**, 125.
69. Kandarakis, J., 1986. *International Dairy Federation Bulletin*, **202**, 118.
70. Casalis, J., 1975. *Rev. Lait. Franc.*, **332**, 403.
71. O'Connor, P. & Fox, P.F., 1979. *J. Dairy Res.*, **44**, 607.
72. Portmann, A., Pierre, Al. & Vedrenne, P., 1968. *Industrie Laitière*, **97**, 2251.
73. Portmann, A., Pierre, Al. & Vedrenne, P., 1967. *Congr. Intern. Du Froid.*, **12**, 10.
74. Cargouet, L. & Serin, C., 1973. *Techn. Lait.*, **767**, 19.
75. Ramos, M. & Juarez, M., 1986. *International Dairy Federation Bulletin*, **202**, 175.
76. Ledda, S., 1987. *Sci. Tech. Lat-Cas.*, **38**, 54.
77. Baltadjieva, M., Kalantzopoulos, G., Stamenova, V. & Sfakianos, A., 1985. *Le Lait*, **65**, 221.
78. Le Jaouen, J.C., (ed) 1987. *Fabrication du Fromage de Chevre Fermier*. ITOVIC, Paris.
79. Le Jaouen, J.C., 1972. *II Seminario Nac. en Ovinos y Caprinos*, Venezuela.
80. Damalas, N.M., 1987. *A. Konstantinidis*, Athens, 30.
81. Rontaboul, M., 1981. *Commic. Inter. Lab. Soc. des Caves de Roquefort*.
82. Anifantakis, E., Kalantzopoulos, G., Voutsinas, L. & Massouras, T., 1985. *Proc. 1st Natl. Congr. Food Technol.*, Athens, p. 311.
83. Krause, J. & Belitz, D., 1985. *Lebensm. Chemie Gericht Chimie*, p. 33.
84. Addeo, F., Chianese, L. & Moio, L., 1991. Pers. comm.
85. Addeo, F., Anelli, G. & Chianese, L., 1986. *International Dairy Federation Bulletin*, **202**, 191.

86. Schutze, A., 1970. Cited from Hanson and Johanson, *Infektion Krankh*, **36**, 5.
87. Perez Florez, F., 1968. *Alimentaria*, **4**, 7.
88. Durand, M., Meusnier, M., Delahaye, J. & Prunot, P., 1974. *Bull. Ac. Vet.*, **47**, 247.
89. Assenat, L., 1975. *Conte-Rendu Reunion Laboratoire*, Tours, France.
90. Radfort, D.V., Than, Y.T. & McPhillips, J., 1981. *Aust. J. Dairy Technol.*, **35**, 114.
91. Rispoli, S. & Saugues, R., 1989. *Lait*, **69**, 211.
92. Rispoli, S., Rivmale, M. & Saugues, R., 1991. *Lait*, **71**, 501.

Appendix: Preliminary Inventory of Cheeses made from Ewes' and Goats' Milk

(Prof. G. Kalantzopoulos & Mme E. Kombaraki, Greece, IDF Group A7.)

Name of cheese	Milk			Country of origin	Characteristics
	Ewe	Goat	Cow		
Abaza	+	+	+	TR	no information
Abertam	+	+	(+)	CS	hard
Alcobaça	+	–	–	PT	soft, fresh
Alicante	–	+	–	ES	fresh
Altenburger	–	+	–	DE	soft
Anejo	(+)	–	–	MX	hard, very salted
Annot	(+)	+	–	FR	pressed, uncooked curd
Aravis	–	+	–	FR	soft cheese with internal blue mould
Aragon	+	+	–	ES	semi hard cheese; coagulation with *Cynora scolymus* extract
Armavir	+	–	–	SU	no information
Ardéchois	+	+	(+)	FR	lactic curd
Asadero	–	+	+	MX	soft, pasta filata type
Asco	+	(+)	(+)	FR (Corsica)	soft
Aragatski	+	(+)	(+)	SU	no information
Axavi	+	+	–	SY	soft, white cheese
Azeitão (AOC)[a]	+	–	–	PT	soft cheese (coagulated with cardo extract)
Azul de Oveja	+	–	–	TN	no information
Banon	+	–	–	FR	soft cheese wrapped in chestnut leaves soaked in spirit
Bagiotto	–	+	+	IT	semi ripened
Belley	–	+	+	FR	lactic curd
Bernode	–	(+)	+	IT	no information
Bevaz-Pevnir	+	+	+	TR	soft, white cheese
Bevos	+	+	–	ES	hard
Bgug-Pevnir	+	–	–	TR	cheese flavoured with aromatic herbs or with maple or vine leaves
Biza	+	–	–	IQ	fresh cheese flavoured with onions and garlic

[a]AOC = Appelation d'Origine Contrôlée

Blue vein	+	—	—	IL	soft cheese, Roquefort type
Bonassai	+	—	—	IT	soft
Bossons macérés	—	+	—	FR	matured in pots with olive oil, white brandy and herbs
Bougon	—	+	—	FR	also designated 'Motte-Bougon'; exterior moulds
Bouton de culotte	—	+	—	FR	soft
Bracki Sir	+	—	—	YU	extra hard curd
Branza	+	—	—	RO	soft cheese stored in vats filled with milk with alternative layers of coarse salt
Branza de Burdouf	—	+	—	RO	same as above
Brique du Forez	—	+	(+)	FR	soft with exterior moulds
Brique du Livradois	—	+	—	FR	same type as Brique du Forez, also called 'Cabrion'
Brinza	+	—	—	IL	white, soft cheese
Brocciu (AOC)	+	+	—	FR (Corsica)	fresh or semi dry cheese
Brousse du Rove	—	+	—	FR	fresh, flavoured
Brousse de la Vésubie	+	+	—	FR	consumed fresh
Brucialepre	—	+	—	IT	soft
Bryzna	+	—	—	HU	piquant flavour
Brynza	+	—	(+)	SU	white, dry cheese in brine
Brynza	+	—	—	IL	white, soft cheese
Burgos	+	—	—	ES	fresh, soft cheese
Cabecou	+	+	—	FR	soft
Cabicou	—	+	—	FR	similar to Cabecou
Cabrales	+	+	+	ES	also called 'Picon' with blue moulds
Cabreiro	+	+	—	PT	soft cheese, served fresh or after maturing for a few months
Cabrion or Cabriou	—	+	—	FR	same type as Brique du Forez
Cachat or Tomme du Mont Ventoux	—	+	—	FR	fresh, rindless, salted cheese
Cachcaval	+	—	—	RO	no information
Cache	+	—	—	RO	white, piquant flavour
Cacio Fiore	+	—	—	IT	renneted with vegetable rennet, coloured with saffron, soft, exquisite flavour
Caciotta	+	—	—	IT	soft or semi hard
Caillebotte d'Aunis	+	—	—	FR	fresh, unsalted
Cajassous	—	+	—	FR	eaten fresh locally

(continued)

APPENDIX: PRELIMINARY INVENTORY OF CHEESES MADE FROM EWES' AND GOATS' MILK — *contd.*

(Prof. G. Kalantzopoulos & Mme E. Kombaraki, Greece, IDF Group A7.)

Name of cheese	Milk			Country of origin	Characteristics
	Ewe	Goat	Cow		
Cajeta	—	+	—	MX	cooked and sweetened curd
Calcagno	+	—	—	IT	hard
Calenzana	+	+	—	FR (Corsica)	hard
Camerano	—	+	—	ES	hard
Canestrato Pugliese	+	—	—	IT	matured, uncooked, hard curd
Canestrato Sardo	+	+	—	IT	cooker curd, hard
Canestrato Siciliano	+	—	—	IT	cooked curd, hard
Canestrato Toscano	+	—	—	IT	cooked curd, hard
Canestrono	+	—	—	IT	also called Cotronese, Moliterno
Caprino Piemontese	—	+	—	IT	soft
Caprino di Rimello	—	+	—	IT	soft
Caprino Sardo	—	+	—	IT	soft
Casciotta d'Urdino	+	—	(+)	IT	soft, or semi hard with 20% cows' milk
Castelo Branco	+	+	—	PT	soft
Cervera	+	—	—	ES	fresh
Cevrindi Coazze	—	+	—	IT	fresh or matured
Chabichou du Poitou (AOC)	—	+	—	FR	matured, soft cheese with mould
Charolais	—	+	(+)	FR	soft
Chanakh	+	+	—	SU	soft rind, dry
Chavignol (AOC)	—	+	—	FR	soft
Chèvre à la feuille	—	+	—	FR	soft
Chevret	—	+	—	FR	also called 'Tome de Belley'
Chevreton d'Ambert	—	+	—	FR	another designation for Brique du Forez
Chevrette des Bauges	—	+	—	FR	uncooked, pressed curd, washed rind
Chevrotin des Aravis	—	+	(+)	FR	uncooked, pressed curd with blue mould
Chevrotin du Bourbonnais	—	+	—	FR	soft, eaten fresh or dry
Chevroton	—	+	—	FR	curd matured in brandy

(continued)

Name	Description	Country			
Chiclosos	cooked, sweetened curd	MX	–	+	–
Chichota	semi hard	MX	–	+	–
Civil-Peyniri	no information	TR	+	+	+
Cordillera	semi hard	CL	–	+	–
Cotronese	similar to Holiterno, hard cooked and curd	IT	–	–	+
Couhé-Vérac	wrapped in maple leaves	FR	–	+	–
Crnogorski Masni	Brinza or Feta type	YU	–	–	+
Crottin de Chavignol (AOC)	soft (identical to Chavignol)	FR	–	+	–
Cubjac	soft	FR	(+)	+	–
Dil-Peynir	fresh	TR	+	+	+
Djanus	hard	SY	–	+	+
Danni	white, soft	EG	–	–	+
Edelpilz	semi hard, sometimes soft	DE	–	–	+
Edirne	fresh	TR	–	–	+
Erivani	no information	SU	–	–	+
Evora	hard	PT	–	–	+
Feta	semi hard	IT	–	+	+
Feuille (chèvre à la)	white, soft	GR	–	+	–
Fiore Sardo (AOC)	ripened in chestnut leaves	FR	–	–	+
Foggiano	soft, slightly pressed	IT	–	+	+
Formagini	similar to Cotronese	IT	–	+	–
Fromage blanc á la samure	fresh	CH	–	+	+
Gaiskäsli	white bovine cheese made from ewes', goats' and cows' milk	DE/CH	(+)	–	–
Galil	soft	IL	–	+	+
Gamonedo	blue veined, semi hard	ES	–	+	+
Gien	semi hard	FR	–	–	–
Gilard	soft, matured in wood ash or in chestnut leaves, or rind	IL	–	+	+
Gjetost	solid curd	NO	–	+	–
Getost	semi hard, whey cheese	ES	–	+	–
Gorbea	semi hard	ES	–	–	+

APPENDIX: PRELIMINARY INVENTORY OF CHEESES MADE FROM EWES' AND GOATS' MILK — *contd.*

(Prof. G. Kalantzopoulos & Mme E. Kombaraki, Greece, IDF Group A7.)

Name of cheese	Ewe	Goat	Cow	Country of origin	Characteristics
Gol-Pamir	+	−	−	IR	flavoured with flower petals
Gratoron	−	+	−	FR	uncooked, pressed curd
Grazalena	+	−	−	ES	hard
Grulinskii	+	+	−	SU	soft and dry rind
Gudbrandsdalsost	−	+	+	NO	semi hard
Haloumi	+	−	−	CY	semi hard
Hrudka	+	−	−	CS	no information
Huelva	+	+	−	ES	semi hard or hard
Hullom	+	+	−	SY	semi hard
Idiazabal (AOC)	+	−	−	ES	salted or smoked with delicate mountain herbs flavour
Ifravi	+	+	−	SY	soft
Jamoncillo	+	+	−	MX	cooked, sweetened
Javorski Sir	−	+	−	YU	type of white cheese
Jbane	−	+	−	MA	hard
Jibneh	+	−	−	JO	fresh
Joehberg	−	+	+	AT	no information
Jonchée Niortaise	−	+	+	FR	fresh, unsalted
Kaser or Kaser Peynir	+	+	+	TR	pressed, semi hard, eaten fresh or matured
Kascaval	+	−	+	HU	semi hard
Katschkawalj	+	+	+	BG	hard
Kautt	+	+	+	Central Asia	no information
Kefalotiri	+	+	−	GR	hard
Kelle-Peyniri	+	+	+	TR	fresh
Kobiisli	+	−	+	SU	hard
Koppen	−	+	−	CS	strong flavour

(continued)

Cheese	Country				Description
Laruns	FR	–	–	+	pressed curd
Lescin	SU	–	–	+	similar to Touloumatyri
Licki	YU	–	–	+	Gruyère type
Lighvan Panir	IR	–	–	+	white
Liptauer	HU	–	–	+	no information
Ljulin	BG	–	–	+	hard
Lor	TR	–	–	+	fresh
Lour	IQ	–	–	+	fresh
Lusignan	FR	–	–	–	soft
Maconnais	FR	–	+	+	another name for Bouton de Culotte
Maille	SU	–	–	+	no information
Maille Pener	SU	–	–	+	no information
Majorero	ES	–	+	–	hard
Makedonski White	YU	+	–	+	similar to Feta
Malaga	ES	–	+	–	white, slightly pressed
Manchego	ES	–	–	+	hard, pressed, kept in olive oil
Manchegó	MX	–	+	–	soft, matured in 15 days
Meira	IQ	–	–	+	no information
Mesitra	SU	–	–	+	soft, eaten fresh
Mesost	SE	+	+	+	no information
Mich	EG	–	+	+	no information
Mihalic	TR	–	–	+	fresh, firm curd
Mizthra	GR	–	–	+	soft, made from Feta whey
Moliterno	IT	–	+	+	no information
Monastorer	RO	–	–	+	no information
Motal	SU	–	–	+	no information
Mothais	FR	+	+	–	soft
Mrsav	YU	–	–	+	another name for Sir Posny
Natillos	MX	–	+	+	cooked, sweetened
Niolo	FR (Corsica)	–	+	+	soft
Nisa	PT	–	+	+	semi hard
Oriental	TN	–	–	+	white, soft

APPENDIX: PRELIMINARY INVENTORY OF CHEESES MADE FROM EWES' AND GOATS' MILK — contd.

(Prof. G. Kalantzopoulos & Mme E. Kombaraki, Greece, IDF Group A7.)

Name of cheese	Milk			Country of origin	Characteristics
	Ewe	Goat	Cow		
Ordúa	+	—	—	ES	semi hard
Oropesa	+	—	—	ES	hard
Ossau-Iraty (AOC)	+	—	—	FR	uncooked, pressed curd
Oschtjepek	+	—	—	CS	similar to Italian Caciocavallo
Osetinskii	+	—	+	SU	hard
Otlu	+	—	—	TR	with herbs
Ourda	+	—	—	RO	white cheese flavoured with herbs
Ovciji	+	—	—	YU	white cheese flavoured with herbs
Olipski	+	—	—	YU	hard, Pecorino type
Pago	+	—	—	YU	no information
Panela	—	+	—	MX	fresh
Panir-Kopéi	+	—	—	IR	no information
Parenica	+	—	—	CS	Caciocavallo type
Parenitza	+	—	—	HU	Caciocavallo type
Parils	—	+	—	IR	kept in goat or lamb skins
Paški	+	—	—	YU	hard
Pecorino Dolce	+	—	—	IT	curd, coloured with annatto
Pecorino Romano*	+	—	—	IT	pressed, white, firm, cooked curd
Pecorino Siciliano*	+	—	—	IT	hard
Pecorino Toscano*	+	—	—	IT	smaller than Pecorino Romano
Pedroches	+	—	—	ES	semi hard
Pepato Sardo	+	—	—	IT	pressed, uncooked curd
Pepato Siciliano	+	—	—	IT	pressed, uncooked curd
Pécardon	—	+	—	FR	soft
Pena Santa	+	—	—	FR (Corsica)	Roquefort analogue
Peneteleu	+	—	—	RO	type of Caciocavallo
Pérail	+	—	—	FR	eaten fresh

(continued)

Name				Country	Description
Persillé des Aravis	—	+	—	FR	soft with internal mould
Peynir	+	—	+	TR	white
Picodon Drôme-Ardèche (AOC)	—	+	+	FR	soft
Picadou	+	+	—	FR	Cabecou type, wrapped in walnut leaves
Pigouille	+	+	+	FR	soft
Piotski Katschkavalj	+	+	—	YU	no information
Pouligny St Pierre (AOC)	—	+	+	FR	soft
Pucol	+	—	+	ES	fresh
Pustagnacq	+	+	—	FR	flavoured, fresh
Pyramide	—	+	+	FR	soft with surface moulds
Quesucos	+	+	—	ES	semi hard
Queso de cabra	—	+	—	VE	no information
Queso de cabra conchile	—	+	—	MX	soft, fresh
Queso de pina	—	+	—	PH	no information
Rabaçal	(+)	+	+	PT	fresh or matured
Raviggiolo	(+)	(+)	—	IT	soft, uncooked curd
Reino	—	+	—	BR	similar to Serra da Estrela
Ricota Salata	+	+	—	IT	soft, semi matured
Resenge Birge	—	+	—	CS	soft
Rola	—	+	—	IT	soft, strong flavour
Rocamadour	+	+	+	FR	soft
Rogeret des Cévennes	—	+	+	FR	soft
Roncal (AOC)	+	—	+	ES	pressed curd
Rodos	+	—	—	IQ	matured in sheep's skins
Roquefort (AOC)	+	+	+	FR	soft, blue veined throughout
Saint Félicien	—	+	—	FR	soft
Saint Maixent	—	+	+	FR	soft
Sainte Maure de Touraine (AOC)	—	+	+	FR	soft
Salamvrali Tulum Peynirl	+	+	+	TR	white
Sancerre	—	+	—	FR	soft
San Maurin	—	+	—	IT	no information

APPENDIX: PRELIMINARY INVENTORY OF CHEESES MADE FROM EWES' AND GOATS' MILK — contd.

(Prof. G. Kalantzopoulos & Mme E. Kombaraki, Greece, IDF Group A7.)

Name of cheese	Ewe	Goat	Cow	Country of origin	Characteristics
Sartenais	+	−	−	FR (Corsica)	pressed, uncooked curd
Sassenage	−	+	+	FR	soft, slightly pressed, with interior moulds
Selles-sur-Cher (AOC)	−	+	−	FR	soft
Semicotto	+	−	−	IT	semi hard
Serena	+	−	−	ES	hard
Serpa	+	−	−	PT	soft or semi hard
Serra da Estrela (AOC)	+	−	−	PT	soft or semi hard
Ser iz Micsine	+	−	−	YU	fresh or matured in a fresh goat's skin
Ser Nastni	+	+	−	YU	fat cheese, hard
Ser Posni	+	−	−	YU	another name for Tord and Mrsav
Smolenskii	+	+	+	SU	hard
Sore	+	−	−	IT	soft
Soria	−	+	−	ES	soft
Stambuli	−	+	−	SY	white, soft
Suluguni	+	+	+	SU	hard
Susilia	+	−	−	TN	pressed curd
Sulal	−	+	−	SY	semi hard
Tchanakh	+	+	+	SU	cheese in brine
Teleme	+	+	−	GR	white, soft
Texel	+	−	−	NL	no information
Tomar	+	−	−	PT	no information
Tome d'Annot	−	+	−	FR	uncooked, pressed curd
Tome d'Arles or Tome de Camargue	+	−	−	FR	fresh, flavoured with thyme or laurel powder
Tome de Belley or Chevret	−	+	(+)	FR	soft
Tome de Brach	+	−	−	FR	soft, blue veined

Name				Country	Description
Tome de Romans	–	(+)	–	FR	soft
Toscanello	+	–	+	IT	very hard
Touchinski	+	–	+	SU	cheese in brine
Toucy	–	+	–	FR	soft
Tournon-St Pierre	–	+	–	FR	soft
Travnicki	+	–	+	YU	similar to Feta
Tracanik	+	+	–	YU	similar to Feta
Tschill	+	+	+	SU, Armenia	no information
Tulum Peyniri	+	+	+	TR	pressed, matured in sheep's skin
Tuma Piemontese (AOC)	+	–	+	IT	soft, uncooked
Tydr Sir	–	–	+	YU	hard
Val de Blore	+	+	–	FR	uncooked, pressed
Valdetaeja	–	+	–	ES	semi hard
Valençay	–	–	+	FR	soft, rind with blue and white moulds
Venaco or Venacais	–	+	+	FR (Corsica)	washed, soft rind cheese
Villalon	+	–	+	ES	soft
Voujnyi	+	–	+	SU	no information
Vizes	–	–	+	GR	hard
Ziegenkäse	+	–	+	DE	no information
Zlatiborski Sir	+	–	+	YU	white cheese
Zomma	+	–	+	TR	Katschkawalj type
Zwirn	+	+	+	SU, Armenia	another name for Tschill

Country Codes

AT — Austria
BG — Bulgaria
BR — Brazil
CH — Switzerland
CL — Chile
CS — Czechoslovakia
CY — Cyprus
DE — Germany, former FR

EG — Egypt
ES — Spain
FR — France
GR — Greece
HU — Hungary
IL — Israel
IQ — Iraq
IR — Iran

IT — Italy
JO — Jordon
MA — Morocco
MX — Mexico
NL — Netherlands
NO — Norway
PH — Philippines
PT — Portugal

RO — Romania
SE — Sweden
SU — USSR
SY — Syria
TN — Tunisia
TR — Turkey
VE — Venezuela
YU — Yugoslavia

Index

Note: Figures and Tables are indicated by *italic page numbers.*

555

DATE DUE

JUN 2 3 1994	APR 8 2001
APR 2 9 1996	OCT 0 1 2001
	APR 1 9 2003
JUN 0 3 1997	
OCT 1 4 1997	
APR 3 0 1998	
MAY 1 9 1999	
FEB 1 9 2002	